Objective Bayesian Inference

Objective Bayesian Inference

James O Berger
Duke University, USA

Jose M Bernardo
University of Valencia, Spain

Dongchu Sun
East China Normal University, China

World Scientific

NEW JERSEY • LONDON • SINGAPORE • BEIJING • SHANGHAI • HONG KONG • TAIPEI • CHENNAI • TOKYO

Published by

World Scientific Publishing Co. Pte. Ltd.
5 Toh Tuck Link, Singapore 596224
USA office: 27 Warren Street, Suite 401-402, Hackensack, NJ 07601
UK office: 57 Shelton Street, Covent Garden, London WC2H 9HE

Library of Congress Control Number: 2024003974

British Library Cataloguing-in-Publication Data
A catalogue record for this book is available from the British Library.

Images on cover: Thomas Bayes (public domain), Pierre-Simon Laplace (public domain), Harold Jeffreys (copyright credit: Walter Stoneman, American Geophysical Union, courtesy AIP Emilio Segrè Visual Archives)

First published 2024 (hardcover)
Reprinted 2026 (in paperback edition)
ISBN 978-981-98-3318-4 (pbk)

OBJECTIVE BAYESIAN INFERENCE

ISBN 978-981-12-8490-8 (hardcover)
ISBN 978-981-12-8491-5 (ebook for institutions)
ISBN 978-981-12-8492-2 (ebook for individuals)

For any available supplementary material, please visit
https://www.worldscientific.com/worldscibooks/10.1142/13640#t=suppl

Preface

Background: Bayesian analysis is today understood to be an extremely powerful method of statistical analysis, as well an approach to statistics that is particularly transparent and intuitive. It is thus being extensively and increasingly utilized in virtually every area of science and society that involves analysis of data.

A widespread misconception is that Bayesian analysis is a more subjective theory of statistical inference than what is now called classical statistics.

This is true neither historically nor in practice. Historically, the first Bayesians — the Reverend Thomas Bayes (see Bayes (1763)) and the great scientist Pierre-Simon Laplace (see Laplace (1812)) — initiated the objective Bayesian approach to statistical inference, which dominated the statistical landscape for the next 150 years (long before 'classical' statistics was developed). Likewise, Bayesian practice today makes heavy (often exclusive) use of objective Bayesian methodology. Yet, curiously, most modern introductions to Bayesian analysis focus on the subjective Bayesian approach, and most popular descriptions of the approach emphasize it's subjective side. The origin of this book was a desire to address this imbalance.

Motivation and Readers: This book has four specific purposes, relating to different potential types of readers:

1. To give an introduction to objective Bayesian enference for non-statisticians; this includes an introduction to Bayesian inference itself for readers who have not previously been exposed to the approach, although some familiarity with probability and statistics is presumed.

2. To give an overview of the development and current state of objective Bayesian analysis and its relationship to subjective Bayesian analysis and classical statistics.

3. To present a careful development of the particular objective Bayesian approach that we recommend, the *reference prior* approach.

4. To present as much practical objective Bayesian methodology as possible for statisticians and scientists primarily interested in practical use of the approach.

Those interested primarily in an introduction to objective Bayesian analysis can concentrate on Chapter 1 (Sections 1.1 through 1.5 and 1.8), Chapter 2 (Sections 2.1 and 2.2), and the beginnings of Chapters 3, 4 and 8. These parts of the book are kept at an introductory level, presuming no previous exposure to Bayesian analysis.

Objective Bayesian analysis has a fascinating history, and has been a subject of intense interest to a multitude of statisticians, mathematicians, philosophers, and scientists, many of whom were motivated by the elusive goal of establishing a unified framework for objectively learning from data. While we do not attempt a systematic review of this work (dozens of volumes would be needed), we do seek to provide enough background and context to understand the current state and potential for objective Bayesian analysis. Chapters 2 through 8 focus on this overall understanding.

The *reference prior* approach is described in Chapters 8 through 10. As this is the objective Bayesian approach that we recommend, we present the theoretical basis of the approach as well as most of the technical details needed for its implementation. This material will primarily be of interest to those who seek to develop objective Bayesian methodology in new situations.

Chapters 11 through 14 describe generalizations of the reference prior approach and its application to important problems in statistics. These chapters tend to have a more methodological orientation, while addressing limitations of the original reference prior approach.

There is a considerable difference between developing and using objective Bayesian methodology. Developing the methodology can be difficult, but its use is often very straightforward. Thus many parts of the book are oriented towards users who simply want to know and apply the best objective Bayesian methodology. As an example, Section 12.6 presents recommended objective Bayesian methodology for handling any hierarchical normal model.

Philosophical Position: Statistics is perhaps the oldest science which does not yet have an agreed upon foundation. Indeed, there are virtually dozens of competing foundations that have been proposed over the last three centuries, the most prominent of which are (in the order of their introduction) objective Bayesian analysis, frequentist analysis, Fisherian/likelihood analysis, and subjective Bayesian analysis. Our philosophical position is rather eclectic, in that we view each of these as providing useful illumination on the complex problem of inference based on data and, hence, as being part of the overall foundation for statistics. However, from an applied and pedagogical perspective, we are strong proponents of the objective Bayesian approach as providing the best working framework for statistical analysis.

What is not covered: As emphasized above, this book is concerned with objective Bayesian *inference*, which refers to statistical problems such as estimation, development of confidence sets, and prediction. Quite different areas of objective Bayesian analysis are model selection, hypothesis testing, and decision analysis. While we include some comments about these areas in the book, they are of such a different character that space precludes extensive coverage. See Consonni et al. (2018) for a recent review article covering these areas

There are two other areas of Bayesian statistics that are highly related to objective Bayesian analysis, namely *robust Bayesian analysis* (see Berger 1994, Berger et al. 1996, Ríos & Ruggeri 2012, and references therein) and *nonparametric Bayesian analysis* (see Ghosal & Van der Vaart 2017, and references therein). Again, space also precludes discussing these approaches.

Acknowledgements: We are indebted to the extensive objective Bayesian community; our interactions over five decades with this community shaped our current views. We are also grateful to Sophia Hallstein for designing the cover of the book and for numerous suggestions to improve the presentation.

Durham NC, USA; Columbia MO, USA; and Valencia, Spain
September 2023

Contents

Chapter 0

Overview and Notation

0.1 Overview

It is often convenient to consider statistical analysis as consisting of multiple phases: a design phase, a data-analytic phase, and an inference or decision phase. These phases of analysis do not necessarily occur in strict sequential order, and some iteration is quite common. The inference or decision phase of analysis is often posed as the problem of what to do when the data has been obtained and a probabilistic model for the data has been determined. This book primarily considers this final phase, focusing on statistical inference within a given model.

The Bayesian approach to statistical inference has been motivated by a wide variety of axiomatic foundations, but neither a formal description of such axioms nor a derivation of their consequences is attempted in this book. One consequence common to the various foundations is that all uncertainties present in the problem should be described by probability distributions. In particular, the parameters of a statistical model should have a joint probability distribution describing the available *prior* information about their values, and uncertainty concerning the parameters after seeing the data should also be represented by a probability distribution, called the *posterior* distribution. As opposed to classical statistics, parameters are thus treated as random variables within the Bayesian paradigm. This is not a description of their variability (parameters are typically fixed unknown quantities, but rather a description of the uncertainty about their true values.

Bayesian analysis itself is divided into a variety of schools, primarily dependent on the type of prior information that is utilized. A relatively recent school of Bayesian analysis is the *subjective* school, in which prior information is viewed as representing individual beliefs prior to seeing the data. The oldest school of Bayesian analysis is the *objective* school, in which prior distributions are chosen in a conventional fashion to reflect either a real lack of prior information or a desire to present an objective analysis, objective in the sense that the analysis depends exclusively on the accepted statistical model and the observed data. The description, derivation and comparison of objective Bayesian methods is the main objective of this book.

0.2 Probabilistic Notation and Distributions

To simplify the exposition, it will be assumed that all probability distributions can be described by probability density functions [1], and no distinction is made between a random quantity and the particular values that it may take. Roman fonts are used for *observable* random quantities (typically data) and for known constants, while Greek fonts are used for *unobservable* random quantities (typically parameters). Bold face is used to denote row vectors, and $\boldsymbol{x}'$ is used to denote the transpose of the vector $\boldsymbol{x}$. Lowercase is used for variables and uppercase for their domains. The standard mathematical convention of referring to *functions*, say f and g of $\boldsymbol{x} \in \boldsymbol{\mathcal{X}}$, respectively, by $f(\boldsymbol{x})$ and $g(\boldsymbol{x})$, will often be used. In particular, $p(\boldsymbol{x} \mid C)$ and $p(\boldsymbol{y} \mid C)$ will, respectively, represent general *probability densities* of the *observable* random vectors $\boldsymbol{x} \in \boldsymbol{\mathcal{X}}$ and $\boldsymbol{y} \in \boldsymbol{\mathcal{Y}}$ under conditions C, without any suggestion that the random vectors $\boldsymbol{x}$ and $\boldsymbol{y}$ have the same distribution. Similarly, $\pi(\boldsymbol{\theta} \mid C)$ and $\pi(\boldsymbol{\omega} \mid C)$ will, respectively, represent general probability densities of the *unobservable* parameter vectors $\boldsymbol{\theta} \in \boldsymbol{\Theta}$ and $\boldsymbol{\omega} \in \boldsymbol{\Omega}$ under conditions C. Thus, $p(\boldsymbol{x} \mid C) \geq 0$, $\int_{\boldsymbol{\mathcal{X}}} p(\boldsymbol{x} \mid C)\, d\boldsymbol{x} = 1$, and $\pi(\boldsymbol{\theta} \mid C) \geq 0$, $\int_{\boldsymbol{\Theta}} \pi(\boldsymbol{\theta} \mid C)\, d\boldsymbol{\theta} = 1$. If the random vectors are discrete, these are probability mass functions, and integrals over their values become sums.

Objective Bayesian analysis often make use of *formal prior functions* for parameters, intended to describe situations where no relevant information about their values is available. These are positive functions which do not necessarily integrate to one and will also be denoted by $\pi(\boldsymbol{\theta})$. A function $\pi(\boldsymbol{\theta}) \geq 0$, $\boldsymbol{\theta} \in \boldsymbol{\Theta}$, such that $\int_{\boldsymbol{\Theta}} \pi(\boldsymbol{\theta})\, d\boldsymbol{\theta} = \infty$ is called an *improper prior function* or *improper prior* for short.

We use $\mathrm{E}[\boldsymbol{x} \mid C]$ and $\mathrm{E}[\boldsymbol{\theta} \mid C]$ to denote the expected values of $\boldsymbol{x}$ and $\boldsymbol{\theta}$, respectively, under conditions C. Similarly, $\mathrm{Var}[\boldsymbol{x} \mid C] = \mathrm{E}[\boldsymbol{x}'\boldsymbol{x} \mid C] - \mathrm{E}[\boldsymbol{x}' \mid C]\,\mathrm{E}[\boldsymbol{x} \mid C]$ and $\mathrm{Var}[\boldsymbol{\theta} \mid C] = \mathrm{E}[\boldsymbol{\theta}'\boldsymbol{\theta} \mid C] - \mathrm{E}[\boldsymbol{\theta}' \mid C]\,\mathrm{E}[\boldsymbol{\theta} \mid C]$ denote the corresponding variance-covariance matrices. The same operation-oriented notation is used for other properties of the distributions, thus $\mathrm{Mo}[\boldsymbol{\theta} \mid C]$ denotes the mode of the distribution of $\boldsymbol{\theta}$ under conditions C and, for a real variable θ, $\mathrm{Me}[\theta \mid C]$ and $\mathrm{Q}_p[\theta \mid C]$ denote the median and p-quantile, respectively, of the distribution of θ. We use $\Pr(\boldsymbol{x} \in A \mid C)$ to denote the probability that $\boldsymbol{x}$ belongs to a region A under conditions C. For instance, $\Pr(\boldsymbol{\theta} \in A \mid \boldsymbol{x}, C) = \int_A p(\boldsymbol{\theta} \mid \boldsymbol{x}, C)\, d\boldsymbol{\theta}$ is the probability that the parameter $\boldsymbol{\theta}$ belongs to A, given data $\boldsymbol{x}$ and conditions C.

---

[1] Formally, probability density functions are assumed to be defined with respect to some σ-finite dominating measure, typically the counting measure in discrete problems and Lebesgue measure in continuous problems; however, knowledge of measure theory is not required.

Specific density functions are denoted by appropriate abbreviations. Thus, if x is a univariate random quantity having a Normal distribution with mean μ and variance σ^2, its probability density function will be denoted $\mathrm{N}(x \mid \mu, \sigma^2)$; if θ has a Beta distribution with parameters a and b, its density function will be denoted $\mathrm{Be}(\theta \mid a, b)$. Table 1 describes the notation used for the more common distributions. The definitions and main properties of all distributions used in the book are collected together in Appendix A.

Table 1: Notation for common probability distributions.

Distribution	Notation
Beta	$\{\mathrm{Be}(x \mid \alpha, \beta),\quad 0 < x < 1,\quad \alpha > 0, \beta > 0\}$ $\mathrm{Be}(x \mid \alpha, \beta) = \frac{\Gamma(\alpha+\beta)}{\Gamma(\alpha)\Gamma(\beta)}\, x^{\alpha-1}(1-x)^{\beta-1}$
Binomial	$\{\mathrm{Bi}(x \mid n, \theta),\quad x \in \{0, \dots, n\}, n \in \mathcal{N} \equiv \{1, 2, \dots\},\quad 0 < \theta < 1\}$ $\mathrm{Bi}(x \mid n, \theta) = \binom{n}{x}\, \theta^x (1-\theta)^{n-x}$
Exponential	$\{\mathrm{Ex}(x \mid \theta),\quad x > 0,\quad \theta > 0\}$ $\mathrm{Ex}(x \mid \theta) = \theta\, e^{-\theta x}$
Gamma	$\{\mathrm{Ga}(x \mid \alpha, \beta),\quad x > 0,\quad \alpha > 0, \beta > 0\}$ $\mathrm{Ga}(x \mid \alpha, \beta) = \frac{\beta^\alpha}{\Gamma(\alpha)}\, x^{\alpha-1} e^{-\beta x}$
Negative Binomial	$\{\mathrm{Nb}(x \mid r, \theta),\quad x \in \mathcal{N},\quad r \in \mathcal{N}, 0 < \theta < 1\}$ $\mathrm{Nb}(x \mid r, \theta) = \binom{r+x-1}{r-1}\, \theta^r (1-\theta)^x$
Normal	$\{\mathrm{N}(x \mid \mu, \sigma^2)\quad x \in \mathbb{R},\quad \mu \in \mathbb{R}, \sigma > 0\}$ $\mathrm{N}(x \mid \mu, \sigma^2) = \frac{1}{\sigma\sqrt{2\pi}} \exp\left[-\frac{1}{2}\left(\frac{x-\mu}{\sigma}\right)^2\right]$
Poisson	$\{\mathrm{Pn}(x \mid \lambda),\quad x \in \mathcal{N},\quad \lambda > 0\}$ $\mathrm{Pn}(x \mid \lambda) = e^{-\lambda}\, \frac{\lambda^x}{x!}$
Student	$\{\mathrm{St}(x \mid \mu, \sigma, \alpha),\quad x \in \mathbb{R},\quad \mu \in \mathbb{R}, \sigma > 0, \alpha > 0\}$ $\mathrm{St}(x \mid \mu, \sigma, \alpha) = \frac{\Gamma\{(\alpha+1)/2)\}}{\Gamma(\alpha/2)} \frac{1}{\sigma\sqrt{\alpha\pi}} \left[1 + \frac{1}{\alpha}\left(\frac{x-\mu}{\sigma}\right)^2\right]^{-(\alpha+1)/2}$

0.3 Statistical Notation and Models

Whenever a *statistical model* for some data $\boldsymbol{x}$ has to be fully specified, the notation $\{p(\boldsymbol{x} \mid \boldsymbol{\theta}), \boldsymbol{x} \in \boldsymbol{\mathcal{X}}, \boldsymbol{\theta} \in \boldsymbol{\Theta}\}$ will be used, and it will be assumed that $p(\boldsymbol{x} \mid \boldsymbol{\theta})$ is a probability density function (a probability mass function in the discrete case) so that $p(\boldsymbol{x} \mid \boldsymbol{\theta}) \geq 0$ for all $\boldsymbol{x} \in \boldsymbol{\mathcal{X}}$ and for all $\boldsymbol{\theta} \in \boldsymbol{\Theta}$ and $\int_{\boldsymbol{\mathcal{X}}} p(\boldsymbol{x} \mid \boldsymbol{\theta})\, d\boldsymbol{x} = 1$ for all $\boldsymbol{\theta} \in \boldsymbol{\Theta}$. The *parameter* $\boldsymbol{\theta}$ will be assumed to be a vector $\boldsymbol{\theta} = (\theta_1, \dots, \theta_m)$ of finite dimension $m \geq 1$, so that $\boldsymbol{\Theta} \subset \mathbb{R}^m$. The sets $\boldsymbol{\mathcal{X}}$ and $\boldsymbol{\Theta}$ will be referred to as the *sampling space* and the *parameter space*, respectively. Occasionally, if there is no possibility of confusion, the model will simply be written $p(\boldsymbol{x} \mid \boldsymbol{\theta})$, without mentioning the sampling space and the parameter space.

Considered as a function of $\boldsymbol{\theta}$, $\text{Lik}(\boldsymbol{\theta} \mid \boldsymbol{x}) = p(\boldsymbol{x} \mid \boldsymbol{\theta})$, the probability density of $\boldsymbol{x}$, is referred to as the *likelihood function* corresponding to the data $\boldsymbol{x}$. Whenever it exists, a maximum of the likelihood function will be denoted by $\hat{\boldsymbol{\theta}} = \hat{\boldsymbol{\theta}}(\boldsymbol{x})$; this is called a *maximum likelihood estimate* or mle of $\boldsymbol{\theta}$.

The complete set of available data will typically be denoted by $\boldsymbol{x}$. In many examples this will be a random sample $\boldsymbol{x} = (x_1, \ldots, x_n)$ from a model of the form $\{p(x \mid \boldsymbol{\theta}), x \in \mathcal{X}, \boldsymbol{\theta} \in \boldsymbol{\Theta}\}$ for a single datum x, that is, a set of n independent, identically distributed (iid) observations, in which case the probability density will be of the form $p(\boldsymbol{x} \mid \boldsymbol{\theta}) = \prod_{i=1}^{n} p(x \mid \boldsymbol{\theta})$ and the sampling space will be $\boldsymbol{\mathcal{X}} = \mathcal{X}^n$, but this may not always be the case. The notation $\boldsymbol{t} = \boldsymbol{t}(\boldsymbol{x})$, $\boldsymbol{t} \in \boldsymbol{\mathcal{T}}$, will be used to refer to a general function of the data; often, but not necessarily, this will be a sufficient statistic (see Subsection 1.2.3).

Example 0.1 *The likelihood for an exponential sample.* Five independent observations are taken from an $\text{Ex}(x \mid \theta)$ distribution where θ is unknown. The result is $\boldsymbol{x} = (x_1, \ldots, x_5) = (4.2, 5.4, 3.6, 2.9, 7.3)$. The likelihood function is then

$$\text{Lik}(\theta \mid \boldsymbol{x}) = p(\boldsymbol{x} \mid \theta) = \prod_{i=1}^{5} \theta\, e^{-x_i\, \theta} = \theta^5 e^{-23.4\, \theta}\,.$$

The mle is found by maximizing this likelihood function over θ; simple calculus yields that this maximum is attained at $\hat{\theta} = 4.68$. In general, for a sample $\boldsymbol{x} = (x_1, \ldots, x_n)$ from an $\text{Ex}(x \mid \theta)$ distribution, the likelihood function is

$$p(\boldsymbol{x} \mid \theta) = \theta^n e^{-n\, \overline{x}\, \theta},$$

and the mle is $\hat{\theta}(\boldsymbol{x}) = \overline{x}$.

We will only use the notation $\text{Lik}(\boldsymbol{\theta} \mid \boldsymbol{x})$ when we are emphasizing that it is a function of $\boldsymbol{\theta}$. Usually we will just abuse notation, however, and call $p(\boldsymbol{x} \mid \boldsymbol{\theta})$ the likelihood function; it is then to be inferred that we are viewing it as a function of $\boldsymbol{\theta}$ with $\boldsymbol{x}$ given.

We will occasionally refer to a (possibly infinite) sequence of replications $\{\boldsymbol{x}_1, \boldsymbol{x}_2, \ldots\}$ of the mechanism which has generated the complete set $\boldsymbol{x}$ of available data; $\boldsymbol{x}^{(k)} = \{\boldsymbol{x}_1, \ldots, \boldsymbol{x}_k\}$ will be used to denote k such replications. Thus, $\boldsymbol{x}^{(k)}$ consists of a random sample of size k from the original statistical model $\{p(\boldsymbol{x} \mid \boldsymbol{\theta}), \boldsymbol{x} \in \boldsymbol{\mathcal{X}}, \boldsymbol{\theta} \in \boldsymbol{\Theta}\}$.

Part I

The Objective Bayesian Paradigm

Chapter 1

Basics of Bayesian Analysis

1.1 Measuring Conditional Uncertainty

A central element of the Bayesian paradigm is the use of probability distributions to describe *all* relevant unknown quantities, interpreting the probability of an event involving an unknown quantity as a measure of the uncertainty of the occurrence of the event, on a $[0,1]$ scale, and typically conditional on some specific conditions. Thus $\Pr(E \mid C)$ is a measure of the belief in the occurrence of the event E under conditions C. The extreme values, 0 and 1, represent impossibility and certainty of the occurrence of the event, respectively. As detailed below, this interpretation of probability is meant to include other interpretations and, most importantly, to correspond to the common intuition behind everyday use of the concept of probability.

It is important to stress that probability is a function of two arguments, the event E whose uncertainty is being measured, and the conditions C under which the measurement takes place. In typical applications, one is interested in the probability of some event E given the available *data* $\boldsymbol{x}$, the set of *assumptions* A which one is prepared to make about the mechanism that has generated the data, and the relevant contextual *knowledge* K about E that might be known from other sources. Thus, $\Pr(E \mid \boldsymbol{x}, A, K)$ is to be interpreted as a measure of rational belief in the occurrence of the *event* E, given data $\boldsymbol{x}$, assumptions A and any other available knowledge K; intuitively, this is often viewed as a measure of how "likely" is the occurrence of E under these conditions. Sometimes, the probability of an event under given conditions can be associated with the relative frequency of "similar" events under the given or similar conditions. Thus the frequency concept of probability defines one route to assessment of probabilities. Many other routes are possible, however, and we do not specify a preference here.

For notational convenience, it is common to omit A, the mechanism that has generated the data, and K, the relevant contextual knowledge, from probability statements, and we will also do so. In the following illustrations of the use of probability as a conditional measure of uncertainty, however, we will try to emphasize both A and K, when relevant. These examples focus on the interpretation of the probability statements, and not on their derivation; their derivation requires Bayes theorem, which is not discussed until the next section. We begin with a stylized version of medical diagnosis that involves only discrete variables.

Example 1.1 *Medical diagnosis.* A human population is known to contain 0.2% of people infected by a particular virus. A person, *randomly selected* from that population, is subject to a test which is known from laboratory data to yield positive results in 98% of infected people and in 1% of non-infected; thus, if V denotes the event that a person carries the virus, $\overline{V}$ its complement, and $+$ denotes a positive result, we have $\Pr(V) = 0.002$, $\Pr(+ \mid V) = 0.98$ and $\Pr(+ \mid \overline{V}) = 0.01$. Suppose that the result of the test turns out to be positive for the person selected. One is then interested in $\Pr(V \mid +)$, the probability that the person carries the virus, given the positive test result. Bayes theorem (see the next section) can be used to show that $\Pr(V \mid +) = 0.164$. The probabilities $\Pr(+ \mid V)$, $\Pr(+ \mid \overline{V})$, and $\Pr(V \mid +)$ are actually conditional on A, the probability mechanism generating the test results, and $\Pr(V)$ and $\Pr(V \mid +)$ are actually conditional on the available knowledge K of the prevalence of the infection in the population, but these conditional features have been suppressed in the notation, as is common.

Note that the four probabilities involved in this problem can be given either uncertainty or frequency interpretations. For instance, $\Pr(V \mid +)$ is both a measure of the uncertainty associated with the event that the particular person who tested positive is actually infected, and an estimate of the proportion of people in the population (about 16.4%) that would prove to be infected, among those which yielded a positive test. Note also that $\Pr(V \mid +)$ is very different than $\Pr(+ \mid V)$; it is common for people to confuse the two, which can be quite problematic.

In the next example, where the proportion of items in a population with a given characteristic has to be estimated from a random sample, the data are still discrete but the quantity of interest is a continuous parameter.

Example 1.2 *Estimation of a proportion.* A survey is conducted to estimate the proportion θ of individuals in a population who share a given property. A random sample of n individuals are studied, r of which are found to possess the property. One is then often interested in specifying intervals in $[0, 1]$, in which the unknown value of θ may plausibly be expected to lie; this is usefully summarized by probabilities of the form $\Pr(a < \theta < b \mid r, n, A, K)$, a conditional measure of the uncertainty about the event that θ belongs to (a, b) given the information provided by the data $\{r, n\}$, the assumptions A made on the behavior of the mechanism which has generated the data (a random sample of n Bernoulli trials), and any relevant knowledge K on the values of θ that might be available from other sources. In practice, A will be omitted from the notation (unless the data-generating probability mechanism is uncertain), and K will be represented by the prior distribution for θ.

As an example, consider a political survey in which 720 citizens out of a random sample of 1500 have declared to be in favor of a particular political measure. With A being the Bernoulli data-generating mechanism, and K being an objective prior distribution for θ, one might conclude (see the next section) that $\Pr(\theta < 0.5 \,|\, 720, 1500, A, K) = 0.934$, indicating a degree of belief of 0.934 that a referendum on that issue would be lost. As another example, after a screening test for an infection where 100 people have been tested, none of which has turned out to be infected, one might conclude that $\Pr(\theta < 0.01 \,|\, 0, 100, A, K) = 0.844$, a conditional measure of uncertainty of 0.844 that the proportion of infected people is smaller than 1%. Note that, in these examples, the concluding probability statements do not have a frequency interpretation; the unknown θ values either are, or are not, in the indicated intervals, so the probabilities only refer to degree of belief.

A large proportion of routine statistical analyses performed in laboratories are variations on the problem of estimating the true value of an unknown quantity based on a finite number of measurements subject to error.

Example 1.3 *Measurement of a physical constant.* A team of scientists, intending to establish the unknown value of a physical constant μ, obtain data $\boldsymbol{x} = (x_1, \ldots, x_n)$ which are considered to be measurements of μ subject to error. Of interest are often probabilities of the form

$$\Pr(a < \mu < b \,|\, x_1, \ldots, x_n, A, K),$$

the probability that the unknown value of μ (fixed in nature, but unknown to the scientists) lies within an interval (a, b) given the information provided by the data $\boldsymbol{x}$, the assumptions A made on the behavior of the measurement mechanism, and whatever other knowledge K might be available on the value of the constant μ.

If the setting is, instead, a classroom experiment to measure the gravitational field g in the school by means of a pendulum, a student might be instructed to ignore other knowledge about g (*e.g.*, knowledge available from the scientific literature). In a Bayesian context this could take the form of an instruction to use only an objective prior distribution for g in any probability computation. After experimentation, the student might report (in $\mathrm{m/s^2}$) that $\Pr(9.788 < g < 9.829 \,|\, \boldsymbol{x}, A, K) = 0.95$, where K now refers to the objective background assumption and A refers to the probability model for the data. This probability of 0.95 is again only a conditional uncertainty measure associated with the event that g is between 9.788 and 9.829, on a [0,1] scale. Although this is not inherently a frequency-based probability, it sometimes can be embedded in scenarios where a frequency-based interpretation is possible. For instance, suppose that the same statistical procedure was used by many other students with similarly obtained data sets. Under certain conditions on the objective prior distribution used in the analysis (see Chapter 5), it is the case that the students' reported intervals would actually cover the true value of g in approximately 95% of the cases.

In many applications, the quantity of interest is not the value of an unknown parameter, but the value of a future observation. Indeed, it is often desired to predict a future observation, *i.e.*, a future realization of the probability mechanism that has generated the available data. First consider a discrete example.

Example 1.4 *Future discrete observations.* Periodically, an experiment is made to count the number of times, x, that an event E takes place in a well defined situation. Replications of the experiment result in the data that E occurs x_i times in replication i, $i = 1, \ldots, n$. For instance, the quality assurance engineer of a firm which produces automobile airbags might observe that $x_i = 0$ complaints from customers were received in each of $n = 10$ consecutive months. It is desired to forecast the number of times, x, that E will take place in a future, similar situation, *e.g.*, the number of complaints that will be received next month.

This is a *prediction* problem on the value of an *observable* quantity x (here discrete), given the information provided by data $\boldsymbol{x} = (x_1, \ldots, x_n)$, accepted assumptions A on the probability mechanism which have generated the data (often a Poisson distribution), and any other available relevant knowledge K. What is required is the computation of the probabilities $\Pr(x \,|\, x_1, \ldots, x_n, A, K)$ for each non-negative integer x. For example, the quality assurance engineer might report, in the above situation with $\boldsymbol{x} = (0, \ldots, 0)$, that $\Pr(x = 0 \,|\, \boldsymbol{x}, A, K) = 0.953$. This should be regarded as a measure, on a $[0, 1]$ scale, of the conditional uncertainty (given observed data, accepted assumptions and contextual knowledge) associated with the event that no airbag complaint will come from next month's production.

Predictive measures also have an inherent frequency interpretation; for instance, 0.953 above would also be an estimate of the proportion of future months that the company's production would be expected to share this desirable property (assuming that conditions remain constant). There is a new subtlety here, however: this is really a degree-of-belief frequency interpretation, and there is no guarantee that, in the long run, the company will actually experience 95.3% of complaint-free months. However, it will be seen (again in Chapter 5) that, with appropriate choice of objective prior distributions, the long-run frequency interpretation is also valid.

Here is a prediction example with continuous observables.

Example 1.5 *Future continuous observations.* From analysis of the breaking strengths $\boldsymbol{x} = (x_1, \ldots, x_{10})$ of 10 randomly chosen safety belt webbings, the quality assurance engineer of the firm has to verify whether or not the material satisfies a requirement from the client that individual webbings have strength above 26 kN. The engineer may then be able to conclude that $\Pr(x > 26 \,|\, \boldsymbol{x}, A, K) = 0.9987$. This should be regarded as a measure, on a $[0, 1]$ scale, of the conditional uncertainty (given observed data, accepted assumptions and contextual knowledge) associated with the event that a randomly chosen safety belt webbing will support no less than 26 kN. If production conditions remain constant, it will also be an estimate of the proportion of safety belts that will conform to this particular specification (with a long-run frequency interpretation if appropriately chosen objective prior distributions are used).

Often, covariates are present in prediction problems.

Example 1.6 *Regression.* In regression, observation y_i is observed at covariate value v_i, for $i = 1, \ldots, n$. For instance, the pollution at the center of a coastal city, measured as the particle density y_i (in μgr/m^3), arises in part from open discharge of coal from ships in the harbor. The pollution is highly dependent on speed and direction of the wind at the moment when a ship is discharged; this covariate can be summarized as the component of wind speed v_i coming from the harbor (the absolute speed in m/sec times the cosine of the separation angle) and was recorded in n episodes of similar ship discharges (along with the corresponding y_i) .

A new environmental directive requires the density of particles present in the air to be below $y^* = 90$ μgr/m^3. Thus, of primary interest is the probability $\Pr(y > y^* \,|\, v, \boldsymbol{x}, A, K)$, which, as usual, is a measure of the uncertainty associated with the event that the pollution levels would be higher than the required limit, conditional on the data $\boldsymbol{x} = \{(y_1, v_1), \ldots, (y_n, v_n)\}$, the probabilistic models used for the data, and background knowledge about the pollution levels. In addition, this probability is conditional on the covariate v, *i.e.*, the wind conditions present. Knowledge of this probability can be used to manage the timing of coal discharges from ships. Moreover, the probability is an estimate of the proportion of days with these wind conditions in which, if a similar ship were discharged, the contamination levels would not be acceptable.

The preceding interpretations of probability, in terms of (objective) degree-of-belief, directly appeals to intuition, and there are numerous formal derivations of such probability based on sensible axiom systems (Bernardo and Smith, 1994, Ch. 2, and references therein). This is not to say that other interpretations of probability are not useful – in particular, the frequency notion of probability was seen in the above examples to also arise naturally. Furthermore, the frequency notion is often the simplest notion of probability to use in defining statistical models. For instance, saying that data are independent realizations from a Bernoulli distribution, with θ being the probability of a success, is most easily given meaning in a frequentist sense: thus, θ is the long-fun frequency of successes in repeated trials from the distribution, and the Bernoulli probabilities themselves can be thought of as the long-run probabilities of success and failure, given θ.

Statistical models can be defined using degree-of-belief considerations. For instance, de Finetti (1937) shows that one can derive the Bernoulli distribution, and give an interpretation for θ solely based on a subjectivist interpretation of probability. But, in practice, most have found it much easier to define models using frequentist thinking, based on considering 'what would have been the probabilities of other possible data.'

While many objective Bayesians believe that models are best defined using frequentist reasoning, virtually all feel that, given the model and the observed data, degree-of-belief (posterior) probabilities are best for inference. There are two primary reasons for the conclusion that degree-of-belief probability should be used for inferential statements. The first is purely practical – most statistical users interpret probability in a degree-of-belief sense, and so giving inference in another language (*e.g.*, the frequentist language) is simply a prescription for confusion. Fundamentally, however, there is also motivation for Bayesian inference based on the need to provide satisfactory 'conditional inference' statements; see Section 2.1.3 and the references therein for further discussion. In any case, these should only be thought of as general guidelines; both frequentist and degree-of-belief probabilities will be relevant throughout the discussion, and we will often argue that it is optimal to simultaneously utilize both.

1.2 Parametric Inference

Let $\boldsymbol{x}$ be the *data* and $\{p(\boldsymbol{x} \mid \boldsymbol{\theta}), \boldsymbol{x} \in \boldsymbol{\mathcal{X}}, \boldsymbol{\theta} \in \boldsymbol{\Theta}\}$ be a statistical model, assumed to describe, for some (unknown) value of the *parameter* $\boldsymbol{\theta}$, the probabilistic mechanism that has generated $\boldsymbol{x}$; thus $p(\boldsymbol{x} \mid \boldsymbol{\theta}) > 0$ for all $\boldsymbol{x} \in \boldsymbol{\mathcal{X}}$, and $\int_{\boldsymbol{\mathcal{X}}} p(\boldsymbol{x} \mid \boldsymbol{\theta})\, d\boldsymbol{\theta} = 1$ for all $\boldsymbol{\theta} \in \boldsymbol{\Theta}$, the *parameter space*. There is no theoretical limitation on the structure of the parameter $\boldsymbol{\theta}$, but in this book we restrict attention to parameter vectors of finite dimension, so that $\boldsymbol{\theta} = (\theta_1, \ldots, \theta_m)$.

1.2.1 The Learning Process

Let $\pi(\boldsymbol{\theta})$ be the density function of a probability distribution which describes the current state of knowledge (or uncertainty) about the value of $\boldsymbol{\theta}$, and suppose that new information is acquired in the form of the data $\boldsymbol{x}$. Bayes theorem shows how to incorporate this additional information with that represented by $\pi(\boldsymbol{\theta})$, resulting in the *posterior* density $\pi(\boldsymbol{\theta} \mid \boldsymbol{x})$, which describes the uncertainty about $\boldsymbol{\theta}$ in light of the combined information.

Proposition 1.1 (Bayes Theorem) *Let $\{p(\boldsymbol{x} \mid \boldsymbol{\theta}), \boldsymbol{x} \in \boldsymbol{\mathcal{X}}, \boldsymbol{\theta} \in \boldsymbol{\Theta}\}$ be a statistical model, $\pi(\boldsymbol{\theta})$ be the prior probability density for $\boldsymbol{\theta}$, and $\boldsymbol{x}$ be the actual data. Then the* posterior density *(the conditional probability density of $\boldsymbol{\theta}$ given $\boldsymbol{x}$) is* *given by*

$$\pi(\boldsymbol{\theta}\,|\,\boldsymbol{x}) = \frac{p(\boldsymbol{x}\,|\,\boldsymbol{\theta})\,\pi(\boldsymbol{\theta})}{\int_{\boldsymbol{\Theta}} p(\boldsymbol{x}\,|\,\boldsymbol{\theta})\,\pi(\boldsymbol{\theta})\,d\boldsymbol{\theta}}\,, \tag{1.1}$$

and this encapsulates the available information about $\boldsymbol{\theta}$ *from the data and the prior knowledge. If the parameter space is* $\boldsymbol{\Theta} = \{\boldsymbol{\theta}_1, \boldsymbol{\theta}_2, \ldots\}$, *so that* $\boldsymbol{\theta}$ *is a* discrete *parameter and* $\pi(\boldsymbol{\theta}_i) = \Pr(\boldsymbol{\theta}_i)$ *is the prior probability that* $\boldsymbol{\theta}_i$ *occurs, the posterior probability of* $\boldsymbol{\theta}_i$, *given the data* $\boldsymbol{x}$, *is*

$$\pi(\boldsymbol{\theta}_i\,|\,\boldsymbol{x}) \equiv \Pr(\boldsymbol{\theta}_i\,|\,\boldsymbol{x}) = \frac{p(\boldsymbol{x}\,|\,\boldsymbol{\theta}_i)\,\Pr(\boldsymbol{\theta}_i)}{\sum_j p(\boldsymbol{x}\,|\,\boldsymbol{\theta}_j)\,\Pr(\boldsymbol{\theta}_j)}\,, \quad i = 1, 2, \ldots. \tag{1.2}$$

The proof of Bayes theorem is elementary in the discrete case. For a detailed technical discussion of its expression in the continuous case, see Hartigan (1983, Ch. 3). The systematic use of Bayes theorem to incorporate the information provided by the data is what justifies the adjective *Bayesian* by which the paradigm is usually known.

It is immediate from Bayes theorem that any value of $\boldsymbol{\theta}$ with zero prior density will have zero posterior density, so that no amount of evidence can overcome this initial belief. It is thus often recommended to choose prior densities so that $p(\boldsymbol{\theta}) > 0$ for all $\boldsymbol{\theta} \in \boldsymbol{\Theta}$. As Savage (1961) put it, keep the mind open, or at least ajar. It is also useful for various technical purposes to have strictly positive prior densities. When operating with a single prior density, this can always be achieved by appropriate restriction of the parameter space $\boldsymbol{\Theta}$ (simply redefine $\boldsymbol{\Theta}$ to be the set of $\boldsymbol{\theta}$ values for which $p(\boldsymbol{\theta}) > 0$).

Example 1.7 *Medical diagnosis (continued).* Consider, again, the simple situation of Example 2.4, where a particular test, designed to detect a virus, is known from laboratory research to give a positive result (+) in 98% of the infected people and in 1% of the non-infected. Let V denote presence of the infection; in the above notation, the unknown parameter can take the two values V and $\overline{V}$ or 'not V.' Let $p = \Pr(V)$ be the prior probability of a person being infected (the *prevalence* of the infection in the population under study), so that $\Pr(\overline{V}) = 1 - p$. Using (1.2), the posterior probability that a person who tested positive is infected is given by

$$\Pr(V\,|\,+) = \frac{p(+\,|\,V)\,\Pr(V)}{p(+\,|\,V)\,\Pr(V) + p(+\,|\,\overline{V})\,\Pr(\overline{V})} = \frac{0.98\,p}{0.98\,p + 0.01\,(1-p)}\,.$$

Figure 1.1 graphs this as a function of $p = \Pr(V)$.

As one would expect, the posterior probability is only zero if the prior probability is zero (so that it is known that the population is free of infection) and it is only one if the prior probability is one (so that it is known that the population is universally infected). Note that, if the infection is rare, then the posterior probability of a randomly chosen person being infected is relatively low even if the test is positive. Indeed, for say $\Pr(V) = 0.002$,

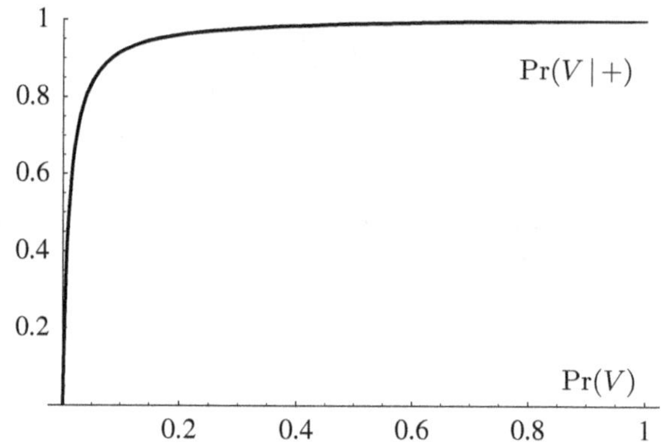

Figure 1.1: Posterior probability of infection $\Pr(V \mid +)$, given a positive test, as a function of the prior probability of infection $\Pr(V)$.

one finds $\Pr(V \mid +) = 0.164$; thus, in a population where only 0.2% of individuals are infected, only 16.4% of those testing positive within a random sample would actually prove to be infected: most positives would actually be *false positives.*

Determination of posterior densities is often facilitated by noting that Bayes theorem can be expressed as

$$\pi(\boldsymbol{\theta} \mid \boldsymbol{x}) \propto p(\boldsymbol{x} \mid \boldsymbol{\theta})\, \pi(\boldsymbol{\theta}), \tag{1.3}$$

where $\propto$ stands for 'proportional to.' The missing proportionality constant, $[\int_{\boldsymbol{\Theta}} p(\boldsymbol{x} \mid \boldsymbol{\theta})\, \pi(\boldsymbol{\theta})\, d\boldsymbol{\theta}]^{-1}$, can always be determined from the fact that $\pi(\boldsymbol{\theta} \mid \boldsymbol{x})$, a probability density, must integrate to one. Hence, to identify the form of a posterior distribution it often suffices to identify a *kernel* of the corresponding probability density, that is a function $k(\boldsymbol{\theta}, \boldsymbol{x})$ such that $\pi(\boldsymbol{\theta} \mid \boldsymbol{x}) = c(\boldsymbol{x})\, k(\boldsymbol{\theta}, \boldsymbol{x})$, for some $c(\boldsymbol{x})$ which does not involve $\boldsymbol{\theta}$. Then one knows that the posterior distribution is that represented by the kernel, without needing to explicitly compute the normalizing constant. In the examples which follow, this technique will frequently be used.

1.2.2 The Likelihood Principle

Recall that, considered as a function of $\boldsymbol{\theta}$, $\text{Lik}(\boldsymbol{\theta} \mid \boldsymbol{x}) = p(\boldsymbol{x} \mid \boldsymbol{\theta})$ is referred to as the *likelihood function* corresponding to the data $\boldsymbol{x}$. Thus, using (1.3), Bayes theorem can be expressed in words by the statement that *the posterior is proportional to the likelihood times the prior.*

Now consider two different *observed* data sets $\boldsymbol{x}_1$ and $\boldsymbol{x}_2$, from two possibly different statistical models

$$\{p_1(\boldsymbol{x}_1 \mid \boldsymbol{\theta}), \boldsymbol{x}_1 \in \boldsymbol{\mathcal{X}}_1, \boldsymbol{\theta} \in \boldsymbol{\Theta}\}, \quad \{p_1(\boldsymbol{x}_2 \mid \boldsymbol{\theta}), \boldsymbol{x}_2 \in \boldsymbol{\mathcal{X}}_2, \boldsymbol{\theta} \in \boldsymbol{\Theta}\}$$

sharing the same parameter $\boldsymbol{\theta} \in \boldsymbol{\Theta}$, which yield *proportional* likelihood functions, so that $\mathrm{Lik}_2(\boldsymbol{\theta} \mid \boldsymbol{x}_2) = c\,\mathrm{Lik}_1(\boldsymbol{\theta} \mid \boldsymbol{x}_1)$ for all $\boldsymbol{\theta}$ values and some constant c. It immediately follows from (1.3) that, provided the *same* prior $\pi(\boldsymbol{\theta})$ is used, the two data sets, $\boldsymbol{x}_1$ or $\boldsymbol{x}_2$, will produce identical posterior distributions for $\boldsymbol{\theta}$.

This consequence of Bayes theorem is often phrased as "the evidence about (the common) $\boldsymbol{\theta}$ that is conveyed by $\boldsymbol{x}_1$ or $\boldsymbol{x}_2$ is the same when their likelihood functions are proportional." This has been proposed as an independent principle, called the *likelihood principle*, and was shown by Birnbaum (1962) to be equivalent to two other commonly accepted principles, the conditionality principle (see Section 2.1.3) and the sufficiency principle (see below). It is thus seen by many as an obvious requirement for reasonable statistical inference. The main limitation of the likelihood principle is that it does not say how to use the likelihood function to make inferences. In Berger and Wolpert (1988) and in Berger, Liseo and Wolpert (1999), it is argued that the most natural way of using the likelihood function for inference is through objective Bayesian analysis. The likelihood principle thus provides one route towards justification of the approach taken in this book.

One consequence of the likelihood principle (or Bayes theorem) is that inferences (or the posterior distribution) depend only on the data $\boldsymbol{x}$ actually observed, and not on the sampling space $\boldsymbol{\mathcal{X}}$. Thus frequentist evaluations (see Section 2.1.3) are not naturally compatible with the likelihood principle (or Bayesian analysis).

Note, however, that the likelihood principle only applies to inferences about $\boldsymbol{\theta}$ after the data have been obtained. Consideration of the sampling space is essential in other aspects of statistical analysis; for instance, in model criticism, in the design of experiments, in prediction problems and, as will later become apparent, in the construction of many objective Bayesian procedures. (There might appear to be a contradiction in this latter statement; we return to this discussion in Section 7.5.3.)

1.2.3 Sufficiency

Three Definitions: For a given statistical model, a *statistic* (*i.e.*, a function of the data) $\boldsymbol{t} = \boldsymbol{t}(\boldsymbol{x})$ is *sufficient* for $\boldsymbol{\theta}$ if $\boldsymbol{t}(\boldsymbol{x})$ contains all the information about $\boldsymbol{\theta}$ that is available from $\boldsymbol{x}$. There are several ways of formally defining sufficiency. One, following from the likelihood principle, is to state that a statistic is sufficient if the likelihood function can be written as $p(\boldsymbol{x} \mid \boldsymbol{\theta}) = f(\boldsymbol{\theta}, \boldsymbol{t})g(\boldsymbol{x})$, for some functions f and g; since the likelihood principle states that proportional likelihoods convey the same information, this would imply that $f(\boldsymbol{\theta}, \boldsymbol{t})$ conveys the same information as the original likelihood. The natural definition of sufficiency from a Bayesian perspective is in terms of the posterior distribution:

Definition 1.1 Sufficient statistic. *A function of the data* $\boldsymbol{t} = \boldsymbol{t}(\boldsymbol{x})$ *is sufficient if (and only if), for any prior* $\pi(\boldsymbol{\theta})$, *the posterior distribution of* $\boldsymbol{\theta}$ *only depends on the data* $\boldsymbol{x}$ *through* $\boldsymbol{t}(\boldsymbol{x})$.

If $\boldsymbol{t} = \boldsymbol{t}(\boldsymbol{x})$ is sufficient, then the posterior distribution of $\boldsymbol{\theta}$ can be directly computed in terms of the sampling distribution of $\boldsymbol{t}$, so that

$$\pi(\boldsymbol{\theta} \mid \boldsymbol{x}) = \pi(\boldsymbol{\theta} \mid \boldsymbol{t}) \propto p(\boldsymbol{t} \mid \boldsymbol{\theta})\, \pi(\boldsymbol{\theta}).$$

The most common definition of a sufficient statistic is the frequentist definition, that $\boldsymbol{t} = \boldsymbol{t}(\boldsymbol{x})$ is sufficient if (and only if) the conditional distribution of $\boldsymbol{x}$, given $\boldsymbol{t}(\boldsymbol{x})$, does not depend on $\boldsymbol{\theta}$. Under very mild conditions, all three definitions of sufficiency are equivalent. From a practical perspective, the first definition is the most useful, in that, by simply looking at the likelihood function, one can often identify a useful sufficient statistic. Sufficient statistics are particularly useful when they are of much lower dimension that the original data $\boldsymbol{x}$.

Examples and Associated Objective Inferences.

Example 1.8 *Inference for a binomial parameter.* Consider a sample of n iid Bernoulli observations $\boldsymbol{x} = (x_1, \ldots, x_n)$, $x_i \in \{0, 1\}$, with parameter $\theta = \Pr(x_i = 1)$. If r of the x_i are observed to equal 1, it is clear that $p(\boldsymbol{x} \mid \theta, n) = \theta^r (1-\theta)^{n-r}$; thus, from the likelihood version of sufficiency, it is clear that $\boldsymbol{t}(\boldsymbol{x}) = (r, n)$ is a sufficient statistic.

It will later be established (Chapters 4 and 9) that, in this problem, lack of prior information about θ is appropriately described by the (proper) Beta density

$$\pi(\theta) = \mathrm{Be}(\theta \mid \tfrac{1}{2}, \tfrac{1}{2}) \propto \theta^{-1/2}(1-\theta)^{-1/2},$$

which is called the *Jeffreys-rule* prior for a binomial parameter, proposed in Jeffreys (1961). (Some, *e.g.*, Bayes (1763), prefer using a uniform density to represent a lack of prior information about θ. This will be discussed in Chapter 3; see, especially, Subsection 3.5.)

Using the proportional form of Bayes theorem in (1.3), the posterior density of θ is then

$$\begin{aligned}\pi(\theta \mid r, n) &\propto \theta^r (1-\theta)^{n-r}\, \theta^{-1/2}(1-\theta)^{-1/2} \\ &\propto \theta^{r-1/2}(1-\theta)^{n-r-1/2}\,,\end{aligned}$$

which is recognizable from Table 1 as being the kernel of another Beta distribution, namely the $\mathrm{Be}(\theta \mid r + \frac{1}{2}, n - r + \frac{1}{2})$ density. Note that it was not necessary to involve the normalizing constants of the prior density and the posterior density, in carrying out this analysis.

As an application, suppose a random survey of $n = 1500$ citizens is conducted, out of which $r = 720$ declare that they would vote in favor of a proposed political measure in a referendum. Using the above results, it follows that the posterior density of θ, given the data, is $\mathrm{Be}(\theta \mid 720.5, 780.5)$. This posterior density is plotted in the top part of Figure 1.2. It can be verified that $\Pr(\theta < 0.5 \mid 720, 1500) = 0.934$ (the area of the shaded region

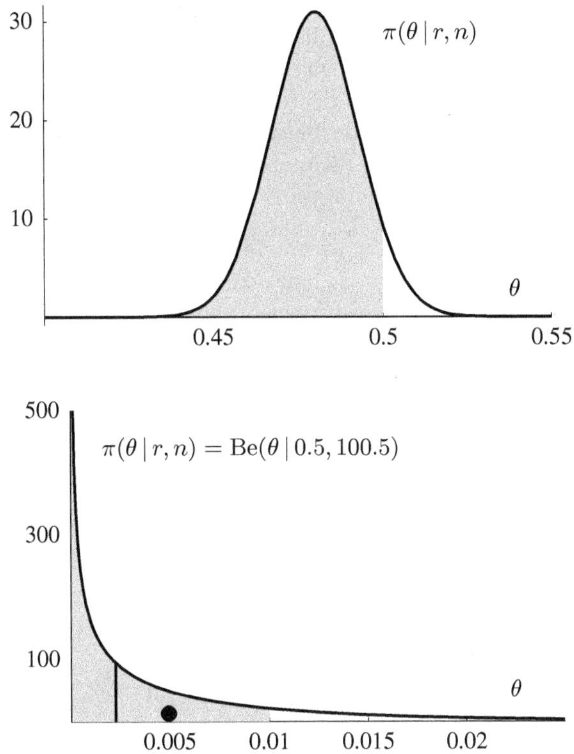

Figure 1.2: Above: Objective posterior density for the proportion θ of citizens that would vote in favor of a referendum. Below: Objective posterior density of the proportion of infected people, based on a random sample of 100 individuals, all of whom tested negative.

in the top part of Figure 1.2). Thus, conditional on the data and the given information, the probability that the referendum will be lost is 0.934.

As another application, with a different flavor, suppose that $n = 100$ randomly selected people were tested for an infection and all tested negative, so that $r = 0$. Suppose that no prior information is available as to the value of θ, the proportion of infected people in the population. (It might also be the case that regulatory requirements preclude use of prior information). As discussed above, the objective prior leads to the (objective) posterior density $\pi(\theta \mid r, n) = \text{Be}(\theta \mid r + \frac{1}{2}, n - r + \frac{1}{2})$ which, for the illustration with $n = 100$ and $r = 0$, is the $\text{Be}(\theta \mid 0.5, 100.5)$ posterior density.

This posterior density is graphed in the lower part of Figure 1.2. Various interesting features of this posterior density: (i) The percentage of infected people is almost surely smaller than 5% (the event $\theta > 0.05$ has posterior probability 0.0013); (ii) the probability that θ is smaller than 0.01 is 0.844

(the area of the shaded region in the lower part of Figure 1.2); (iii) the median of the posterior density is 0.00227 (at the location of the vertical line in the figure), so that it is equally likely for θ to be above or below this value; (iv) the mean of the posterior density is $\mathrm{E}[\theta \mid r, n] = 0.00495$, (represented in the figure by the black circle), which has the additional interpretation of being a good estimate of the probability that a person randomly chosen from the population is infected.

Example 1.9 *Inference for the parameters of a Normal model.* Consider a random sample $\boldsymbol{x} = (x_1 \ldots, x_n)$ of size n from a normal distribution $\mathrm{N}(x \mid \mu, \sigma^2)$, so that the likelihood function is

$$p(\boldsymbol{x} \mid \mu, \sigma) = \prod_{i=1}^{n} \mathrm{N}(x_i \mid \mu, \sigma^2) \propto \sigma^{-n} \exp\left\{-\frac{n}{2\sigma^2}\left[(\overline{x} - \mu)^2 + s^2\right]\right\},$$

where $\overline{x} = n^{-1}\sum_{i=1}^{n} x_i$ and $s^2 = n^{-1}\sum_{i=1}^{n}(x_i - \overline{x})^2$ are, respectively, the sample mean and the sample variance. It follows from the likelihood version of sufficiency that $\boldsymbol{t}(\boldsymbol{x}) = (\overline{x}, s^2)$ is a sufficient statistic.

It will later be established that, in this problem, lack of prior information about either μ or σ is appropriately described by the (improper) prior function $\pi(\mu, \sigma) = \sigma^{-1}$. Using the proportional form of Bayes theorem in (1.3), the joint posterior density of (μ, σ) is then

$$\pi(\mu, \sigma \mid \boldsymbol{x}) = \pi(\mu, \sigma \mid \overline{x}, s^2) \propto \sigma^{-(n+1)} \exp\left\{-\frac{n}{2\sigma^2}\left[(\overline{x} - \mu)^2] + s^2\right]\right\}.$$

Usually, the mean μ is the parameter of interest. It's posterior density is just the marginal posterior density found by integrating $\pi(\mu, \sigma \mid \boldsymbol{x})$ over σ, resulting in

$$\begin{aligned}
\pi(\mu \mid \boldsymbol{x}) &= \int_0^\infty \pi(\mu, \sigma \mid \boldsymbol{x})\, d\sigma \\
&\propto \int_0^\infty \sigma^{-(n+1)} \exp\left\{-\frac{n}{2\sigma^2}\left[(\overline{x} - \mu)^2] + s^2\right]\right\} d\sigma \\
&\propto \left[s^2 + (\mu - \overline{x})^2\right]^{-n/2},
\end{aligned}$$

which can be recognized as the kernel of the Student t distribution, centered at $\overline{x}$, with scale $s/\sqrt{n-1}$, and with $n-1$ degrees of freedom. Thus

$$\pi(\mu \mid \boldsymbol{x}) = \mathrm{St}(\mu \mid \overline{x}, s/\sqrt{n-1}, n-1). \tag{1.4}$$

Similarly, the marginal posterior density of σ is

$$\begin{aligned}
\pi(\sigma \mid \boldsymbol{x}) &= \int_{-\infty}^{\infty} \pi(\mu, \sigma \mid \boldsymbol{x})\, d\mu \\
&\propto \sigma^{-n} \exp\left\{-\frac{ns^2}{2\sigma^2}\right\} \int_{-\infty}^{\infty} \sigma^{-1} \exp\left\{-\frac{n}{2\sigma^2}(\mu - \overline{x})^2)\right\} d\mu \\
&\propto \sigma^{-n} \exp\left\{-\frac{ns^2}{2\sigma^2}\right\}.
\end{aligned}$$

(Change variables in the integral to $y = (\mu - \overline{x})/\sigma$ to see that the integral is a constant.) The change of variables $\sigma \to \sigma^2$ has Jacobian $\frac{d\sigma}{d\sigma^2} = \frac{d\sqrt{\sigma^2}}{d\sigma^2} = \frac{1}{2\sqrt{\sigma^2}}$, so the marginal posterior density of σ^2 is

$$\pi(\sigma^2 \,|\, \boldsymbol{x}) \propto [\sigma^2]^{-n/2} \exp\left\{-\frac{ns^2}{2\sigma^2}\right\} \frac{1}{\sqrt{\sigma^2}} = [\sigma^2]^{-(n+1)/2} \exp\left\{-\frac{ns^2}{2\sigma^2}\right\}.$$

This is clearly the kernel of the Inverted Gamma density (see Appendix A),

$$\pi(\sigma^2 \,|\, \boldsymbol{x}) = \mathrm{Ig}(\sigma^2 \,|\, \tfrac{1}{2}(n-1), \tfrac{1}{2}\, ns^2)\,,$$

which is thus the posterior distribution of σ^2.

Exponential Families. This is a technical section and can be skipped. Sufficiency can always be usefully applied whenever the statistical model belongs to the *exponential family*, which includes many of the more frequently used statistical models.

A probability density (or mass function) $p(x \,|\, \boldsymbol{\theta})$, $x \in \mathcal{X}$, $\boldsymbol{\theta} \in \Theta \subseteq \mathbb{R}^m$, is said to belong to the m-parameter exponential family if it is of the form

$$p(x \,|\, \boldsymbol{\theta}) = \mathrm{Ef}_m(x \,|\, f, g, \boldsymbol{h}, \boldsymbol{\phi}, \boldsymbol{\theta}, \boldsymbol{c}) = f(x)\, g(\boldsymbol{\theta}) \exp\left\{\sum_{i=1}^{m} c_i\, \phi_i(\boldsymbol{\theta})\, h_i(x)\right\},$$

where $\boldsymbol{h} = (h_1, \ldots, h_m)$, $\boldsymbol{\phi}(\boldsymbol{\theta}) = (\phi_1, \ldots, \phi_m)$ and, given the functions $f, \boldsymbol{h}, \boldsymbol{\phi}$, and the constants c_i,

$$\frac{1}{g(\boldsymbol{\theta})} = \int_{\mathcal{X}} f(x) \exp\left\{\sum_{i=1}^{m} c_i\, \phi_i(\boldsymbol{\theta}) h_i(x)\right\} dx < \infty.$$

The family is called regular if $\mathcal{X}$ does not depend on $\boldsymbol{\theta}$ and non-regular otherwise.

Given a random sample $\boldsymbol{x} = (x_1, \ldots, x_n)$ from a regular m-parameter exponential family model $\mathrm{Ef}_m(x_i \,|\, f, g, \boldsymbol{h}, \boldsymbol{\phi}, \boldsymbol{\theta}, \boldsymbol{c})$, the function

$$\boldsymbol{t} = \boldsymbol{t}(x_1, \ldots, x_n) = \left[n, \sum\nolimits_{i=1}^{n} h_1(x_i), \ldots, \sum\nolimits_{i=1}^{n} h_m(x_i)\right]$$

is a sufficient statistic of fixed dimension m.

For example, the Bernoulli distribution $p(x \,|\, \theta) = \mathrm{Br}(x \,|\, \theta) = \theta^x (1-\theta)^{1-x}$, $x \in (0, 1)$, $\theta \in [0, 1]$, is a one parameter exponential family model with $f(x) = 1$, $g(\theta) = 1 - \theta$, $h(x) = x$, $\phi(\theta) = \log \theta/(1-\theta)$ and $c = 1$. The corresponding sufficient statistic is $\boldsymbol{t}(\boldsymbol{x}) = \{n, r\}$, $r = \sum_{i=1}^{n} x_i$.

When working with exponential families, a noticeable notational simplification is obtained by defining $\boldsymbol{\psi} = (\psi_1, \ldots, \psi_m)$, where $\psi_i = c_i\, \phi_i(\boldsymbol{\theta})$, and $\boldsymbol{y} = (y_1, \ldots, y_m)$, where $y_i = h(x_i)$. Then the density can be rewritten

$$p(\boldsymbol{y} \,|\, \boldsymbol{\psi}) = a(\boldsymbol{y}) \exp\left\{\boldsymbol{y}^t \boldsymbol{\psi} - b(\boldsymbol{\psi})\right\}, \quad \mathbf{y} \in \mathcal{Y},$$

where

$$b(\boldsymbol{\psi}) = -\log\left(g\left(\phi_1^{-1}\left(\frac{\psi_1}{c_1}\right),\ldots,\phi_m^{-1}\left(\frac{\psi_m}{c_m}\right)\right)\right)$$

and $a(\boldsymbol{y})$ is found by changing variables from $\boldsymbol{x}$ to $\boldsymbol{y}$. (In applying Bayes theorem, $a(\boldsymbol{y})$ is irrelevant.) This is referred to as the *canonical* (or natural) form of the exponential family.

For more details on exponential families, see Brown (1986), Bernardo and Smith, (1994, Ch. 4), and references therein. Objective priors for the exponential family are given in Consonni et al. (2004).

1.2.4 Conjugate Priors

Definition: To obtain an analytical expression for the posterior distribution given by Bayes theorem

$$\pi(\boldsymbol{\theta}\,|\,\boldsymbol{x}) = \frac{p(\boldsymbol{x}\,|\,\boldsymbol{\theta})\,\pi(\boldsymbol{\theta})}{\int_{\Theta} p(\boldsymbol{x}\,|\,\boldsymbol{\theta})\,\pi(\boldsymbol{\theta})\,d\boldsymbol{\theta}}\,,$$

the integral in the denominator must be evaluated. This is often easy if the prior distribution $\pi(\boldsymbol{\theta})$ is chosen from the class of distributions with the same 'structure' as the likelihood function $p(\boldsymbol{x}\,|\,\boldsymbol{\theta})$, when this is viewed as a function of $\boldsymbol{\theta}$.

Definition 1.2 Conjugate prior family. *A conjugate family of prior densities (proper or improper) for $\boldsymbol{\theta} \in \Theta$, with respect to a likelihood $p(\boldsymbol{x}\,|\,\boldsymbol{\theta})$, is a class $\{\pi(\boldsymbol{\theta}\,|\,\boldsymbol{\tau}), \boldsymbol{\tau} \in \mathcal{T}\}$, such that the resulting posterior is proper and in the class for any $\boldsymbol{x}$ and any $\pi(\boldsymbol{\theta}\,|\,\boldsymbol{\tau})$.*

Example 1.10 *Bernoulli data.* Suppose $\boldsymbol{x} = (x_1, \ldots, x_n)$ is a random sample of n Bernoulli observations, so that $p(\boldsymbol{x}\,|\,\theta) = \theta^r(1-\theta)^{n-r}$, where $r = \sum_{i=1}^n x_i$. The prior densities having the same structure as this likelihood function are clearly of the form $\pi(\theta\,|\,\boldsymbol{\tau}) \propto \theta^{\tau_1}(1-\theta)^{\tau_2}$. The posterior is proper (for all r and n) if $\tau_1 > 0$ and $\tau_2 > 0$. This defines the class of $\mathrm{Be}(\theta\,|\,\tau_1, \tau_2)$ densities, so the beta densities form a conjugate family of priors for the binomial sampling model. The posterior density is recognizable as the $\mathrm{Be}(\theta\,|\,r+\tau_1, n-r+\tau_2)$ density, so it is not necessary to compute the posterior normalizing constant. Note that this is not always the case, however. For instance the class of all prior distributions is clearly also a conjugate family of priors but, for most priors in that class, the normalizing constant would need to be computed numerically.

Conjugate Priors for Models in the Exponential Family. If the model belongs to the exponential family class and this is parametrized in its natural form, $p(\boldsymbol{y}\,|\,\boldsymbol{\psi}) = a(\boldsymbol{y})\exp\{\boldsymbol{y}^t\boldsymbol{\psi} - b(\boldsymbol{\psi})\}$, the prior densities of the same structural form are, with $\boldsymbol{\tau} = (\tau_1, \ldots, \tau_m)$,

$$p(\boldsymbol{\psi}\,|\,\tau_0, \boldsymbol{\tau}) \propto \exp\left\{\tau_0(\boldsymbol{\tau}^t\boldsymbol{\psi} - b(\boldsymbol{\psi}))\right\}, \quad \boldsymbol{\psi} \in \Psi\,.$$

The conjugate family would be defined by the set of all $(\tau_0, \boldsymbol{\tau})$ for which the posterior is proper for a sample of size one, *i.e.*, for which

$$\int_{\boldsymbol{\Psi}} \exp\left\{(\boldsymbol{y}^t + \tau_0\boldsymbol{\tau}^t)\boldsymbol{\psi} - (\tau_0+1)b(\boldsymbol{\psi})\right\}\, d\boldsymbol{\psi} < \infty \quad \text{for all } \boldsymbol{y}\,.$$

The class of such prior distributions is often called the *natural conjugate family*, to distinguish it from other conjugate classes.

If $(\boldsymbol{y}_1, \ldots, \boldsymbol{y}_n)$ is a random sample of size n from the canonical exponential family model, the posterior density corresponding to the conjugate prior $p(\boldsymbol{\psi} \mid \tau_0, \boldsymbol{\tau})$ is clearly given by

$$p(\boldsymbol{\psi} \mid \tau_0, \boldsymbol{\tau}, \boldsymbol{y}_1, \ldots, \boldsymbol{y}_n) \propto \exp\left\{\left(\sum_{i=1}^{n} \boldsymbol{y}_i^t + \tau_0\boldsymbol{\tau}^t\right)\boldsymbol{\psi} - (\tau_0+n)b(\boldsymbol{\psi})\right\},$$

which can be recognized as being the $p\left(\boldsymbol{\psi} \mid \tau_0+n,\ [n\overline{\boldsymbol{y}}_n^t + \tau_0\boldsymbol{\tau}^t]/(\tau_0+n)\right)$ density from the natural conjugate family, where $\overline{\boldsymbol{y}}_n = \sum_{i=1}^n \boldsymbol{y}_i/n$. The hyperparameter τ_0 is often interpreted as the equivalent sample size for the prior information, since it plays the same role in the posterior as does n.

Example 1.11 *Bernoulli data (continued).* In canonical form, the Bernoulli model is $p(x \mid \theta) = \exp\{x \log(\theta/(1-\theta)) + \log(1-\theta)\}$ and, hence, the natural conjugate priors are of the form

$$p(\theta \mid \tau_0, \tau_1) \propto \exp\{\tau_0[\tau_1 \log(\theta/(1-\theta)) + \log(1-\theta)]\}\,,$$

with the canonical form obtained by setting $y = x$, $\psi = \log(\theta/(1-\theta))$, $a(y) = 1$ and $b(\psi) = \log(1+e^{\psi})$. The posterior distribution of the canonical parameter ψ is given by

$$p(\psi \mid \tau_0, \tau_1, y_1, \ldots, y_n) \propto \exp\left[(\tau_0+n)\left\{\left[\frac{n\overline{y}_n + \tau_0\tau_1}{\tau_0+n}\right]\psi - b(\psi)\right\}\right].$$

As the last example demonstrates, the canonical form may not be as intuitively appealing to work with as the original parameterization. However, the canonical form is very useful for establishing general properties of inference within exponential families; see Diaconis & Ylvisaker (1979) for such results.

1.2.5 Sequential Updating

As one would expect from the coherence induced by probability theory, if data $\boldsymbol{x} = (\boldsymbol{x}_1, \ldots, \boldsymbol{x}_n)$ are sequentially presented, the final result will be the same whether data are globally or sequentially processed. (An example of the latter is clinical trials, where the data from subjects often arises sequentially.) Indeed,

$$\pi(\boldsymbol{\theta} \mid \boldsymbol{x}_1, \ldots, \boldsymbol{x}_{i+1}) \propto p(\boldsymbol{x}_{i+1} \mid \boldsymbol{x}_1, \ldots, \boldsymbol{x}_i, \boldsymbol{\theta})\, \pi(\boldsymbol{\theta} \mid \boldsymbol{x}_1, \ldots, \boldsymbol{x}_i),$$

for $i = 1, \ldots, n-1$, so that the 'posterior' at a given stage can be viewed as the 'prior' at the next. In most (but not all) sequential settings, the new posterior, $\pi(\boldsymbol{\theta} \mid \boldsymbol{x}_1, \ldots, \boldsymbol{x}_{i+1})$, will be more concentrated around the true value of $\boldsymbol{\theta}$ than the previous posterior, $\pi(\boldsymbol{\theta} \mid \boldsymbol{x}_1, \ldots, \boldsymbol{x}_i)$.

To illustrate these ideas and to demonstrate the use of simulated data to study the behavior of a procedure, consider the analysis of data simulated from a Poisson distribution.

Example 1.12 *Simulated Poisson data.* Let $\boldsymbol{x} = (x_1, \ldots, x_n)$ be a random sample of size n from a Poisson distribution $\text{Pn}(x \mid \lambda)$, so that

$$p(\boldsymbol{x} \mid \lambda) = \prod_{i=1}^{n} \frac{\lambda^{x_i} e^{-\lambda}}{x_i!} \propto e^{-n\lambda} \lambda^t ,$$

where $t = \sum_{j=1}^{n} x_j$ is a sufficient statistic. It will be shown later (Chapter 4) that an appropriate objective prior function for this problem is the improper prior function $\pi(\lambda) = \lambda^{-1/2}$. By Bayes theorem, the corresponding posterior is

$$\pi(\lambda \mid \boldsymbol{x}) \propto p(\boldsymbol{x} \mid \lambda)\, \pi(\lambda) \propto e^{-n\lambda} \lambda^{t-1/2},$$

which is the kernel of the Gamma distribution $\text{Ga}(\lambda \mid t + 1/2, n)$, this distribution thus being the objective posterior distribution. Note that the posterior mean of λ is $(t + 1/2)/n = \overline{x} + 1/(2n)$, which is not the sample mean $\overline{x}$.

The effect of the prior is especially appreciable with few or extreme observations. For instance, if $t = 0$, so that no events have been observed in n random Poisson observations, the corresponding posterior mean of λ is $1/(2n)$. While close to zero, this is positive, and arguably superior to the maximum likelihood estimate $\hat{\lambda} = 0$.

For a numerical illustration, suppose that no earthquakes have been observed in $n = 100$ consecutive years at a particular location. Assuming that the number of earthquakes in a year has a Poisson distribution with parameter λ, the posterior distribution of λ is $\text{Ga}(\lambda \mid 1/2, 100)$, represented in Figure 1.3. The expected number of earthquakes per year in that location is then $1/200 = 0.005$, but their median (shown with a dashed line) is only 0.0023.

If, as is often the case with Poisson observations, data arrived sequentially, the objective posterior after the first i observations would be

$$\pi(\lambda \mid x_1, \ldots, x_i) = \text{Ga}(\lambda \mid t_i + 1/2, i) ,$$

where $t_i = \sum_{j=1}^{i} x_j$. With a new observation x_{i+1}, the posterior is

$$\pi(\lambda \mid x_1, \ldots, x_i, x_{i+1}) = \text{Ga}(\lambda \mid t_{i+1} + 1/2, , i + 1).$$

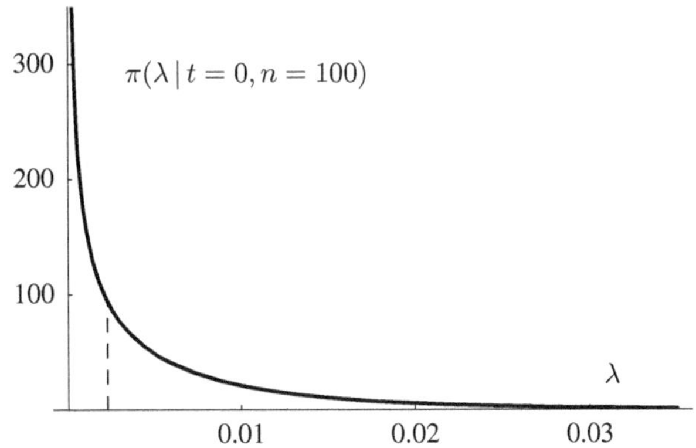

Figure 1.3: Objective posterior density $\pi(\lambda \mid \boldsymbol{x})$ of the parameter of a Poisson distribution, given $t = 0$ and $n = 100$.

This could also be obtained by using Bayes theorem to directly incorporate the last observation. Indeed, noting that x_{i+1} is assumed to be independent of the previous observations,

$$\pi(\lambda \mid x_1, \ldots, x_{i+1}) \propto p(x_{i+1} \mid \lambda)\, \pi(\lambda \mid x_1, \ldots, x_i) \propto e^{-(i+1)\lambda} \lambda^{t_i + x_{i+1} - 1/2},$$

which is the kernel of the Gamma density $\text{Ga}(\lambda \mid t_{i+1} + 1/2, i+1)$, with $t_{i+1} = t_i + x_{i+1}$. The result after all observations have been included will be the final posterior distribution obtained before, $\pi(\lambda \mid \boldsymbol{x}) = \text{Ga}(\lambda \mid t + 1/2, n)$.

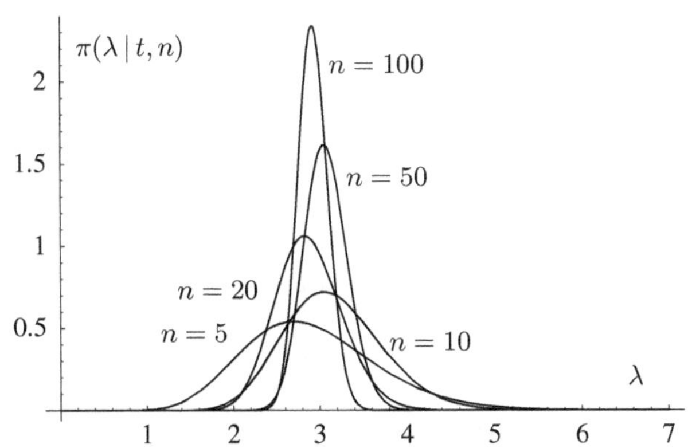

Figure 1.4: Objective posterior distributions $\pi(\lambda \mid t, n)$ of the parameter of a Poisson distribution, derived from partial samples of sizes $n = 5, 10, 20, 50$ and 100, extracted from a sample of size 100 obtained by simulation from a Poisson distribution with parameter $\lambda = 3$.

For a numerical illustration, we simulated a sample of size $n = 100$ from a Poisson distribution with parameter $\lambda = 3$, resulting in a sufficient statistic $t = \sum_j x_j = 292$. The corresponding objective posterior density is $\text{Ga}(\lambda \mid 292.5, 100)$, given in Figure 1.4. We also took the subsamples corresponding to the first $5, 10, 20$, and 50 observations, computed the corresponding posteriors and plotted them in Figure 1.4 for comparison. The progressive, intuitive convergence towards the value $\lambda = 3$, from which the sample was simulated, is immediately apparent.

In the preceding example, and in most situations, the posterior distribution is 'narrower' than the prior. However, this is not always the case: occasionally, a 'surprising' observation will increase, rather than decrease, the uncertainty about the value of $\boldsymbol{\theta}$. For instance, in probabilistic diagnosis, a sharp posterior probability distribution for the possible cause $\theta \in \{\delta_1, \ldots, \delta_s\}$ of a syndrome, *i.e.*, a posterior distribution with a large probability for disease δ_i (describing, a 'clear' diagnosis of the particular disease δ_i), would typically update to a less concentrated posterior probability distribution over $\{\delta_1, \ldots, \delta_s\}$ if a new clinical study yielded data which were unlikely under δ_i.

1.2.6 Sensitivity Analysis

Typically, different priors lead to different posteriors. Indeed, Bayes theorem can be described as a data-driven probability transformation machine which maps prior distributions (describing prior knowledge) into posterior distributions (representing combined prior and data knowledge). It is important to analyze whether or not modest changes in the prior would induce noticeable changes in the posterior; this subject is called *sensitivity analysis.* Posterior distributions based on objective priors play a central role in sensitivity analysis, since they indicate 'what the data alone has to say.' If a posterior resulting from an objective prior differs markedly from that arising from a subjective prior, it behooves the analyst to make sure that the subjective information is reliable and was properly encoded in the prior.

Example 1.13 *Simple sensitivity analysis.* Let x be one observation from a normal $\text{N}(x \mid \mu, 1)$ distribution, and consider three possible priors: a standard Cauchy $\pi_1(\mu) = \text{Ca}(\mu \mid 0, 1)$, a Normal $\pi_2(\mu) = \text{N}(\mu \mid 0, 1.48^2)$ and the objective (improper) uniform prior function $\pi_3(\mu) = 1$. The scales of the Normal and the Cauchy priors were chosen so that $\Pr(-1 < \mu < 1) = \frac{1}{2}$; these two densities are the lower two light gray curves, on the left, in Figure 1.5. The corresponding posterior distributions are immediately obtained by Bayes theorem as $\pi_i(\mu \mid x) \propto \text{N}(x \mid \mu, 1)\, \pi_i(\mu)$, where numerical integration is required in the case of the Cauchy prior to obtain the proportionality constant.

Suppose, first, that $x = 0$ is observed, a value compatible with all three priors. The posterior distributions, all centered at zero, are plotted in Figure 1.5. The objective posterior (solid line) is, as one would expect, the most disperse of the posteriors, as it uses only the information in the data;

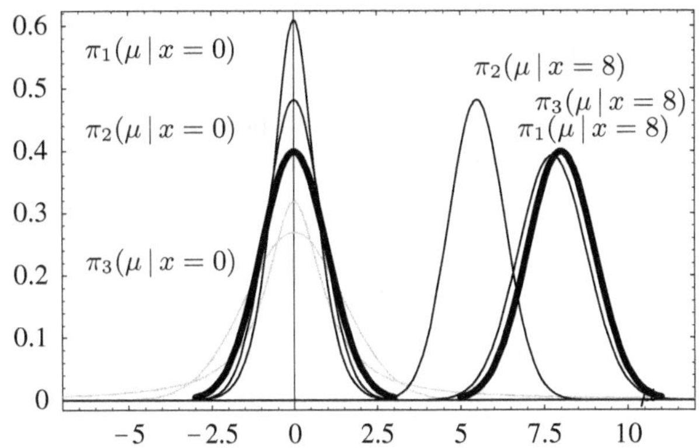

Figure 1.5: Cauchy and normal prior densities – the two low light grey densities on the left. The other curves are posterior distributions for the mean μ of a Normal density, when the observation is $x = 0$ (left) or $x = 8$ (right). $\pi_1(\mu \mid x)$ is the posterior from the Cauchy prior; $\pi_2(\mu \mid x)$ is the posterior from the normal prior; and $\pi_3(\mu \mid x)$ is the posterior from the objective constant prior.

the Cauchy and normal priors yield more peaked, and narrower, posteriors. Considering that we only have one observation, it is not unreasonable to view these posteriors as being compatible, *i.e.*, the sensitivity to the prior is modest.

Suppose next that $x = 8$ is observed, a value obviously compatible with the model and the objective prior, but quite unlikely under the Cauchy prior and almost completely incompatible with the Normal prior. The corresponding posteriors are also plotted in Figure 1.5. Interestingly, the objective posterior and the posterior from the Cauchy prior virtually coincide, but the normal prior produces a posterior shifted far away (3 sample standard deviations) from the region of μ values supported by the data. Thus comparison with the objective posterior suggests that the Cauchy encoding of prior information — that the prior median is 0 and the prior quartiles are ± 1 — is much less sensitive than is the normal prior encoding of that information. Unfortunately, the normal prior encoding is much simpler computationally, because it is a conjugate prior; but the sensitivity analysis suggests that use of the Cauchy encoding is much better.

Table 1.1: Posterior medians.

x	0	2	4	6	8	10	12
Normal	0.00	1.37	2.75	4.12	5.50	6.87	8.25
Cauchy	0.00	1.21	3.44	5.65	7.74	9.80	11.83
Objective	0.00	2.00	4.00	6.00	8.00	10.00	12.00

The posterior medians which correspond to each of the three priors, for values of x ranging from 0 to 12, are shown in Table 1.1. Note that the Cauchy posterior medians are always reasonably compatible with the objective posterior medians, while the normal posterior medians diverge from the objective posterior medians as soon as x becomes incompatible with the normal prior.

1.2.7 Nuisance Parameters

Often only some components of the parameter vector are of interest. In the normal Example 1.9 for instance, the mean μ will often be the *parameter of interest*, while σ (or σ^2) is not of direct interest; parameters that are not of direct interest are called *nuisance parameters*. In that example, it was possible to eliminate the nuisance parameter by simply considering the marginal posterior density of μ (1.4), obtained by integrating the joint posterior density of μ and σ over σ. This is the general theme of Bayesian treatment of nuisance parameters: find the joint posterior distribution of the parameters and then integrate out the nuisance parameters.

It is useful to note how to deal with the more general situation in which some function $\boldsymbol{\phi} = \boldsymbol{\phi}(\boldsymbol{\theta})$ (of possibly lower dimension than $\boldsymbol{\theta}$) is of interest. Doing so is just basic probability theory, but we record the results in terms of the posterior distribution for later use.

Suppose we have the posterior distribution, $\pi(\boldsymbol{\theta} \,|\, \boldsymbol{x})$, and $\boldsymbol{\phi} = \boldsymbol{\phi}(\boldsymbol{\theta}) \in \boldsymbol{\Phi}$ is of interest, so the posterior distribution, $\pi(\boldsymbol{\phi} \,|\, \boldsymbol{x})$, of $\boldsymbol{\phi}$ is desired. This is just the marginal posterior distribution of $\boldsymbol{\phi}$, but to find it one must reparameterize by a one-to-to one transformation and use probability calculus. Thus choose some other function $\boldsymbol{\omega} = \boldsymbol{\omega}(\boldsymbol{\theta}) \in \boldsymbol{\Omega}$, of the original parameters, in such a way that $\boldsymbol{\psi}(\boldsymbol{\theta}) = (\boldsymbol{\phi}, \boldsymbol{\omega})$ is a one-to-one transformation of $\boldsymbol{\theta}$. Let $J(\boldsymbol{\psi}) = (\partial\boldsymbol{\theta}/\partial\boldsymbol{\psi})$ be the Jacobian matrix of the inverse transformation $\boldsymbol{\theta} = \boldsymbol{\psi}^{-1}((\boldsymbol{\phi}, \boldsymbol{\omega}))$ (the (i, j) element of the Jacobian being $\partial\boldsymbol{\theta}_i/\partial\boldsymbol{\psi}_j$). Of course, introducing $\boldsymbol{\omega}$ is not necessary if $\boldsymbol{\phi}(\boldsymbol{\theta})$ is itself a one-to-one transformation of $\boldsymbol{\theta}$. Standard change-of-variables in probability yields that posterior density of $(\boldsymbol{\phi}, \boldsymbol{\omega})$ is

$$\pi(\boldsymbol{\phi}, \boldsymbol{\omega} \,|\, \boldsymbol{x}) = |J(\boldsymbol{\psi})| \Big[\pi(\boldsymbol{\theta} \,|\, \boldsymbol{x})\Big]_{\boldsymbol{\theta}=\boldsymbol{\psi}^{-1}((\boldsymbol{\phi}, \boldsymbol{\omega}))} .$$

The desired posterior of $\boldsymbol{\phi}$ is the marginal posterior density, obtained by integration over the nuisance parameter $\boldsymbol{\omega}$,

$$\pi(\boldsymbol{\phi} \,|\, \boldsymbol{x}) = \int_{\boldsymbol{\Omega}} \pi(\boldsymbol{\phi}, \boldsymbol{\omega} \,|\, \boldsymbol{x})\, d\boldsymbol{\omega}.$$

1.2.8 Restricted Parameter Spaces

Sometimes, the range of possible values of $\boldsymbol{\theta}$ is restricted by contextual considerations. For instance, it could be known that a normal mean is positive.

Suppose the posterior density $\pi(\boldsymbol{\theta}\,|\,\boldsymbol{x})$ is known over the unconstrained space $\boldsymbol{\Theta}$, but one then learns that $\boldsymbol{\theta}$ actually belongs to $\boldsymbol{\Theta}_c \subset \boldsymbol{\Theta}$. Treating this as new knowledge, it is immediate from Bayes theorem that the restricted posterior density is given by

$$\pi(\boldsymbol{\theta}\,|\,\boldsymbol{x}, \boldsymbol{\theta} \in \boldsymbol{\Theta}_c) = \begin{cases} \dfrac{\pi(\boldsymbol{\theta}\,|\,\boldsymbol{x})}{\int_{\boldsymbol{\Theta}_c} \pi(\boldsymbol{\theta}\,|\,\boldsymbol{x})}, & \text{if } \boldsymbol{\theta} \in \boldsymbol{\Theta}_c \\ 0 & \text{otherwise} \end{cases}.$$

Thus, to incorporate a restriction on the possible values of the parameters, it suffices to *renormalize* the unrestricted posterior distribution to the set $\boldsymbol{\Theta}_c \subset \boldsymbol{\Theta}$ of parameter values which satisfy the required condition. This simple renormalization provides another example of the conceptually simple solutions that Bayesian methods provide.

Example 1.14 *Normal Data. Example 1.9 continued.* For a random sample from a $\mathrm{N}(x\,|\,\mu, \sigma^2)$ distribution, with mean and variance unknown, the objective (marginal) posterior density of μ was seen to be $\pi(\mu\,|\,\boldsymbol{x}) = \mathrm{St}(\mu\,|\,\overline{x}, s/\sqrt{n-1}, n-1)$. In the physics classroom experiment to measure the gravitational field g with a pendulum, a student has obtained $n = 20$ measurements of g yielding (in m/sec^2) a mean $\overline{x} = 9.8087$, and a standard deviation $s = 0.0428$. The corresponding (objective) posterior distribution is $\pi(g\,|\,\boldsymbol{x}) = \mathrm{St}(g\,|\,9.8087, 0.0098, 19)$ represented in the upper part of Figure 1.6. Computation yields $\Pr(9.788 < g < 9.829\,|\,\boldsymbol{x}) = 0.95$; thus, the gravitational field at the location of the laboratory can be expected to lie between 9.788 and 9.829 with probability (rational degree of belief) 0.95.

This is actually not formally correct, because we know that $g > 0$ and the parameter space should be restricted accordingly. However, as is immediately obvious from the figure, this would not have had any appreciable effect because the likelihood function is so concentrated on positive g values. Suppose, however, that the student is further instructed to incorporate into the analysis the fact that the value of the gravitational field g at the laboratory is known to lie between 9.7803 m/sec^2 (average value of g at the equator) and 9.8322 m/sec^2 (average value at the poles). The updated posterior distribution will then be

$$p(g\,|\,\boldsymbol{x}, g \in G_c) = \frac{\mathrm{St}(g\,|\,9.8087, 0.0098, 19)}{\int_{g \in G_c} \mathrm{St}(g\,|\,9.8087, 0.0098, 19)}, \quad g \in G_c,$$

where $G_c = \{g;\, 9.7803 < g < 9.8322\}$, represented in the lower part of Figure 1.6. One-dimensional numerical integration yields $\Pr(g > 9.792\,|\,\boldsymbol{x}, g \in G_c) = 0.95$. Thus, adding the knowledge that g must lie in $[9.7803, 9.8322]$, results in the conclusion that the gravitational field at the location of the laboratory lies between 9.792 and 9.8322, with probability (rational degree of belief) 0.95.

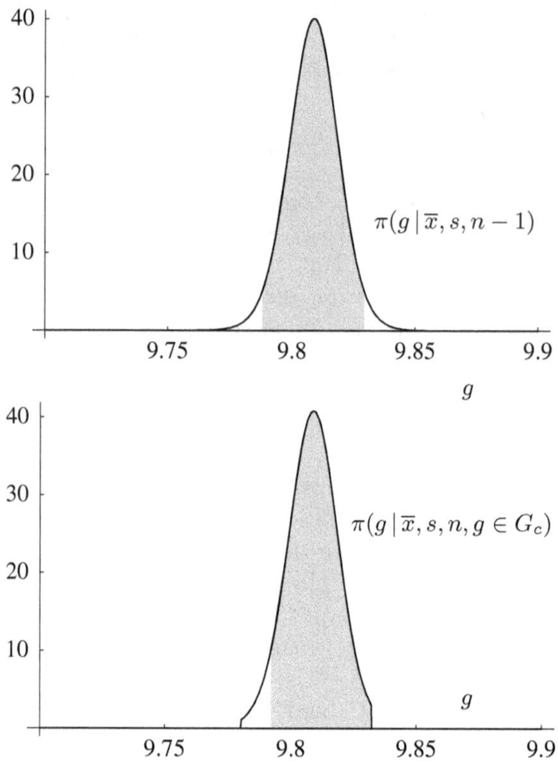

Figure 1.6: Posterior density $\pi(g \mid m, s, n-1)$ of the value g of the gravitational field, given $n = 20$ normal measurements with $\overline{x} = 9.8087$ and $s = 0.0428$, with no additional information (above) and with the value of g restricted to $G_c = \{g;\, 9.7803 < g < 9.8322\}$ (below). Shaded areas correspond to regions of 0.95 probability contents.

1.3 Inference Summaries

From the Bayesian viewpoint, all information about *any* unknown quantity is contained in the corresponding posterior distribution. Thus, given data $\boldsymbol{x}$ and conditions C, *all* that can be said about any function $\boldsymbol{\omega}$ of the parameters which govern the model is contained in the posterior distribution $\pi(\boldsymbol{\omega} \mid \boldsymbol{x}, C)$. [1]

In one or two dimensions, a graph of the posterior density of the quantity of interest (or the probability mass function in the discrete case) immediately conveys an intuitive summary of the location of the quantity of interest. In-

[1] In fact (Bernardo, 1979a), Bayesian inference may technically be described as a decision problem where one has to choose a posterior probability distributions of the quantity of interest encapsulating all available information. The action space is the class of probability distributions of the quantity of interest which are compatible with available information, and the utility function is a measure of the value of the information contained in the posterior.

deed, this readily interpreted graphical summary of the situation is a strength of the Bayesian approach. For instance, all important conclusions about the value of the gravitational field in Example 1.14 are qualitatively available from Figure 1.6.

In more than two dimensions, one can still proceed graphically by presenting the marginal posterior distributions, $\pi(\theta_i \mid \boldsymbol{x})$, for all parameters, and the bivariate marginal distributions, $\pi(\theta_i, \theta_j \mid \boldsymbol{x})$, of all pairs of parameters. If there are many parameters, this can be a lot to process, however, and it is then common to, instead, use numerical summaries of the posterior distribution. Also, many users of statistics have been trained to utilize numerical summaries, such as point estimates or confidence intervals, and so it is of interest to also provide such inferences in Bayesian analysis.

1.3.1 Point Estimation

In statistical analysis it is often of interest to provide point estimates of relevant quantities. It will later be shown (Section 1.6) that Bayesian point estimation is best described as a *decision problem* where one has to choose a particular value $\tilde{\boldsymbol{\theta}}$ of the parameter of interest as a proxy for the actual, unknown value of $\boldsymbol{\theta}$. A formal definition of a Bayesian point estimate is therefore postponed to that section.

At an intuitive level, any appropriate location measure of the posterior distribution $\pi(\boldsymbol{\theta} \mid \boldsymbol{x})$ can be used as a point estimator. Common point estimators include the posterior mean, $\mathrm{E}[\boldsymbol{\theta} \mid \boldsymbol{x}]$, and the posterior mode, $\mathrm{Mo}[\boldsymbol{\theta} \mid \boldsymbol{x}]$, given (when they exist) by

$$\mathrm{E}[\boldsymbol{\theta} \mid \boldsymbol{x}] = \int_{\Theta} \boldsymbol{\theta}\, \pi(\boldsymbol{\theta} \mid \boldsymbol{x})\, d\boldsymbol{\theta}, \qquad \mathrm{Mo}[\boldsymbol{\theta} \mid \boldsymbol{x}] = \arg \sup_{\boldsymbol{\theta} \in \Theta} \pi(\boldsymbol{\theta} \mid \boldsymbol{x}),$$

respectively. Indeed, the posterior mean is by far the most commonly used point estimator in Bayesian inference, partly for theoretical reasons and partly for computational reasons.

One issue with the posterior mean and mode is that they are not typically invariant under reparameterization. Thus, when either of them is used, the point estimator $\tilde{\boldsymbol{\psi}}$ of a one-to-one transformation $\boldsymbol{\psi} = \boldsymbol{\psi}(\boldsymbol{\theta})$ of the original parameter $\boldsymbol{\theta}$ will generally not be equal to $\boldsymbol{\psi}(\tilde{\boldsymbol{\theta}})$. For instance, in the Normal Example 1.9, the posterior mean or mode of σ^2 will not be the square of the posterior mean or mode of σ. While the differences are typically slight and, hence, not much of a problem in practice, the issue is of philosophical concern.

In one-dimensional continuous problems, the posterior median,

$$\mathrm{Me}[\theta \mid \boldsymbol{x}] = q\,; \qquad \int_{\{\theta \leq q\}} \pi(\theta \mid \boldsymbol{x})\, d\theta = \tfrac{1}{2},$$

is an attractive alternative. First, it is easy to show that it is *is invariant* to reparameterization. (The posterior median of σ^2 will be the square of the posterior median of σ.) Second, the posterior median exists and is unique under far wider conditions than either the posterior mean or the posterior mode. Finally the posterior median is typically less sensitive than the mean or mode to small variations in the data. As a consequence, the posterior median is often considered to be the best 'automatic' choice in one-dimensional problems.

In multiparameter problems, one could certainly compute the posterior median $\text{Me}[\theta_i \mid \boldsymbol{x}]$, of each parameter, from its marginal posterior density, $\pi(\theta_i \mid \boldsymbol{x})$, and define the posterior median of $\boldsymbol{\theta}$ to be $\text{Me}[\boldsymbol{\theta} \mid \boldsymbol{x}] = (\text{Me}[\theta_1 \mid \boldsymbol{x}], \ldots, \text{Me}[\theta_m \mid \boldsymbol{x}])$. This seems very sensible and attractive, but is not common in Bayesian practice.

Example 1.15 *Point estimation of a Poisson parameter.* Continuing with Example 1.12, the top portion of Table 1.2 reproduces the sample sizes and sufficient statistics of the samples simulated in Example 1.12, from a Poisson distribution with parameter $\lambda = 3$. The middle portion of the table contains the posterior mean, the posterior mode, and the posterior median of the corresponding posterior distributions $\text{Ga}(\lambda \mid t_j + 1/2, n_j)$, where $t_j = \sum_{j=1}^{n_j} x_j$ (graphed in Figure 1.4). While there is little difference in the estimates for large samples, the differences for small samples could matter.

Table 1.2: Point Estimates for a Poisson parameter.

n_j	5	10	20	50	100
t_j	14	31	57	153	292
$\text{E}[\lambda \mid t_j, n_j]$	2.900	3.150	2.875	3.070	2.925
$\text{Mo}[\lambda \mid t_j, n_j]$	2.700	3.050	2.825	3.050	2.915
$\text{Me}[\lambda \mid t_j, n_j]$	2.834	3.117	2.858	3.063	2.922
$\exp\{\text{E}[\log \lambda \mid t_j, n_j]\}$	2.801	3.100	2.850	3.060	2.920
$\exp\{\text{Mo}[\log \lambda \mid t_j, n_j]\}$	2.900	3.150	2.875	3.070	2.925
$\exp\{\text{Me}[\log \lambda \mid t_j, n_j]\}$	2.834	3.117	2.858	3.063	2.922

Alternatively, one might decide to work on a logarithmic scale, in terms of the parameter $\phi = \phi(\lambda) = \log(\lambda)$. Changing variables, the corresponding objective posterior distribution for ϕ is

$$\pi(\phi \mid t, n) \propto e^{\phi}\, \text{Ga}(e^{\phi} \mid t + 1/2, n)\,.$$

The corresponding posterior mean, posterior mode and posterior median provide point estimates of ϕ. But, since $\lambda = e^{\phi}$ these estimates can be transformed back to the original λ scale. The results are shown in the bottom portion of the table. The median, of course, remains invariant, while the mean and mode are not. It is curious that the posterior means of λ equal the estimates obtained by exponentiating the posterior modes of $\log \lambda$; this can be analytically verified.

1.3.2 Credible Regions

Another useful summary of a posterior distribution is to provide a region of the parameter space having specified probability content under $\pi(\boldsymbol{\theta} \mid \boldsymbol{x})$. In one or two dimensions, the identification of regions containing 50%, 90%, 95%, or 99% of the probability under the posterior may be sufficient to convey the general quantitative messages implicit in $\pi(\boldsymbol{\theta} \mid \boldsymbol{x})$.

Definition 1.3 Posterior credible region. *A subset* $\boldsymbol{\Theta}_c$ *of the parameter space* $\boldsymbol{\Theta}$, *such that*

$$\int_{\boldsymbol{\Theta}_c} \pi(\boldsymbol{\theta} \mid \boldsymbol{x})\, d\boldsymbol{\theta} = 1 - \alpha\,,$$

is called a posterior $(1-\alpha)$*-credible region for* $\boldsymbol{\theta}$ *or, equivalently, a* $100(1-\alpha)\%$*-credible region for* $\boldsymbol{\theta}$.

Thus, if $\boldsymbol{\Theta}_c$ is a posterior $(1-\alpha)$-credible region for $\boldsymbol{\theta}$, then the true value of $\boldsymbol{\theta}$ is expected to lie outside $\boldsymbol{\Theta}_c$ with probability α. For instance, in Examples 1.8 and 1.14 above (Figures 1.2 and 1.6), the shaded regions (with area $1-\alpha$) identify $1-\alpha$ credible intervals. Notice that this provides direct, immediately intuitive statements about the unknown quantity of interest in probability terms: given all available information, the quantity of interest is asserted to be in $\boldsymbol{\Theta}_c$ with a (presumably rational) degree of belief $1-\alpha$.

For any given α, there are generally infinitely many $(1-\alpha)$-credible regions, when $\boldsymbol{\theta}$ is a continuous parameter. An important property of a credible region is that it is invariant under reparameterization; thus, for any $(1-\alpha)$ credible region $\boldsymbol{\Theta}_c$ for $\boldsymbol{\theta}$, $\phi(\boldsymbol{\Theta}_c)$ (the set obtained by mapping each point of $\boldsymbol{\Theta}_c$ by ϕ) is an $(1-\alpha)$-credible region for $\boldsymbol{\phi} = \phi(\boldsymbol{\theta})$.

Sometimes, credible regions are selected to have minimum size (length, area, volume), resulting in what are called the *highest probability density* (HPD) regions, where all points in the region have larger probability density than all points outside. Unfortunately, HPD regions are *not* invariant under reparameterization: the image $\phi(\boldsymbol{\Theta}_c)$ of an HPD region $\boldsymbol{\Theta}_c$ will be a credible region of the same probability for $\boldsymbol{\phi}$, but will not generally be HPD in the transformed space, as the following example from Berger (1985) dramatically shows.

Example 1.16 *Lack of Invariance of HPD credible sets.* Suppose x has density $p(x \mid \theta) = e^{-(x-\theta)}$ on (θ, ∞), where $\theta > 0$, and the prior density is the Cauchy density $\text{Ca}(\theta \mid 0, 1)$. Since e^{-x} is constant in θ, the posterior density then satisfies $\pi(\theta \mid x) \propto (1+\theta^2)^{-1} e^{\theta}$ on the interval $(0, x)$. Differentiation shows that this is increasing in θ over the interval, so any HPD credible set must be an interval of the form $(b(x), x)$. For instance, if $x = 2$, numerical integration shows that the 75%-credible interval for θ is $(0.55, 2)$.

Consider now the reparameterization $\eta = e^{\theta}$. Changing variables yields, as the posterior distribution of η, $\pi(\eta \mid x) \propto (1 + [\log \eta^2])^{-1}$ on the interval $(1, e^x)$. This is clearly a decreasing function of η on the interval, so any

HPD credible interval for η must be of the form $(1, a(x))$. Since $\theta = \log(\eta)$, this interval transforms to $(0, \log(a(x))$ as the credible interval for θ. For instance, if $x = 2$, numerical integration shows that the 75%-credible interval for η is $(1, 4.63)$, which transforms into the 75%-credible interval for θ, $(0, 1.53)$. So, depending on the parameterization used to compute the HPD credible set, the resulting 75% credible interval for θ could be either $(0, 1.53)$ or $(0.55, 2)$, a rather extreme difference.

In one-dimensional continuous problems, posterior quantiles are often used to derive credible intervals. Thus, if $\theta_q = Q_q[\theta \,|\, \boldsymbol{x}]$ is the q-quantile of the posterior distribution of θ, then the interval $\Theta_{1-\alpha} = \{\theta;\ \theta \leq \theta_{1-\alpha}\}$ is a $(1-\alpha)$-credible region, and it is invariant under reparameterization. Quantile based credible intervals, which are also simpler to compute than HPD intervals, are often considerer the best 'automatic' choices in one-dimensional continuous problems.

Of particular interest are *equal-tailed* $(1-\alpha)$-credible intervals of the form

$$\Theta_{1-\alpha} = \{\theta;\ \theta_{\alpha/2} \leq \theta \leq \theta_{1-\alpha/2}\},$$

which are typically unique and are invariant under reparameterization. They also convey the additional useful information that the probability of 'missing' θ on either side is $(1-\alpha/2)$. The equal-tailed credible regions are thus, arguably, the best choice for a credible set.

There is a class of problems where Bayesian credible regions are also frequentist confidence intervals. A prime example is provided by the Normal mean.

Example 1.17 *Credible intervals for a Normal mean.* Consider again the situation described in Example 1.9. The objective posterior density of the mean of a normal distribution, when both parameters are unknown, was found to be $\pi(\mu \,|\, \boldsymbol{x}) = \mathrm{St}(\mu \,|\, \overline{x}, s/\sqrt{n-1}, n-1)$. Notice that this implies that the objective *posterior* distribution of the standardized quantity

$$\tau = \tau(\mu \,|\, \overline{x}, s, n) = \sqrt{n-1}\,(\mu - \overline{x})/s$$

is a standard Student-t distribution, *i.e.*, $\pi(\tau \,|\, \overline{x}, s, n) = \mathrm{St}(\tau \,|\, 0, 1, n-1)$. Thus, the equal-tailed $(1-\alpha)$-credible interval for μ is

$$\{\mu;\ \ \mu \in \overline{x} \pm q_{\{n-1, 1-\frac{\alpha}{2}\}}\, s/\sqrt{n-1}\},$$

where $q_{\{n-1,1-\frac{\alpha}{2}\}}$ is the $1-\alpha/2$ quantile of a standard Student distribution with $n-1$ degrees of freedom.

Note however that, for fixed μ, the expression $\sqrt{n-1}\,(\mu - \overline{x})/s$ may also be analyzed as the function of the data

$$t = t(\overline{x}, s \,|\, \mu, n) = \sqrt{n-1}\,(\mu - \overline{x})/s.$$

(Gosset 1908) showed that the *sampling* distribution of this statistic is *also* a standard Student $p(t \,|\, \mu, n) = \mathrm{St}(t \,|\, 0, 1, n-1)$ with the same degrees of freedom implying that, in this example, objective Bayesian credible intervals will *also* be *exact* frequentist confidence intervals.

The *exact* agreement between Bayesian credible intervals and frequentist confidence intervals illustrated in the last example is the exception, not the norm. However, for large samples, one may generally expect an *approximate* agreement. This will be discussed in Chapter 5 and will be used to provide a frequentist calibration for objective Bayesian methods.

An exact equivalence between Bayesian credible intervals and frequentist confidence intervals is typically impossible when the data are discrete. Indeed, exact (non randomized) frequentist confidence intervals do not exist in that case for most confidence levels. The computation of Bayesian credible regions for continuous parameters is, however, precisely the same whether the data is discrete or continuous.

Example 1.18 *Credible intervals for a Poisson parameter.* Consider again Example 1.12. For each of the samples simulated in Example 1.12 from a Poisson distribution with parameter $\lambda = 3$, Table 1.3 gives the equal-tailed posterior 95%-credible intervals of the corresponding objective posterior distributions $\text{Ga}(\lambda \mid t_j + 1/2, n_j)$ (graphed in Figure 1.4), where $t_j = \sum_{j=1}^{n_j} x_j$.

Table 1.3: Credible Intervals for a Poisson parameter.

n_j	5	10	20	50	100
t_j	14	31	57	153	292
$\text{Q}_{0.025}[\lambda \mid t_j, n_j]$	1.605	2.148	2.180	2.604	2.599
$\text{Q}_{0.975}[\lambda \mid t_j, n_j]$	4.572	4.341	3.664	3.574	3.270

These are derived by solving (numerically, or in terms of the incomplete gamma function) the equations $\int_0^q \text{Ga}(\lambda \mid t_j + 1/2, n_j)\, d\lambda = q$, for $q = \alpha/2$ and $q = 1 - \alpha/2$.

1.4 Prediction

1.4.1 Posterior Predictive Distributions

Let data $\boldsymbol{x} = (x_1, \ldots, x_n)$, $x_i \in \mathcal{X}$, consist of a set of somewhat 'homogeneous' observations, and suppose it is desired to predict the value of a future observation $x \in \mathcal{X}$ generated by the same random mechanism that has generated the others. It follows from the foundations arguments mentioned in Section 1.1, that the solution to this prediction problem *must* have the form of a probability distribution $p(x \mid \boldsymbol{x})$ describing the uncertainty on the value that the new observation x will take, given the information provided by $\boldsymbol{x}$ and any other available knowledge. This is called the *posterior predictive* distribution of x. To be able to proceed, it is necessary to specify the precise sense in which the x_i's are judged to be homogeneous.

Many statistical analyses directly assume that the data $\boldsymbol{x} = (x_1, \ldots, x_n)$ consist of a random sample of n conditionally independent (iid) observations from some specified statistical model, $\{p(x \,|\, \boldsymbol{\theta}), x \in \mathcal{X}, \boldsymbol{\theta} \in \boldsymbol{\Theta}\}$, so that their likelihood function is of the form

$$p(\boldsymbol{x} \,|\, \boldsymbol{\theta}) = p(x_1, \ldots, x_n \,|\, \boldsymbol{\theta}) = \prod\nolimits_{j=1}^{n} p(x_j \,|\, \boldsymbol{\theta}).$$

If this is the case, determination of the posterior predictive distribution is straightforward, once a prior distribution $\pi(\boldsymbol{\theta})$ has been specified for the parameter vector $\boldsymbol{\theta}$.

Proposition 1.2 *Let $\boldsymbol{x} = (x_1, \ldots, x_n)$, $x_i \in \mathcal{X}$, be a random sample from the statistical model $\{p(x \,|\, \boldsymbol{\theta}), x \in \mathcal{X}, \boldsymbol{\theta} \in \boldsymbol{\Theta}\}$, and let $\pi(\boldsymbol{\theta})$ be a prior distribution describing available knowledge (or lack thereof) on the value of $\boldsymbol{\theta}$. Then, the posterior predictive distribution*

$$p(x \,|\, \boldsymbol{x}) = p(x \,|\, x_1, \ldots, x_n) = \int_{\boldsymbol{\Theta}} p(x \,|\, \boldsymbol{\theta})\, \pi(\boldsymbol{\theta} \,|\, \boldsymbol{x})\, d\boldsymbol{\theta}$$

encapsulates all available knowledge about the outcome of any future observation $x \in \mathcal{X}$ from the same model.

Proof. By the total probability theorem, $p(x \,|\, \boldsymbol{x}) = \int_{\boldsymbol{\Theta}} p(x \,|\, \boldsymbol{\theta}, \boldsymbol{x})\, \pi(\boldsymbol{\theta} \,|\, \boldsymbol{x})\, d\boldsymbol{\theta}$ but, by the assumed conditional independence, $p(x \,|\, \boldsymbol{\theta}, \boldsymbol{x}) = p(x \,|\, \boldsymbol{\theta})$. □

The observable values $x \in \mathcal{X}$ can be either discrete of continuous. The next example illustrates prediction in a discrete case.

Example 1.19 *Prediction in a Poisson process.* Consider the situation of Example 1.12. Let data $\boldsymbol{x}$ be a random sample $(x_1, \ldots, x_n)$ from a Poisson distribution $\mathrm{Pn}(x \,|\, \lambda)$. We have already discussed that the corresponding objective posterior density of λ is $\pi(\lambda \,|\, \boldsymbol{x}) = \mathrm{Ga}(\lambda \,|\, t + 1/2, n)$, where $t = \sum_{i=1}^{n} x_i$. Using Proposition 1.2 and the Gamma integral, the corresponding predictive distribution is immediately found to be the Poisson-Gamma distribution with probability mass function

$$\begin{aligned} p(x \,|\, \boldsymbol{x}) = \Pr[x \,|\, t, n] &= \int_0^\infty \mathrm{Pn}(x \,|\, \lambda)\, \mathrm{Ga}(\lambda \,|\, t + \tfrac{1}{2}, n)\, d\lambda \\ &= \frac{n^{t+1/2}}{\Gamma(t + 1/2)} \frac{1}{x!} \frac{\Gamma(x + t + 1/2)}{(1+n)^{x+t+1/2}} \,. \end{aligned}$$

When $t = 0$, the posterior predictive distribution becomes

$$\Pr(x \,|\, t = 0, n) = \frac{n^{1/2}}{(1+n)^{x+1/2}} \frac{\Gamma(x + 1/2)}{x! \sqrt{\pi}} \,;$$

in particular, $\Pr(x = 0 \,|\, t = 0, n) = \sqrt{n/(n+1)}$. As a specific example, suppose a firm producing automobile restraint systems has received no complaints from its clients in each of 10 consecutive months. With no additional information on the average number λ of complaints per month, the quality assurance department of the firm reports that the probability that x complaints will be received in the next month of production is given by the Poisson-Gamma distribution above, with $t = 0$ and $n = 10$. In particular, $\Pr[0 \mid t, n] = 0.953$, $\Pr[1 \mid t, n] = 0.043$, and $\Pr[2 \mid t, n] = 0.003$. Many other situations can be described with the same model. For instance, if meteorological conditions remain similar in a given area, $\Pr[0 \mid t, n] = 0.953$ would describe the chances of, say, no flash flood next year, given 10 years without flash floods in that area.

As the prime simple example of predicting future continuous observations, we consider the ubiquitous problem of predicting future normal observations.

Example 1.20 *Prediction of Normal measurements.* Consider now prediction of a continuous variable in the context of Example 1.9. The data $\boldsymbol{x}$ is a random sample $(x_1, \ldots, x_n)$ from a normal distribution $\mathrm{N}(x \mid \mu, \sigma^2)$, and suppose one is interested in predicting the value of a future observation $x \in \mathbb{R}$ to be obtained under the same conditions as the x_j's already observed. As mentioned before, absence of initial information on the values of both μ and σ is formally described by the *improper* prior function $\pi(\mu, \sigma) = \sigma^{-1}$ and, in terms of the precision $\lambda = \sigma^{-2}$, this leads to the joint posterior

$$\pi(\mu, \lambda \mid \boldsymbol{x}) = \mathrm{N}(\mu \mid \overline{x}, (n\lambda)^{-1}) \, \mathrm{Ga}(\lambda \mid (n-1)/2, ns^2/2).$$

Using Proposition 1.2 and the Gamma integral, the corresponding posterior predictive distribution is

$$\begin{aligned} \pi(x \mid \boldsymbol{x}) &= \int_0^\infty \int_{-\infty}^\infty \mathrm{N}(x \mid \mu, \lambda^{-1}) \, \pi(\mu, \lambda \mid \boldsymbol{x}) \, d\mu \, d\lambda \\ &\propto [(1+n)s^2 + (\mu - \overline{x})^2]^{-n/2}, \end{aligned}$$

which is recognizable as the kernel of the Student t density

$$p(x \mid \boldsymbol{x}) = \mathrm{St}(x \mid \overline{x}, s\sqrt{\frac{n+1}{n-1}}, n-1) \,. \tag{1.5}$$

If μ is known to be positive, the appropriate objective prior is

$$\pi(\mu, \sigma) = \begin{cases} \sigma^{-1} & \text{if } \mu > 0 \\ 0 & \text{otherwise} \end{cases} \,. \tag{1.6}$$

However, the resulting posterior predictive distribution will still, for all practical purposes, be the Student t density above, provided the likelihood function $p(\boldsymbol{x} \mid \mu, \sigma)$ is concentrated on positive μ values.

Suppose, for example, that a firm producing automobile restraint system observes breaking strengths for $n = 10$ randomly chosen safety belt

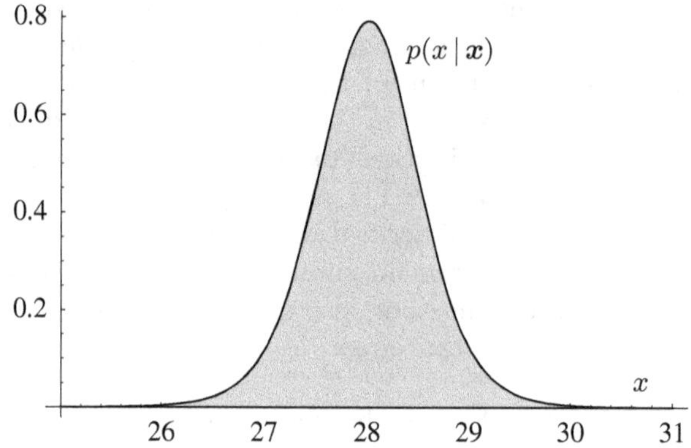

Figure 1.7: Objective posterior predictive density for the breaking strength of future webbing samples.

webbings, the resulting sample mean and variance being $\overline{x} = 28.011$ kN and $s^2 = 0.443^2$ kN^2, respectively. Engineering specifications require the breaking strengths to be larger than 26 kN. Assuming the data is a random sample from the normal distribution and the recommended objective prior is to be used, the posterior predictive distribution of the strength of a new webbing is $p(x \mid \boldsymbol{x}) = \text{St}(x \mid 28.011, 0.490, 9)$, plotted in Figure 1.7. It follows that the quality engineer will be able to claim that a safety belt randomly chosen from the same batch as the sample tested would satisfy the required specification with probability

$$\Pr(x > 26 \mid \boldsymbol{x}) = \int_{26}^{\infty} \text{St}(x \mid 28.011, 0.490, 9)\, dx = 0.9987,$$

the shaded area in the figure. Note that is also implies that, if production conditions remain constant, 99.87% of the future safety belt webbings produced can be expected to have acceptable breaking strengths.

1.4.2 Prediction in Regression

In prediction problems, one often has additional information provided by covariates $\boldsymbol{v}$ (*cf.* Example 1.6). The data typically consists of a collection of pairs $\boldsymbol{x} = \{(\boldsymbol{y}_1, \boldsymbol{v}_1), \ldots (\boldsymbol{y}_n, \boldsymbol{v}_n)\}$ where $\boldsymbol{y}_i$ is the observation that was made when the covariate was $\boldsymbol{v}_i$; both the $\boldsymbol{y}_i$'s and the $\boldsymbol{v}_i$'s can be multidimensional quantities. The goal here is to predict a new observation $\boldsymbol{y}$ under the associated covariate $\boldsymbol{v}$. Formally, what it is required is the posterior predictive distribution $p(\boldsymbol{y} \mid \boldsymbol{v}, \boldsymbol{x})$.

The first step is to establish a model $\{p(\boldsymbol{y} \mid \boldsymbol{v}, \boldsymbol{\theta}), \boldsymbol{y} \in \boldsymbol{\mathcal{Y}}, \boldsymbol{\theta} \in \boldsymbol{\Theta}\}$ which makes precise the relationship presumed to exist between the quantity of interest $\boldsymbol{y}$ and the covariates $\boldsymbol{v}$. Simplest is a linear relationship, the univariate version of which is discussed here.

Univariate Linear Regression. Let $y \in \mathcal{Y}$ be a real-valued observable to be predicted on the basis of covariates $\{v_1, \ldots, v_m\}$ and suppose that one expects a linear relationship

$$y = \beta_0 + v_1\beta_1 + \ldots + v_m\beta_m + \varepsilon$$

between y and the covariates, where $\boldsymbol{\beta} = (\beta_0, \beta_1, \ldots \beta_m)$ is a vector of unknown *regression coefficients* (β_0 reflecting an overall unknown constant level) and ε is a normal error $\mathrm{N}(\varepsilon \mid 0, \sigma^2)$, for some unknown $\sigma^2 > 0$. This can be compactly written $p(y \mid \boldsymbol{v}, \boldsymbol{\beta}, \sigma^2) = \mathrm{N}(y \mid \boldsymbol{v}\boldsymbol{\beta}', \sigma^2)$, where $\boldsymbol{v} = (1, v_1, v_2, \ldots, v_m)$.

Let the data $\boldsymbol{x}$ consist of n pairs $\{(y_1, \boldsymbol{v}_1), \ldots, (y_n, \boldsymbol{v}_n)\}$ of observations and their corresponding covariates. This can more compactly be expressed by re-defining $\boldsymbol{x} = (\boldsymbol{y}, \boldsymbol{V})$, where $\boldsymbol{y}$ is the column vector $\boldsymbol{y} = (y_1, \ldots, y_n)'$ and $\boldsymbol{V}$ is the $n \times (m+1)$ matrix having the $\boldsymbol{v}_i$'s as rows. The likelihood function is then $p(\boldsymbol{y} \mid \boldsymbol{V}, \boldsymbol{\beta}, \sigma^2) = \mathrm{N}_n(\boldsymbol{y} \mid \boldsymbol{V}\boldsymbol{\beta}', \sigma^2 \boldsymbol{I}_n)$, where $\boldsymbol{I}_n$ is the $n \times n$ identity matrix.

It will later be established (Example 10.9) that the appropriate objective prior for this problem is $\pi(\boldsymbol{\beta}, \sigma) = \sigma^{-1}$. What is required is $p(x \mid \boldsymbol{v}, \boldsymbol{y}, \boldsymbol{V})$, the posterior predictive density of a new observation y, having covariates $\boldsymbol{v}$, given the available data $\boldsymbol{x} = (\boldsymbol{y}, \boldsymbol{V})$. Using Proposition 1.2, it may be established—see *e.g.*, Zellner (1971) or Geisser (1993)—that the desired posterior predictive distribution is the Student t distribution with $n - m - 1$ degrees of freedom,

$$p(y \mid \boldsymbol{v}, \boldsymbol{y}, \boldsymbol{V}) = \mathrm{St}\left(y \,\middle|\, \boldsymbol{v}\hat{\boldsymbol{\beta}}',\ s\, f(\boldsymbol{v}, \boldsymbol{V}) \sqrt{\frac{n}{n-m-1}}, n-m-1\right),$$

$$\hat{\boldsymbol{\beta}}' = (\boldsymbol{V}'\boldsymbol{V})^{-1}\boldsymbol{V}'\boldsymbol{y}, \quad ns^2 = \sum_{i=1}^{n} (\boldsymbol{y} - \boldsymbol{v}\hat{\boldsymbol{\beta}})'(\boldsymbol{y} - \boldsymbol{v}\hat{\boldsymbol{\beta}}),$$

$$f(\boldsymbol{v}, \boldsymbol{V}) = \sqrt{1 + \boldsymbol{v}(\boldsymbol{V}'\boldsymbol{V})^{-1}\boldsymbol{v}'}\,.$$

If there is only one covariate v, then $p(y \mid v, \boldsymbol{\beta}, \sigma^2) = \mathrm{N}(y \mid \beta_0 + \beta_1 v, \sigma^2)$. It can be shown that a sufficient statistic is $\boldsymbol{t} = \{\overline{v}, \overline{y}, s_{vy}, s_{vv}\}$, where $\overline{v} = n^{-1}\Sigma v_j$ and $\overline{y} = n^{-1}\Sigma y_j$ are the sample means, and $s^2_{vv} = [n^{-1}\Sigma v_j^2 - \overline{v}^2]$ and $s^2_{yv} = [n^{-1}\Sigma v_j y_j - \overline{v}\,\overline{y}]$ are the sample variance and covariance (between the y_i and v_i). The result above then reduces to

$$p(y \mid v, \boldsymbol{t}) = \mathrm{St}\left(y \,\middle|\, \hat{\beta}_0 + \hat{\beta}_1 v,\ s\, f(v, \boldsymbol{t}) \sqrt{\frac{n}{n-2}}, n-2\right)$$

$$\hat{\beta}_0 = \overline{y} - \hat{\beta}_1 \overline{v}, \quad \hat{\beta}_1 = \frac{s^2_{vy}}{s^2_{vv}}, \quad f(v, \boldsymbol{t}) = \sqrt{1 + \frac{1}{n}\frac{(v - \overline{v})^2 + s^2_{vv}}{s^2_{vv}}},$$

$$ns^2 = \sum\nolimits_{j=1}^{n} (y_j - \hat{\beta}_1 - \hat{\beta}_2 v_j)^2.$$

Example 1.21 *Ship contamination (continuation of Example 1.6).* As an illustration, consider the problem of predicting the level of contamination, y

(in μgr/m^3), induced by a ship discharge as a function of the speed of wind proceeding from the harbor, v (in m/s). The data consist of the $n = 12$ pairs of observations contained in Table 1.4 and represented in the upper portion of Figure 1.8.

Table 1.4: Pollution levels in the city center and wind speeds from the harbor.

y_j (in μgr/m^3)	1212	836	850	446	1164	601
v_j (in m/s)	4.8	3.3	3.1	1.7	4.7	2.1
y_j (in μgr/m^3)	1074	284	352	1064	712	976
v_j (in m/s)	3.9	0.9	1.4	4.3	2.9	3.4

Using the formulas just described, one can compute the predictive probability density for the contamination level y which corresponds to any wind speed v. Those Student t distributions are centered on the regression line,

$$y = 53.9 + 244.5\, v \equiv c(v)\,,$$

have a scale parameter, $s(v)$, which increases as one moves away from the mean $\overline{v}$ of the v values, and have $n-2 = 10$ degrees of freedom. The upper part of Figure 1.8 contains the regression line and the (slightly divergent) curves defined by the 0.95-equal-tailed credible intervals for the value of y. The lower portion of Figure 1.8 represents the predictive distributions which correspond to winds of $v = 0, 1, 3$ and 5 m/s. The means and scale parameters of these predictive densities are given in Table 1.5.

Table 1.5: Posterior predictive densities $\mathrm{St}(y \,|\, v, c(v), s(v), n-2)$ of contamination levels y at various wind speeds v.

v	$c(v)$	$s(v)$	$n-2$
0.0	53.9	62.4	10
1.0	298.4	56.7	10
3.0	787.4	51.6	10
5.0	1276.4	56.3	10

Since contamination cannot be negative, the resulting distributions should be renormalized (see Section 1.2.8) to positive values when predicting contamination levels for small wind speeds. This is really only an issue for $v = 0$, since only the corresponding Student t density would have had appreciable negative mass; this renormalized predictive density is what is actually graphed in Figure 1.8. As a specific example, suppose there is concern if the contamination exceeds 50 μgr/m^3. The probability of this, when $v = 0$, is $p(y > 50 \,|\, v = 0, y \geq 0, \boldsymbol{t}) = 0.66$, and is given by the shaded area in the density in Figure 1.8.

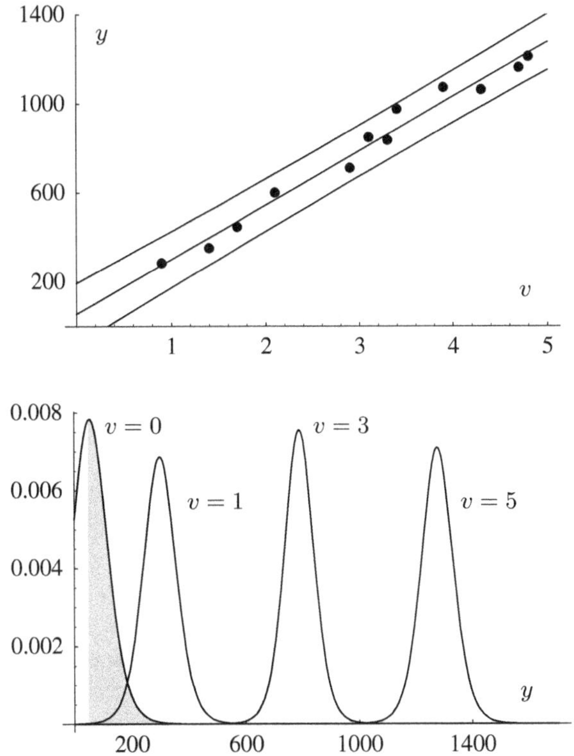

Figure 1.8: Top: the 12 data points, the regression line, and the (slightly divergent) curves defined by the 0.95-equal-tailed credible intervals for the predicted value of y at each possible v. Bottom: Predictive posterior densities $p(y \mid v, \boldsymbol{t})$ of the pollution level y based on simple linear regression on the wind speed, v, for no wind ($v = 0$), and wind speeds of 1, 3 and 5 m/s.

1.5 Hierarchical Modeling and Analysis

1.5.1 Hierarchical Models

An extremely powerful Bayesian modeling tool involves building models at different levels or 'hierarchies.'

Example 1.22 *Hierarchical Normal model.* Blood hemoglobin levels are measured at n regular intervals for a random set of individuals, $i = 1, \ldots m$, from a population. For individual i, the data is thus $\boldsymbol{x}_i = (x_{i1}, \ldots, x_{in})$, and the x_{ij} are assumed to be independent normal observations, $x_{ij} \sim \mathrm{N}(x_{ij} \mid \mu_i, \sigma^2)$, with μ_i being the unknown average hemoglobin level of individual i and (for simplicity here) the variance σ^2 being assumed to be known. By sufficiency, the information in the data for individual i is contained in $\overline{x}_i$, and the sampling distribution of $\overline{x}_i$ is $\mathrm{N}(\overline{x}_i \mid \mu_i, \sigma^2/n)$.

The individual average hemoglobin levels are themselves assumed to be independently normally distributed, *i.e.*, $\mu_i \sim \mathrm{N}(\mu_i \mid \xi, \tau^2)$, with ξ and τ^2 also being unknown. Thus the collection of unknown parameters in the model are $\boldsymbol{\theta} = (\mu_1, \ldots, \mu_m, \xi, \tau^2)$, but note that only the μ_i relate immediately to the data, while ξ and τ^2 are parameters of the distribution of other unknown parameters (the μ_i). From a Bayesian perspective, $\mathrm{N}(\mu_i \mid \xi, \tau^2)$ is the prior distribution of the μ_i, for given ξ and τ^2.

Algebra shows that the posterior density of μ_i, given $\overline{x}_i$, ξ, and τ^2, is

$$\begin{aligned}\pi(\mu_i \mid \overline{x}_i, \xi, \tau^2) &\propto \mathrm{N}\left(\overline{x}_i \mid \mu_i, \sigma^2/n\right) \mathrm{N}(\mu_i \mid \xi, \tau^2) \\ &\propto \mathrm{N}\left(\mu_i \;\middle|\; \frac{\sigma^2 \xi + n\tau^2 \overline{x}_i}{n\tau^2 + \sigma^2}, \frac{\tau^2\sigma^2}{n\tau^2 + \sigma^2}\right). \qquad (1.7)\end{aligned}$$

Distributions at the level of the data (*e.g.*, the $\mathrm{N}(\overline{x}_i \mid \mu_i, \sigma^2)$ distributions in the above example) are called *first level* distributions, while the $\mathrm{N}(\mu_i \mid \xi, \tau^2)$ distribution above is a *second level* distribution. Models that can be viewed as consisting of several levels, such as this, are called *hierarchical models*, and parameters that occur at the higher levels (such as ξ and τ^2) are called *hyperparameters*. While this is the standard terminology in Bayesian analysis, other names are often used for these quantities. In classical statistics, hierarchical models are often called random effects models; the μ_i in the above example would be the random effects, and $\mathrm{N}(\mu_i \mid \xi, \tau^2)$ would be called the distribution of the random effects. Other names for hierarchical models include multilevel models, variance component models, mixed models, random coefficient models, credibility models, and empirical Bayes models.

Sometimes it is the hyperparameters of a hierarchical model that are of interest. Thus in Example 1.22, the goal might be to determine the population parameters ξ and τ^2, since these describe the overall distribution of individual average hemoglobin in the population. Inference about hyperparameters can often be addressed by looking at the *marginal density* of the data, given the hyperparameters, found by integrating the joint density over the μ_i. The resulting function of the hyperparameters is called the *marginal likelihood function.*

Example 1.23 *Hierarchical Normal model (continued).* Consider again Example 1.22. Writing $\overline{x}_i = \mu_i + \epsilon_i$, where $\epsilon_i \sim \mathrm{N}(\epsilon_i \mid 0, \sigma^2/n)$, and $\mu_i = \xi + \delta_i$, where $\delta_i \sim \mathrm{N}(\delta_i \mid 0, \tau^2)$,it is clear that $\overline{x}_i = \xi + \epsilon_i + \delta_i$. It is well known that the sum of zero mean normal random variables is itself normal with zero mean and variance equal to the sum of the variances, so it can be concluded that, marginally, the distribution of $\overline{x}_i$ is $\mathrm{N}(\overline{x}_i \mid \xi, \tau^2 + \sigma^2/n)$. Furthermore, the $\overline{x}_i$ are independent (because the $\overline{x}_i$ given μ_i were independent and the μ_i were independent) so that the overall marginal density of $\overline{\boldsymbol{x}} = (\overline{x}_1, \ldots, \overline{x}_m)$ is the product of these normal densities, which algebra shows to be

$$p(\overline{\boldsymbol{x}} \mid \xi, \tau^2) = \frac{1}{(2\pi)^{m/2}(\tau^2+\sigma^2/n)^{m/2}} \exp\left\{-\frac{\sum_{i=1}^m (\overline{x}_i - \xi)^2}{2(\tau^2+\sigma^2/n)}\right\}$$
$$\propto \frac{1}{(\tau^2+\sigma^2/n)^{m/2}} \exp\left\{-\frac{m(\bar{\bar{x}}-\xi)^2 + s^2}{2(\tau^2+\sigma^2/n)}\right\}, \quad (1.8)$$

where $\bar{\bar{x}}$ is the mean of the means and $s^2 = \sum_{i=1}^m (\overline{x}_i - \bar{\bar{x}})^2$. When considered as a function of (ξ, τ^2) for given $\overline{\boldsymbol{x}}$, this is also the marginal likelihood function of ξ and τ^2.

It is clear from (1.8) that considerable information is available about the hyperparameters from the data. Hierarchical Bayesian models have this characteristic, that information about at least part of the prior is available from the data.

At other times, the primary parameters of interest will be the μ_i themselves, with ξ and τ^2 being viewed as nuisance parameters. Objective Bayesian analysis in both these situations is illustrated in the next subsection.

1.5.2 Hierarchical Bayesian Analysis

In principle, Bayesian analysis for hierarchical models is no different than general Bayesian analysis; one simply chooses a prior distribution for all unknown parameters, and computes the posterior distribution of quantities of interest. The only difference is that the prior distributions of many of the parameters are already specified as part of the hierarchical model; indeed, one needs only to specify the prior distribution of the hyperparameters.

Example 1.24 *Hierachical Normal model (continued).* Continuing with Examples 1.22 and 1.23, and writing $\boldsymbol{\mu} = (\mu_1, \ldots, \mu_m)$ and $\mathbf{1} = (1, \ldots, 1)$, the hierarchical model has already stated that $\pi(\boldsymbol{\mu} \mid \xi, \tau^2)$ is $\mathrm{N}_m(\boldsymbol{\mu} \mid \xi\mathbf{1}, \tau^2\boldsymbol{I})$. It will be seen in Section 12.6 that the recommended objective prior for (ξ, τ^2) is the reference prior

$$\pi^R(\xi, \tau^2) = \frac{1}{\tau^2 + \sigma^2/n}\,. \quad (1.9)$$

Hence the overall prior density is

$$\pi(\boldsymbol{\mu}, \xi, \tau^2) = \pi(\boldsymbol{\mu} \mid \xi, \tau^2)\,\pi(\xi, \tau^2) = \mathrm{N}_m(\boldsymbol{\mu} \mid \xi\mathbf{1}, \tau^2\boldsymbol{I})(\tau^2+\sigma^2/n)^{-1},$$

and the resulting posterior density is simply proportional to the product of this and the normal likelihood function from the data. Utilizing probability theory, (1.7), and (1.8), the joint posterior density of all parameters is

$$\pi(\boldsymbol{\mu}, \xi, \tau^2 \mid \overline{\boldsymbol{x}}) \propto \pi(\boldsymbol{\mu} \mid \xi, \tau^2, \overline{\boldsymbol{x}})\, p(\overline{\boldsymbol{x}} \mid \xi, \tau^2)\, \pi^R(\xi, \tau^2)$$
$$\propto \mathrm{N}_m\left(\boldsymbol{\mu} \,\middle|\, \frac{\sigma^2\xi\mathbf{1} + n\tau^2\overline{\boldsymbol{x}}}{n\tau^2+\sigma^2}, \frac{\tau^2\sigma^2}{n\tau^2+\sigma^2}\boldsymbol{I}\right)$$
$$\times \frac{1}{(\tau^2+\sigma^2/n)^{(m+2)/2}} \exp\left\{-\frac{m(\bar{\bar{x}}-\xi)^2+s^2}{2(\tau^2+\sigma^2/n)}\right\}. \quad (1.10)$$

Integrating this joint posterior over $\boldsymbol{\mu}$ shows immediately that the marginal posterior distribution of the hyperparameters is given by

$$\pi(\xi, \tau^2 \,|\, \overline{\boldsymbol{x}}) \propto \frac{1}{(\tau^2 + \sigma^2/n)^{(m+2)/2}} \exp\left\{ -\frac{m(\bar{\bar{x}} - \xi)^2 + s^2}{2(\tau^2 + \sigma^2/n)} \right\}.$$

From this, it follows that the marginal posterior distribution of τ^2 is

$$\begin{aligned} \pi(\tau^2 \,|\, \overline{\boldsymbol{x}}) &\propto \int \frac{1}{(\tau^2 + \sigma^2/n)^{(m+2)/2}} \exp\left\{ -\frac{m(\bar{\bar{x}} - \xi)^2 + s^2}{2(\tau^2 + \sigma^2/n)} \right\} d\xi \\ &\propto \frac{1}{(\tau^2 + \sigma^2/n)^{(m+1)/2}} \exp\left\{ -\frac{s^2}{2(\tau^2 + \sigma^2/n)} \right\}. \end{aligned} \tag{1.11}$$

Finally, we record that the marginal posterior distribution of a particular μ_i can be found by first integrating (1.10) over the other μ_j (leaving only (1.7) as the first term on the right hand side), and then integrating over ξ resulting, after some algebra, in

$$\begin{aligned} \pi(\mu_i \,|\, \boldsymbol{x}) &\propto \int \mathrm{N}\left(\mu_i \,\middle|\, \frac{\sigma^2\xi + n\tau^2\overline{x}_i}{n\tau^2 + \sigma^2}, \frac{\tau^2\sigma^2}{n\tau^2 + \sigma^2} \right) \\ &\quad \times \frac{1}{(\tau^2 + \sigma^2/n)^{(m+2)/2}} \exp\left\{ -\frac{m(\bar{\bar{x}} - \xi)^2 + s^2}{2(\tau^2 + \sigma^2/n)} \right\} d\xi d\tau^2. \\ &\propto \int_0^\infty \mathrm{N}\left(\mu_i \,\middle|\, \frac{\sigma^2\bar{\bar{x}} + n\tau^2\overline{x}_i}{n\tau^2 + \sigma^2}, \frac{\sigma^2(mn\tau^2 + \sigma^2)}{mn(n\tau^2 + \sigma^2)} \right) \\ &\quad \times \frac{1}{(\tau^2 + \sigma^2/n)^{(m+1)/2}} \exp\left\{ -\frac{s^2}{2(\tau^2 + \sigma^2/n)} \right\} d\tau^2. \end{aligned}$$

Determination (and computation with) this posterior distribution requires one-dimensional numerical integration.

1.6 Decision Making

1.6.1 Definition of a Decision Problem and its Analysis

A decision problem exists whenever there are two or more possible courses of action; let $\mathcal{A}$ be the class of possible actions. Moreover, for each $a \in \mathcal{A}$, let $\boldsymbol{\Theta}$ be the set of *relevant events* which may affect the result of choosing a, and let $c(a, \boldsymbol{\theta})$ be the *consequence* of having chosen action a when event $\boldsymbol{\theta}$ takes place. In this context, $\boldsymbol{\theta}$ if often referred to as the *parameter of interest.* Without loss of generality, it may be assumed that the possible actions are mutually exclusive.

Different sets of principles have been proposed to capture a minimum collection of logical rules that could sensibly be required for 'rational' decision-making. These all consist of axioms with a strong intuitive appeal; examples include the *transitivity* of preferences – if $a_1 \succ a_2$ given C (to be read as a_1 is preferred to a_2 given conditions C) and $a_2 \succ a_3$ given C, then $a_1 \succ a_3$ given C; and the *sure-thing principle* – if $a_1 \succ a_2$ given C and E, and $a_1 \succ a_2$ given C and not E, then $a_1 \succ a_2$ given C. Note that these rules are not intended as a description of actual human decision-making, but as a *normative* set of principles to be followed by someone who aspires to coherent decision-making.

The vast majority of axiom systems for decision theory lead to basically the same set of conclusions, namely:

• The consequences of wrong actions should be evaluated in terms of a real-valued *loss function* $\ell(a, \boldsymbol{\theta})$ which specifies, on some numerical scale, their undesirability. Alternatively, one may evaluate the desirability of each possible consequence with a real-valued *utility function* $u(a, \boldsymbol{\theta})$ (essentially the negative of the loss function).

• The uncertainty of relevant events should be measured with a set of *probability* distributions $\{\pi(\boldsymbol{\theta} \,|\, C, a),\ \boldsymbol{\theta} \in \boldsymbol{\Theta},\ a \in \mathcal{A}\}$ describing their plausibility given the conditions C under which the decision must be taken.

• The relative undesirability of available actions $a \in \mathcal{A}$ is then measured by their *expected loss*

$$\bar{\ell}(a \,|\, C) = \int_{\boldsymbol{\Theta}} \ell(a, \boldsymbol{\theta})\, \pi(\boldsymbol{\theta} \,|\, C, a)\, d\boldsymbol{\theta}.$$

If a utility function is specified, the relative desirability of each action is then measured by their *expected utility* $\overline{u}(a \,|\, C) = \int_{\boldsymbol{\Theta}} u(a, \boldsymbol{\theta})\, \pi(\boldsymbol{\theta} \,|\, C, a)\, d\boldsymbol{\theta}$.

Note, in particular, that these axiomatic conclusions indicate that uncertainty about all relevant unknown quantities (the actual values of the $\boldsymbol{\theta}$'s) should be quantified by probability distributions. These distributions are conditional on the circumstances, C, under which the decision is to be taken, which typically, but not necessarily, include the results $\boldsymbol{x}$ of some relevant experimental or observational data.

There are several ways to embed a degree of 'objectivity' into decision analysis. If presented with a subjectively elicited loss function, one could still utilize an objective prior distribution, find the posterior distribution (if data $\boldsymbol{x}$ is available) and then minimize the expected loss over a. There are also certain loss functions that are viewed as being objective (or at least are conventionally used) and using them with objective posteriors is common. Several such loss functions are discussed in this section.

Statistical inference is often thought of as different from decision analysis. There are, however, two powerful arguments for treating inference problems as decision problems. First, a problem of statistical inference is typically considered worth analyzing because it *may* eventually help to make sensible decisions; as Ramsey (1931) put it, "a lump of arsenic is poisonous because it *may* kill someone, not because it has actually killed someone." Second, it can be shown (Bernardo, 1979*a*) that statistical inference on $\boldsymbol{\theta}$ actually *has* the mathematical structure of a decision problem, where the class of alternatives is the functional space of the conditional probability distributions of $\boldsymbol{\theta}$ given the data, and the utility function is a measure of the amount of information about $\boldsymbol{\theta}$ which the data may be expected to provide. In particular, direct application of formal decision theory to statistical inference provides a systematic procedure to derive both point and region estimates that can be argued to be optimal in a well-defined sense. The intuitive estimates discussed in Section 1.3 are found to be particular cases of this general procedure.

1.6.2 Formal Point Estimation

Let $\boldsymbol{x}$ be the available data, which are assumed to have been generated by a statistical model $\{p(\boldsymbol{x} \,|\, \boldsymbol{\theta}), \boldsymbol{\theta} \in \boldsymbol{\Theta}\}$, and let $\boldsymbol{\phi} = \boldsymbol{\phi}(\boldsymbol{\theta}) \in \boldsymbol{\Phi}$ be the quantity of interest. A *point estimator* of $\boldsymbol{\phi}$ is some function of the data $\tilde{\boldsymbol{\phi}} = \tilde{\boldsymbol{\phi}}(\boldsymbol{x})$ which could be regarded as an appropriate proxy for the actual, unknown value of $\boldsymbol{\phi}$. Without loss of generality, the model can be explicitly specified in terms of the quantity of interest, in the form $\{p(\boldsymbol{x} \,|\, \boldsymbol{\phi}, \boldsymbol{\omega}),\ \boldsymbol{\phi} \in \boldsymbol{\Phi}, \boldsymbol{\omega} \in \boldsymbol{\Omega}\}$, where $(\boldsymbol{\phi}, \boldsymbol{\omega})$ is a one-to-one transformation of $\boldsymbol{\theta}$.

Formally, choosing a point estimate for $\boldsymbol{\phi}$ is a *decision problem* where the action space is the class $\boldsymbol{\Phi}$ of possible $\boldsymbol{\phi}$ values. As prescribed by the foundations of decision theory described above, to solve this decision problem it is necessary to specify a *loss function* $\ell(\tilde{\boldsymbol{\phi}}, \boldsymbol{\phi})$ measuring the consequences of acting *as if* the true value of the quantity of interest was $\tilde{\boldsymbol{\phi}}$, when it is actually $\boldsymbol{\phi}$. The expected loss if $\tilde{\boldsymbol{\phi}}$ were used is then given by

$$\bar{\ell}[\tilde{\boldsymbol{\phi}} \,|\, \boldsymbol{x}] = \int_{\boldsymbol{\Phi}} \ell(\tilde{\boldsymbol{\phi}}, \boldsymbol{\phi})\, \pi(\boldsymbol{\phi} \,|\, \boldsymbol{x})\, d\boldsymbol{\phi},$$

and the corresponding *Bayes estimator*, denoted $\boldsymbol{\phi}^*$, is that function of the data, $\boldsymbol{\phi}^* = \boldsymbol{\phi}^*(\boldsymbol{x})$, which minimizes in $\boldsymbol{\Phi}$ the expected loss $\bar{\ell}[\tilde{\boldsymbol{\phi}} \,|\, \boldsymbol{x}]$.

For any given model and data, the Bayes estimator obviously depends on the chosen loss function. The loss function is context specific, and should be chosen in terms of the anticipated uses of the estimate; however, a number of conventional loss functions have been suggested for those situations where no particular uses are envisaged. These loss functions produce estimates which can be regarded as stylized descriptions of the *location* of the posterior distribution.

Proposition 1.3 Conventional estimators.
(i) If the quadratic loss function $\ell(\tilde{\boldsymbol{\phi}}, \boldsymbol{\phi}) = (\tilde{\boldsymbol{\phi}} - \boldsymbol{\phi})^t(\tilde{\boldsymbol{\phi}} - \boldsymbol{\phi})$ *is used, the Bayes estimator is the* posterior mean $\boldsymbol{\phi}^* = E[\boldsymbol{\phi} \,|\, \boldsymbol{x}]$, *assuming that it exists.*
(ii) If the loss function is a zero-one function, so that $\ell(\tilde{\boldsymbol{\phi}}, \boldsymbol{\phi}) = 0$ *if* $\tilde{\boldsymbol{\phi}}$ *belongs to a ball of radius* ϵ *centered in* $\boldsymbol{\phi}$ *and* $\ell(\tilde{\boldsymbol{\phi}}, \boldsymbol{\phi}) = 1$ *otherwise, then the Bayes estimator* $\boldsymbol{\phi}^*$ *tends to the* posterior mode *as the ball radius* ϵ *tends to zero, assuming that a unique mode exists.*

In one dimensional estimation problems, quantiles can be seen to be Bayes estimators with respect to a linear loss function.

Proposition 1.4 One dimensional quantiles. *If* ϕ *is univariate and continuous* *and the loss function is linear, so that*

$$\ell(\tilde{\phi}, \phi) = \begin{cases} c_1(\tilde{\phi} - \phi) & \text{if } \tilde{\phi} \geq \phi \\ c_2(\phi - \tilde{\phi}) & \text{otherwise,} \end{cases} ,$$

then the Bayes estimator is the posterior quantile *of order* $c_2/(c_1 + c_2)$*, so that* $\Pr[\phi < \phi^*] = c_2/(c_1 + c_2)$*. In particular, if* $c_1 = c_2$*, the Bayes estimator is the* posterior median.

The last result clearly illustrates the fact that *any* possible parameter value may turn out be the Bayes estimator; it all depends on the loss function describing the consequences of the anticipated uses of the estimate.

1.6.3 Formal Region Estimation

Any loss function can be used to formally define credible regions, by simply choosing those points which minimize the expected loss, given $\boldsymbol{x}$, and have desired probability q.

Definition 1.4 *A* Bayes q-credible region $\boldsymbol{\Phi}^*_q(\boldsymbol{x}) \subset \boldsymbol{\Phi}$ *is a q-credible region where any value within the region has a smaller posterior expected loss than any value outside the region, so that*

$$\forall \phi_1 \in \boldsymbol{\Phi}^*_q(\boldsymbol{x}),\ \forall \phi_2 \notin \boldsymbol{\Phi}^*_q(\boldsymbol{x}),\quad \bar{\ell}(\phi_1 \,|\, \boldsymbol{x}) \leq \bar{\ell}(\phi_2 \,|\, \boldsymbol{z}).$$

The quadratic loss function yields credible regions which contain those ϕ values which are closest to the posterior expectation in Euclidean distance. The zero-one loss function leads to highest posterior density (HPD) credible regions.

1.6.4 Kullback-Leibler (KL) Divergence

In many sections of the book — starting with this section — extensive use will be made of Kullback-Leibler (KL) divergence, a general measure of the goodness of the approximation of a true probability density $p(\boldsymbol{x})$ by another density $q(\boldsymbol{x})$.

Definition 1.5 *The* KL divergence *of a probability density* $q(\boldsymbol{x})$*, of the random vector* $\boldsymbol{x} \in \boldsymbol{\mathcal{X}}$*, from its true probability density* $p(\boldsymbol{x})$*, is*

$$\kappa\{q \,|\, p\} = \int_{\boldsymbol{\mathcal{X}}} p(\boldsymbol{x}) \log \frac{p(\boldsymbol{x})}{q(\boldsymbol{x})}\, d\boldsymbol{x}, \tag{1.12}$$

provided the integral (or the sum if $\boldsymbol{x}$ is discrete) exists.

The properties of $\kappa\{q \,|\, p\}$ have been extensively studied; see *e.g.*, Gibbs (1876), Shannon (1948), Kullback & Leibler (1951), Chernoff (1956), Kullback (1968), Csiszar (1985) and Rosenkrantz (1989). In particular, it can be shown that the KL divergence is non-negative and is zero if, and only if, $q(\boldsymbol{x}) = p(\boldsymbol{x})$

almost everywhere. Moreover, the KL divergence is invariant under one-to-one transformations of $\boldsymbol{x}$.

KL divergence has an interesting information-theoretical interpretation. Indeed (see *e.g.*, Csiszar (1985)) $\kappa\{q \mid p\}$ is the amount of information which would be necessary to recover the true distribution $p(\boldsymbol{x})$ from its approximation $q(\boldsymbol{x})$. Thus KL divergence is not defined on an arbitrary scale but, rather, is a measure in information units (*nits* if natural logarithms are used, *bits* if base 2 logarithms are used instead).

KL divergence has an interesting statistical interpretation in terms of the likelihood ratios which result when comparing two models $\{p_1(\boldsymbol{x} \mid \boldsymbol{\theta}_1), \boldsymbol{\theta}_1 \in \boldsymbol{\Theta}_1\}$ and $\{p_2(\boldsymbol{x} \mid \boldsymbol{\theta}_2), \boldsymbol{\theta}_2 \in \boldsymbol{\Theta}_2\}$. If, say, p_2 is the true model, then $\kappa\{p_1 \mid p_2\}$ is clearly the expectation (under p_2) of the log-likehood ratio in favor of the true model. Thus, a KL divergence of, say $\log[100] \approx 4.6$, suggests likelihood ratios for the true model of about 100, making discrimination very easy. On the other hand, a KL divergence of, say $\log[1.01] \approx 0.01$, suggests likelihood ratios for the true model of about 1.01, making discrimination very hard. More generally, since $\log(1+\epsilon) \approx \epsilon$ for all small ϵ, small discrepancies directly suggest likelihood ratios for the true model of only $1+\epsilon$, thus providing a transparent measure of how similar the compared models are.

Example 1.25 *Poisson approximation of a Binomial distribution.* One of the first approximations introduced in many elementary probability courses is the approximation of a Binomial distribution $p_1(x) = \mathrm{Bi}(x \mid n, \theta)$ by the Poisson distribution $p_2(x) = \mathrm{Pn}(x \mid n\theta)$. Clearly

$$\kappa\{p_2 \mid p_1\} = \sum_{x=0}^{n} \mathrm{Bi}(x \mid n, \theta) \log \frac{\mathrm{Bi}(x \mid n, \theta)}{\mathrm{Pn}(x \mid n\theta)}.$$

The result is plotted in Figure 1.9 as a function of θ for several values of n. As one might expect, the KL divergence converges to zero as θ decreases and as n increases, but it is apparent from the graph that the important condition for the approximation to work is that θ has to be small. For instance, with $n = 10$, the KL divergence is smaller than 0.001 for all $\theta < 0.06$.

The arithmetic average of the log-probability ratios $\log\{\mathrm{Bi}(x \mid 10, 0.05)/\mathrm{Pn}(x \mid 0.5)\}$ in favor of the Binomial, which could be obtained from Binomial $\mathrm{Bi}(x \mid 10, 0.05)$ samples, will cluster around the value $\kappa\{\mathrm{Pn}(\cdot \mid 0.5) \mid \mathrm{Bi}(\cdot \mid 10, 0.05)\} = 0.00067$, which corresponds to a probability ratio for the Binomial when it is true of only 1.00067.

1.6.5 Intrinsic Loss

While important foundationally, this section of the book is primarily theoretical and can be skipped by those interested primarily in methodology.

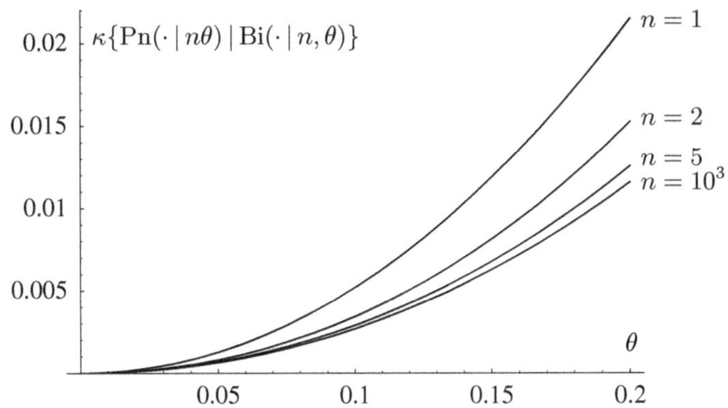

Figure 1.9: KL divergence between a Binomial distribution $\text{Bi}(x \mid n, \theta)$ and a Poisson distribution $\text{Pn}(x \mid n\theta)$ as a function of θ, for $n = 1, 2, 5$ and 1000.

Definition of Intrinsic Loss: Bayes procedures with conventional loss functions are typically *not* invariant under one-to-one transformations. Thus, the Bayes estimator of a variance under quadratic loss is not the square of the Bayes estimator of the standard deviation under quadratic loss. This can be unappealing, and leads to consideration of loss functions that are invariant, so that $\ell\{\tilde{\boldsymbol{\phi}}, \boldsymbol{\phi}\} = \ell\{g(\tilde{\boldsymbol{\phi}}), g(\boldsymbol{\phi})\}$ for any one-to-one transformation g. The most natural way to achieve this is to use a loss function that measures the discrepancy between the *models* $p(\boldsymbol{x} \mid \tilde{\boldsymbol{\phi}})$ and $p(\boldsymbol{x} \mid \boldsymbol{\phi})$, rather than the discrepancy between $\tilde{\boldsymbol{\phi}}$ and $\boldsymbol{\phi}$.

KL divergence (see Definition 1.5) is an invariant measure of the discrepancy between two distributions, but it has the sometimes problematical property of not being symmetric, *i.e.*, $\kappa\{q \mid p\} \neq \kappa\{p \mid q\}$. It can be symmetrized, however.

Definition 1.6 *The* intrinsic discrepancy $\delta\{p, q\}$ *between two probability densities* $p(\boldsymbol{x})$ *and* $q(\boldsymbol{x})$ *of* $\boldsymbol{x}$ *over* $\boldsymbol{\mathcal{X}}$ *is*

$$\delta\{p, q\} = \min[\kappa\{p \mid q\}, \kappa\{q \mid p\}] \,.$$

Besides being non-negative and symmetric, the intrinsic discrepancy inherits the KL divergence invariance under one-to-one transformations of $\boldsymbol{x} \in \boldsymbol{\mathcal{X}}$. Also, it is defined even if the supports of p and q are strictly nested, thus allowing its use to measure the dissimilarity between nonregular models, or between a density and its compact approximation. From its definition, the intrinsic discrepancy $\delta\{p, q\}$ may be interpreted as the minimum amount of information required to discriminate between p and q. The use of the logarithm makes this loss additive, in that, if $\boldsymbol{x}$ is a random sample $(x_1, \ldots, x_n)$, then $\delta\{p(\boldsymbol{x}), q(\boldsymbol{x})\} = n\,\delta\{p(x), q(x)\}$.

Estimation Under the Intrinsic Loss. The intrinsic discrepancy can be used to define an appropriate, invariant loss function for estimation problems.

Definition 1.7 *The* intrinsic loss *of using $\tilde{\boldsymbol{\phi}}$ as a proxy for $\boldsymbol{\phi}$ within the class of models $\{p(\boldsymbol{x} \mid \boldsymbol{\phi}, \boldsymbol{\omega}), \boldsymbol{x} \in \mathcal{X}, \boldsymbol{\phi} \in \boldsymbol{\Phi}, \boldsymbol{\omega} \in \boldsymbol{\Omega}\}$ is the intrinsic discrepancy between the true and estimated models,*

$$\mathcal{L}(\tilde{\boldsymbol{\phi}}, (\boldsymbol{\phi}, \boldsymbol{\omega})) = \delta\{p(\cdot \mid \boldsymbol{\phi}, \boldsymbol{\omega}),\, p(\cdot \mid \tilde{\boldsymbol{\phi}}, \boldsymbol{\omega})\}.$$

Given data $\boldsymbol{x}$ assumed to have been generated from $p(\boldsymbol{x} \mid \boldsymbol{\phi}, \boldsymbol{\omega})$, the expected intrinsic loss if $\tilde{\boldsymbol{\phi}}$ were used as a proxy for $\boldsymbol{\phi}$ is

$$\overline{\mathcal{L}}(\tilde{\boldsymbol{\phi}}, (\boldsymbol{\phi}, \boldsymbol{\omega}) \mid \boldsymbol{x}) = \int_{\boldsymbol{\Phi}} \int_{\boldsymbol{\Omega}} \mathcal{L}(\tilde{\boldsymbol{\phi}}, (\boldsymbol{\phi}, \boldsymbol{\omega}))\, \pi(\boldsymbol{\phi}, \boldsymbol{\omega} \mid \boldsymbol{x})\, d\boldsymbol{\phi}\, d\boldsymbol{\omega}\,.$$

Minimizing this over $\tilde{\boldsymbol{\phi}}$ results in the Bayes estimate under the intrinsic loss.

Example 1.26 *Normal variance.* Let $\boldsymbol{x} = (x_1, \ldots, x_n)$ be a random sample from a normal $\mathrm{N}(x \mid \mu, \sigma)$ distribution, and suppose that a point estimate of the variance σ^2 (or any one-to-one function of this, such as σ or $\log(\sigma)$) is required. Thus $\boldsymbol{\phi}$ above is σ^2 and $\boldsymbol{\omega}$ is μ. Let $\overline{x}$ and s^2 be the sample mean and variance.

To find the intrinsic discrepancy between the $\mathrm{N}(x \mid \mu, \tilde{\sigma})$ and $\mathrm{N}(x \mid \mu, \sigma)$ distributions, note that their KL divergence is

$$\begin{aligned}
\kappa\{\mathrm{N}(\cdot \mid \mu, \tilde{\sigma}) \mid \mathrm{N}(\cdot \mid \mu, \sigma)\} &= \int \mathrm{N}(x \mid \mu, \sigma) \log \frac{\mathrm{N}(x \mid \mu, \sigma)}{\mathrm{N}(x \mid \mu, \tilde{\sigma})}\, dx \\
&= \int \mathrm{N}(x \mid \mu, \sigma) \left[\log \frac{\tilde{\sigma}}{\sigma} + \left(\frac{1}{2\tilde{\sigma}^2} - \frac{1}{2\sigma^2}\right)(x - \mu)^2\right] dx \\
&= \frac{1}{2}\left[\log \frac{\tilde{\sigma}^2}{\sigma^2} + \frac{\sigma^2}{\tilde{\sigma}^2} - 1\right].
\end{aligned}$$

Thus, the intrinsic discrepancy between the $\mathrm{N}(x \mid \mu, \tilde{\sigma})$ and $\mathrm{N}(x \mid \mu, \sigma)$ distributions is given by the following expression, where $\theta = \tilde{\sigma}^2/\sigma^2$:

$$\begin{aligned}
\delta\{\mathrm{N}(\cdot \mid \mu, \tilde{\sigma}) \mid \mathrm{N}(\cdot \mid \mu, \sigma)\} &= \min\{\kappa\{\mathrm{N}(\cdot \mid \mu, \tilde{\sigma}) \mid \mathrm{N}(\cdot \mid \mu, \sigma)\},\, \kappa\{\mathrm{N}(\cdot \mid \mu, \sigma) \mid \mathrm{N}(\cdot \mid \mu, \tilde{\sigma})\}\} \\
&= \delta(\theta) = \begin{cases} \frac{1}{2}\left[\log \theta^{-1} + \theta - 1\right], & \text{if } \theta < 1 \\[2ex] \frac{1}{2}\left[\log \theta + \theta^{-1} - 1\right], & \text{otherwise} \end{cases}.
\end{aligned}$$

In Example 1.9, it was shown that the appropriate objective posterior distribution of σ^2 is the $\mathrm{Ig}(\sigma^2 \mid (n-1)/2, ns^2/2)$ distribution. The objective posterior distribution of $\theta = \tilde{\sigma}^2/\sigma^2$ is then the gamma density $\mathrm{Ga}[\theta \mid (n-1)/2, ns^2/(2\tilde{\sigma}^2)]$. Hence, the intrinsic estimator of the normal variance is that value, σ^2_{int}, of $\tilde{\sigma}^2$ which minimizes the expected posterior loss

$$\overline{\mathcal{L}}(\tilde{\sigma}^2 \mid n, s^2) = \int_0^\infty \delta(\theta)\, \mathrm{Ga}[\theta \mid (n-1)/2, ns^2/(2\tilde{\sigma}^2)]\, d\theta,$$

which may be rewritten, in terms of the chi-squared density, as

$$\overline{\mathcal{L}}(\tilde{\sigma}^2 \mid n, s^2) = \overline{\mathcal{L}}(a \mid n) = \int_0^\infty \delta(a\,\tau)\, \chi^2(\tau \mid n-1)\, d\tau, \qquad a = \frac{\tilde{\sigma}^2}{ns^2}\,.$$

Thus, the intrinsic estimator is of the form $\sigma^2_{int}(n, s^2) = c(n)\, s^2$, $c(n) = n\, a^*_n$, where a^*_n is the value of a which minimizes $\mathcal{L}(a \,|\, n)$ above. The exact value of a^*_n can be found by one-dimensional numerical integration followed by numerical optimization; some results are $\sigma^2_{int}(2, s^2) \approx 5\, s^2$, $\sigma^2_{int}(3, s^2) \approx 2.4\, s^2$, and $\sigma^2_{int}(4, s^2) \approx 1.8\, s^2$. For larger n, $c(n)$ is reasonably well approximated by its asymptotic limit, $n/(n-2)$, yielding

$$\sigma^2_{int}(n, s^2) \approx \frac{n\, s^2}{n-2}.$$

This is between the maximum likelihood estimate s^2 and the posterior mean $ns^2/(n-3)$.

Since intrinsic estimation is consistent under one-to-one reparameterizations, the intrinsic estimator of the standard deviation is just σ_{int}, the squared root of σ^2_{int}, and the intrinsic estimator of, say, $\log \sigma$ is simply $\log \sigma_{int}$. For more details on this historically important example, see Bernardo (2006).

Credible Sets Under the Intrinsic Loss. Again, many methods of deriving Bayes credible regions are *not* invariant under one-to-one transformations. This was dramatically shown for HPD regions in Example 1.16. As in the case of point estimation, invariant Bayes credible regions can be obtained by using an invariant loss function, as introduced in Bernardo (2005*a*). Here is an example.

Example 1.27 *Uniform model.* Let $\boldsymbol{x} = (x_1, \ldots, x_n)$ be a random sample from a uniform distribution $\text{Un}(x \,|\, 0, \theta)$ on $(0, \theta)$, $\theta > 0$, and suppose that one is interested in an interval estimate of θ, or any one-to-one function of θ. Notice that this is not a regular model, since the sampling space $\mathcal{X} = (0, \theta)$ depends on the parameter.

It will later be seen that an appropriate objective prior for this model is $\pi(\theta) = \theta^{-1}$. After $\boldsymbol{x}$ has been observed this leads to the Pareto posterior density $\pi(\theta \,|\, \boldsymbol{x}) = \pi(\theta \,|\, t, n) = n\, t^n \theta^{-(n+1)}$, with support on (t, ∞), where $t = \max\{x_1, \ldots, x_n\}$ is a sufficient statistic. The corresponding q-posterior quantile is $\theta_q = t\,(1-q)^{-1/n}$; in particular the median is $t\, 2^{1/n}$.

Using the definition of intrinsic loss for n independent observations of the uniform model immediately yields $\mathcal{L}(\tilde{\theta}, \theta) = n \log(\tilde{\theta}/\theta)$, if $\tilde{\theta} \geq \theta$, and $n \log(\theta/\tilde{\theta})$ otherwise. The expected posterior intrinsic loss from using $\tilde{\theta}$ as a proxy for θ (with $\tilde{\theta} > t$ since, given the data, θ is known to be larger than t) is

$$\overline{\mathcal{L}}(\tilde{\theta} \,|\, \boldsymbol{x}) = \int_t^\infty \mathcal{L}(\tilde{\theta}, \theta)\, \pi(\theta \,|\, \boldsymbol{x})\, d\theta = 2\,r - \log r - 1, \quad r = (t/\tilde{\theta})^n.$$

This is a positive convex function of r with support on $(0, 1)$ which tends to ∞ as $r \to 0$, has unique minimum at $r = 1/2$ and takes the value 1 at $r = 1$. As a function of $\tilde{\theta}$, this is minimized at $\theta^* = t\, 2^{1/n}$, which is therefore the intrinsic estimator (and, as mentioned above, is also the posterior median).

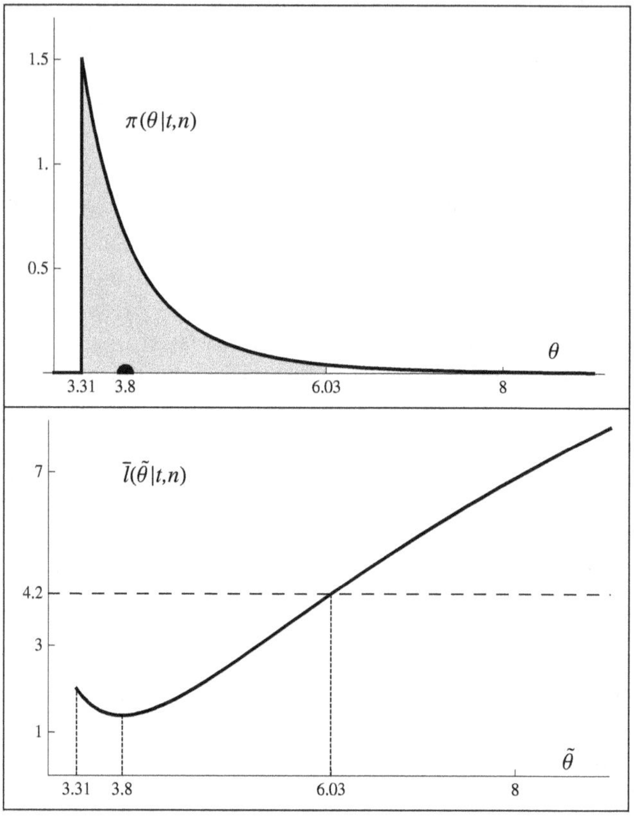

Figure 1.10: Top: the objective posterior distribution for θ arising from a sample of size 5 from the $\text{Un}(x\,|\,0,5)$ distribution; the half dot is the intrinsic estimate of θ and the shaded region is the intrinsic 95% credible set, consisting of those values that minimize the expected intrinsic loss given in the bottom panel and have posterior probability 0.95.

An intrinsic q-credible region will consist of the set of θ values with lower expected loss which have a posterior probability equal to q. It follows from the shape of $\overline{\mathcal{L}}(\tilde{\theta}\,|\,\boldsymbol{x})$ that, for sufficiently large q, these regions will be intervals of the form $R_q = (t, \theta_q)$, where $\theta_q = t\,(1-q)^{-1/n}$ is the q-quantile of $\pi(\theta\,|\,t, n)$.

As an illustration, a small sample of size $n = 5$ was simulated from a uniform distribution on $(0, 5)$, and the sample had maximum value $t = 3.31$. The corresponding objective posterior density is plotted in the upper panel of Figure 1.10. The figure includes the intrinsic estimator $\theta^* = 3.80$ (indicated by half a solid dot) and the intrinsic 0.95-credible region $(t, 6.03)$

(indicated as a shaded area). The expected intrinsic loss $\overline{\mathcal{L}}(\tilde{\theta}\,|\,\boldsymbol{x})$ from using $\tilde{\theta}$ as a proxy for θ is plotted in the lower panel of Figure 1.10. Indeed the intrinsic estimator corresponds to the minimum value of the expected loss, and all values within the intrinsic credible region have smaller expected loss than all values outside the region.

1.7 Large Sample Behavior

Let $\mathcal{M} = \{p(\boldsymbol{x}\,|\,\boldsymbol{\theta}), \boldsymbol{x} \in \boldsymbol{\mathcal{X}}, \boldsymbol{\theta} \in \boldsymbol{\Theta}\}$ be a parametric model, and consider a sequence $\boldsymbol{z}_n = (\boldsymbol{x}_1, \ldots, \boldsymbol{x}_n)$ of n independent observations from $\boldsymbol{\mathcal{X}}$, so that $\boldsymbol{z}_n$ is one observation from $\{p(\boldsymbol{z}\,|\,\boldsymbol{\theta}) = \prod_{i=1}^{n} p(\boldsymbol{x}_i\,|\,\boldsymbol{\theta}), \boldsymbol{z} \in \boldsymbol{\mathcal{X}}^n, \boldsymbol{\theta} \in \boldsymbol{\Theta}\}$. In this section, the behavior of the corresponding posterior distribution, $p(\boldsymbol{\theta}\,|\,\boldsymbol{z}_n)$, is discussed as $n \to \infty$. The two crucial features of this limit are (i) the result does not depend on the prior $\pi(\boldsymbol{\theta})$, since the information about $\boldsymbol{\theta}$ contained in the prior is dominated by the information provided by the data, and (ii) the posterior $p(\boldsymbol{\theta}\,|\,\boldsymbol{z}_n)$ will become increasingly concentrated around the true value of $\boldsymbol{\theta}$ as n increases.

These large sample results are of considerable interest, both theoretically and practically. Theoretically, they establish that Bayesian procedures are consistent, since the posterior distribution concentrates around the true value as the sample size increases. Practically, the results provide a useful approximation to the posterior distribution when n is large. Last, but not least, these asymptotic results are important for objective Bayesian analysis since, as will later be demonstrated, the recommended objective prior for the model parameters only depend on the large sample behavior of the model.

1.7.1 Discrete Parameters

Suppose first that the parameter space is discrete, so that $\boldsymbol{\Theta} = \{\boldsymbol{\theta}_1, \boldsymbol{\theta}_2, \ldots\}$ has a countable, possibly finite, number of elements, and where the true parameter value $\boldsymbol{\theta}_t$ is distinguishable from the others, in the sense that the KL divergences $\kappa\{p(\cdot\,|\,\boldsymbol{\theta}_i)\,|\,p(\cdot\,|\,\boldsymbol{\theta}_t)\}$ are strictly positive for all $i \neq t$. Using the strong law of large numbers, one can then prove that the posterior distribution converges to a probability distribution which gives probability one to the true parameter value.

Proposition 1.5 *Let $\boldsymbol{z}_n = (\boldsymbol{x}_1, \ldots, \boldsymbol{x}_n)$ be a random sample from $p(\boldsymbol{x}\,|\,\boldsymbol{\theta})$, $\boldsymbol{x} \in \boldsymbol{\mathcal{X}}$, $\boldsymbol{\theta} \in \boldsymbol{\Theta}$, where $\boldsymbol{\Theta} = \{\boldsymbol{\theta}_1, \boldsymbol{\theta}_2, \ldots\}$, and let the the prior distribution be $\pi(\theta_i) = p_i$, where $p_i > 0$ and $\sum_i p_i = 1$. Suppose that $\boldsymbol{\theta}_t \in \boldsymbol{\Theta}$ is the true value of $\boldsymbol{\theta}$ and that, for all $i \neq t$, $\kappa\{p(\cdot\,|\,\boldsymbol{\theta}_i)\,|\,p(\cdot\,|\,\boldsymbol{\theta}_t)\} > 0$. Then*

$$\lim_{n\to\infty} \pi(\boldsymbol{\theta}_t\,|\,\boldsymbol{z}_n) = 1, \quad \lim_{n\to\infty} \pi(\boldsymbol{\theta}_i\,|\,\boldsymbol{z}_n) = 0 \ \textit{for } i \neq t.$$

This result can be extended to more general forms of $\boldsymbol{\Theta}$; for details, see Berk (1966), Berk (1970). In particular, it can be shown that, if the true $\boldsymbol{\theta}_t$ is *not* in $\boldsymbol{\Theta}$, then the posterior degenerates onto the value in $\boldsymbol{\Theta}$ which gives the parametric model closest in KL divergence to the true model.

1.7.2 Continuous Parameters

Consider now the continuous case, where $\boldsymbol{\Theta} \subset \mathbb{R}^m$, and the prior distribution $\pi(\boldsymbol{\theta})$ has a strictly positive, continuous, 'smooth' probability density on $\boldsymbol{\Theta}$. Expanding $\log p(\boldsymbol{\theta} \mid \boldsymbol{z}_n)$ in a Taylor series around the mle $\hat{\boldsymbol{\theta}}_n$ and taking limits, one can prove that, under suitable regularity conditions, the posterior distribution $\pi(\boldsymbol{\theta} \mid \boldsymbol{z}_n)$ converges to an m-variate normal distribution.

Proposition 1.6 *Let $\boldsymbol{z}_n = (\boldsymbol{x}_1, \ldots, \boldsymbol{x}_n)$ be a random sample from $p(\boldsymbol{x} \mid \boldsymbol{\theta})$, $\boldsymbol{x} \in \boldsymbol{\mathcal{X}}$, $\boldsymbol{\theta} \in \boldsymbol{\Theta}$, where $\boldsymbol{\Theta} \subset \mathbb{R}^m$, and assume that the prior distribution $\pi(\boldsymbol{\theta})$ has a strictly positive, sufficiently regular probability density on $\boldsymbol{\Theta}$. Let $\hat{\boldsymbol{\theta}}_n$ be the value of $\boldsymbol{\theta}$ with maximizes the likelihood function $p(\boldsymbol{z}_n \mid \boldsymbol{\theta})$, and let $I(\boldsymbol{\theta} \mid \boldsymbol{z}_n)$ be the observed information matrix, with (i, j) element*

$$I_{ij}(\boldsymbol{\theta} \mid \boldsymbol{z}_n) = -\frac{\partial^2}{\partial\theta_i \partial\theta_j} \log p(\boldsymbol{z}_n \mid \boldsymbol{\theta}).$$

Then, under appropriate regularity conditions, $\pi(\boldsymbol{\theta} \mid \boldsymbol{z}_n)$ converges to the m-variate normal distribution, $N_m[\boldsymbol{\theta} \mid \hat{\boldsymbol{\theta}}_n, I^{-1}(\hat{\boldsymbol{\theta}}_n \mid \boldsymbol{z}_n)]$.

Under further regularity conditions on the likelihood function, the law of large numbers can be used to prove that $I(\boldsymbol{\theta} \mid \boldsymbol{z}_n)$ converges to $n\, I(\boldsymbol{\theta})$, where $I(\boldsymbol{\theta})$ is the Fisher information matrix having (i, j) element

$$I_{ij}(\boldsymbol{\theta}) = -\int_{\boldsymbol{\mathcal{X}}} p(\boldsymbol{x} \mid \boldsymbol{\theta}) \, \frac{\partial^2 \log p(\boldsymbol{x} \mid \boldsymbol{\theta})}{\partial\theta_i \partial\theta_j} \, d\boldsymbol{x} \cdot$$

Proposition 1.7 (Corollary). *Under sufficient regularity conditions, $\pi(\boldsymbol{\theta} \mid \boldsymbol{z}_n)$ converges to the m-variate normal distribution, $N_m[\boldsymbol{\theta} \mid \hat{\boldsymbol{\theta}}_n, n^{-1}\, I^{-1}(\hat{\boldsymbol{\theta}}_n)]$.*

There is a huge literature on the regularity conditions required to justify the results described above, starting with the pioneering work of Laplace (1812) and Jeffreys (1961, Ch. 4). For more recent accounts, see Chen (1985), Bernardo and Smith (1994, Ch. 5) and the references therein.

If a different parameterization is used for the model, the posterior is still asymptotically normal and is given as follows. (See Mendoza (1994) for the required regularity conditions.)

Proposition 1.8 *If $\boldsymbol{\theta}$ has an asymptotic normal distribution $N_m(\boldsymbol{\theta} \mid \hat{\boldsymbol{\theta}}_n, \boldsymbol{\Sigma}_n)$ and $\boldsymbol{g}(\boldsymbol{\theta})$ is a one-to-one smooth transformation such that $\boldsymbol{J}_{\boldsymbol{g}}(\boldsymbol{\theta}) = \partial\boldsymbol{g}(\boldsymbol{\theta})/\partial\boldsymbol{\theta}$ is non-singular with continuous entries, then, $\boldsymbol{g}(\boldsymbol{\theta})$ has an asymptotic normal distribution with mean $\boldsymbol{g}(\hat{\boldsymbol{\theta}}_n)$ and covariance matrix $\boldsymbol{J}_{\boldsymbol{g}}(\hat{\boldsymbol{\theta}}_n)\boldsymbol{\Sigma}_n \boldsymbol{J}_{\boldsymbol{g}}'(\hat{\boldsymbol{\theta}}_n)$.*

Example 1.28 *Ratio of Normal means.* Consider two independent random samples $\boldsymbol{x} = (x_1, \ldots, x_n)$ from $\mathrm{N}(x \mid \mu_1, 1)$ and $\boldsymbol{y} = (y_1, \ldots, y_n)$ from $\mathrm{N}(y \mid \mu_2, 1)$, and suppose that the quantity of interest is the ratio of the means $\phi_1 = \mu_1/\mu_2$. Using Theorem 1.7, the asymptotic posterior distribution of $\boldsymbol{\mu} = (\mu_1, \mu_2)$ is bivariate normal with mean $(\bar{x}_n, \bar{y}_n)$ and covariance matrix $n^{-1}\boldsymbol{I}_2$, where $\boldsymbol{I}_2$ is the identity matrix of dimension 2. The one-to-one transformation $\boldsymbol{\phi}(\boldsymbol{\mu}) = (\phi_1, \phi_2)$, with $\phi_2 = \mu_2$, has Jacobian

$$\mathbf{J}_{\boldsymbol{\phi}}(\boldsymbol{\mu}) = \begin{pmatrix} \partial\phi_1/\partial\mu_1 & \partial\phi_1/\partial\mu_2 \\ \partial\phi_2/\partial\mu_1 & \partial\phi_2/\partial\mu_2 \end{pmatrix} = \begin{pmatrix} \mu_2^{-1} & -\mu_1\mu_2^{-2} \\ 0 & 1 \end{pmatrix}.$$

Using Theorem 1.8, the asymptotic posterior distribution of $\boldsymbol{\phi}$ is thus bivariate normal with mean $(\bar{x}_n/\bar{y}_n, \bar{y}_n)$ and covariance matrix

$$\frac{1}{n\bar{y}_n^2}\begin{pmatrix} 1 + (\bar{x}_n/\bar{y}_n)^2 & -\bar{x}_n \\ -\bar{x}_n & \bar{y}_n^2 \end{pmatrix}.$$

Hence, the posterior distribution of $\phi_1 = \mu_1/\mu_2$ is asymptotically normal with mean $\bar{x}_n/\bar{y}_n$ and variance $(1 + (\bar{x}_n/\bar{y}_n)^2)/n\bar{y}_n^2$.

Different parameterizations can have differing speeds of convergence to asymptotic normality. The fastest is often attained with a transformation for which the covariance matrix does not depend on the parameter. In the one-dimensional case, and under appropriate regularity conditions, this is approximately achieved by $g(\theta) = \int I(\theta)^{1/2}\, d\theta$, with $I(\theta)$ being the one-dimensional Fisher information.

Example 1.29 *Normal approximations to Bernoulli data.* As an illustration, consider n with parameter θ, and the objective prior $\pi(\theta) = \text{Be}(\theta \mid 1/2, 1/2)$. If r denotes the number of observed successes, the exact posterior for θ is $\text{Be}(\theta \mid r + 1/2, n - r + 1/2)$, the mle is $\hat{\theta} = r/n$, the Fisher information is $n\theta^{-1}(1-\theta)^{-1}$ and, using Theorem 1.7, the direct normal approximation is $\pi(\theta \mid r, n) \approx \text{N}[\theta \mid \hat{\theta}, n^{-1}\hat{\theta}(1-\hat{\theta})]$.

The upper panel of Figure 1.11 contains the exact posterior densities for θ (continuous lines) and the corresponding normal approximations (dotted lines) arising from $n = 20$ and $n = 100$ simulated observations with $\theta = 0.2$, with $r = 2$ and $r = 19$ successes, respectively. The normal approximation is not very good (especially in the left tail) for $n = 20$, but is reasonable for $n = 100$.

The function which approximately stabilizes the variance is

$$\phi(\theta) = \int \sqrt{I(\theta)}\, d\theta = \int \theta^{-1/2}(1-\theta)^{-1/2}\, d\theta = 2\text{ArcSin}\sqrt{\theta}.$$

Changing variables, the objective prior $\pi(\theta)$ becomes $\pi(\phi) = 1$ on $(0, \pi)$. Using Theorem 1.8, the posterior distribution of ϕ is approximately normal $\text{N}(\phi \mid \phi(\hat{\theta}), n^{-1})$. The lower panel of Figure 1.11 contains the exact posteriors for ϕ, and the corresponding normal approximations, for the same samples used in the upper panel. The approximation is vastly improved by the transformation.

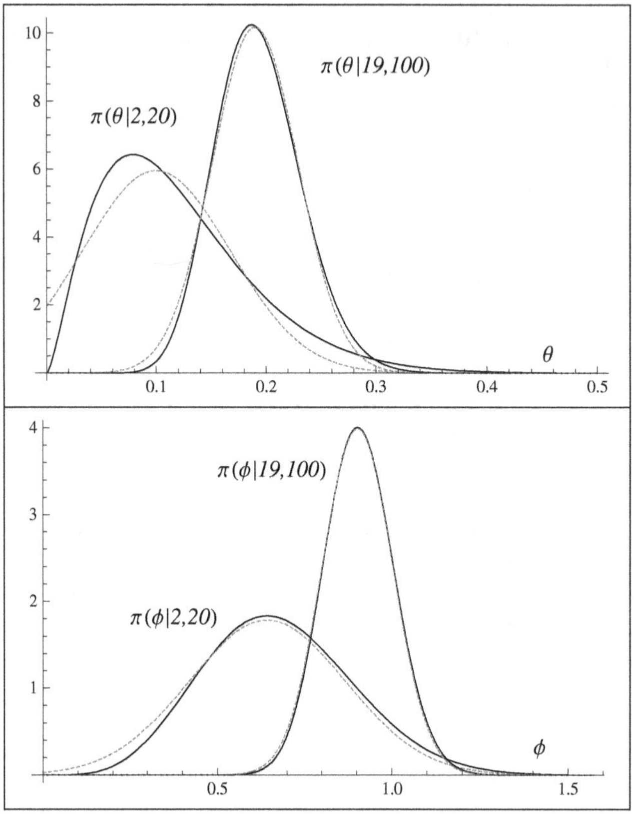

Figure 1.11: Normal approximations to the posterior densities arising from Bernoulli, $\text{Br}(x \mid \theta)$, observations. Upper panel: exact and normal approximation of $\pi(\theta \mid r, n)$. Lower panel: exact and normal approximation of $\pi(\phi \mid r, n)$, for the transformed parameter $\phi(\theta) = 2\text{ArcSin}\sqrt{\theta}$.

1.8 Generating Samples from the Posteriors

This section provides a brief survey of the the most prominent methods of simulating (taking draws of the unknown parameter $\boldsymbol{\theta}$) from the posterior distribution of $\boldsymbol{\theta}$. This is not meant to be a 'how to do' simulation treatise; the subject is far too complicated. The intent is to, instead, provide a brief introduction into simulation methods so that, when such methods are referred to later in the text (for a given model and objective prior, we will often mention particularly appropriate computational methods), a ready reference is available as to what is meant. For in-depth study of simulation, see Robert & Casella (2004).

The section begins with the most general method (but least accurate) and ends with the least general (but most accurate) methods. The most general method is the Metropolis-Hastings algorithm, discussed in Section 1.8.1; it can virtually always be applied. Section 1.8.3, discusses Gibbs Sampling, a powerful method for situations in which conditional posterior distributions (distributions of parameters given other parameters) are available.

Both of these methods yield dependent draws from the posterior distribution, and the dependence can be so severe as to make any inferences resulting from the methods questionable. In the other sections, methods are described that guarantee independent draws from the posterior; with such draws one can efficiently obtain desired estimates and accurately assess the error in the estimates. Unfortunately, these accurate methods are only applicable in certain situations.

1.8.1 Simulation via the Metropolis-Hastings Algorithm

The Metropolis-Hastings algorithm is a Markov chain Monte Carlo (MCMC) method for obtaining a sequence of (dependent) samples from a probability distribution such as the posterior density, $\pi(\boldsymbol{\theta} \mid \boldsymbol{x})$, for which direct sampling could be difficult. This sequence can be used to approximate the target density or to compute an integral (such as an expected value). The algorithm was named after Metropolis *et al.* (1953), and used for symmetrical proposal distributions (see below). Hastings (1970) extended the algorithm to general proposal distributions. Metropolis-Hastings and other MCMC algorithms are often used for sampling from multi-dimensional distributions, especially when the dimension is high.

The *Metropolis-Hastings* algorithm proceeds as follows,

Step 0. Choose a starting value $\boldsymbol{\theta}_0$ (*e.g.*, the mle), and set $t = 1$.

Step 1. Draw $\boldsymbol{\theta}^*$ from a *proposal* distribution q_t with density $q_t(\boldsymbol{\theta}^* \mid \boldsymbol{\theta}_{t-1})$.

Step 2. Calculate the ratio

$$r_t = \frac{\pi(\boldsymbol{\theta}^* \mid \boldsymbol{x})/q_t(\boldsymbol{\theta}^* \mid \boldsymbol{\theta}_{t-1})}{\pi(\boldsymbol{\theta}_{t-1} \mid \boldsymbol{x})/q_t(\boldsymbol{\theta}_{t-1} \mid \boldsymbol{\theta}^*)} \,.$$

Note that $p(\boldsymbol{x})$, the marginal density of $\boldsymbol{x}$, will cancel in the numerator and denominator and so it need not be computed.

Step 3. Let $\boldsymbol{\theta}_t = \boldsymbol{\theta}^*$ with probability $\min\{r_t; 1\}$; else, set $\boldsymbol{\theta}_t = \boldsymbol{\theta}_{t-1}$.

Step 4. Set $t = t + 1$ and return to *Step 1.*

Discard the initial M values (called 'burn-in' values). The output from the last T iterations: $\{\boldsymbol{\theta}_t,\ t = M + 1, ..., M + T\}$ is then treated as a sample (not independent) from the posterior density. This sample can then be used to approximate the posterior density (through a histogram, say) and to approximate the posterior mean of $h(\boldsymbol{\theta})$ given $\boldsymbol{x}$, using

$$\mathrm{E}[h(\boldsymbol{\theta}) \mid \boldsymbol{x}] = \int h(\boldsymbol{\theta})\, \pi(\boldsymbol{\theta} \mid \boldsymbol{x})\, d\boldsymbol{\theta} \approx \frac{1}{T} \sum\nolimits_{j=1}^{T} h(\boldsymbol{\theta}_{M+j})\,. \tag{1.13}$$

The Metropolis algorithm is the special case when $q_t(\boldsymbol{\theta}^* \mid \boldsymbol{\theta}_{t-1}) = q_t(\boldsymbol{\theta}_{t-1} \mid \boldsymbol{\theta}^*)$, which results in $r_t = \pi(\boldsymbol{\theta}^* \mid \boldsymbol{x})/\pi(\boldsymbol{\theta}_{t-1} \mid \boldsymbol{x})$.

For illustration, we give an example of a one-dimensional distribution. A relatively simple example is chosen because this can also be used to illustrate the other computational methods which will later be discussed.

Example 1.30 *Genetic linkage.* One hundred animals are distributed into three categories, with $\boldsymbol{x} = (x_1, x_2, x_3)$ being the number of animals in each category. The data arise from a multinomial genetic linkage model, $\text{Mu}_2(p_1, p_2 \mid n = 100)$, where $p_1 = \theta/4$, $p_2 = (1-\theta)/2$, (and, therefore, $p_3 = (1 - p_1 - p_2) = (2+\theta)/4)$ for some unknown parameter $\theta \in (0,1)$. Observing $\boldsymbol{x} = (16, 19, 65)$ and choosing a uniform $\text{Un}(\theta \mid 0, 1)$ prior yields the posterior

$$\pi(\theta \mid \boldsymbol{x}) \propto f(\theta) = \theta^{x_1}(1-\theta)^{x_2}(2+\theta)^{x_3} = \theta^{16}(1-\theta)^{19}(2+\theta)^{65}; \quad (1.14)$$

again, the normalizing constant for the posterior is not needed.

The posterior mean of θ will be approximated using the Metropolis-Hastings algorithm. In this simple problem, there is no need to choose a proposal distribution for θ^* that depends on θ_{t-1}. Instead, one can just choose $q(\theta^*)$ to be $\text{Un}(\theta^* \mid 0, 1)$, *i.e.*, just sample independently from the uniform distribution; thus the proposal acceptance probability is

$$r_t = \frac{\pi(\theta^* \mid \boldsymbol{x})}{\pi(\theta_{t-1} \mid \boldsymbol{x})} = \frac{(\theta^*)^{16}(1-\theta^*)^{19}(2+\theta^*)^{65}}{\theta_{t-1}^{16}(1-\theta_{t-1})^{19}(2+\theta_{t-1})^{65}} \, .$$

Using the starting value $\theta_0 = 0.5$, $M = 500$ and $T = 2000$, applying the Metropolis-Hastings algorithm, and using (1.13), the estimated posterior mean of θ is 0.629.

The samples produced by the Metropolis-Hastings algorithm can be directly used to create a posterior histogram of θ; this is displayed in Figure 1.12, part (a), with the continuous line representing the true posterior density of θ. The posterior histogram is a reasonably accurate portrayal of the posterior density.

The sample path of the last T outputs (simply a plot of the last T outputs) of θ is displayed in Figure 1.13, part (a). This can be used to assess how well the algorithm 'mixes' which is a combination of how often it moves (notice that it does periodically stay at the same value of θ) and how much it moves throughout the parameter space. In this problem, the algorithm usually makes big moves (when it moves). Also, the probability of moving (the average of the $\min\{r_t; 1\}$) is about 0.2332, so it moves reasonably often.

The success of Metropolis-Hastings here is because the posterior distribution of θ is quite diffuse, so that use of the uniform proposal distribution $q(\theta^*)$ is reasonable. Typically, the posterior distribution will be quite concentrated, and very different (and cleverly chosen) proposal distributions are needed.

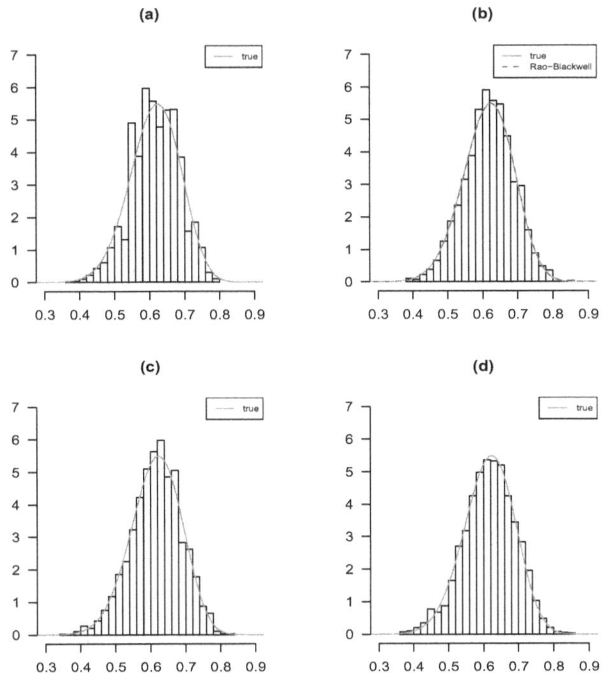

Figure 1.12: Posterior histograms of θ based on $T = 2000$ cycles from (a) the Metropolis-Hastings algorithm, (b) Gibbs sampling, (c) the rejection method, and (d) the ratio-of-uniforms method, for the genetic linkage problem.

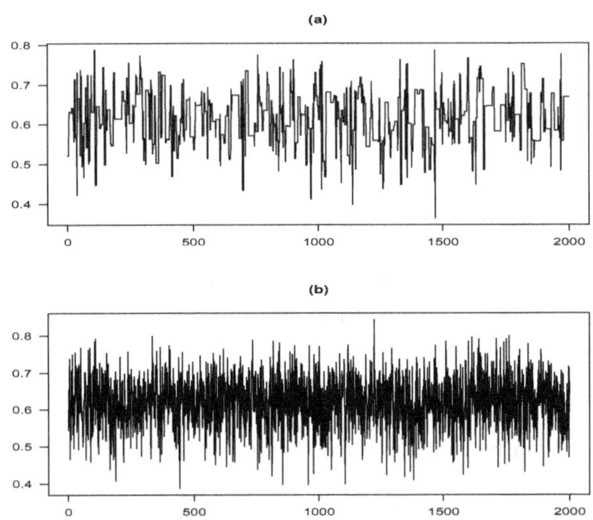

Figure 1.13: Sample paths of θ from (a) the Metropolis-Hastings algorithm and (b) Gibbs sampling, based on $T = 2000$ cycles for the genetic linkage problem.

1.8.2 Importance Sampling

Closely related to the Metropolis-Hastings algorithm is *importance sampling*, which is usually focused on estimating integrals such as $\int h(\boldsymbol{\theta})\, \pi(\boldsymbol{\theta} \mid \boldsymbol{x})\, d\boldsymbol{\theta}$. Importance sampling begins with a proposal distribution $q(\boldsymbol{\theta})$, chosen so as try to mimic the posterior distribution (cf. Liu (2001)). One then takes T draws $\{\boldsymbol{\theta}_1, \ldots, \boldsymbol{\theta}_T\}$ from this proposal distribution, and approximates the integral by

$$\int h(\boldsymbol{\theta})\, \pi(\boldsymbol{\theta} \mid \boldsymbol{x}) d\boldsymbol{\theta} \approx \frac{1}{T} \sum_{i=1}^{T} h(\boldsymbol{\theta}_i)\, \frac{\pi(\boldsymbol{\theta}_i \mid \boldsymbol{x})}{q(\boldsymbol{\theta}_i)} \, .$$

Often, $p(\boldsymbol{x}) = \int p(\boldsymbol{x} \mid \boldsymbol{\theta})\, \pi(\boldsymbol{\theta})\, d\boldsymbol{\theta}$ (the denominator of $\pi(\boldsymbol{\theta} \mid \boldsymbol{x})$) is not available in closed form, in which case importance sampling can also be used to approximate this marginal distribution. Indeed, the approximation to the posterior expectation of $h(\boldsymbol{\theta})$ would then be

$$\begin{aligned} \int h(\boldsymbol{\theta})\, \pi(\boldsymbol{\theta} \mid \boldsymbol{x})\, d\boldsymbol{\theta} &= \int h(\boldsymbol{\theta})\, \frac{p(\boldsymbol{x} \mid \boldsymbol{\theta})\, \pi(\boldsymbol{\theta})}{p(\boldsymbol{x})}\, d\boldsymbol{\theta} \\ &\approx \frac{\sum_{i=1}^{T} h(\boldsymbol{\theta}_i)\, p(\boldsymbol{x} \mid \boldsymbol{\theta}_i)\, \pi(\boldsymbol{\theta}_i)/q(\boldsymbol{\theta}_i)}{\sum_{i=1}^{T} p(\boldsymbol{x} \mid \boldsymbol{\theta}_i)\, \pi(\boldsymbol{\theta}_i)/q(\boldsymbol{\theta}_i)} \, . \end{aligned}$$

Example 1.31 *Genetic linkage continued.* Consider again Example 1.30. Choosing $q(\theta)$ to be $Un(\theta \mid 0, 1)$ and $T = 2000$, the above approximation yields 0.6177 as an estimate of the posterior mean.

1.8.3 Simulation via Gibbs Sampling

Let $\pi(\boldsymbol{\theta} \mid \boldsymbol{x})$ be the posterior density for $\boldsymbol{\theta} = (\theta_1, \ldots, \theta_m) \in \boldsymbol{\Theta}$. A *full conditional posterior distribution* is the posterior distribution of one of the θ_i, given values of all the other parameters and $\boldsymbol{x}$, *i.e.*,

$$\pi_i(\theta_i \mid \theta_1, \ldots, \theta_{i-1}, \theta_{i+1}, \ldots, \theta_m, \boldsymbol{x}), \qquad i = 1, \ldots, m.$$

For situations where all the full conditional posterior distributions are available for sampling, Geman & Geman (1984) introduced another algorithm to produce (dependent) samples from the posterior distribution; the algorithm is commonly referred to as the *Gibbs sampler.*

The algorithm is a Markovian updating scheme and proceeds as follows.

Step 0. Begin with any initial value $\boldsymbol{\theta}_0 = (\theta_{0,1}, \theta_{0,2}, \ldots, \theta_{0,m})$.

Step 1. Generate $\theta_{1,1}$ from $\pi_1(\theta_1 \mid \theta_{0,2}, \theta_{0,3}, \ldots, \theta_{0,m}, \boldsymbol{x})$;

Step 2. Generate $\theta_{1,2}$ from $\pi_2(\theta_2 \mid \theta_{1,1}, \theta_{0,3}, \ldots, \theta_{0,m}, \boldsymbol{x})$;

...

Step m. Generate $\theta_{1,m}$ from $\pi_m(\theta_m \,|\, \theta_{1,1}, \theta_{1,2}, \ldots, \theta_{1,m-1}, \boldsymbol{x})$.

The above loop results in a single iterate (or a cycle) of the Gibbs sampler resulting in a 'draw' $\boldsymbol{\theta}_1 = (\theta_{1,1}, \theta_{1,2}, \ldots, \theta_{1,m})$. Then repeat the process, using $\boldsymbol{\theta}_1 = (\theta_{1,1}, \theta_{1,2}, \ldots, \theta_{1,m})$ as the initial values, resulting in the new 'draw' $\boldsymbol{\theta}_2 = (\theta_{2,1}, \theta_{2,2}, \ldots, \theta_{2,m})$. Continue iterating and throw out the first M 'burn in' samples, resulting in the final sample $\{\boldsymbol{\theta}_{M+i}, i = 1, \ldots, T\}$.

This can be viewed as a (dependent) sample from the posterior distribution and (under mild conditions), the average

$$\frac{1}{T}\sum\nolimits_{j=1}^{T} h(\boldsymbol{\theta}_{M+j,1}, \boldsymbol{\theta}_{M+j,2}, \ldots, \boldsymbol{\theta}_{M+j,m})$$

will converge to $\int h(\boldsymbol{\theta})\, \pi(\boldsymbol{\theta} \,|\, \boldsymbol{x})\, d\boldsymbol{\theta}$ as T increases. Furthermore, $\pi(\theta_i \,|\, \boldsymbol{x})$, the marginal posterior density of θ_i can be estimated by the finite mixture density

$$\frac{1}{T}\sum\nolimits_{j=1}^{T} \pi_i(\theta_i \,|\, \theta_{j,1}, \ldots, \theta_{j,i-1}, \theta_{j,i+1}, \ldots, \theta_{j,m}). \tag{1.15}$$

This is called "Rao-Blackwellization" in Gelfand & Smith (1990). For further discussion of the Gibbs sampler, see Robert & Casella (2004) and Bolstad (2010).

Example 1.32 *Genetic linkage continued.* Consider again Example 1.30. In order to use the Gibbs Sampler to simulate from the posterior of θ, we introduce a variable z that is called a 'latent variable.' Indeed, we consider the extended model $\boldsymbol{x}^* = (x_1, x_2, z) \sim \text{Mu}_3(\boldsymbol{x}^* \,|\, p_1, p_2, p_3, n = 100)$, where

$$p_1 = \theta/4, \quad p_2 = (1-\theta)/2, \quad p_3 = 1/2, \qquad \theta \in (0,1).$$

(Note that, here, $x_4 = 1 - x_1 - x_2 - z = x_3 - z$ and $p_4 = \theta/4$.) It is easy to show that that marginal distribution of (x_1, x_2, x_3) is the same as before, so the introduction of the latent z has not changed the distribution in terms of inference concerning θ.

With a uniform prior $Un(\theta \,|\, 0, 1)$ for θ, the joint posterior of (z, θ) is

$$\pi(z, \theta \,|\, \boldsymbol{x}) \propto \frac{2^z}{z!(x_3 - z)!}\theta^{x_1 + x_3 - z}(1-\theta)^{x_2}. \tag{1.16}$$

Computation shows that the full conditional posterior distributions of (z, θ), given $\boldsymbol{x}$, are

$$\begin{aligned} p(z \,|\, \theta, \boldsymbol{x}) &= \text{Bi}(z \,|\, x_3, 2/(2+\theta)); \\ p(\theta \,|\, z, \boldsymbol{x}) &= \text{Be}(\theta \,|\, x_1 + x_3 - z + 1, x_2 + 1). \end{aligned}$$

The Gibbs sampling was run with $M = 50$, $T = 2000$, and a starting value of $\theta_0 = 0.5$. The resulting estimated posterior mean for θ is 0.612. Note that (1.15) becomes

$$\hat{\pi}(\theta \,|\, \boldsymbol{x}) = \frac{1}{T}\sum\nolimits_{j=1}^{T} \text{Be}(\theta \,|\, x_1, x_3 - z_{M+j} + 1, x_2 + 1), \tag{1.17}$$

where $\{z_j, j = M+1, \ldots, M+T\}$ are the draws from the Gibbs sampling. The dashed line, in Figure 1.12, part (b), is this estimated posterior density, and it is indistinguishable from the true posterior density (continuous line). The posterior histogram, based on the draws, is also displayed, and is inferior to the Rao-Blackwell estimate.

The sample path of the draws of θ is displayed in Figure 1.13, part (b). The mixing appears much better than with the Metropolis-Hastings algorithm, with changes always happening and with almost as many large changes. This gain was achieved because of the efficiency arising from knowing the full conditional distributions; if these had not been available in recognizable closed form, one could not have done Gibbs sampling so efficiently.

1.8.4 Simulation via Rejection Sampling

A general discussion of rejection sampling can be found in Casella, Robert, Wells et al. (2004). This is the first of the methods to be discussed that will directly yield an independent sample from the posterior $\pi(\boldsymbol{\theta} \mid \boldsymbol{x})$.

Write $\pi(\boldsymbol{\theta} \mid \boldsymbol{x}) = C_1 f(\boldsymbol{\theta})$; the constant C_1 can be ignored, so that one does not need to be able to compute the normalization constant of the posterior. One then tries to find a proposal function, $q(\boldsymbol{\theta})$, such that $C_2 q(\boldsymbol{\theta})$ is a proper density, for some constant C_2, and for which

$$f(\boldsymbol{\theta}) \leq M\, q(\boldsymbol{\theta}), \quad \boldsymbol{\theta} \in \boldsymbol{\Theta}, \tag{1.18}$$

for some constant $0 < M < \infty$. Rather than the term 'proposal function,' it is more common to call q an *envelope function* in rejection sampling.

One can then obtain independent samples from the posterior by simulating (independent) $\boldsymbol{\theta}^*$ from $C_2 q(\boldsymbol{\theta})$, and accepting each $\boldsymbol{\theta}^*$ as an observation from $\pi(\boldsymbol{\theta} \mid \boldsymbol{x})$ with probability $f(\boldsymbol{\theta}^*)/(M\, q(\boldsymbol{\theta}^*))$. An algorithm to do this is as follows.

Step 1. Independently generate u from $\text{Un}(u \mid 0, 1)$, and $\boldsymbol{\theta}^*$ from $C_2 q(\boldsymbol{\theta})$.

Step 2. If $0 < u < f(\boldsymbol{\theta}^*)/(M\, g(\boldsymbol{\theta}^*))$, report $\boldsymbol{\theta}^*$ as an independent draw from the posterior distribution; otherwise, return to *Step 1* and try again.

Clearly, if a finite M cannot be found, the algorithm cannot be used. But, even if M is found, the algorithm will be essentially useless if M is too large (the acceptance probabilities will be near zero.) Hence, again, the need is to find a clever $q(\boldsymbol{\theta})$, from which one can generate samples and which leads to a reasonable M.

A useful class of densities for which such a q can be found is the class of log-concave densities (see Devroye (1984)). Even for such densities, one may have to find q by adaptive methods, starting with some q and continually refining it to make M smaller (Gilks and Wild, 1992). The function *rARS* from the R package *AdapSamp* can be used to do this.

Example 1.33 *Genetic linkage continued.* Consider again Example 1.32. It follows from (1.14) that, for any $\theta \in (0, 1)$,

$$\frac{d^2}{d\theta^2} \log[f(\theta)] = -\frac{x_1}{\theta^2} - \frac{x_2}{(1-\theta)^2} - \frac{x_3}{(2+\theta)^2} < 0,$$

so the posterior density of θ is log-concave. The R function *rARS* was used to simulate 2000 observations from this posterior. The posterior histogram of θ is displayed in Figure 1.12, part (c), where the continuous line is the true posterior density. The histogram is quite accurate because it is based on 2000 *independent* observations from the posterior.

1.8.5 Ratio-of-Uniforms Method

When θ is one-dimensional, an alternative scheme to generate an independent sample from $\pi(\theta \mid \boldsymbol{x}) \propto f(\theta)$ is the ratio-of-uniforms method. Here f is an integrable function with support (θ_0, θ_1), the endpoints not necessarily being finite. This method is different than the previous methods in that it does not utilize a proposal distribution. Instead, it is based on the fact (Leydold (2003)) that, if a bivariate random variable (u, v) is uniform on

$$\mathcal{A} = \left\{(u, v) : 0 < u < \sqrt{f(\xi + v/u)}, \quad \theta_0 < \xi + v/u < \theta_1\right\}, \tag{1.19}$$

where ξ could be any number, then $\theta = \xi + v/u$ is a random draw from $\pi(\theta \mid \boldsymbol{x})$.

The method for $\xi = 0$ was introduced in Kinderman & Monahan (1977). The value of ξ can be adjusted to improve the efficiency of the algorithm; often, a reasonable choice is the mode of the posterior distribution of θ.

To convert this fact into an algorithm, Wakefield and Gelfand (1991) find the 'minimal bounding rectangle' for $\mathcal{A}$ (and hence the most efficient region in which to generate uniform random variables)

$$\overline{\theta} = \sup_\theta \sqrt{f(\theta)}, \quad \overline{v} = \sup_\theta \left[(\theta - \xi)\sqrt{f(\theta)}\right], \quad \underline{v} = \inf_\theta \left[(\theta - \xi)\sqrt{f(\theta)}\right].$$

The algorithm is then as follows.

Step 1. Independently simulate u from $\text{Un}(u \mid 0, \overline{\theta})$ and v from $\text{Un}(v \mid \underline{v}, \overline{v})$.

Step 2. If $0 < u < \sqrt{f(\xi + v/u)}$, report $\xi + v/u$; otherwise, return to *Step 1.*

Extension to an m-dimensional posterior density, $\pi(\boldsymbol{\theta} \mid \boldsymbol{x}) \propto f(\boldsymbol{\theta})$, can be found in Wakefield and Gelfand (1991). Indeed, if $(u, \boldsymbol{v})$ are uniform on

$$\mathcal{A} = \{(u, \boldsymbol{v});\ \boldsymbol{v} \in \mathbb{R}^m,\ 0 < u < f(\boldsymbol{\xi} + \boldsymbol{v}/u^r)^{1/[r\,m+1]}\}, \tag{1.20}$$

for some $\boldsymbol{\xi} \in \mathbb{R}^m$, then $\boldsymbol{\theta} = \boldsymbol{\xi} + \boldsymbol{v}/u^r$ has the posterior density $\pi(\boldsymbol{\theta} \mid \boldsymbol{x})$. When $r = 1$, this yields the standard ratio-of-uniforms method, which is actually defined for any dimension m (see Hörnmann *et al.*, 2004); when $r = 0$, this is the *slice sampler* of Neal (2003).

Example 1.34 *Genetic linkage continued.* Consider again Example 1.32. The ratio-of-uniforms method was employed to sample 2000 points from the posterior of θ, first using the algorithm with $\xi = 0$ and then with $\xi = 0.615$ (the posterior mode). These sampled points, together with the boundaries of the acceptance regions, are plotted and compared in Figure 1.14. When $\xi = 0$, the acceptance probability is about 0.1410; when $\xi = 0.615$, it becomes 0.7430, yielding a much more efficient algorithm.

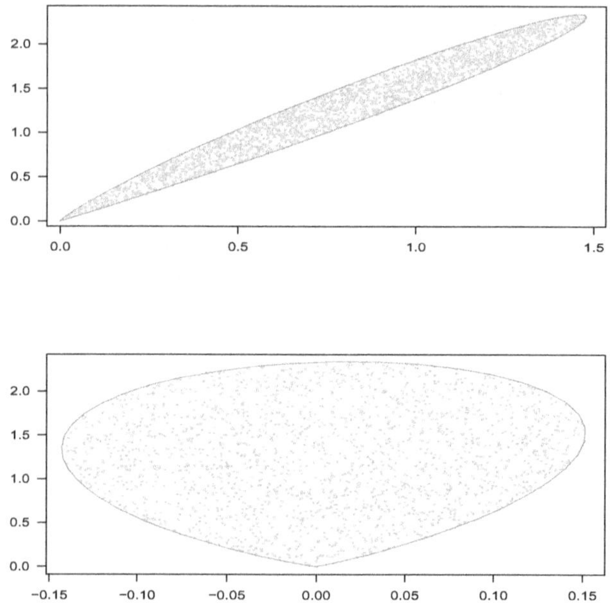

Figure 1.14: Acceptance regions for the ratio of uniforms method when $\xi = 0$ (upper frame) and when $\xi = \text{Mo}[\theta \mid \boldsymbol{x}] = 0.615$ (lower frame).

The posterior histogram of θ is displayed in Figure 1.12, part (d), where the continuous line is the true posterior density. Again, since this was based on an independent sample of size 2000, the histogram is quite accurate.

1.8.6 Constructive Random Posteriors

Sometimes the posterior density is quite complicated — apparently necessitating use of one of the above sampling schemes — but an explicit expression is available to generate directly from the posterior; we call such an expression a *constructive random posterior.* In this section, x^* will be used to denote a draw from some specific density $p(x)$.

Example 1.35 *Normal mean posterior continued.* Consider again Example 1.9. The marginal posterior of μ is $\pi(\mu \mid \boldsymbol{x}) = \mathrm{St}(\mu \mid \overline{x}, s/\sqrt{n-1}, n-1)$. Draws from this posterior can be obtained by either of the following procedures:

Option 1. Let t^*_{n-1} be one random observation from $\mathrm{St}(t \mid 0, 1, n-1)$, a standard t random variable with $n-1$ degrees of freedom. Then, a constructive random posterior draw, μ^*, from $\pi(\mu \mid \boldsymbol{x})$, is given by

$$\mu^* = \bar{x} + \frac{s}{\sqrt{n-1}}\, t^*_{n-1}. \tag{1.21}$$

Option 2. Let z^* be a draw from the standard normal distribution, and, independently, let χ^{2*}_{n-1} be a draw from the chi-squared distribution with $n-1$ degrees of freedom. Then, another constructive random posterior draw, μ^*, from $\pi(\mu \mid \boldsymbol{x})$, is given by

$$\mu^* = \bar{x} + \frac{s}{\sqrt{n-1}} \frac{z^*}{\sqrt{\chi^{2*}_{n-1}}}.$$

Considerably more sophisticated examples of constructive random posteriors are given in Section 10.3.3.

1.9 Discussion and Additional References

1.9.1 Formal Definition of Probability

As discussed in Section 1.1, we described the probability $\Pr(E \mid C)$ as a measure of the belief in the occurrence of the event E under conditions C; however, no formal definition was provided, since this is not necessary to understand and implement the methods we propose. Formal definitions of probability are, however, available from a variety of viewpoints and are briefly discussed below.

We favor an *operational* definition of probability, in the sense of Percy Bridgman (1927). Since probability is a *measure* of uncertainty, some type of *unit* of uncertainty is needed to obtain probabilities by appropriate comparison. This can be done by defining a set of standard events of known probabilities and using these to make comparisons (see, *e.g.*, Bernardo and Smith, 1994, Ch. 2, and references therein).

An alternative approach to the definition of probability as a degree of belief, without explicit use of the concept of utility, was pioneered by de Finetti (1937), who proved that coherent degrees of belief satisfy the rules of probability, and developed probability theory from this point of view (de Finetti 1970). Related arguments were given by Freedman & Purves (1969), Cornfield (1969), Heath & Sudderth (1972, 1978) and Buehler (1976). Bertrand Russell (1948) provided an early, provocative, discussion on the relationship between 'individual' and 'scientific' knowledge.

The first approach to a formal axiom system for degrees of belief was given by Wrinch and Jeffreys (1919, 1921), and its crucial implications developed in Jeffreys (1961). Both R. T. Cox (1946, 1961) and Jaynes (1958, 2003) developed axiom systems to prove that probability constitutes the only consistent extension of ordinary logic in which degrees of belief are measured by real numbers. Other important contributions to this general topic include Keynes (1929), Koopman (1940) and Carnap (1962).

Formal mathematical axioms for probability, with no specific reference to degrees of belief, were first given in definitive form by Andréi Kolmogórov (1956); those are today an important part of measure theory. A number of extensions and generalizations have been published since; in particular, Rényi (1955) provided an axiomatic theory for probability motivated by the need to include unbounded measures; see also Taraldsen & Lindqvist (2010). Other relevant pointers are Reichenbach (1949), Popper (1957) and Suppes (1973).

1.9.2 Normative Approach to Foundations

There are many normative decision theories, which involve a unified treatment of both probability and utility, that lead to the expected utility criterion described in Section 1.6. The earlier axiomatic approach to decision making under uncertainty was outlined in Ramsey (1931), but the first formal theory appeared in Savage (1954), with many variations and extensions published in the following three decades. Good pointers to the large relevant bibliography are Fishburn (1981, 1982, 1986), French (1982), Jeffrey (1983), Amarante (2009) and references therein.

The word 'normative' here reflects that the theories describe how decisions should ideally be made. The theories are not descriptive, in the sense of explaining how people actually make decisions. Indeed, empirical evidence suggests that people often violate the normative theories in making decisions; famous examples include Allais (1953, 1990) and Ellsberg (1961), suggesting that considerable training might be required for an appropriate management of uncertainty. The concept of probability and its relationship with human behavior has received a great deal of attention from both philosophers of science and psychologists. Some relevant pointers to this huge literature are Stäel von Holstein (1974), Slovic & Tversky (1974), Kahneman et al. (1982), Machina (1992), Kaplan (1998), Talbott (2001), Oaksford & Chater (2007), Luce (2012), Giocoli (2013), Hajec (2019), Hacking (1965). See also Savage (1980).

1.9.3 Bayes Theorem

Bayes theorem was posthumously published Bayes (1763) through the initiative of his friend Richard Price, but some recent research (Hooper 2013) suggests that Price himself may have played a major role in its conception.

In the discrete case, the theorem is a simple corollary of probability laws, but its statement in the continuous case requires an elaborate proof (Hartigan

(1983), Ch. 3). Formally, this requires the use of proper priors, but extensions are available to allow its careful use with improper priors (Rényi (1955), Taraldsen & Lindqvist (2010), McCullagh & Han (2011)). A Bayes Hilbert space is a linear space of equivalence classes of proportional σ-finite measures, including probability measures; addition is then given by Bayes rule and substraction by RadonNikodym derivatives (Van den Boogaart et al. 2014).

Bayes theorem may be presented in a ratio form where the posterior probabilities relative to the prior probabilities equals the conditional probability of data relative to the marginal probability of data. Rouder & Morey (2019) argue that this may provide a unified view of estimation and model comparison.

For a simple, philosophy oriented introduction to Bayes theorem, see Joyce (2003) and references therein. McGrayne 2011 and Efron 2013 describe the ability of Bayesian statistics to analyze complex problems and to connect disparate inferences on complicated interrrelated questions.

The applications of Bayesian methods have exponentially grown within new areas in the last years. These include, for instance, law and criminology (Fenton et al. 2016, Wollert 2006, Jowett 2001, Salo et al. 2013), dispute resolution (Jones & Yarn 2003), marketing (Michaeli & Simon 2008), computing (Xu et al. 2013), graph theory (Pawlak 2003), urban planning (Abdullahi et al. 2015), electroencephalography (Kopp 2007), big data analysis (Balamurugan & Rajaram 2009), astrobiology (Catling et al. 2018), astrophysics (Hee et al. 2017), climate change (Cook & Lewandowsky 2016), human reasoning (Oaksford & Chater 2007), forensic science (Gutiérrez-Redomero et al. 2008), paleontology (Byers & Roberts 2003), geophysics (Hu et al. 2011, Wang et al. 2014), classification (Hand & Yu 2001), behavioral science (Mossman 2000, Pete Blair & Rossmo 2010), physics (Casadei 2012) and Bayesian networks (Pearl 1988, de Melo & Sanchez 2008).

Chapter 2

Basics of Objective Bayesian Analysis

2.1 Motivation for Objective Bayesian Methods

2.1.1 Ease of Interpretation

One of the positive aspects of Bayesian analysis is the easy interpretation of resulting inferences. Thus, saying that $(2.2, 3.4)$ is a 95% credible interval for θ has the direct interpretation that it is believed, with probability 0.95, that the unknown θ lies between 2.2 and 3.4. In contrast, the frequentist interpretation of confidence sets is rather tortured, in that it refers to a long sequence of repetitive uses of the confidence procedure, and not to the actual interval obtained in a given situation.

The intuitive 'degree of belief' interpretation of probability raises one big question: *whose* degree of belief? In subjective Bayesian analysis, the probability statement nominally refers to the beliefs, given the data, of those who chose the model and prior distribution. If others prefer different models and/or priors, they are welcomed to do their own subjective Bayesian analysis to determine their own beliefs. While this keeps everything neat and tidy in the abstract, it is rare to see a fully subjective Bayesian analysis done for complex problems. Even rarer is to find users of complex statistical analyses who are willing or able to redo the analysis with their own models and/or priors. Such users are also often leery of simply trusting the beliefs of the original analyst. Subjective Bayesians are working hard to overcome these obstacles, but elicitation and computational limitations that prevent full subjective Bayesian analysis in complex problems will always be present.

In objective Bayesian analysis, the prior distribution is chosen to represent 'minimal information' in some sense, so that the resulting Bayesian probability statements refer to degree of belief of an individual who accepts the statistical model, but who has no additional information about the unknown parameters of the model. Individuals with strong prior opinions will recognize that the objective Bayesian answers are not necessarily their own beliefs, but it is useful even for strong subjectivists to know what the 'data alone (and model)' have to say (if, for nothing else, to provide a reference for the effect of their subjective prior distribution).

2.1.2 Objectivity and Conventionality

There is an undeniable attraction to the idea that statistical analysis of data could be objective, and part of the broad appeal of objective Bayesian analysis is the extent to which it can claim to be so. It would, indeed, be wonderful if there were a universally accepted way to define objective prior distributions.

The earliest effort in this direction was due to Laplace (1812), who used, for any unknown parameter $\boldsymbol{\theta}$, the constant prior distribution $\pi(\boldsymbol{\theta}) = 1$. This approach, which came to be known as *inverse probability* dominated statistical inference for over 150 years. It was eventually recognized, however, that continual use of a constant prior density was not logically sound, in that the answer to a statistical analysis can then be seen to depend on the (often arbitrary) choice of a parameterization of the problem. This issue and the history of inverse probability is extensively discussed in Section 3.3.

Since Laplace, there have been many efforts to define objective prior distributions. We will review these efforts in the other chapters of Part II of the book. Our own belief is that – except for certain special cases – the effort is a misguided search for the holy grail. There is no unambiguous objectivity; one can at best seek prior distributions that represent minimal information in a clearly defined technical sense, and which have a host of desirable operational properties. It is our position that, once found, such priors can become the *conventional* basis for objective Bayesian analysis In other words, we professionally agree to define certain priors as the objective priors corresponding to a specified inference problem, rather than demanding logical proof that a prior is objective.

This last position is consistent with the views of most statisticians regarding objectivity; few actually believe that any statistical analysis is really objective. For instance, choice of the model will typically have a much greater effect on the answer than will the choice of the prior distribution, and it is virtually impossible to eliminate subjectivity in the development of the model. See Berger & Berry (1988) for discussion of this issue.

In this light, it may seem odd that we used the word *Objective* in the title of the book. Some arguments for the choice follow.

- It is the natural linguistic complement to *subjective* Bayesian analysis.
- We will provide considerable evidence that our recommended version of objective Bayesian analysis — use of the 'reference prior' — provides a statistical analysis that is as objective as is possible, once the model has been chosen.

- The appearance of objectivity is often important in statistical analysis – *e.g.*, in regulatory situations,
- Other names that have been proposed for the approach are less appealing.

Among the many alternative names that have been suggested, and used, are *noninformative*, *reference*, *default*, *conventional*, and *non-subjective* analysis. *Reference* analysis has now become associated with the particular technical approach to developing objective priors that we recommend, and so is not a suitable general name. The other names are themselves either misleading or unappealing (in that they sound unscientific).

One formal approach to achieving objectivity should be mentioned: *robust Bayesian analysis*. If there are a variety of plausible models and priors under consideration, the robust Bayesian approach seeks to determine if the Bayesian answer remains essentially the same over the plausible range of models and priors; see, *e.g.*, Berger (1984, 1994), Ríos and Ruggeri (2012), and references therein). If so, the answer is deemed to be objective, in the sense that all will agree as to its use. While this is a nice ideal to keep in mind, it is difficult in practice to carry out robust Bayesian analyses that are of a refined enough nature to yield usable objective conclusions.

2.1.3 Conditioning and Frequentist Performance

While objective Bayesian analysis shares many of the desirable properties of subjective Bayesian analysis, it is not as firmly grounded in axiomatic foundations (although many – *e.g.*, Jeffreys (1961) – did provide a number of reasonable foundational arguments). We discuss foundational questions in Chapter 7 but, for now, it suffices to simply admit that objective Bayesian analysis is not unassailable from a foundational perspective. This means that its justification must arise, in part, from extrinsic operational properties of the approach. Perhaps the two most significant of these are the success of the objective Bayesian approach in ensuring proper conditioning, and its success in achieving optimal frequentist performance. We begin discussion of these two issues here.

Conditioning. Conditional inference is one of the most important concepts in statistics, but is often not taught in statistics courses or even graduate programs. In part this is because conditioning is automatic in the Bayesian paradigm, and hence not a subject of particular methodological interest to Bayesians, while, in the frequentist paradigm, there is no established general theory as to how to condition. Frequentists do automatically condition in various circumstances, such as in the following example, which is a version of the famous Cox (1958) example.

Example 2.1 *Cox example.* An assay is sometimes run with a sample of size $n = 10$ and other times with a sample of size $n = 20$. If the choice of sample size does not depend on the unknowns under consideration in the assay (*e.g.*, if it depends only on whether an employee is home sick or not), then virtually everyone would condition on the sample size, rather than, say, report an error probability that is the average of the error probabilities one would obtain for the two sample sizes. The principle that one should condition on the experiment actually performed, rather than providing results averaged over all experiments that might have been performed, is called the *conditionality principle* in statistics.

To be precise as to the type of conditioning we will discuss, it is useful to begin with a simple example, taken from Berger & Wolpert (1988) (which also discusses conditioning in general; see also Reid (1995*a*) and Bjornstad (1996).)

Example 2.2 Two observations, x_1 and x_2, are to be taken, where

$$x_i = \begin{cases} \theta+1 & \text{with probability } \frac{1}{2} \\ \theta-1 & \text{with probability } \frac{1}{2} \end{cases}.$$

Consider the frequentist confidence set, for the unknown θ, defined by

$$C(x_1,x_2) = \begin{cases} \text{the point } \{\frac{1}{2}(x_1+x_2)\} & \text{if } x_1 \neq x_2 \\ \text{the point } \{x_1-1\} & \text{if } x_1 = x_2. \end{cases}$$

The (unconditional) frequentist coverage of this confidence set can easily be shown to be

$$\Pr(C(x_1,x_2) \text{ contains } \theta \,|\, \theta) = 0.75.$$

This is not at all a sensible report, once the data is at hand. To see this, observe that, if $x_1 \neq x_2$, then we know for sure that their average is equal to θ, so that the confidence set is then actually 100% accurate. On the other hand, if $x_1 = x_2$, we do not know if θ is the data's common value plus one or their common value minus one, and each of these possibilities is equally likely to have occurred.

To obtain sensible frequentist answers here, one must define a conditioning statistic such as $s = |x_1 - x_2|$, which can be thought of as measuring the 'strength of evidence' in the data ($s = 2$ reflecting data with maximal evidential content and $s = 0$ being data of minimal evidential content). Then one defines frequentist coverage conditional on the strength of evidence s. For the example, an easy computation shows that this conditional confidence equals, for the two distinct cases,

$$\Pr(C(x_1,x_2) \text{ contains } \theta \,|\, s=2,\theta) = 1$$
$$\Pr(C(x_1,x_2) \text{ contains } \theta \,|\, s=0,\theta) = \tfrac{1}{2}.$$

It is important to realize that conditional frequentist measures are fully frequentist and (to most people) clearly better than unconditional frequentist measures. They have the same unconditional property (*e.g.*, in the above example one will report 100% confidence half the time and 50% confidence half the time, resulting in an 'average' of 75% confidence, as must be the case for a frequentist measure), yet give much better indications of the accuracy for the type of data that one has actually encountered.

Finding good conditioning statistics is usually very difficult — so much so that the conditional frequentist theory of statistics is quite underdeveloped. In contrast, the objective Bayesian approach automatically conditions appropriately.

Example 2.3 *Example 2.2 continued.* The natural objective prior for this example is $\pi(\theta) = 1$ (since θ is a location parameter — see Chapter 6). Application of Bayes theorem trivially yields that, if $x_1 \neq x_2$, then the posterior distribution for the unknown θ gives probability one to the point $(x_1+x_2)/2$ while, if $x_1 = x_2$, then the posterior distribution gives probability $1/2$ each to the common value of the data plus 1 and the common value minus 1. It is immediate that the objective Bayesian confidence statements for $C(x_1, x_2)$ are 1 and 0.5 for the two cases, respectively, which is the optimal answer.

Noteworthy in the above example is that the objective prior Bayesian analysis trivially reproduced the correct (conditional) frequentist analysis. In other words, not only did the objective Bayesian analysis introduce needed conditioning, but the resulting Bayesian posterior probability statements were actually identical with the corresponding (conditional) frequentist probability statements. The Bayesian and frequentist probability statements have different meanings, of course, but the reported numbers are the same.

We will see that this is not a coincidence. Quite generally, objective Bayesian procedures yield answers that are also (nearly) optimal (conditional) frequentist answers, and often the latter cannot be obtained in any other way.

Frequentist Performance. In addition to its conditional benefits, Bayesian analysis with objective priors is being increasingly recognized as a method for classical statisticians to develop good unconditional frequentist procedures. For instance, the *frequentist-matching* approach to developing objective priors (see Chapter 5) is based on ensuring that one has Bayesian credible sets with good frequentist properties. *Reference priors* will also be seen to be excellent in this regard. Indeed, simple Bayesian credible sets from objective priors typically outperform — in a frequentist sense — much more involved confidence sets found classically.

Example 2.4 *Medical diagnosis, continuation of Example 2.4.* Consider a population with a prevalence of a disease D, $p_0 = \Pr(\text{Disease } D)$, and for which a diagnostic test results in either a Positive (+) or Negative (-) reading. Let $p_1 = \Pr(+ \mid D)$ and $p_2 = \Pr(+ \mid \text{not } D)$. By Bayes theorem,

$$\theta = \Pr(D \mid +) = \frac{p_0\, p_1}{p_0\, p_1 + (1 - p_0)\, p_2}.$$

In practice, the p_i are typically unknown but, for $i = 0, 1, 2$, there are available (independent) data, x_i, having $\text{Bi}(x_i \mid n_i, p_i)$ densities. It is desired to find a $100(1-\alpha)\%$ confidence set for θ that has good conditional and frequentist properties.

This problem has been studied in the classical literature, using standard 'log odds' and 'delta' method procedures to develop confidence sets, as well as more sophisticated approaches such as the Gart & Nam (1988) procedure. For a description of these methods, as applied to this problem, see Mossman & Berger (2001).

A simple objective Bayesian approach to this problem is to utilize the Jeffreys-rule $\text{Be}(p_i \mid \frac{1}{2}, \frac{1}{2})$ priors (motivated in Example 4.3) for each of the p_i, and compute the $100(1-\alpha)\%$ equal-tailed posterior credible set. This can be easily done using the following simulation algorithm, utilizing the results from Example 4.3.

- Draw random p_i from the $\text{Be}(p_i \mid x_i + \frac{1}{2}, n_i - x_i + \frac{1}{2})$ posterior distributions, $i = 0, 1, 2$.
- Compute the associated $\theta = \dfrac{p_0\, p_1}{p_0\, p_1 + (1-p_0)\, p_2}$.
- Repeat this process 10,000 times.
- The $\alpha/2$ and $1-\alpha/2$ quantiles of these 10,000 generated θ form the desired confidence limits. (In other words, simply order the 10,000 values of θ, and let the confidence interval be the interval between the $10^4\,(\alpha/2)$-th and $10^4\,(1-\alpha)/2$-th values.)

Table 2.1: The probability that the nominal 95% interval misses the true value of θ on the left and on the right, for the indicated parameter values, and when $n_0 = n_1 = n_2 = 20$; ideal, would be entries of 0.025.

(p_0, p_1, p_2)	O-Bayes	Log Odds	Gart-Nam	Delta
$(\frac{1}{4}, \frac{3}{4}, \frac{1}{4})$	.0286, .0271	.0153, .0155	.0277, .0257	.0268, .0245
$(\frac{1}{10}, \frac{9}{10}, \frac{1}{10})$	.0223, .0247	.0017, .0003	.0158, .0214	.0083, .0041
$(\frac{1}{2}, \frac{9}{10}, \frac{1}{10})$	.0281, .0240	.0004, .0440	.0240, .0212	.0125, .0191

Table 2.1 gives an indication of the frequentist performance of the confidence sets developed by the four methods mentioned above. It is based on a simulation that repeatedly generates data from binomial distributions with sample sizes 20 and the indicated values of the parameters (p_0, p_1, p_2). For each generated triplet of data, the 95% confidence interval is computed by the studied procedures, and it is noted whether the interval contains the true θ or misses to the left or right. The entries in the table are the

long run proportion of misses to the left or right. Ideally, these proportions should be 2.5% and, at the least, their sum should be 5%.

Clearly the objective Bayesian interval has better performance in this regard than any of the classically derived confidence intervals. Furthermore, it can be seen that the objective Bayes intervals are, on average, smaller than the classically derived intervals. (See Mossman & Berger (2001), for more extensive computations.) This combination of small confidence sets with correct frequentist coverage has been repeatedly shown to happen when the objective Bayesian approach is used to derive confidence sets. (See Section 7.6 for more discussion.)

Another classical use of objective Bayesian procedures is as a replacement of asymptotics. In classical statistics, asymptotics is often used to develop statistical procedures, with the procedures then being used with small or moderate sample sizes (with perhaps a cursory investigation of small sample size properties of the procedures). The objective Bayesian approach is a very attractive alternative to this practice, for two reasons.

First, Bayesian procedures typically automatically have the right asymptotic properties. (For instance, the central limit theorem was first developed by Laplace (1812) as a normal approximation to the posterior distribution; the approximation is identical to the frequentist version commonly taught today.) This is true even in problems with irregular asymptotics. It should be emphasized that Bayesians need not do the asymptotics; the Bayesian procedures will automatically yield the correct asymptotic answer, assuming the sample size is large enough that the asymptotic answer would apply.

The second motivation for using objective Bayesian analysis, in lieu of asymptotics, is that the objective Bayesian answers are still likely to be sensible when the sample size is too small for the asymptotics to apply. Example 2.4 provides a good illustration of this, in that the sample sizes of 20 are too small to trust asymptotics.

2.1.4 General Methodological Advantages

Here are some general methodologically oriented advantages of using objective Bayesian analysis as a tool.

Simultaneous study of multiple models. A wide variety of similar models can be simultaneously considered in terms of robustness. For instance, suppose $x_1, \ldots, x_n$ are iid from a location model, *i.e.*, a probability model with density of the form $p(x - \theta)$. Then $\pi(\theta) = 1$ is the usual objective prior, no matter what the functional form of p, and a wide variety of possible p can easily be simultaneously considered and evaluated.

All inferences can be done simultaneously. Estimation, confidence sets, prediction ... can all be done with essentially the same methodology, and usually (but not always) the same objective priors.

Analysis of restricted parameter spaces is easy. As an example, observe – in Example 1.14 – that going from an unrestricted parameter space to a restricted parameter space required only a minor modification in the analysis. (See Berger & Chen (1993) for an involved example.) In contrast, classical statistical inference becomes considerably more difficult when restricted parameter spaces are considered.

Elimination of nuisance parameters is easy. In Section 1.2.7, it was shown how easy it is to focus on parameters of interest in Bayesian analysis, by integrating out nuisance parameters. This ease in eliminating nuisance parameters was especially apparent in the hierarchical setting of Example 1.24, and will be sharply contrasted with the difficulty of dealing with nuisance parameters via other methods in Section 2.2.5. See Berger et al. (1999) for numerous other examples and references.

Sequential problems pose no additional difficulties. In some contexts (e.g., many medical studies) the data $x_1, x_2, \ldots$ arrive sequentially in time, allowing the possibility of 'interim analysis' at any point. In frequentist statistics, the standard methodology to deal with such a situation is to pre-specify a 'stopping rule.'

> **Example 2.5** Suppose the data are independent $\mathrm{N}(x \mid \mu, \sigma^2)$ observations, and one wants to test $H_0 : \mu = 0$ versus $H_1 : \mu \neq 0$. The data arrive sequentially, with a maximum of 20 observations being possible. The stopping rule chosen is the following: at the 10th observation, if the p-value is less than α, stop experimentation and reject H_0; if the p-value is greater than α, take the next ten observations and reject H_0 if the p-value, for the combined data, is less than α.

Interestingly, in frequentist analysis, one must take the stopping rule into account in analyzing the experiment. For instance, in the above example, one might think that one could report error probability α upon rejecting (at either time point), but this is not so. One must compute the error probability for the entire procedure, and it will be bigger than α (essentially because one had two chances to reject). Sequential analysis in frequentist statistics is typically much harder than non-sequential analysis.

In contrast, Bayesian analysis is completely unaffected by sequential data (also see Section 1.2.5). At any point of time, a Bayesian analysis of the current data can be performed, and any warranted conclusion drawn. There is no need to adjust for a stopping rule, or even to specify a stopping rule. For discussion, see Berger (1985), Berger & Wolpert (1988).

One caveat should be mentioned: in application of formal objective Bayesian methodologies, the objective prior itself can depend on the stopping rule used. We defer discussion of this until Section 14.4.

Dealing with censored data poses only modest additional difficulties. Censoring arises when the the observation x is not fully reported; instead only some feature of x is reported. For instance, suppose a measuring instrument has an upper limit of 100; when measuring x one will then only observe $\min\{x, 100\}$.

In principle, censoring poses no difficulty to Bayesian analysis; one just finds the posterior distribution conditioned on what one actually observes: $\min\{x, 100\}$ in the above situation. If $x < 100$, the posterior distribution is unchanged. But, if 100 is observed, all that is really known is that $x \geq 100$, and so this is the event that must be conditioned upon in computing the posterior, namely

$$\pi(\boldsymbol{\theta} \,|\, x \geq 100) = \frac{\Pr(x \geq 100 \,|\, \boldsymbol{\theta})\pi(\boldsymbol{\theta})}{\int \Pr(x \geq 100 \,|\, \boldsymbol{\theta})\, \pi(\boldsymbol{\theta})\, d\boldsymbol{\theta}}\,.$$

Posterior distributions conditioned on censored observations can be more difficult to work with, but there are many tricks that can make them easy to work with (see *e.g.*, Ferguson & Phadia (1979), Chaloner (1991), Zhang et al. (2008), Yu (2010)).

2.1.5 Objective and Subjective Bayesian Analysis

We are fans of subjective Bayesian analysis, when it is needed and when it can be done. But this is a book on objective Bayesian analysis, and so we must highlight a number of the practical advantages of objective Bayesian analysis over subjective Bayesian analysis.

Allowing focus on model choice. Selection of a statistical model for the data is usually the most important subjective component of an analysis. It is convenient to have available objective prior distributions for unknown parameters of a proposed model, since one can then easily perform Bayesian inference for models under consideration, allowing more time to be spent on the 'first order' issue of model choice, than the 'second order' issue of prior selection. Indeed, if a wide variety of models were under consideration, performing subjective elicitations for the unknown parameters of all models would be essentially impossible.

Determining if subjective elicitation is needed. An objective Bayesian analysis can be run initially, to assess if subjective priors are even needed. (Perhaps the data will 'swamp the prior.')

Example 2.6 *Example 1.17 continued.* Suppose the 95% objective Bayesian credible interval for the normal mean μ is the interval $(5.23, 5.36)$, and this is small enough to give a good answer to the question of interest and is also compatible with subjective prior beliefs. There is then no practical reason to go through a careful elicitation of the prior beliefs about μ; it would be a waste of time, as one knows the answer will change very little.

When subjective analysis is not possible. Frequently, elicitation of subjective prior distributions is impossible, because of time or cost limitations, or resistance or lack of training of clients. Or one may be in a situation where appearance of objectivity is important (*e.g.*, when dealing with a regulatory agency). Objective prior distributions are needed in both these situations.

Objective Bayes as a reference. Subjective elicitation can easily result in poor prior distributions, because of systematic elicitation bias (*e.g.*, the almost universal tendency to underestimate the actual amount of uncertainty about unknowns) and the fact that elicitation typically yields only a few features of the prior, with the rest of the prior (*e.g.*, its functional form) being chosen in a convenient, but possibly inappropriate, way. It is thus good practice to compare answers from a subjective analysis with answers from an objective prior analysis. If there are substantial differences, it is important to check that the differences are due to features of the prior that are trusted, and not due to either unelicited 'convenience' features of the prior or suspect elicitations (*e.g.*, too-small prior variances).

Example 2.7 *Cepheid variable stars.* In Barnes III et al. (2003), astronomical data was analyzed with the goal of determining the distance to Cepheid variable stars. As is common in astronomy, the observations $x_1, \ldots, x_n$ were assumed to be independent and distributed as $\text{N}(x \mid \mu, \sigma_i^2)$, with the σ_i^2 being given; *i.e.*, each observation comes attached with its own known variance, nominally arising from knowledge about the astronomical instruments used to obtain the data. This is — in some sense — an extreme of 'subjective elicitation,' assuming exactly known variances.

To study this elicitation, Barnes III et al. (2003) instead considered, for x_i, the model $\text{N}(x \mid \mu, \tau^2\sigma_i^2)$, where τ^2 was unknown and assigned the objective prior $\pi(\tau^2) = 1/\tau^2$. Non-surprisingly the posterior distribution of τ^2 was centered at about 3, indicating that the elicited σ_i^2 were on the order of three times too small.

Use for nuisance parameters. In high dimensional problems, the best one can typically hope for is to develop subjective priors for the important parameters, with the unimportant or 'nuisance' parameters being given objective priors. In Barnes III et al. (2003), for instance, there were 227 nuisance parameters that were handled with formal objective priors; 2 parameters that were handled with informal constant priors on intervals (with sensitivity with respect to the choice of the interval studied); and subjective priors were assessed for 2 parameters.

Possible inappropriateness of standard priors. Through study of objective priors, one can obtain insight into possibly bad behavior of standard (*e.g.*, conjugate) subjective priors.

Example 2.8 *Covariance matrices.* Consider iid multivariate normal data $(\boldsymbol{x}_1, \ldots, \boldsymbol{x}_n)$, where each column vector $\boldsymbol{x}_i$ arises from the $\mathrm{N}_k(\boldsymbol{x} \mid \mathbf{0}, \boldsymbol{\Sigma})$ density, with $\boldsymbol{\Sigma}$ unknown. The sufficient statistic for $\boldsymbol{\Sigma}$ is easily seen to be $\boldsymbol{S} = \sum_{i=1}^{n} \boldsymbol{x}_i \, \boldsymbol{x}_i'$. A commonly used subjective prior for $\boldsymbol{\Sigma}$ is the inverse Wishart prior (for subjectively specified a and b)

$$\pi(\boldsymbol{\Sigma}) \propto |\boldsymbol{\Sigma}|^{-a/2} \exp\{-\tfrac{1}{2}\mathrm{tr}[b\,\boldsymbol{\Sigma}^{-1}]\}. \tag{2.1}$$

A frequently used objective version of this prior (choosing $a = k + 1$ and $b = 0$) is the Jeffreys-rule prior

$$\pi^J(\boldsymbol{\Sigma}) \propto |\boldsymbol{\Sigma}|^{-(k+1)/2}, \tag{2.2}$$

found by applying (4.4) after computing the Fisher information for $\boldsymbol{\Sigma}$.

To see the concern with these priors, consider the spectral decomposition $\boldsymbol{\Sigma} = \boldsymbol{O}\boldsymbol{D}\boldsymbol{O}'$, where $\boldsymbol{O}$ is an orthogonal matrix and $\boldsymbol{D}$ is a diagonal matrix with diagonal entries $d_1 > d_2 > \cdots > d_k$. A change of variables (see Yang & Berger (1994)) yields

$$\pi(\boldsymbol{\Sigma})\, d\boldsymbol{\Sigma} = |\boldsymbol{D}|^{-a/2} \exp\{-\tfrac{1}{2}\mathrm{tr}[b\,\boldsymbol{D}^{-1}]\} \prod_{i<j} (d_i - d_j) \cdot I_{[d_1 > \cdots > d_k]}\, d\boldsymbol{D}\, d\boldsymbol{O},$$

where $I_{[d_1 > \cdots > d_k]}$ denotes the indicator function on the given set. Since $\prod_{i<j}(d_i - d_j)$ is near zero when any two eigenvalues are close, it follows that the conjugate priors (and the Jeffreys-rule prior) tend to force apart the eigenvalues of the covariance matrix.

This is contrary to typical prior beliefs. Indeed, often in modelling multivariate normal data, one is debating between assuming an exchangeable covariance structure (and hence equal eigenvalues) or a more general structure. When one is contemplating whether or not to assume equal eigenvalues, it is clearly inappropriate to use a prior distribution that gives essentially no weight to equal eigenvalues, and instead forces them apart. Use of such priors can also have a detrimental effect on inferences, as will be shown when we return to this example in Sections 7.2 and 12.6.

The point here is that proper priors that are commonly used by subjectivists can have hidden and highly undesirable features. Objective Bayesian tools can expose these features and allow for development of better subjective priors; see Berger et al. (2020*a*) for development of better subjective priors for covariance matrices.

2.2 Informal Objective Bayesian Solutions

As mentioned above, complete subjective Bayesian analysis is rarely done in complex problems. Also, formal objective Bayesian analysis is not as common as one would wish, in part because there have not been consensus professional recommendations as to objective priors for complex problems. (Providing such recommendations is, in part, the purpose of this book.) Thus what has developed in practice is a rather *ad hoc* collection of informal default methods of choosing a prior. These methods are briefly reviewed here — along with some of their problems — so as to indicate the need for approaching objective Bayesian analysis seriously.

Use of a constant prior distribution is by far the most common adhoc objective Bayesian method and deserves extensive discussion; this is delayed until Chapter 3. Numerous examples will be presented in which the constant prior is markedly inferior, even resulting in improper posterior distributions.

2.2.1 Truncation of the Parameter Space

It is a common misconception that, to avoid potential difficulties with improper priors, one need only choose (extreme) bounds on the parameter space and confine analysis to this bounded space (in which the prior will presumably be proper). For instance, a common attempt to avoid possible posterior impropriety when using the constant prior, is to choose the prior to be constant over some (large) bounded region of $\boldsymbol{\Theta}$. This will not solve the problem, however, in that, if the posterior resulting from the constant prior were improper, then the ensuing inferences will often be highly dependent on the actual bounds that were used. (The answers obtained by truncating at $\pm K$ could then be very different than the answers obtained by truncating at $\pm 2K$.) At the very least, this approach should only be used if a very careful sensitivity study is done with respect to these bounds, with bounds for different parameters varying independently in the sensitivity study.

Example 2.9 A model has three parameters (μ, ξ, η) with domains $(-\infty, \infty)$ and three parameters (σ, τ, γ) with domains $(0, \infty)$. If truncation of the parameter space is to be employed, a *minimally acceptable* sensitivity study would be to consider the truncated parameter space

$$\boldsymbol{\Theta} = (-K_\mu, K_\mu)\times(-K_\xi, K_\xi)\times(-K_\eta, K_\eta)\times(1/K_\sigma, K_\sigma)\times(1/K_\tau, K_\tau)\times(1/K_\gamma, K_\gamma),$$

and independently vary the truncation constants, K_μ, K_ξ, K_η, K_σ, K_τ, and K_γ, in a variety of ways to see if the Bayesian answer varies appreciably. One can construct examples where this type of sensitivity analysis fails to detect a problem that does exist, but the sensitivity analysis would be a reasonable effort.

2.2.2 Vague Proper Priors

Use of vague proper priors has become very popular, especially because of their use in programs such as BUGS. If a vague proper prior is a good approximation to an optimal objective prior, then its use is typically fine.

Example 2.10 *Normal mean.* Suppose the statistical problem is estimation of a normal mean θ, with known variance. The natural (and good) objective prior is $\pi(\theta) = 1$. It is common to utilize instead a $\mathrm{N}(\theta \mid 0, K)$ prior distribution, with a very large value of K, and it is possible to show that this will yield essentially the same answer (in estimation problems).

If, on the other hand, a vague proper prior is an approximation to a suboptimal or bad default prior, then use of the vague proper prior is also bad, and indeed often merely hides – not solves – the problem. For instance, consider the following common situation:

Example 2.11 *Hierarchical model continued.* Consider Example 1.24 and recall that the recommended hyperprior for (ξ, τ^2) (the second level mean and variance) was $\pi^R(\xi, \tau^2) = (\tau^2 + \sigma^2/n)^{-1}$. (Recall that σ^2 was assumed to be known.) A common error is to instead use the prior $\pi(\xi, \tau^2) = 1/\tau^2$, which is the independence Jeffreys prior for a mean and variance at the data-level of a model. The resulting marginal posterior density of τ^2 is (contrast with (1.11))

$$\pi(\tau^2 \mid \overline{\boldsymbol{x}}) \propto \frac{1}{\tau^2(\tau^2 + \sigma^2/n)^{(m-1)/2}} \exp\left\{-\frac{s^2}{2(\tau^2 + \sigma^2/n)}\right\}.$$

The problem with this posterior is what happens when $\tau^2 \to 0$. Indeed,

$$\lim_{\tau^2 \to 0} \pi(\tau^2 \mid \overline{\boldsymbol{x}}) \propto \frac{C}{\tau^2}, \quad \text{where} \quad C = \frac{1}{(\sigma^2/n)^{(m-1)/2}} \exp\left\{-\frac{ns^2}{2\sigma^2}\right\}$$

is a constant. But C/τ^2 is not integrable at 0, so this posterior distribution is not proper – a disaster!

Note that this error is often missed, because of the use of MCMC methods of analysis, which often do not give obvious indications of a problem (as would, say, common numerical integration methods). Indeed, the common Gibbs sampling analysis (*cf.* Section 1.8.3) for this problem would be a simulation consisting of generating from the full-conditionals (the conditional posterior density of a parameter, given the data and the other parameters). Each of the full-conditionals is easily seen to be a proper distribution. In particular, the full-conditional of τ^2 is

$$\pi(\tau^2 \mid \bar{\boldsymbol{x}}, \xi, \boldsymbol{\mu}) \propto \frac{1}{\tau^{(m+2)}} \exp\Big\{ - \frac{\sum_{i=1}^m (\mu_i - \xi)^2}{2\tau^2} \Big\},$$

which is recognizable as the inverse-gamma $\text{Ig}(\tau^2 \mid \frac{1}{2}m, \frac{1}{2}\sum_{i=1}^{m}(\mu_i - \xi)^2)$ distribution; this is a quite concentrated proper distribution (given the other parameters) and is easily sampled.

The main indication of a problem, that could be observed in the Gibbs sampling analysis, is that the values of τ^2 would increasingly pile up near 0 as the sampling proceeds. This could, however, be missed in short runs of the Gibbs sampler, or even in long runs if the parameter τ^2 is being viewed as a nuisance parameter and is not carefully monitored.

A 'solution' to this problem that one often sees is to replace the improper prior $1/\tau^2$ with the proper prior proportional to $\tau^{-2(1+\epsilon)}\exp(-\epsilon/\tau^2)$ (which is the $\text{Ig}(\tau^2 \mid \epsilon, \epsilon)$ distribution), with ϵ small. The argument is that this is a 'vague proper prior' that approximates the original improper prior, and its propriety will supposedly fix the problem.

This argument is nonsense. It is true that the ensuing posterior distribution will be proper, but the practical problem does not disappear. The resulting 'proper' posterior distribution will be highly concentrated at values of τ^2 near zero, exactly as would have happened with use of Gibbs sampling with the improper posterior. Furthermore, this will have a dramatic (and highly misleading) effect on resultant inferences. For instance, the posterior distributions of the μ_i will be highly concentrated near $\bar{\bar{x}}$, since $\tau^2 \approx 0$ is equivalent to saying *a priori* that all μ_i are approximately equal. This will happen even when the data indicate that the μ_i differ dramatically.

Indeed, the situation with a vague proper prior is typically much worse than that with an improper prior, in that it is much easier to miss the fact that a ridiculous answer was obtained. In contrast, when an improper prior leads to an improper posterior, one has a good chance of catching the problem with numerical methods, and a decent chance of catching it when using MCMC (if careful diagnostic monitoring is employed).

2.2.3 Weakly Informative Priors

Gelman et al. (2013) recommends using weakly informative priors, which are related to vague proper priors, except that the 'vagueness' of the prior is conservatively subjectively chosen. For instance, the scientist involved might be able to say that θ is for sure between 5 and 10, and one can then choose a prior compatible with this statement. In a purely subjective approach, the scientist would try to pin down exactly what is apriori known about θ, rather than just making the above conservative statement.

There is nothing unreasonable about this approach, but it does require some subjective input and is, hence, not an automatic procedure. Also, it's implementation is unclear in multiparameter problems, as little to nothing may be known about some parameters. Finally, as with truncating priors or the vague proper prior approach, one needs to verify that the answer is not sensitive to the rather arbitrarily chosen weak information.

2.2.4 Transformation of the Parameter

This approach consists of first transforming θ to a bounded interval; for instance, by defining $\psi = g(\theta)$, where $g : \Theta \to [0, 1]$ is a $1-1$ differentiable transformation, and then placing an objective (but proper) prior on ψ, say $\pi(\psi) = 1$. This will typically ensure that the posterior is proper.

Note, however, that choosing g is essentially equivalent to choosing the original prior $\pi(\theta)$. For instance, a simple change of variables shows that (under mild conditions) use of $\pi(\psi) = 1$ is equivalent to use of $\pi(\theta) = |g'(\theta)|$, where $g'(\theta)$ is the derivative of $g(\theta)$. This approach is thus essentially equivalent to choosing a subjective proper prior, but is guilty of hiding the fact that this has been done!

2.2.5 Data-Dependent Priors

Sensible objective priors can depend on the stochastic model for the data and covariates relevant to the model. For instance, the recommended objective prior in (1.9), for (μ, τ^2) in a normal hierarchical model, depends on the sample size n. This is unavoidable in use of Jeffreys priors (Chapter 4) and reference priors (Chapter 8).

A variety of devices have been proposed for choosing priors that are based, not only on the model and covariates under consideration, but also on the actual data. These include empirical Bayes priors, vague proper priors specifically chosen to 'span the range of the likelihood function,' and objective priors chosen in a data-dependent fashion. We briefly introduce and discuss these ideas here.

Empirical Bayes Priors. Empirical Bayes analysis applies to hierarchical models, such as those discussed in Section 1.5. Instead of placing an objective prior distribution on the hyperparameters, the empirical Bayes approach consists of directly estimating the hyperparameters, and using these estimates in the ensuing Bayesian analysis; the name comes from the fact that one is, in a sense, estimating the prior distribution before doing the Bayesian analysis.

Example 2.12 Consider again Example 1.23. Here the hyperparameters are ξ and τ^2 (remember that σ^2 is known), and they can be directly estimated from the marginal likelihood function in (1.8). The obvious estimate of ξ is $\hat{\xi} = \bar{x}$. For τ^2, two classical estimates are the 'unbiased estimate,' which can be shown to be $\hat{\tau}^2_U = [s^2/(m-1) - \sigma^2/n]$, and the mle, which is easily calculated to be $\hat{\tau}^2_{mle} = \max\{0, s^2/m - \sigma^2/n\}$ (replace ξ by $\bar{\bar{x}}$ in (1.8) and then maximize the resulting expression over τ^2). The unbiased estimate of τ^2 has the rather severe difficulty that it can be negative — and a negative variance estimate is ridiculous. Hence we focus on use of the mle, which has become the standard empirical Bayes estimate (although see Morris (1983)).

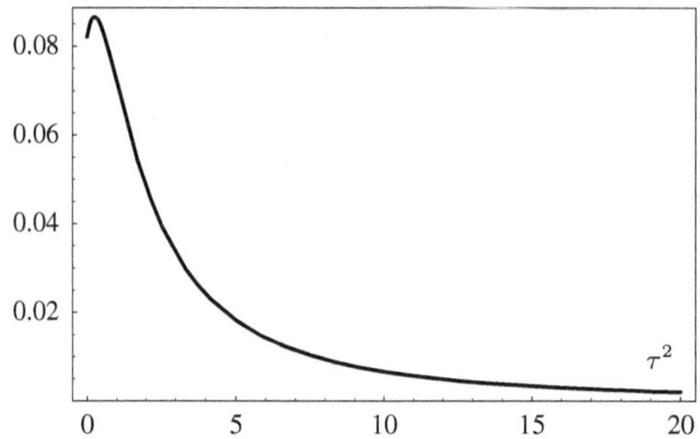

Figure 2.1: Marginal likelihood function of τ^2 when $p = 5$, $n = 1$, $\sigma^2 = 1$, and $s^2 = 5$ is observed.

Even $\hat{\tau}^2_{mle}$ is problematical. This is particularly clear if $s^2/m < \sigma^2/n$, in which case the mle would be $\hat{\tau}^2_{mle} = 0$. While a value of $s^2/m < \sigma^2/n$ is somewhat unusual here (if, for instance, $m = 5$, $n = 1$ and $\tau^2 = \sigma^2$, then $\Pr(s^2/5 < \sigma^2) = 0.264$), it is quite common in more complicated hierarchical models to have at least one mle variance estimate equal to 0. See Berger & Bernardo (1992*c*) and Belzile & Angers (1995) for handling multiple variance components.

That having an mle variance estimate equal to 0 is a problem is most clearly indicated by looking at the marginal likelihood function for τ^2 in such a situation, given by integrating (1.8) over ξ (as in the derivation of (1.11)), yielding

$$p(\overline{\boldsymbol{x}} \mid \tau^2) \propto (\tau^2 + \sigma^2/n)^{-(m-1)/2} \exp\left\{ - \frac{s^2}{2(\tau^2 + \sigma^2/n)} \right\}. \tag{2.3}$$

For illustration, Figure 2.1 graphs this marginal likelihood function in the 'borderline' situation when $m = 5$, $n = 1$, $s^2 = 5$ and $\sigma^2 = 1$. This marginal likelihood of τ^2 is mostly decreasing away from 0, but not particularly quickly, clearly indicating that there is considerable uncertainty as to the true value of τ^2, even though the mle was $\hat{\tau}^2_{mle} = 0$. (The mle for the integrated likelihood (2.3) is slightly bigger than 0, reflecting the well-known fact that, if one must do empirical Bayes analysis, it is better to use mle's from integrated likelihoods.)

While it is always dangerous in statistical analysis to simply replace unknown parameters by their estimates, this can be particularly dangerous in hierarchical settings because of the fact that estimates can easily be on the boundaries of parameter space. In the above example, for instance, setting τ^2 to 0 is equivalent to stating that all the μ_i are exactly equal to each other. This is clearly unreasonable in light of the fact that there is actually great uncertainty about τ^2, as reflected in Figure 2.1. Since the maximum of the likelihood (1.8) is occurring at the boundary of the parameter space, it is also very difficult to utilize likelihood or frequentist methods to attempt to incorporate uncertainty about τ^2 into the analysis. Thus we strongly prefer full objective hierarchical Bayesian analysis to empirical Bayesian analysis.

Data-Dependent Vague Proper Priors. The other common data-dependent choice of prior is using a vague proper prior specifically chosen to 'span the range of the likelihood function.' The rationale for doing this is typically a desire to eliminate the influence of the prior distribution; by choosing a prior that 'lives near the likelihood function,' the hope is that one has a reasonable objective Bayesian analysis.

It is not really possible to evaluate this strategy. On the one hand, it may often be better than using an arbitrarily chosen objective prior (e.g. a constant prior), but it corresponds to an extremely disturbing double use of the data; an extreme version of this is Aitkin (1991), which uses the likelihood function itself as the prior, essentially resulting in the data being counted twice. We certainly do not recommend such strategies in choosing the prior.

Data-Dependent Objective Priors. There have been some developments of objective priors which explicitly recommend priors that depend on the actual data (as well as the model). A prime example of this is Fraser et al. (2010), which argues that data-dependent priors can yield better frequentist performance than other objective priors. These are interesting developments but have not been extensively studied.

2.3 Permissible Improper Priors

2.3.1 Introduction

In Section 1.2, improper priors were introduced and were utilized in a number of the subsequent examples. Establishing the legitimacy of using improper priors in the formal version of Bayes theorem can done by showing that the answer so obtained is the same as that which would be obtained in the limit from an approximating sequence of proper prior distributions. Establishing this result is the primary purpose of this section. Most of this material was first developed in Berger et al. (2009*a*).

The approximating prior distributions that will be considered are the improper prior truncated to a sequence of increasingly large compact subsets of the parameter space. This choice of the sequence of proper priors has an additional purpose (beyond establishment of the formal version of Bayes theorem), based on a common viewpoint concerning the finiteness of reality. We frequently use $(-\infty, \infty)$ as the parameter space for a normal mean, but do not really believe that the parameter space is infinite in a real problem. Rather, the real parameter space might be $(-K, K)$, for some large K, and $(-\infty, \infty)$ is simply being used as an approximation. This is only justified, however, if the answers obtained using $(-\infty, \infty)$ are essentially the same as those obtained using $(-K, K)$. Thus the larger purpose of this section is to justify using (suitable) improper priors with infinite parameter spaces, to approximate finite reality, when the finite limits are unknown.

To begin, we define the notion of convergence of distributions that we will use; this notion of convergence will also have a number of other uses throughout the book. Note that this is a theoretical section, that can be skipped by the reader interested primarily in methodology.

2.3.2 Kullback-Leibler (KL) Convergence

We now define the first notion of convergence that will be used to verify that improper priors can be limits of proper priors. It is based on KL divergence between two distributions, as defined in Definition 1.5.

Definition 2.1 KL Convergence. *A sequence of probability distributions, defined by their density functions $\{p_i(\boldsymbol{x})\}_{i=1}^{\infty}$, is said to KL converge to a probability distribution with density $p(\boldsymbol{x})$ whenever*

$$\lim_{i\to\infty} k\{p_i \,|\, p\} = 0,$$

that is, whenever the KL divergence between $p_i(x)$ and $p(x)$ converges to zero.

As an example of KL convergence, consider the approximation of a Student t distribution with large degrees of freedom by a normal distribution.

Example 2.13 *Student t convergence to normality.* The KL divergence between a standard Normal distribution and a standard Student t distribution with α degrees of freedom is

$$\delta(\alpha) = \kappa\{\mathrm{St}(\cdot \,|\, 0, 1, \alpha) \,|\, \mathrm{N}(\cdot \,|\, 0, 1)\} = \int_{-\infty}^{\infty} \mathrm{N}(x \,|\, 0, 1) \log \frac{\mathrm{N}(x \,|\, 0, 1)}{\mathrm{St}(x \,|\, 0, 1, \alpha)}\, dx\,.$$

This can be analytically evaluated in terms of the Hypergeometric and Beta functions which, for large α, can be approximated by Stirling's formula to obtain

$$\delta(\alpha) = \int_{-\infty}^{\infty} \mathrm{N}(x \,|\, 0, 1) \log \frac{\mathrm{N}(x \,|\, 0, 1)}{\mathrm{St}(x \,|\, 0, 1, \alpha)}\, dx \approx \frac{4}{(\alpha + 2)(4\alpha + 1)}\,,$$

a function which rapidly converges to zero as α grows.

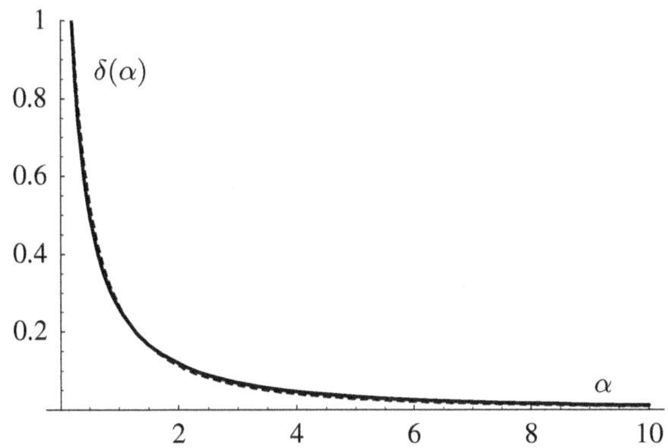

Figure 2.2: KL divergence $\delta(\alpha)$ between a Normal distribution, $\mathrm{N}(x \mid 0, 1)$, and a standard Student t distribution, $\mathrm{St}(x \mid 0, 1, \alpha)$, as a function of the degrees of freedom α (solid line) and its Stirling approximation (dashed line).

The exact value of $\delta(\alpha)$ and its approximation are both given in Figure 2.2. Note that the Stirling approximation to $\delta(\alpha)$ is extremely good, even for small values of the degrees of freedom α. Particular values are $\delta(20) = 0.0034$, and $\delta(39) = 0.001$, substantiating the usual claim that Student t distributions with more than 20 degrees of freedom are fairly well approximated by normals, and that the approximation is extremely good for $\alpha \geq 40$.

2.3.3 Justifying Posteriors from Improper Priors

Justification as a KL Limit of Proper Priors. When $\pi(\boldsymbol{\theta})$ is improper, Bayes theorem no longer applies, and so it is not obvious how to justify use of the posterior density

$$\pi(\boldsymbol{\theta} \mid \boldsymbol{x}) = \frac{p(\boldsymbol{x} \mid \boldsymbol{\theta})\, \pi(\boldsymbol{\theta})}{\int_{\boldsymbol{\Theta}} p(\boldsymbol{x} \mid \boldsymbol{\theta})\, \pi(\boldsymbol{\theta})\, d\boldsymbol{\theta}} \,. \tag{2.4}$$

For this expression to define a probability density, it is clearly necessary that

$$p(\boldsymbol{x}) = \int_{\boldsymbol{\Theta}} p(\boldsymbol{x} \mid \boldsymbol{\theta})\, \pi(\boldsymbol{\theta})\, d\boldsymbol{\theta} < \infty \,, \tag{2.5}$$

and we take this as a minimal requisite condition for use of the improper prior to make sense (but see Lindqvist & Taraldsen (2018) for counterarguments).

This section presents a strong justification for use of (2.4) with improper priors, showing that $\pi(\boldsymbol{\theta} \mid \boldsymbol{x})$ is the limit of proper posteriors arising from proper priors that approximate the improper $\pi(\boldsymbol{\theta})$. (See Akaike (1980*b*) for a different approach.) To this end, for an improper prior $\pi(\boldsymbol{\theta})$, consider an increasing sequence of subsets of the parameter space $\{\boldsymbol{\Theta}_1, \boldsymbol{\Theta}_2 \ldots\}$, such that

$$\boldsymbol{\Theta}_i \subset \boldsymbol{\Theta}_{i+1}, \int_{\boldsymbol{\Theta}_i} \pi(\boldsymbol{\theta}) < \infty, \; \forall\, i \geq 1, \quad \text{and} \quad \lim_{i\to\infty} \boldsymbol{\Theta}_i = \boldsymbol{\Theta}. \tag{2.6}$$

Then, $\pi(\boldsymbol{\theta})$ can be normalized within each of the $\boldsymbol{\Theta}_i$'s to produce the sequence of *proper* priors $\{\pi_1(\boldsymbol{\theta}), \pi_2(\boldsymbol{\theta}), \ldots\}$ defined by

$$\pi_i(\boldsymbol{\theta}) = \begin{cases} \dfrac{\pi(\boldsymbol{\theta})}{\int_{\boldsymbol{\Theta}_i} \pi(\boldsymbol{\theta})\, d\boldsymbol{\theta}} & \text{if } \boldsymbol{\theta} \in \boldsymbol{\Theta}_i \\ 0 & \text{otherwise.} \end{cases}$$

Since the $\pi_i(\boldsymbol{\theta})$'s are all proper probability distributions, Bayes theorem can be applied, given the data $\boldsymbol{x}$, to obtain the corresponding sequence of posterior densities $\{\pi_1(\boldsymbol{\theta} \,|\, \boldsymbol{x}), \pi_2(\boldsymbol{\theta} \,|\, \boldsymbol{x}), \ldots\}$ defined by

$$\pi_i(\boldsymbol{\theta} \,|\, \boldsymbol{x}) = \begin{cases} \dfrac{p(\boldsymbol{x} \,|\, \boldsymbol{\theta})\, \pi_i(\boldsymbol{\theta})}{\int_{\boldsymbol{\Theta}_i} p(\boldsymbol{x} \,|\, \boldsymbol{\theta})\, \pi_i(\boldsymbol{\theta})\, d\boldsymbol{\theta}} & \text{if } \boldsymbol{\theta} \in \boldsymbol{\Theta}_i \\ 0 & \text{otherwise.} \end{cases}$$

The goal now might seem to be to show that $\{\pi_i(\boldsymbol{\theta} \,|\, \boldsymbol{x})\}$ is KL convergent to $\pi(\boldsymbol{\theta} \,|\, \boldsymbol{x})$ at $\boldsymbol{x}$, but that is not possible because $\pi(\boldsymbol{\theta} \,|\, \boldsymbol{x})$ has larger support than $\pi_i(\boldsymbol{\theta} \,|\, \boldsymbol{x})$, so $\kappa\{\pi_i(\boldsymbol{\theta} \,|\, \boldsymbol{x}) \,|\, \pi(\boldsymbol{\theta} \,|\, \boldsymbol{x})\}$ does not exist. Also, the $\pi_i(\boldsymbol{\theta} \,|\, \boldsymbol{x})$ are being viewed as the valid posteriors, since they arose from a proper application of Bayes theorem; indeed, $\pi(\boldsymbol{\theta} \,|\, \boldsymbol{x})$ is being viewed as their appoximation. Both issues are resolved by using a different definition of convergence.

Definition 2.2 Reverse KL Convergence. *A sequence of probability distributions, defined by their density functions* $\{p_i(\boldsymbol{x})\}_{i=1}^{\infty}$, *is said to* reverse KL converge *to a probability distribution with density* $p(\boldsymbol{x})$ *whenever*

$$\lim_{i\to\infty} k\{p \,|\, p_i\} = 0,$$

that is, whenever the KL divergence between $p(x)$ *and* $p_i(x)$ *converges to zero.*

Theorem 2.1 *Let* $\pi(\boldsymbol{\theta})$ *be an improper prior on* $\boldsymbol{\Theta}$, *having a sequence of subsets satisfying (2.6). If* $\int_{\boldsymbol{\Theta}} p(\boldsymbol{x} \,|\, \boldsymbol{\theta})\, \pi(\boldsymbol{\theta})\, d\boldsymbol{\theta} < \infty$ *at* $\boldsymbol{x}$, *then* $\{\pi_i(\boldsymbol{\theta} \,|\, \boldsymbol{x})\}_{i=1}^{\infty}$ *is reverse KL convergent to* $\pi(\boldsymbol{\theta} \,|\, \boldsymbol{x})$ *at* $\boldsymbol{x}$.

Proof. We must show that

$$\lim_{i\to\infty} \int_{\boldsymbol{\Theta}_i} \pi_i(\boldsymbol{\theta} \,|\, \boldsymbol{x}) \log \frac{\pi_i(\boldsymbol{\theta} \,|\, \boldsymbol{x})}{\pi(\boldsymbol{\theta} \,|\, \boldsymbol{x})}\, d\boldsymbol{\theta} = 0.$$

Let $c_i = \int_{\boldsymbol{\Theta}_i} \pi(\boldsymbol{\theta})\, d\boldsymbol{\theta}$, so that $\pi_i(\boldsymbol{\theta}) = c_i^{-1} \pi(\boldsymbol{\theta})$, with support on $\boldsymbol{\Theta}_i$, and define $p_i(\boldsymbol{x}) = \int_{\boldsymbol{\Theta}_i} p(\boldsymbol{x} \,|\, \boldsymbol{\theta})\, \pi_i(\boldsymbol{\theta})\, d\boldsymbol{\theta}$ and $p(\boldsymbol{x}) = \int_{\boldsymbol{\Theta}} p(\boldsymbol{x} \,|\, \boldsymbol{\theta})\, \pi(\boldsymbol{\theta})\, d\boldsymbol{\theta}$ (which has been assumed to be finite). Using, for the posteriors, the expressions provided by Bayes theorem yields

$$\int_{\boldsymbol{\Theta}_i} \pi_i(\boldsymbol{\theta} \,|\, \boldsymbol{x}) \log \frac{\pi_i(\boldsymbol{\theta} \,|\, \boldsymbol{x})}{\pi(\boldsymbol{\theta} \,|\, \boldsymbol{x})} \, d\boldsymbol{\theta} = \int_{\boldsymbol{\Theta}_i} \pi_i(\boldsymbol{\theta} \,|\, \boldsymbol{x}) \log \frac{p(\boldsymbol{x})\, \pi_i(\boldsymbol{\theta})}{p_i(\boldsymbol{x})\, \pi(\boldsymbol{\theta})} \, d\boldsymbol{\theta}$$

$$= \int_{\boldsymbol{\Theta}_i} \pi_i(\boldsymbol{\theta} \,|\, \boldsymbol{x}) \log \frac{p(\boldsymbol{x})}{p_i(\boldsymbol{x})\, c_i} \, d\boldsymbol{\theta} = \log \frac{p(\boldsymbol{x})}{p_i(\boldsymbol{x})\, c_i} = \log \frac{\int_{\boldsymbol{\Theta}} p(\boldsymbol{x} \,|\, \boldsymbol{\theta})\, \pi(\boldsymbol{\theta})\, d\boldsymbol{\theta}}{\int_{\boldsymbol{\Theta}_i} p(\boldsymbol{x} \,|\, \boldsymbol{\theta})\, \pi(\boldsymbol{\theta})\, d\boldsymbol{\theta}} \,.$$

The last expression converges to zero if, and only if,

$$\lim_{i \to \infty} \int_{\boldsymbol{\Theta}_i} p(\boldsymbol{x} \,|\, \boldsymbol{\theta})\, \pi(\boldsymbol{\theta})\, d\boldsymbol{\theta} = \int_{\boldsymbol{\Theta}} p(\boldsymbol{x} \,|\, \boldsymbol{\theta})\, \pi(\boldsymbol{\theta})\, d\boldsymbol{\theta},$$

and this follows from the monotone convergence theorem. □

This theorem is remarkable in its generality. As long as the improper prior has a proper posterior at $\boldsymbol{x}$, it is a good approximation to the proper prior formed by restriction to *any* sufficiently large compact subset of the parameter space.

Example 2.14 *Sequential compact approximation to the posterior.* For a random sample $\boldsymbol{x} = (x_1, \ldots, x_n)$ from a $\mathrm{N}(x \,|\, \mu, \sigma^2)$ distribution with known variance σ^2, the natural objective prior density is $\pi(\mu) = 1$, leading to the *proper* $\mathrm{N}(\mu \,|\, \overline{x}, \sigma/\sqrt{n})$ objective posterior distribution. For instance, $n = 9$ observations were simulated from a normal distribution with mean $\mu = 3$ and variance $\sigma^2 = 16$, resulting in a sample average $\overline{x} = 2.11$ and the objective posterior distribution $\mathrm{N}(\mu \,|\, 2.11, (4/3)^2)$, the density of which is given by the solid line in Figure 2.3.

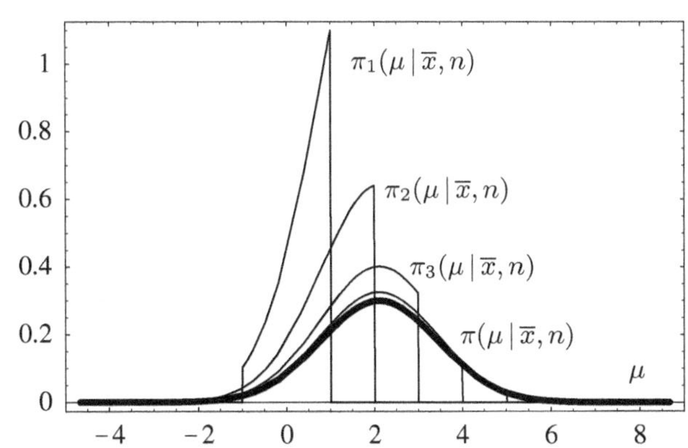

Figure 2.3: The approximating sequence of compact subset posteriors to the objective posterior distribution, for a sample of 9 normal observations from the $\mathrm{N}(x \,|\, 3, 16)$ distribution.

Consider the increasing sequence of intervals $\{\Theta_i\}_{i=1}^{\infty}$, with $\Theta_i = [-i, i]$, converging to $\mathrm{I\!R}$. The relevant sequence of proper priors would then be

$\pi_i(\mu) = (2i)^{-1}$ for $\mu \in \Theta_i$, and zero otherwise. The corresponding sequence of posteriors consist of the truncated, renormalized normal densities,

$$\pi_i(\mu \,|\, \overline{x}, n) = \begin{cases} \dfrac{\exp[-n(\overline{x}-\mu)^2/(2\sigma^2)]}{\int_{-i}^{i} \exp[-n(\overline{x}-\mu)^2/(2\sigma^2)]\, d\mu}\,, & \text{if } \mu \in \Theta_i \\ 0 & \text{otherwise} \end{cases} \tag{2.7}$$

presented in Figure 2.3 for $i = 1, 2, 3$ and 8 (the light line that is almost indistinguishable from the dark line). Note that $\pi(\mu \,|\, \overline{x}, n)$ is an excellent approximation to $\pi_i(\mu \,|\, \overline{x}, n)$ for all $i > 8$.

Justification as an Expected KL Limit of Proper Priors. Reverse KL convergence of posteriors is a pointwise property; it holds at each $\boldsymbol{x}$. The following stronger notion of convergence will also be needed.

Definition 2.3 Expected Reverse KL Convergence. *Let $\pi(\boldsymbol{\theta})$ be an improper prior on $\boldsymbol{\Theta}$, which has subsets satisfying (2.6), and $\int_{\boldsymbol{\Theta}} p(\boldsymbol{x} \,|\, \boldsymbol{\theta})\, \pi(\boldsymbol{\theta})\, d\boldsymbol{\theta} < \infty$ for any $\boldsymbol{x} \in \boldsymbol{\mathcal{X}}$. $\{\pi_i(\boldsymbol{\theta} \,|\, \cdot)\}_{i=1}^{\infty}$ is* expected reverse KL convergent *to $\pi(\boldsymbol{\theta} \,|\, \cdot)$ if the expected KL divergence between $\pi(\boldsymbol{\theta} \,|\, \boldsymbol{x})$ and $\pi_i(\boldsymbol{\theta} \,|\, \boldsymbol{x})$ converges to 0, that is*

$$\lim_{i\to\infty} \int_{\boldsymbol{\mathcal{X}}} p_i(\boldsymbol{x})\, \kappa\{\pi(\cdot \,|\, \boldsymbol{x}) \,|\, \pi_i(\cdot \,|\, \boldsymbol{x})\}\, d\boldsymbol{x} \;=\; 0\,, \tag{2.8}$$

where $p_i(\boldsymbol{x}) = \int_{\boldsymbol{\Theta}_i} p(\boldsymbol{x} \,|\, \boldsymbol{\theta})\, \pi_i(\boldsymbol{\theta})\, d\boldsymbol{\theta}$.

Since $\kappa\{\pi(\cdot \,|\, \boldsymbol{x}) \,|\, \pi_i(\cdot \,|\, \boldsymbol{x})\} \geq 0$ for any $\boldsymbol{x} \in \boldsymbol{\mathcal{X}}$, expected reverse KL convergence implies reverse KL convergence (almost surely).

Example 2.15 Again consider Example 2.14, and let $\boldsymbol{x} = (x_1, \ldots, x_n)$ be a random sample from a $\mathrm{N}(x \,|\, \mu, \sigma^2)$ distribution with known variance σ^2. Without loss of generality, assume that $\sigma = 1$. Under the constant prior $\pi(\mu) = 1$, $\pi(\mu \,|\, \boldsymbol{x})$ is $\mathrm{N}(\mu \,|\, \overline{x}, 1/\sqrt{n})$, $p(x) = \int \mathrm{N}(\mu \,|\, \overline{x}, 1/\sqrt{n})\, d\mu = 1$, $\pi_i(\mu \,|\, \boldsymbol{x})$ is given by (2.7), and

$$p_i(\boldsymbol{x}) = \frac{1}{2i} \int_{-i}^{i} \exp\left[-\frac{1}{2} n(\overline{x}-\mu)^2\right] d\mu = \frac{\Phi[\sqrt{n}(\overline{x}+i)] - \Phi[\sqrt{n}(\overline{x}-i)]}{2i}\,,$$

where Φ is the standard normal cdf. Therefore, the KL divergence between $\pi(\mu \,|\, \boldsymbol{x})$ and $\pi_i(\mu \,|\, \boldsymbol{x})$ is given by

$$\begin{aligned} \kappa\{\pi(\cdot \,|\, \boldsymbol{x}) \,|\, \pi_i(\cdot \,|\, \boldsymbol{x})\} &= \int_{-i}^{i} \pi_i(\mu \,|\, \boldsymbol{x}) \log \frac{\pi_i(\mu \,|\, \boldsymbol{x})}{\pi(\mu \,|\, \boldsymbol{x})}\, d\mu \\ &= \int_{-i}^{i} \pi_i(\mu \,|\, \boldsymbol{x}) \log \frac{1}{2i\, p_i(\boldsymbol{x})}\, d\mu \\ &= -\log\left(\Phi[\sqrt{n}(\overline{x}+i)] - \Phi[\sqrt{n}(\overline{x}-i)]\right)\,. \end{aligned} \tag{2.9}$$

Thus,

$$\int p_i(\boldsymbol{x})\, \kappa\{\pi(\cdot \mid \boldsymbol{x}) \mid \pi_i(\cdot \mid \boldsymbol{x})\}\, d\boldsymbol{x}$$

$$= -\frac{1}{2i} \int_{-\infty}^{\infty} \Big(\Phi[\sqrt{n}(\overline{x}+i)] - \Phi[-\sqrt{n}(\overline{x}-i)] \Big) \times$$

$$\log \Big(\Phi[\sqrt{n}(\overline{x}+i)] - \Phi[-\sqrt{n}(\overline{x}-i)] \Big)\, d\overline{x}$$

$$= -\int_{1}^{\infty} \Big(\Phi(y\sqrt{ni}) - \Phi[(y-2)\sqrt{ni}] \Big) \log \Big([\Phi(y\sqrt{ni}) - \Phi[(y-2)\sqrt{ni}] \Big) dy,$$

the last step using symmetry (half the integral is the same as the integral from 0 to ∞) and making the transformation $y = (\overline{x}+i)/i$. Break this integral into $\int_1^3 + \int_3^\infty$. Since $-v \log v \le e^{-1}$ for $0 \le v \le 1$, the dominated convergence theorem can be applied to the first integral to show that it converges to 0 as $i \to \infty$. For the second integral, the inequality

$$1 - \tfrac{0.5}{v} e^{-\frac{1}{2}v^2} \le \Phi(v) \le 1 - \tfrac{0.3}{v} e^{-\frac{1}{2}v^2}$$

for large v can be used to prove convergence to 0 as $i \to \infty$. (The details are a bit lengthy and will be omitted.) Hence Condition (2.8) is satisfied for any n and $\{\pi_i(\boldsymbol{\theta} \mid \cdot)\}_{i=1}^{\infty}$ is expected reverse KL convergent to $\pi(\boldsymbol{\theta} \mid \cdot)$.

To see why expected reverse KL convergence is important, consider the following famous historical example, in which reverse KL convergence holds, but expected reverse KL convergence does not hold, and the lack of the latter has serious consequences.

Example 2.16 Fraser et al. (1985) and Berger & Bernardo (1992*a*) considered a discrete problem with $\mathcal{X} = \Theta = \{1, 2, 3, \ldots\}$ and

$$p(x \mid \theta) = \tfrac{1}{3} \text{ for } x \in \{[\tfrac{\theta}{2}],\ 2\theta,\ 2\theta+1\}, \tag{2.10}$$

with $[v]$ denoting the integer part of v and $[\frac{1}{2}]$ separately defined as 1. Note that, when x is observed, θ must lie in $\{[\frac{x}{2}],\ 2x,\ 2x+1\}$, and that the likelihood function is constant over this set. It is immediate that, if one used the objective prior $\pi(\theta) = 1$, then

$$\pi(\theta \mid x) = \tfrac{1}{3} \text{ for } \theta \in \{[\tfrac{x}{2}],\ 2x,\ 2x+1\}. \tag{2.11}$$

Take the compact sets to be $\Theta_i = \{1, 2, \ldots, 2i\}$. Then $x \le 4i+1$ and $\pi_i(\theta)$ becomes uniform on Θ_i, so that (2.11) is modified to be

$$\pi_i(\theta \mid x) = \begin{cases} \frac{1}{3} \text{ for } \theta \in \{[\frac{x}{2}], 2x, 2x+1\} & \text{if } x < i \\ \frac{1}{2} \text{ for } \theta \in \{[\frac{x}{2}], 2x\} & \text{if } x = i \\ 1 \text{ for } \theta = [\frac{x}{2}] & \text{if } i < x \le 4i+1\ . \end{cases}$$

It follows that the KL divergence between $\pi(\theta\,|\,\boldsymbol{x})$ and $\pi_i(\theta\,|\,\boldsymbol{x})$ is

$$\begin{aligned}\kappa\{\pi(\cdot\,|\,x)\,|\,\pi_i(\cdot\,|\,x)\} &= \sum_{i=1}^{\infty}\pi_i(\theta\,|\,x)\log\frac{\pi_i(\theta\,|\,x)}{\pi(\theta\,|\,x)}\\ &= \begin{cases}\log(1)=0 & \text{if } x<i\\ \log(3/2) & \text{if } x=i\\ \log(3) & \text{if } i<x\le 4i+1\\ 0 & \text{if } 4i+1<x,\end{cases}\end{aligned}$$

which converges to 0 as $i\to\infty$ for any fixed x. Therefore, $\{\pi_i(\boldsymbol{\theta}\,|\,\boldsymbol{x})\}_{i=1}^{\infty}$ is reverse KL convergent to $\pi(\boldsymbol{\theta}\,|\,\boldsymbol{x})$.

Next, it is easy to see that

$$p_i(x)=\sum_{\theta=1}^{\infty}p(x\,|\,\theta)\pi_i(\theta)=\begin{cases}1/i & \text{if } x<i\\ 2/(3i) & \text{if } x=i\\ 1/(3i) & \text{if } i<x\le 4i+1\\ 0 & \text{if } 4i+1<x.\end{cases}$$

Finally, the expected KL divergence between $\pi(\theta\,|\,\boldsymbol{x})$ and $\pi_i(\theta\,|\,\boldsymbol{x})$ is

$$\sum_{x=1}^{\infty}p_i(x)\,\kappa\{\pi(\cdot\,|\,x)\,|\,\pi_i(\cdot\,|\,x)\}=\frac{2}{3i}\log\left(\frac{3}{2}\right)+\sum_{x=i+1}^{4i+1}\frac{\log(3)}{3i},$$

which converges to $\log(3)$ as $i\to\infty$. Thus, $\{\pi_i(\boldsymbol{\theta}\,|\,\cdot)\}_{i=1}^{\infty}$ is *not* expected reverse KL convergent to $\pi(\boldsymbol{\theta}\,|\,\cdot)$ and this has serious consequences.

Indeed, the posterior (2.11) can yield very unsatisfactory answers, as discussed in Fraser et al. (1985) and Berger & Wolpert (1988). As a simple example of this inadequacy, consider the credible set $C(x)=\{2x,\ 2x+1\}$, which according to (2.11) would have probability 2/3 of containing θ, given x. But it is easy to check that the frequentist coverage probability of $C(x)$, considered as a confidence set, is

$$\Pr(C(x)\text{ contains }\theta\,|\,\theta)=\tfrac{1}{3}\text{ for all }\theta.$$

Thus the Bayesian answer with the constant prior very significantly overstates the confidence in $C(x)$. This is an example of 'strong inconsistency' (see Stone (1976) for other examples) and indicates a serious problem with the constant prior.

It is interesting to note that the noninformative prior $\pi(\theta)=1/\theta$ performs perfectly satisfactorily here, resulting in posterior probabilities and coverage probabilities that are in essential agreement (see Berger & Wolpert (1988)). It is thus quite likely that this prior will result in expected reverse KL convergence for approximating proper priors from subsets, and that would be interesting to demonstrate.

The importance of versions of expected discrepancy convergence has been noted before (Dawid and Stone, 1973, Wallstrom, 2007); however, they worked with discrepancy functions other than KL divergence. Gutiérrez-Peña (1992) used expected KL divergence.

2.3.4 Permissible Priors

Based on the previous considerations, we formally restrict consideration of objective priors to those that satisfy the expected reverse KL convergence criterion, and formally define them as follows.

Definition 2.4 *A positive continuous function* $\pi(\boldsymbol{\theta})$ *is a* permissible prior *for model* $\mathcal{M} = \{p(\boldsymbol{x} \mid \boldsymbol{\theta}),\ \boldsymbol{x} \in \boldsymbol{\mathcal{X}},\ \boldsymbol{\theta} \in \Theta\}$ *if*

(i) *for all* $\boldsymbol{x} \in \boldsymbol{\mathcal{X}}$, $\pi(\boldsymbol{\theta} \mid \boldsymbol{x})$ *is proper, i.e.* $\int_{\boldsymbol{\Theta}} p(\boldsymbol{x} \mid \boldsymbol{\theta})\, \pi(\boldsymbol{\theta})\, d\boldsymbol{\theta} < \infty$;

(ii) *on some approximating compact sequence of subsets of* $\boldsymbol{\Theta}$, *the corresponding posterior sequence is expected reverse KL convergent to* $\pi(\boldsymbol{\theta} \mid \boldsymbol{x}) \propto p(\boldsymbol{x} \mid \boldsymbol{\theta})\, \pi(\boldsymbol{\theta})$.

Example 2.17 Consider the exponential density $p(x \mid \sigma) = \sigma^{-1} e^{-x/\sigma}$, with priors $\pi^a(\sigma) = \sigma^{-a}$. Choose compact sets $C_i = (i^{-1}, i)$, $i = 1, 2 \ldots$, and let π_i^a be the proper prior arising from restricting $\pi^a(\sigma)$ to C_i. Then it can be shown that $\{\pi_i^a(\sigma \mid \cdot)\}_{i=1}^{\infty}$ is expected reverse KL convergent to $\pi^a(\sigma \mid \cdot)$ if and only if $a = 1$. Thus, among the priors $\pi^a(\sigma) = \sigma^{-a}$, only the Jeffreys-rule prior, $\pi^J(\sigma) = 1/\sigma$, is permissible. That the Jeffreys-rule prior is permissible also follows directly from the subsequent Corollary 2.2.

This example shows that expected reverse KL convergence is potentially a serious tool in restricting the choice among improper priors. Stated informally, improper priors cannot be 'too improper.' Here, priors with $a > 1$ are fine as $\sigma \to \infty$ (indeed the right tail of the prior is then proper), but they are 'too improper' as $\sigma \to 0$. The reverse is true for $a < 1$. Only for $a = 1$ is the prior fine in both tails.

Unfortunately, expected reverse KL convergence is usually difficult to verify, and so it has not become a practical tool for identifying potential objective priors. Here is a general result on expected reverse KL convergence for the location family from Berger et al. (2009*a*), where the proof can be found.

Theorem 2.2 *Consider a location family,* $p(x \mid \theta) = f(x - \theta)$, *where* $x \in \mathbb{R}$ *and* $\theta \in \Theta = \mathbb{R}$, *and* f *is a known density. Let* $\pi(\theta) = 1$ *and choose* $\Theta_i = [-i, i]$. *Assume that there is* $\epsilon > 0$, *such that*

$$|x|^{1+\epsilon} f(x) \to 0, \ \text{as}\ |x| \to \infty. \tag{2.12}$$

Then $\{\pi_i = (2i)^{-1} 1_{[-i,i]}(\theta)\}_{i=1}^{\infty}$ *yield a sequence of posteriors that are expected reverse KL convergent to* $\pi(\theta \mid \cdot)$. *Thus* $\pi(\theta) = 1$ *is a permissible prior for a location model for which (2.12) holds.*

When condition (2.12) does not hold, the location model may not exhibit expected reverse KL convergence; indeed, an example of such failure is given below. Thus, interestingly, a prior being 'too improper' is not necessarily solely a property of the tails of the prior; it can also depend on the tails of the model density.

Example 2.18 *A non-permissible constant prior in a location model.* Consider the location model $\mathcal{M} \equiv \{p(x \mid \theta) = f(x - \theta),\, \theta \in \mathbb{R},\, x > \theta + e\}$, where $f(t) = t^{-1}(\log t)^{-2}$, $t > e$. Berger *et al.* (2009*a*) show that, if $\pi(\theta) = 1$, then $\int_{\Theta_0} \kappa\{\pi(\theta \mid x) \mid \pi_0(\theta \mid x)\}\, p_0(x)\, dx = \infty$ for any compact set $\Theta_0 = [a, b]$ with $b - a \geq 1$; thus $\pi(\theta) = 1$ is *not* a permissible prior for $\mathcal{M}$. Note that this model does not satisfy (2.12).

This is an interesting example because it is a location density, so that $\pi(\theta) = 1$ is natural and satisfies numerous nice properties, such as being exact frequentist matching (*i.e.*, a Bayesian $100(1 - \alpha)\%$ credible set will also be a frequentist $100(1 - \alpha)\%$ confidence set). This is in stark contrast to the situation with the Fraser, Monette and Ng example. However, the basic fact remains that posteriors from uniform priors on large compact sets do not seem here to be well approximated (in terms of KL divergence) by a constant prior on the full parameter space. The suggestion is that this is a situation in which assessment of the 'true' bounded parameter space is potentially needed.

Of course, a prior might be permissible for a larger sample size, even if it is not permissible for the minimal sample size. For instance, we suspect that $\pi(\theta) = 1$ is permissible for any location model having two or more independent observations.

The condition in the definition of permissibility that the posterior must be proper is not vacuous, as the following example shows.

Example 2.19 *Mixture model.* Let $\boldsymbol{x} = (x_1, \ldots, x_n)$ be a random sample from the mixture model having density $p(x \mid \theta) = \frac{1}{2}\mathrm{N}(x \mid \theta, 1) + \frac{1}{2}\mathrm{N}(x \mid 0, 1)$, and consider the constant prior $\pi(\theta) = 1$. Since the likelihood function is bounded below by $2^{-n} \prod_{j=1}^{n} \mathrm{N}(x_j \mid 0, 1) > 0$, the integrated likelihood $\int_{-\infty}^{\infty} p(\boldsymbol{x} \mid \theta)\, \pi(\theta)\, d\theta = \int_{-\infty}^{\infty} p(\boldsymbol{x} \mid \theta)\, d\theta$ is infinite. The corresponding formal posterior is thus improper, implying that the uniform prior is *not* a permissible prior for this model. This fact holds quite generally for mixture models, no matter the number of mixture components (as long as there are at least two) and the number of observations.

For this mixture model, it can be shown that the Jeffreys-rule prior (see Chapter 4 for definition) has the shape of an inverted bell, with a minimum value of $1/2$ at $\mu = 0$; since it is thus bounded from below, it also cannot be a permissible prior. This is noteworthy because it is very rare for the Jeffreys-rule prior to yield an improper posterior in univariate problems. It is also of interest because there is no natural objective prior available for the problem. (There are data-dependent objective priors for mixtures: see Wasserman, 2000.)

Theorem 2.2 can easily be modified to apply to models that can be transformed into a location model.

Corollary 2.1 *Consider* $\mathcal{M} \equiv \{p(x \mid \theta),\, \theta \in \Theta,\, x \in \mathcal{X}\}$. *If there are monotone functions* $y = y(x)$ *and* $\phi = \phi(\theta)$ *such that* $p(y \mid \phi) = f(y - \phi)$ *is a location model, and there exists* $\epsilon > 0$ *such that* $\lim_{|t| \to 0} |t|^{1+\epsilon} f(t) = 0$, *then* $\pi(\theta) = |\phi'(\theta)|$ *is a permissible prior for* $\mathcal{M}$.

The most frequent transformation is the log transformation, which converts a scale model into a location model. Indeed, this transformation yields the following direct analogue of Theorem 2.2.

Corollary 2.2 *Consider* $\mathcal{M} = \{p(x \mid \theta) = \theta^{-1} f(|x|/\theta),\, \theta > 0,\, x \in \mathbb{R}\}$, *a scale model where* $f(s)$, $s > 0$, *is a density function. If, for some* $\epsilon > 0$,

$$\lim_{|t| \to \infty} |t|^{1+\epsilon} e^t f(e^t) = 0, \tag{2.13}$$

then $\pi(\theta) = \theta^{-1}$ *is a permissible prior for the scale model* $\mathcal{M}$.

Example 2.20 *Exponential data.* If x is an observation from an exponential density, (2.13) becomes $|t|^{1+\epsilon} e^t \exp(-e^t) \to 0$, as $|t| \to \infty$, which is true. From Corollary 2.2, $\pi(\theta) = \theta^{-1}$ is a permissible prior; indeed, $\{\pi_i(\theta) = (2i)^{-1}\theta^{-1}\ ,\ e^{-i} \le \theta \le e^i\}$ yields posteriors that are expected reverse KL convergent to $\pi(\theta \mid \cdot)$.

Example 2.21 *Uniform data.* Let x be one observation from the uniform distribution $\mathcal{M} = \{\mathrm{Un}(x \mid 0, \theta) = \theta^{-1},\, x \in [0, \theta],\, \theta > 0\}$. This is a scale density, and equation (2.13) becomes $|t|^{1+\epsilon} e^t\, \mathbf{1}_{\{0 < e^t < 1\}} \to 0$, as $|t| \to \infty$, which is indeed true. Thus $\pi(\theta) = \theta^{-1}$ is a permissible prior for $\mathcal{M}$.

The examples showing permissibility were for a single observation. Pleasantly, it is enough to establish permissibility for a single observation or, more generally, for the sample size necessary for posterior propriety of $\pi(\theta \mid \boldsymbol{x})$, because of the following theorem, which shows that expected reverse KL discrepancy is monotonically non-increasing in sample size. For the proof of the theorem, see Berger et al. (2009*a*). (That paper was only for the case of $\boldsymbol{\theta}$ being a scalar, but the proof in the paper goes through in general.)

Theorem 2.3 Monotone expected reverse KL discrepancy. *Consider the parametric model* $\mathcal{M} = \{p(\boldsymbol{x}_1, \boldsymbol{x}_2 \mid \boldsymbol{\theta}) = p(\boldsymbol{x}_1 \mid \boldsymbol{\theta})\, p(\boldsymbol{x}_2 \mid \boldsymbol{x}_1, \boldsymbol{\theta}),\ \boldsymbol{x}_1 \in \boldsymbol{\mathcal{X}}_1,\ \boldsymbol{x}_2 \in \boldsymbol{\mathcal{X}}_2,\ \boldsymbol{\theta} \in \boldsymbol{\Theta}\}$, *and let* $\pi(\boldsymbol{\theta})$ *be a continuous improper prior such that*

$$m(\boldsymbol{x}_1) = \int_{\boldsymbol{\Theta}} p(\boldsymbol{x}_1 \mid \boldsymbol{\theta})\, \pi(\boldsymbol{\theta})\, d\boldsymbol{\theta} < \infty,\ m(\boldsymbol{x}_1, \boldsymbol{x}_2) = \int_{\boldsymbol{\Theta}} p(\boldsymbol{x}_1, \boldsymbol{x}_2 \mid \boldsymbol{\theta})\, \pi(\boldsymbol{\theta})\, d\boldsymbol{\theta} < \infty.$$

For any compact set $\boldsymbol{\Theta}_0 \subset \boldsymbol{\Theta}$, *let* $\pi_0(\boldsymbol{\theta}) = \pi(\boldsymbol{\theta}) \mathbf{1}_{\boldsymbol{\Theta}_0}(\boldsymbol{\theta}) / \int_{\boldsymbol{\Theta}_0} \pi(\boldsymbol{\theta})\, d\boldsymbol{\theta}$, *so that*

$$\pi_0(\boldsymbol{\theta} \mid \boldsymbol{x}_1, \boldsymbol{x}_2) = \frac{p(\boldsymbol{x}_1, \boldsymbol{x}_2 \mid \boldsymbol{\theta})\, \pi(\boldsymbol{\theta})}{m_0(\boldsymbol{x}_1, \boldsymbol{x}_2)}, \quad m_0(\boldsymbol{x}_1, \boldsymbol{x}_2) = \int_{\boldsymbol{\Theta}_0} p(\boldsymbol{x}_1, \boldsymbol{x}_2 \mid \boldsymbol{\theta})\, \pi(\boldsymbol{\theta})\, d\boldsymbol{\theta},$$

$$\pi_0(\boldsymbol{\theta} \mid \boldsymbol{x}_1) = \frac{p(\boldsymbol{x}_1 \mid \boldsymbol{\theta})\, \pi(\boldsymbol{\theta})}{m_0(\boldsymbol{x}_1)}, \quad m_0(\boldsymbol{x}_1) = \int_{\boldsymbol{\Theta}_0} p(\boldsymbol{x}_1 \mid \boldsymbol{\theta})\, \pi(\boldsymbol{\theta})\, d\boldsymbol{\theta}.$$

Then,

$$\begin{aligned}\int\!\!\int_{\boldsymbol{\mathcal{X}}_1\times\boldsymbol{\mathcal{X}}_2} &\kappa\{\pi(\cdot\,|\,\boldsymbol{x}_1,\boldsymbol{x}_2)\,|\,\pi_0(\cdot\,|\,\boldsymbol{x}_1,\boldsymbol{x}_2)\}\,m_0(\boldsymbol{x}_1,\boldsymbol{x}_2)\,d\boldsymbol{x}_1\,d\boldsymbol{x}_2\\ &\le \int_{\boldsymbol{\mathcal{X}}_1}\kappa\{\pi(\cdot\,|\,\boldsymbol{x}_1)\,|\,\pi_0(\cdot\,|\,\boldsymbol{x}_1)\}\,m_0(\boldsymbol{x}_1)\,d\boldsymbol{x}_1, \qquad (2.14)\end{aligned}$$

As an aside, the above result suggests that, as the sample size grows, the convergence of the posterior to normality given in Clarke (1996) is monotone.

2.4 Discussion and Additional References

2.4.1 Conditioning

The *conditionality principle* — that one should draw conclusions conditional on what has been observed, rather that what could have been observed but was not — is automatic in Bayesian statistics, but is typically violated by frequentist procedures, generating considerable controversy. For detailed arguments on the need to appropriately condition see Berger (2014) and references therein. Useful pointers to the huge relevant literature include Durbin (1970), Savage (1970*a*), Kalbfleisch (1975), Kiefer (1977), Godambe (1980, 1984), Liang (1984), Bhapkar (1989), Helland (1995), Reid (1995*b*), Robins & Wasserman (2000), Diciccio et al. (2012), Azriel (2019), Lavine (2020), and references therein.

2.4.2 Frequentist Performance

The frequentist performance of objective Bayesian procedures has received considerable attention. For a review of the main issues involved, see Bayarri & Berger (2004). This will also be extensively discussed in later chapters. Early studies of frequentist performance of Bayesian procedures include Agresti & Min (2005), providing confidence intervals for the difference of proportions, relative risk, and odds ratio in 2×2 contingency tables, and Lecoutre et al. (2010), comparing the success rates of two medical treatments. Other examples include Hartigan (1966), Nicolaou (1993), Datta & Ghosh (1995*b*), Datta, Mukerjee, Ghosh, Sweeting et al. (2000), Cox (2006), Wagenmakers et al. (2008), Yuan et al. (2009), Fraser et al. (2010), Samaniego (2010), Wakefield (2013), Efron (2015), Szabó et al. (2015), Rousseau (2016) and Albers et al. (2018).

2.4.3 Informal Objective Bayesian Procedures

As described in Section 2.2, the uncritical use of *ad hoc* default priors is not recommended. In particular, the use of proper 'vague' priors, obtained by choosing conjugate priors with extreme hyperparameter values, or by truncating the parameter space, should be avoided unless accompanied by a very careful sensitivity analysis with respect to these choices; for further discussion, see Steel (1991), Berger, De Oliveira & Sansó (2001), Hjort et al. (2010), Wang et al. (2013), Pulcini & Liseo (2015), Chang et al. (2017), Manderson et al. (2020), Borowska et al. (2021), Tran et al. (2021), and references therein.

Part II

Select Early Objective Bayesian Developments

Chapter 3

The Constant Prior

The first 180 years of objective Bayes were dominated by use of the constant prior. Here is some of the history.

3.1 Bayes

The first clear Bayesian was, fittingly, the Reverend Thomas Bayes. His paper (published posthumously) *An essay towards solving a problem in the doctrine of chances* (Bayes, 1763) was the first unambiguous statement and use of what is now known as Bayes theorem.

The paper was also a significant work in objective Bayesian analysis. In today's language, the paper considered the basic binomial problem of observing x from $p(x \mid \theta) = \text{Bi}(x \mid n, \theta)$, with θ unknown. While inference concerning θ given an observed x (via Bayes theorem) was one focus of the paper, the other focus was that of coming up with a reasonable objective prior distribution, $\pi(\theta)$, for θ.

Bayes's solution to developing an objective prior for θ was strikingly modern. He did not consider θ directly, but instead considered the marginal (or predictive) distribution of x, namely $p(x) = \int_0^1 \text{Bi}(x \mid n, \theta)\pi(\theta)\, d\theta$. He suggested that an objective prior, $\pi(\theta)$, would be one for which $p(x) = (n+1)^{-1}$, *i.e.*, for which each possible value of x would be equally likely: Bayes (1763, p. 392), "... that concerning such an event I have no reason to think that, in a certain number of trials, it should rather happen any one possible number of times than another." The only prior distribution for which this holds (for all n) can be shown to be $\pi(\theta) = 1$, the uniform density. This is thus the objective prior recommended by Bayes.

3.2 Laplace

The real inventor of objective Bayesian analysis, as a paradigm for statistical analysis, was Simon Laplace (also a great mathematician, astronomer and civil servant). In a series of works, including the amazing book *Théorie Analytique des Probabilités* (Laplace, 1812), he invented a statistical paradigm that was to dominate statistics for well over a century.

Laplace was responsible for many of the most important individual developments in statistics, such as the central limit theorem, the Bayesian version of which is that (in usual situations) the posterior distribution can be well-approximated by a normal distribution for large sample sizes, and this approximation will not depend on the prior distribution. For reviews of the enormous contributions of Laplace, see Stigler (1986*b*), Dale (2012), Hald (1998) and Kass et al. (1991).

The approach of Laplace to statistical inference was essentially objective Bayesian analysis with utilization of a constant prior density for all unknown parameters of statistical models. We are about to criticize the automatic use of constant prior densities, but Laplace's use of such was reasoned.

Originally, he suggested that one should model a problem so that all the possible "cases" of an unknown quantity would reasonably be equally plausible. This later became famous as the *Principle of Insufficient Reason*, first named by de Morgan (1838).

Principle of Insufficient Reason. *If $\boldsymbol{\theta}$ is unknown, and one has no knowledge that certain values of $\boldsymbol{\theta}$ are more likely to occur than others, then the prior density should be chosen to be constant.*

Laplace did not follow this blindly, but rather argued that, if one case were more plausible than another, then one should divide that case up into smaller cases so as to achieve equal plausibility. In today's language, this is roughly a statement that one should choose a parameterization of any unknown quantity so that the unknown has a constant prior density.

Later, with his invention of the central limit theorem, Laplace realized that, with enough data, the initial choice of the prior distribution did not matter. Thus making the simplest choice, a constant density, seemed fine.

Ultimately, however, Laplace was a great pragmatist when it came to statistics. He became involved with the field, and worked in statistics for most of his life because he needed to solve real scientific and societal problems involving data. His invention of objective Bayesian analysis with a constant prior density was a marvelous tool that allowed him to obtain far better practical answers to the problems he encountered than any other methods that existed at the time. Thus it is reasonable to conjecture that his belief in the approach primarily arose from his practical success in its use.

3.3 Inverse Probability

Given a probability distribution, one can deductively determine which possible observations are plausible and, hence, if the actual observed data is reasonable in light of the assumed hypothesis. Laplace was arguably (Hald 1998) the first to formalize this forward approach to statistical analysis, now associated with the frequentist school. The major revolution started by Laplace, however, was going backwards, from the observations to probabilistic inference about unknown features of the probability distribution.

Laplace himself distinguished between the two approaches by using different words for the types of probability. This was rather awkward, however, and never became established. Augustus de Morgan (Morgan, 1838) distinguished between the two types of reasoning by calling reasoning from the data to the model *inverse probability*. This name reigned until roughly 1950, when it became replaced by 'Bayesian analysis' for rather unclear reasons, as extensively discussed in Fienberg (2006).

Inverse probability (*i.e.*, objective Bayesian analysis using a constant prior density) was arguably the dominant mode of formal statistical inference until the 1920's (Dale 2012, Hald 1998). While dominating statistical practice, inverse probability was not without its critics, however. Much of the criticism was ill-placed, but at least some of the philosophical criticisms were legitimate.

Foremost among these criticisms was the fact that the statistical answers with inverse probability will depend on the choice of the parameterization that is used for the statistical model.

Example 3.1 *Inference on a binomial parameter.* Consider again the problem described in Example 1.8. Suppose that we observe x distributed according to the binomial $\mathrm{Bi}(x \mid n, \theta)$ distribution. Then the Principle of Insufficient Reason would lead to the objective prior $\pi(\theta) = 1$. Suppose that someone had, instead, chosen $\psi(\theta) = \frac{2}{\pi}\mathrm{ArcSin}(\sqrt{\theta})$ as the unknown parameter, rather than θ. Then the objective prior would have been $\pi(\psi) = 1$ (note that $0 < \psi < 1$).

To see that this will typically result in different Bayesian answers, make the change of variables from ψ back to θ. With this change of variables, the objective prior $\pi(\psi) = 1$ transforms to

$$\pi^J(\theta) = 1 \times \psi'(\theta) = \pi^{-1}\, \theta^{-1/2}(1-\theta)^{-1/2},$$

which is the $\mathrm{Be}(\theta \mid \frac{1}{2}, \frac{1}{2})$ density (the 'J' superscript because it is the Jeffreys-rule prior, as mentioned in Example 1.8). Clearly the uniform prior and the $\mathrm{Be}(\theta \mid \frac{1}{2}, \frac{1}{2})$ prior will typically give different answers. For instance, the posterior mean for θ corresponding to the uniform prior can be shown to be $(x+1)/(n+1)$, while that corresponding to the $\mathrm{Be}(\theta \mid \frac{1}{2}, \frac{1}{2})$ prior will be $(x+\frac{1}{2})/(n+\frac{1}{2})$.

While Laplace did not specifically address this issue for such continuous parameter problems, recall that his primary motivation for the inverse probability approach was its practical success. In this regard, note for instance that, if $x = 5$ and $n = 30$ in the example, the two posterior means are 0.188 and 0.177 respectively, and the difference is negligible from a practical perspective (since the associated posterior standard errors are on the order of 0.07).

As a prelude to what is to come, consider the development of the Student t distribution, certainly one of the top-five distributions in statistics.

Example 3.2 *The Student t distribution.* Consider again Example 1.9 on the problem of inference about a normal mean μ with unknown variance σ^2. Lüroth (1876) and Edgeworth (1883) were the first to present an exact solution (Laplace used approximations). They did so using the method of inverse probability with a constant prior on the mean and on the precision $\theta = 1/\sigma^2$, showing that the resulting inference should be based on the Student t distribution with n degrees of freedom.

This was almost right. We now know that $n-1$ is the correct degrees of freedom, and this was first obtained by William S. Gosset (better known by his pen name Student) in Gosset (1908), using a frequentist argument. That the correct degrees of freedom could be obtained by an objective Bayesian argument was first shown by Harold Jeffreys in the 1930's, deriving the correct answer by using a constant prior for the parametrization $\theta = \log(\sigma^2)$. (Transforming back to the (μ, σ^2) parameterization yields the prior $\pi(\mu, \sigma^2) = 1/\sigma^2$ or, equivalently, $\pi(\mu, \sigma) = 1/\sigma$, that we have been using for this problem.)

Of course, the difference between n and $n-1$ degrees of freedom is typically practically irrelevant (if n is so small that it matters, the background distributional assumptions are suspect), but the example is interesting as a microcosm of the history of statistics:

(i) Inverse probability provided a practical solution to the problem in 1876.

(ii) Frequentist statistics caught up 30 years later, and produced a slight improvement over the inverse probability answer.

(iii) Objective Bayesian analysis provided the correct objective prior 30 years after that.

In spite of its practical success, inverse probability was being increasingly criticized by mathematicians such as Boole (1854) and Venn (1866), because of the lack of invariance to choice of the parameter. In a sense this was healthy, in that there was a growing desire to put statistical analysis on a logical scientific footing, but there was an unfortunate lack of understanding as to whether the parameterization issue was a major issue or a minor issue.

The period from 1920-1935 is historically recognized as the time when Ronald Fisher and Jerzy Neyman had arrived on the scene and were in the process of fundamentally changing the course of statistics. Yet, even during this period, some of the best statisticians (not named Fisher or Neyman) were still trying to come to grips with how to 'correct' inverse probability.

As an example, consider the efforts of Egon Pearson (he of the fundamental frequentist Neyman-Pearson Lemma) to settle the issue of 'what is the correct objective prior for a binomial proportion?' In Pearson (1925), we see a modern sophisticated 'empirical Bayesian' analysis (long before the name was invented) that tried to address the issue through collection of experimental data. Indeed, Pearson gathered a large number of estimates of proportions θ_i from different real-world binomial situations, $\text{Bi}(x_i \mid \theta_i, n_i)$, and treated these as arising from the marginal (predictive) distributions $p_i(x_i) = \int_0^1 \text{Bi}(x_i \mid \theta_i, n_i)\pi(\theta_i)\, d\theta_i$, corresponding to a fixed prior π, *nature's prior* for binomial situations. Estimating the underlying prior distribution led him to suggest the prior $\pi(\theta) \propto \theta^{-0.662}(1-\theta)^{-0.662}$, something very close to the currently accepted answer, the Jeffreys-rule prior $\pi^J(\theta) \propto \theta^{-1/2}(1-\theta)^{-1/2}$. Besides producing a masterful result in empirical Bayes analysis — many years in advance of the development of that field — it is of considerable historical interest that one of the best statisticians of the time (and one now identified, incorrectly, with the frequentist paradigm) was working on objective Bayesian analysis.

During the 1930's, Harold Jeffreys had begun the modern theory of objective Bayesian analysis, by introducing the notion that the objective prior needs to depend on the model and the parameterization used; this will be discussed in Chapter 4. While inverse probability continued to be used in practice, by 1950 it was fading out, being replaced by other forms of Bayesianism, primarily subjective Bayes and objective Bayes in the school of Jeffreys.

3.4 The Constant Prior after 1950

Curiously, after a period where there was routine avoidance of use of the constant prior (except when justified, as in location parameter problems), use of the constant prior has come back into vogue in some circles. Here are some of the reasons.

(i) The difficulty of subjective Bayesian analysis in complex problems.

(ii) The increasing recognition that the Jeffreys-rule prior can be very problematical in multi-parameter problems (see Section 4.3).

(iii) The difficulty of developing good multi-parameter objective priors (*e.g.*, reference priors).

(iv) With the passage of time, forgetfulness of the potential dangers with the constant prior.

(v) A feeling that using 'vague proper' priors, instead of the constant prior, provides protection against bad things happening.

Some (maybe many) uses of constant priors are fine. After all, Laplace was not wrong. He (necessarily for the time) dealt with situations with low-dimensional unknown parameters, and parameters for which the "principle of insufficient reason" was not unreasonable. If one is today dealing with similar situations and formal objective Bayesian analysis is not possible, use of a constant prior is far from the worst thing that one could do. The difficulty, however, is in judging when it is safe to use a constant prior and when it is dangerous. In Section 7.2, some of the dangers with use of the constant prior are highlighted, problems that are often exacerbated as the dimension of the parameter gets larger. Throughout the text, we will also indicate scenarios in which it is safe to use a constant prior, when that is known.

3.5 Discussion and Additional References

3.5.1 The Historial Controversy on Inverse Probability

The years preceding the second world war saw a raging controversy over the ideas behind inverse probability, as objective Bayesian methods were then called. For some pointers to this interesting part of statistics history, see Fisher (1930), Fisher (1932), Fisher (1933), Haldane (1932), Isserlis (1936), Jeffreys (1938), Wertheimer (1939), Gordon (1940) and Feibleman (1945). For more recent appraisals of this history, from a modern perspective, see Dawid & Stone (1972), Stigler (1986*a*), Zabell (1989), Edwards (1997), Fienberg (1997), Hald (1999), Aldrich (2000), Dale (2012), and references therein.

3.5.2 Modern Advocates of the Use of a Constant Prior

There are advocates for use of a constant prior, in certain situations, particularly for inference in the binomial model, $\text{Bi}(x \mid n, \theta)$. Here are some of the historical arguments for use of the constant prior therein:

- Bayes had the very nice argument that the marginal (predictive) distribution of x should be uniform, and this implied a uniform prior for θ. Why the marginal distribution of x should be uniform is not elaborated upon in his article, but it is certainly a reasonable assumption. See Tuyl et al. (2009) for extensions of this argument.

- Geisser (1984) argued that one should clearly use uniform priors for the (discrete) parameters of hypergeometric and negative hypergeometric distributions, and the limiting binomial distribution should thus inherit the uniform prior (but see Example 11.2 for the reverse argument).

- Gelman et al. (2013) argued that the binomial likelihood $\theta^x(1-\theta)^{n-x}$ is constant when there is no information (*i.e.*, when $x = 0$ and $n = 0$), and so the constant prior is natural to reflect 'no information.'

Of course, while listing these arguments, we must also remember the delightful empirical Bayes analysis of Pearson (1925), which lends considerable support to choice of the Jeffreys-rule prior as 'nature's prior' for binomial problems.

The main recent criticism of the Jeffreys-rule prior for the binomial problem is given in Tuyl et al. (2009). The key issue is the behavior of the posterior from the Jeffreys-rule prior when the observation x is very small, especially $x = 0$. The expressed concern is that the posterior then gives too much mass to θ near zero.

Example 3.3 *Binomial data continued.* Consider again Example 1.8, where x has a $\text{Bi}(x \mid n, \theta)$ distribution and the $\text{Be}(\theta \mid \frac{1}{2}, \frac{1}{2})$ Jeffreys-rule prior is used, so that the posterior distribution is then $\text{Be}(\theta \mid x + \frac{1}{2}, n - x + \frac{1}{2})$. If the uniform prior on θ were used, the posterior distribution would, instead, be $\text{Be}(\theta \mid x + 1, n - x + 1)$.

First note that, for most binomial data, use of either the uniform prior or the Jeffreys-rule prior will give essentially the same answer, so we are not talking about a common practical issue. For instance, the posterior mean and the posterior standard deviation for the Jeffreys-rule prior and the uniform prior are, respectively

$$\frac{x+0.5}{n+1}\,,\ \left[\frac{(x+0.5)(n-x+0.5)}{(n+1)^2(n+2)}\right]^{1/2};\quad \frac{x+1}{n+2}\,,\ \left[\frac{(x+1)(n-x+1)}{(n+2)^2(n+3)}\right]^{1/2}.$$

The most extreme difference between the two occurs when $n = 1$ and $x = 0$, the posterior means being 0.25 versus 0.330 and the associated posterior standard deviations being 0.25 versus 0.236; these are not big differences in practice, especially considering that we have only one observation!

But sometimes one can be very concerned about what happens for θ near 0 (or near 1); a concern expressed in Tuyl et al. (2009) is that, when $x = 0$ is observed, the Jeffreys-rule posterior can be much more concentrated near $\theta = 0$ than the uniform prior posterior. Indeed, a Stirling's approximation argument shows that, if $\epsilon \ll 1/n$, then

$$\Pr^J(0 < \theta < \epsilon \mid x = 0) \approx 1.13\sqrt{n}\sqrt{\epsilon}, \quad \Pr^U(0 < \theta < \epsilon \mid x = 0) \approx (n+1)\epsilon\,.$$

For instance, if $\epsilon = n^{-3/2}$, these probabilities are approximately $1.13\, n^{-1/4}$ and $n^{-1/2}$, respectively. While both probabilities are very small if n is appreciable, it is true that the Jeffreys-rule posterior probability is larger by a factor of $n^{1/4}$. Tuyl et al. (2009) present an example where they feel this difference is a concern.

One of the main arguments in favor of the Jeffreys-rule prior is that it is asymptotically matching (see Chapter 5), meaning that, as $n \to \infty$, one-sided Bayesian $100(1-\alpha)\%$ credible sets have frequentist coverage of $100(1-\alpha)\%$, up to an error term of order $O(1/n)$. In contrast, while other priors also asymptotically have frequentist coverage of $1-\alpha$, their frequentist error term is of order $O(1/\sqrt{n})$, which is significantly larger. Note that this result does not hold true for two-sided credible intervals. Indeed, Hartigan (1965) shows that the $100(1-\alpha)\%$ equal-tailed credible interval is (under standard regularity conditions) matching to order $O(1/n)$ for essentially *any* prior, a rather astonishing result. For the Jeffreys-rule posterior, things would be balanced on each side to order $O(1/n)$, while this would not be so for other posteriors (*e.g.*, the Jeffreys-rule posterior might 'miss' by .05% on both the left and the right, while another prior might miss by 2.1% on the left and -2% on the right, but the misses will mostly cancel so that both will be overall matching to $O(1/n)$).

But one can do different kinds of asymptotics here; Tuyl et al. (2009) consider frequentist coverage as $n \to \infty$, but also with $\theta \to 0$ faster than $1/n$; they show that the uniform prior is better in this type of asymptotics (the matching result for the Jeffreys-rule prior is based on fixed θ) and state "...so one could argue that, near 0 and 1, the argument for the reference prior based on regularity requires too large a sample size to be practically relevant."

What is one to make of all this? Since reasonable arguments can be made in favor of either the Jeffreys-rule prior or the uniform prior for the binomial problem, we do not feel that there is any compelling conclusion that can be reached. And, in the vast majority of applications, the choice makes almost no difference. For consistency with the 'reference prior' philosophy in this book, we will continue to advocate for the Jeffreys-rule prior in the binomial problem, while recognizing that a sensitivity analysis involving the uniform prior is reasonable, especially when $x = 0$ or $x = n$.

For an interesting use of the constant prior for the log-binomial model, see Zhou et al. (2014).

Chapter 4

Jeffreys-Rule Priors

4.1 Development

In the early 1930's, Harold Jeffreys weighed in on the ubiquitous use of the constant prior as follows:

"There is not more need for the later idea [the constant prior] than there is to say that an oven that has once cooked roast beef can never cook anything but roast beef." (Jeffreys, 1961, p. 118).

Being very aware of the inconsistencies that could arise if one used the constant prior, no matter what the parameterization, Jeffreys sought a 'rule' to specify an objective prior which would lead to the same results, whichever parameterization is chosen. In Jeffreys (1946) he developed such a rule using an information-based discrepancy measure between probability distributions (see also Jeffreys, 1961, Section 3.1). In particular he proposed using the square of an element of distance between probability distributions in curvilinear coordinates (which is invariant under one-to-one reparameterizations), and showed that, for any continuous parameter vector $\boldsymbol{\theta} = (\theta_1, \ldots, \theta_m) \in \boldsymbol{\Theta} \subset \mathbb{R}^m$, this has the form

$$|I(\boldsymbol{\theta})|^{1/2} d\theta_1 \ldots d\theta_m \,, \tag{4.1}$$

where $I(\boldsymbol{\theta})$ is the $m \times m$ matrix having elements

$$I_{ij}(\boldsymbol{\theta}) = \mathrm{E}^{\boldsymbol{x}\,|\,\boldsymbol{\theta}}\left[-\frac{\partial^2}{\partial\theta_i \partial\theta_j} \log p(\boldsymbol{x}\,|\,\boldsymbol{\theta}) \right], \tag{4.2}$$

often referred to as *Fishers's information matrix*; here $\mathrm{E}^{\boldsymbol{x}\,|\,\boldsymbol{\theta}}[\cdot]$ denotes the conditional expectation of $\boldsymbol{x}$ given $\boldsymbol{\theta}$, namely

$$\mathrm{E}^{\boldsymbol{x}\,|\,\boldsymbol{\theta}}[f(\boldsymbol{x}, \boldsymbol{\theta})] = \int_{\mathcal{X}} f(\boldsymbol{x}, \boldsymbol{\theta})\, p(\boldsymbol{x}\,|\,\boldsymbol{\theta})\, d\boldsymbol{x} \,. \tag{4.3}$$

This, of course, assumes that the Fisher information matrix exists (that $p(\boldsymbol{x} \mid \boldsymbol{\theta})$ is twice differentiable in the θ_i and that the expectations are finite).

A well known property of the information matrix is that, under a reparameterization, the new information matrix is the old information matrix, pre and post multiplied by the Jacobian of the transformation and its transpose, respectively. Thus the square root of the determinant of the new information matrix is the old information matrix multiplied by the determinant of the Jacobian of the reparameterization. It follows that the rule

$$\pi(\boldsymbol{\theta}) \propto |I(\boldsymbol{\theta})|^{1/2} \tag{4.4}$$

is indeed invariant under one-to-one reparameterization. This is known today as the *Jeffreys-rule* prior. The Jeffreys-rule prior can be proper or improper. When it is improper, it has the rather amazing property of virtually always yielding a proper posterior $\pi(\boldsymbol{\theta} \mid \boldsymbol{x}) \propto p(\boldsymbol{x} \mid \boldsymbol{\theta})\, \pi(\boldsymbol{\theta})$, *i.e.*,

$$\int_{\boldsymbol{\Theta}} p(\boldsymbol{x} \mid \boldsymbol{\theta}) |I(\boldsymbol{\theta})|^{1/2}\, d\boldsymbol{\theta} < \infty$$

for virtually all models. (Example 2.19 is one of the very few examples where this fails). No one has been able to provide even a heuristic argument as to why the Jeffreys-rule prior almost always results in proper posteriors.

A crucial simplification arises when one has n iid observations. Then $I(\boldsymbol{\theta})$ can be easily seen to just be n times the Fisher information for a single observation. Since proportionality constants like n are irrelevant, it follows that one need only compute the Fisher information for one observation.

The famous monograph Jeffreys (1961) explores the consequences of this proposal (*i.e.*, using (4.4) with possible *ad hoc* modifications), in many problems of inference. The Jeffreys-rule prior turns out to be generally excellent (and is our recommended prior) for one parameter problems, as discussed in the next section. Section 4.3 discusses the case of multiple parameter models; unfortunately, the Jeffreys-rule prior is considerably less successful for such problems.

4.2 The Single Parameter Case

In one parameter *regular* continuous models, $\{p(x \mid \theta), \theta \in \Theta \subset \mathbb{R}, x \in \mathcal{X}\}$, where the sampling space $\mathcal{X}$ does not depend on the parameter θ, and the model probability density is twice continuously differentiable with respect to θ, Jeffreys rule (4.4) reduces to

$$\pi(\theta) \propto \sqrt{I(\theta)}, \quad I(\theta) = \mathrm{E}^{x \mid \theta}\left[-\frac{\partial^2}{(\partial\theta)^2} \log p(x \mid \theta)\right]. \tag{4.5}$$

Immediate applications include many of the simple textbook examples of probability models.

Example 4.1 *Jeffreys-rule prior for the parameter of a Poisson model.* Let $\boldsymbol{x} = (x_1, \ldots, x_n)$ be a random sample of size n from a Poisson probability model $\text{Pn}(x \mid \theta) = e^{-\theta}\theta^x/x!$. Recalling that, in the iid case, it suffices to compute the Fisher information matrix for a single observation, and noting that $\text{E}[x \mid \theta] = \theta$, the corresponding information is

$$I(\theta) \quad = \text{E}^{x \mid \theta}\Big[-\frac{\partial^2}{(\partial\theta)^2}\log\frac{e^{-\theta}\theta^x}{x!}\Big] = \text{E}^{x \mid \theta}\Big[\frac{x}{\theta^2}\Big] = \frac{1}{\theta}\,.$$

Thus, using (4.5), the Jeffreys-rule prior for this model is the improper prior $\pi(\theta) \propto \sqrt{I(\theta)} = \theta^{-1/2}$.

In his writings, Jeffreys did not always recommend the use of his own rule; in particular, for the Poisson model, he used an *ad hoc* argument to recommend the use of $\pi(\theta) \propto \theta^{-1}$ (Jeffreys, 1961, p. 136). Thus, we will sometimes have to distinguish between the Jeffreys-rule prior and the Jeffreys-recommended prior. Our own recommendation here is to use the Jeffreys-rule prior, rather than his recommended prior. Not only does it have better properties (*e.g.*, is 'frequentist matching,' as we will discuss), but the prior θ^{-1} has the disconcerting property of yielding an improper posterior if $x = 0$ is observed.

In problems with more than one parameter, Jeffreys univariate rule can obviously be used to provide an objective prior for one of the parameters when the others are known. More generally, Jeffreys rule may be used to provide an objective prior for any one-dimensional function of the parameters, when the other parameters are known. It will later be demonstrated that the construction of these one-dimensional *conditional* objective priors is a crucial element in the derivation of good multivariate objective priors.

Example 4.2 *Normal data.* Let $\boldsymbol{x} = (x_1, \ldots, x_n)$ be a random sample of size n for a normal $\text{N}(x \mid \mu, \sigma^2)$ probability model, and consider the following one-parameter problems:

Variance σ^2 known. If an objective prior is required for the mean μ, the relevant information function is

$$I(\mu) = \text{E}^{x \mid \mu,\sigma}\big[-\frac{\partial^2}{(\partial\mu)^2}\log \text{N}(x \mid \mu, \sigma^2)\big] = \sigma^{-2},$$

a known constant. Hence, as one would expect in a location problem, the Jeffreys-rule prior is the constant prior $\pi(\mu) = 1$. (The Bayesian answers are the same for any constant, so it is common to just choose 1.) Use of Bayes theorem immediately yields

$$\pi(\mu \mid \boldsymbol{x}) = \text{N}(\mu \mid \overline{x}, \sigma^2/n), \quad \overline{x} = n^{-1}\sum\nolimits_{j=1}^{n} x_j, \tag{4.6}$$

which is proper for any $n \geq 1$. Moreover, since use of Jeffreys rule yields priors that are invariant under one-to-one reparameterization, the prior in terms of any one-to-one function of $\psi = \psi(\mu)$ of μ is simply

$$\pi(\psi) \propto \pi(\mu)\left|\frac{\partial\psi}{\partial\mu}\right|^{-1}\Bigg|_{\mu=\mu(\psi)} = \left|\frac{\partial\psi}{\partial\mu}\right|^{-1}\Bigg|_{\mu=\mu(\psi)},$$

where $\mu = \mu(\psi)$ is the inverse function of $\psi(\mu)$. In particular, the Jeffreys-rule prior for the standardized mean $\phi = \mu/\sigma$ is also uniform and, as one would expect, the corresponding posterior distribution is $\pi(\phi \mid \boldsymbol{x}, \sigma^2) = \mathrm{N}(\phi \mid \overline{x}/\sigma, 1/n)$.

Mean μ known. The relevant information function for σ^2 is

$$I(\sigma^2) = \mathrm{E}^{x \mid \mu,\sigma^2}\left[-\frac{\partial^2}{(\partial\sigma^2)^2}\log \mathrm{N}(x \mid \mu, \sigma^2)\right] = \frac{1}{2\sigma^4}.$$

Hence, the Jeffreys-rule prior is $\pi(\sigma^2) = \sigma^{-2}$. Using Bayes theorem, this leads to the posterior

$$\pi(\sigma^2 \mid \boldsymbol{x}, \mu) \propto \frac{1}{\sigma^{(n+2)}}\exp\left(-\frac{ns_\mu^2}{2\sigma^2}\right), \quad s_\mu^2 = n^{-1}\sum\nolimits_{j=1}^n (x_j - \mu)^2,$$

which can be recognized as the inverse gamma $\mathrm{Ig}(n/2, 2/(ns_\mu^2))$ distribution, and is proper for any $n \geq 1$. Again, since Jeffreys rule is invariant under reparameterization, the prior in terms of any one-to-one function $\psi = \psi(\sigma^2)$ of σ^2 is simply

$$\pi(\psi) = \pi(\sigma^2)\left|\frac{\partial\psi}{\partial\sigma^2}\right|^{-1}\Bigg|_{\sigma^2=\sigma^2(\psi)} = \frac{1}{\sigma^2}\left|\frac{\partial\psi}{\partial\sigma^2}\right|^{-1}\Bigg|_{\sigma^2=\sigma^2(\psi)}.$$

In particular, the Jeffreys-rule prior for $\log(\sigma^2)$ (or $\log(\sigma)$) is a constant, and the Jeffreys-rule prior for the precision $\tau^2 = \sigma^{-2}$ has the same functional form $\pi(\tau^2) = 1/\tau^2$, with the corresponding posterior distribution being $\mathrm{Ga}(\tau^2 \mid n/2,\, ns_\mu^2/2)$. Changing variables to σ yields

$$\pi(\sigma \mid \boldsymbol{x}, \mu) = \pi(\sigma^2 \mid \boldsymbol{x}, \mu)\, 2\sigma \propto \frac{1}{\sigma^{(n+1)}}\exp\left(-\frac{ns_\mu^2}{2\sigma^2}\right). \tag{4.7}$$

Finally, the Jeffreys-rule prior for the standardized mean $\phi = \mu/\sigma$, when μ is known, is $\pi(\phi) = \phi^{-1}$ (not constant as it was with μ unknown and σ known). If (say) $\mu > 0$, the corresponding posterior distribution is

$$\pi(\phi \mid \boldsymbol{x}, \mu) \propto \phi^{n-1}\exp\{-ns_\mu^2\phi^2/(2\mu^2)\}, \quad \phi > 0.$$

For a numerical illustration of these results, consider the following data, simulated from a normal density with mean $\mu = 10$ and variance $\sigma^2 = 4$:

$$\boldsymbol{x} = \{9.20, 8.70, 11.02, 13.43, 8.48, 4.78, 9.10, 11.87, 9.85, 7.91\},$$

so that $n = 10$, $\overline{x} = 9.43$ and $s_\mu^2 = 5.34$. The posterior density of σ^2 assuming $\mu = 10$ is known, is the $\mathrm{Ig}(\sigma^2 \,|\, 5, 2/53.4)$ density; the corresponding posterior density of σ, $\pi(\sigma \,|\, \overline{x} = 9.43, \mu = 10)$ is given in Figure 4.1. The very different posteriors of the standardized mean $\phi = \mu/\sigma$, which result from assuming $\sigma^2 = 4$ is known and from assuming $\mu = 10$ is known, are both given in Figure 4.2.

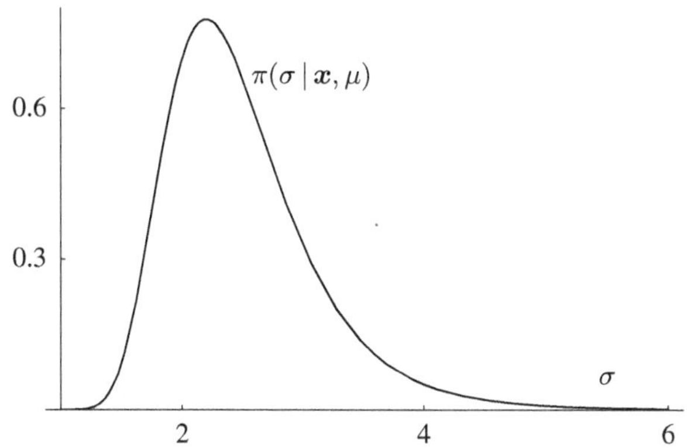

Figure 4.1: Jeffreys posterior density $\pi(\sigma \,|\, \boldsymbol{x}, \mu)$ of the standard deviation σ with $n = 10$ and $s_\mu^2 = 5.34$, assuming $\mu = 10$ is known.

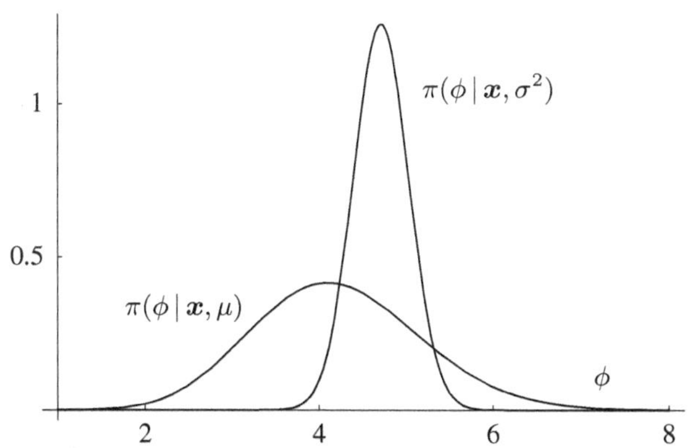

Figure 4.2: Jeffreys posterior densities of the standardized normal mean $\phi = \mu/\sigma$, with $n = 10$, $\overline{x} = 9.43$ and $s_\mu^2 = 5.34$, assuming $\mu = 10$ is known (flatter density), and assuming $\sigma = 2$ known (sharper density).

The Jeffreys-rule prior is a function of the complete probability model $\{p(x \,|\, \theta), x \in \mathcal{X}, \theta \in \Theta\}$, and is not simply a function of the observed likelihood function. As a consequence, two different probability models labeled by the same parameter will generally lead to different Jeffreys-rule priors, even if the two models provide proportional likelihood functions.

Example 4.3 *Bernoulli data.* Let $\boldsymbol{x} = (x_1, \ldots, x_n)$ be a set of n independent Bernoulli observations, so that $p(x \mid \theta) = \theta^x (1-\theta)^{1-x}$, $x \in \{0, 1\}$, $0 < \theta < 1$, and consider the following one-parameter problems:

Direct sampling. Assume that $\boldsymbol{x}$ is a random sample of size n from $p(x \mid \theta)$, so that the number of observations n is a priori fixed. The Fisher information for a single observation, noting that $\mathrm{E}[x \mid \theta] = \theta$, is

$$I(\theta) = \mathrm{E}\left[-\frac{\partial^2}{(\partial\theta)^2} \log[\theta^x(1-\theta)^{1-x}]\right] = \mathrm{E}^{x \mid \theta}\left[\left(\frac{x}{\theta^2} + \frac{1-x}{(1-\theta)^2}\right)\right] = \frac{1}{\theta(1-\theta)};$$

thus, the Jeffreys-rule prior is $\pi(\theta) \propto \theta^{-1/2}(1-\theta)^{-1/2}$, a proper prior, actually the Beta density $\mathrm{Be}(\theta \mid \frac{1}{2}, \frac{1}{2})$. Writing $r = \Sigma_j x_j$ and using Bayes theorem, the corresponding proper posterior is

$$\begin{aligned}\pi(\theta \mid \boldsymbol{x}) &\propto \left[\prod\nolimits_{j=1}^{n} \theta^{x_j}(1-\theta)^{1-x_j}\right] \theta^{-1/2}(1-\theta)^{-1/2} \\ &\propto \theta^{r+\frac{1}{2}-1}(1-\theta)^{n-r+\frac{1}{2}-1} \propto \mathrm{Be}(\theta \mid r+\tfrac{1}{2}, n-r+\tfrac{1}{2}).\end{aligned}$$

The analysis could, instead, have been done in terms of the sampling distribution of the sufficient statistic r, namely $p(r \mid \theta) = \mathrm{Bi}(r \mid n, \theta)$. The same Jeffreys-rule prior and posterior would be obtained.

Note, in passing, that Haldane's prior (Haldane (1948)), $\pi(\theta) \propto \theta^{-1}(1-\theta)^{-1}$, yields a posterior which is improper whenever either $r = 0$ or $r = n$, an unacceptable situation.

Inverse sampling. Assume now that observations are taken until r of the x_i are 1, where r is fixed a priori. Then x, the number of the x_i that equal 0, is distributed according to the negative binomial model

$$p(x \mid r, \theta) = \binom{r+x-1}{r-1} \theta^r (1-\theta)^x .$$

Noting that $\mathrm{E}[x \mid \theta] = r(1-\theta)/\theta$, the Fisher information is

$$I(\theta) = \mathrm{E}^{x \mid \theta}\left[\frac{r}{\theta^2} + \frac{x}{(1-\theta)^2}\right] = \frac{r}{\theta^2(1-\theta)},$$

and, hence, the Jeffreys-rule prior is $\pi(\theta) \propto \theta^{-1}(1-\theta)^{-1/2}$, an improper prior which is different from that derived from direct sampling. Using Bayes theorem, the corresponding posterior is

$$\pi(\theta \mid \boldsymbol{x}) \propto p(x \mid r, \theta)\, \theta^{-1}(1-\theta)^{-1/2} \propto \mathrm{Be}(\theta \mid r, x+\tfrac{1}{2}),$$

which is proper for any data.

The foundational issue here is that the likelihood functions for the negative binomial and binomial densities are proportional to $\theta^{[\#1's]}(1-\theta)^{[\#0's]}$. The Likelihood Principle (see Section 1.2.2) states that inferences drawn from proportional likelihoods should be the same, yet the Jeffreys-rule answers here differ. We defer discussion of this until Section 7.5.3.

Example 4.4 *Location models.* Let $\boldsymbol{x} = (x_1, \ldots, x_n)$ be a random sample from a location model $p(x \mid \theta) = h(x - \theta)$, $x \in \mathbb{R}$, $\theta \in \mathbb{R}$. Jeffreys (1961, Sec. 3.1) used *ad hoc* and historical arguments to recommend the use of a constant prior for all location models, and this is certainly the result of formal application of his rule. Indeed using (4.5) leads to

$$\pi(\theta) \propto \left[\int_{-\infty}^{\infty} \frac{h'(x-\theta)^2 - h(x-\theta)h''(x-\theta)}{h(x-\theta)} \, dx \right]^{1/2},$$

and changing the integration variable to $y = x - \theta$ shows that the integral does not depend on θ and, hence, the resulting prior is proportional to some constant.

Example 4.5 *Scale models.* Consider the model $p(x \mid \theta) = \theta^{-1} g(x/\theta)$. This is a general scale model and can be transformed to the location model $p(y \mid \phi) = g(\exp\{y - \phi\})$, with the change of variables $y = \log x$ and $\phi = \log \theta$. For ϕ, the Jeffreys-rule prior is constant and changing variables back to θ results in $\pi(\theta) \propto \pi(\phi)|d\phi/d\theta| \propto \theta^{-1}$, which is thus the Jeffreys-rule prior for scale models.

As stated earlier, Jeffreys rule cannot be applied to non-regular problems. This includes those situations where the sampling space $\mathcal{X}$ depends on the parameter.

Example 4.6 *Uniform data on* $[0, \theta]$. Let $\boldsymbol{x} = (x_1, \ldots, x_n)$ be a random sample of size n from a uniform distribution on the real interval $[0, \theta]$, so that $p(x \mid \theta) = \theta^{-1}$, if $0 < x < \theta$, and zero otherwise. The sample space $\mathcal{X} = (0, \theta)$ depends on the parameter θ and, therefore, Jeffreys rule (4.5) is not applicable (blind use of the formula yields a negative Fisher information function). This will be considered, again, in Section 9.4, where the reference prior approach will be shown to yield the objective prior $\pi(\theta) = 1/\theta$.

Note that the data does not have to be i.i.d. to determine the Jeffreys-rule prior.

Example 4.7 *Data from different distributions.* Suppose $\boldsymbol{x} = (x_1, x_2)$, with $x_1 \sim \mathrm{N}(x_1 \mid \theta, 1)$, where it is known that $\theta > 0$ and, independently, $x_2 \sim \mathrm{Po}(x_2 \mid \theta)$. Calculation yields $I(\theta) = 1 + \theta^{-1}$, so the Jeffreys-rule prior is $\pi(\theta) = (1 + \theta^{-1})^{1/2}$.

4.3 The Multi-parameter Case

Although Jeffreys rule (4.4) does provide a multivariate invariant objective prior, the results are very often unappealing.

Example 4.8 *Inference on a normal mean, with variance unknown.*

Let $\boldsymbol{x} = (x_1, \ldots, x_n)$ be a random sample of size n from a normal distribution $\mathrm{N}(x \mid \mu, \sigma^2)$, both parameters unknown. Using (4.2), the Fisher information matrix for a single observation is

$$I(\mu, \sigma^2) = \begin{pmatrix} 1/\sigma^2 & 0 \\ 0 & 1/(2\,\sigma^4) \end{pmatrix} \tag{4.8}$$

and, hence, by (4.4), the Jeffreys-rule prior is

$$\pi(\mu, \sigma^2) \propto |I(\mu, \sigma^2)|^{1/2} \propto \sigma^{-3}. \tag{4.9}$$

If the prior is of the form $\pi(\mu, \sigma^2) \propto \sigma^{-\alpha}$, for some $\alpha > 0$, Bayes theorem yields a joint posterior

$$\pi(\mu, \sigma^2 \mid \boldsymbol{x}, \alpha) \propto \sigma^{-(n+\alpha)} \exp\left\{-\frac{n}{2\sigma^2}[s^2 + (\overline{x} - \mu)^2]\right\}, \tag{4.10}$$

where $\overline{x}$ is the sample mean and $s^2 = \sum(x_i - \overline{x})^2/n$. Integrating out σ^2, the corresponding marginal posterior for μ is (provided $n + \alpha > 3$), the Student t density

$$\pi(\mu \mid \boldsymbol{x}, \alpha) = \mathrm{St}(\mu \mid \overline{x},\ s/\sqrt{n + \alpha - 3}, n + \alpha - 3)\,. \tag{4.11}$$

The use of (4.9) ($\alpha = 3$) thus leads to $\pi(\mu \mid \boldsymbol{x}) = \mathrm{St}(\mu \mid \overline{x},\ s/\sqrt{n}, n)$.

Jeffreys (1961, Sec. 3.4) was unhappy with this result; he mentions that this is against the 'standard practice' of removing one degree of freedom from the sample size per estimated parameter (which is the result for $\alpha = 2$). This led Jeffreys to recommend the use of $\alpha = 2$, *i.e.*, of $\pi(\mu, \sigma^2) \propto \sigma^{-2}$, even though it contradicted his multivariate rule. His argument to rationalize this choice was to treat μ and σ independently.

"If the standard error is unknown, its prior must be proportional to $d\sigma/\sigma$ (...) Also we need not suppose that any previous knowledge of μ would tell us anything directly about σ, so that the prior probabilities of μ and σ may be taken independent. Then $\pi(\mu, \sigma) \propto d\mu\, d\sigma/\sigma$." (Jeffreys, 1961, p. 138). Note that use of $d\sigma/\sigma$ is equivalent to the use of $d\sigma^2/\sigma^2$. This Jeffreys-recommended prior has now become known as the *Jeffreys independence prior.*

Another argument for the Jeffreys independence prior is that it is the only choice of α for which the resulting posterior credible sets for μ have exactly matching frequentist coverage probability. This is discussed in Chapter 5.

Mainly motivated by this example, Jeffreys argued that location and scale parameters should *always* be considered independent a priori and, hence, for location-scale models of the form $p(x \mid \mu, \sigma) = \sigma^{-1} f((x - \mu)/\sigma)$ he recommended the prior $\pi(\mu, \sigma) \propto \pi(\mu)\, \pi(\sigma) \propto \sigma^{-1}$. It will later be demonstrated that more general methods for the construction of objective priors also lead to this result.

Definition 4.1 Jeffreys independence prior for location-scale problems. *The Jeffreys independence prior for location-scale models of the form*

$$p(x \mid \mu, \sigma) = \frac{1}{\sigma} f\left(\frac{x-\mu}{\sigma}\right), \quad x \in \mathbb{R}, \quad \theta \in \mathbb{R}, \quad \sigma > 0,$$

is defined to be $\pi(\mu, \sigma) = \sigma^{-1}$. *If* σ^2 *is used instead of* σ, *as the parameter, the corresponding prior density would be* $\pi(\mu, \sigma^2) = \sigma^{-2}$.

Example 4.9 *Neyman-Scott problem.* Consider the situation where two observations are independently obtained from each of m normal distributions, with differing means μ_i but common variance σ^2, so that $\boldsymbol{x} = \{x_{ij}\}$, $i = 1, \ldots, m$, $j = 1, 2$, and

$$p(\boldsymbol{x} \mid \mu_1, \ldots, \mu_m, \sigma^2\} = \prod_{i=1}^{m} \prod_{j=1}^{2} \mathrm{N}(x_{ij} \mid \mu_i, \sigma^2).$$

Suppose that one is interested in inference about the common variance σ^2. This problem was introduced by Neyman and Scott (1948) and has since become a test for all new methods of inference.

The Fisher information matrix is easily derived to be

$$I(\mu_1, \ldots, \mu_m, \sigma^2) = \begin{pmatrix} 2/\sigma^2 & 0 & \ldots & 0 & 0 \\ 0 & 2/\sigma^2 & \ldots & 0 & 0 \\ \ldots & \ldots & \ldots & \ldots & \ldots \\ 0 & 0 & \ldots & 2/\sigma^2 & 0 \\ 0 & 0 & \ldots & 0 & m/\sigma^4 \end{pmatrix}, \tag{4.12}$$

so that $|I(\mu_1, \ldots, \mu_m, \sigma^2)| = m\, 2^m \sigma^{-(2m+4)}$ and, using 4.4, the corresponding Jeffreys-rule prior is $\pi(\mu_1, \ldots, \mu_m, \sigma^2) \propto \sigma^{-(m+2)}$. The joint posterior density, which corresponds to a prior of the form $\pi(\mu_1, \ldots, \mu_m, \sigma^2) = \sigma^{-\alpha}$, is easily derived to be

$$\pi_\alpha(\mu_1, \ldots, \mu_m, \sigma^2 \mid \boldsymbol{x}) \propto \frac{1}{\sigma^{2m+\alpha}} \exp\left\{-\frac{1}{\sigma^2}\left[\frac{S^2}{2} + \sum\nolimits_{i=1}^{m} (\overline{x}_i - \mu_i)^2\right]\right\}, \tag{4.13}$$

where $\overline{x}_i = (x_{i1} + x_{i2})/2$ and $S^2 = \sum_{i=1}^{m} \sum_{j=1}^{2} (x_{ij} - \overline{x}_i)^2$. Integrating out the μ_i results in the marginal posterior density of σ^2 (the parameter of interest)

$$\pi_\alpha(\sigma^2 \mid \boldsymbol{x}) \propto \frac{1}{\sigma^{m+\alpha}} \exp\left\{-\frac{S^2}{2\sigma^2}\right\},$$

which is recognizable as the $\mathrm{Ig}([m + \alpha - 2]/2, S^2/2)$ distribution. This distribution has mean

$$\mathrm{E}[\sigma^2 \mid \boldsymbol{x}] = \frac{S^2}{(m + \alpha - 4)}.$$

If the Jeffreys-rule prior is used (which corresponds to $\alpha = m+2$) this becomes $\mathrm{E}[\sigma^2 \mid \boldsymbol{x}] = S^2/[2m-2]$. Letting σ_T^2 denote the true value of the variance, a computation shows that the the frequentist mean of S^2 is $\mathrm{E}^x[S^2 \mid \sigma_T^2] = m\sigma_T^2$. By the law of large numbers, S^2/m thus converges (in a frequentist sense) to σ_T^2, as m increases. It follows that, for the Jeffreys-rule prior,

$$\lim_{m\to\infty} \mathrm{E}[\sigma^2 \mid \boldsymbol{x}] = \lim_{m\to\infty} \frac{S^2}{2m-2} = \frac{\sigma_T^2}{2},$$

which is only half the true value. Also, the posterior variance will go to 0 as m grows, so the posterior distribution concentrates around a completely wrong value; this is an example of *posterior inconsistency.* The fact that the Jeffreys-rule prior gets worse and worse as the dimension grows conveys a clear warning concerning its use for higher dimensional problems.

In marked contrast, use of the Jeffreys independence prior, which here would be $\pi(\mu_1,\ldots,\mu_m,\sigma^2) \propto \sigma^{-2}$ (which corresponds to $\alpha = 2$), leads to a posterior density of σ^2 with mean $S^2/(m-2)$, which converges to σ_T^2 as m increases. We will later see that this is also the prior which results from the reference prior approach.

Example 4.10 *Multinomial distribution.* Suppose $\boldsymbol{x} = (x_1,\ldots,x_m)$ is multinomial $\mathrm{Mu}(\boldsymbol{x} \mid n,\theta_1,\ldots,\theta_m)$ (suppressing the $(m+1)$st cell count, $x_{m+1} = n - \sum_{j=1}^m x_j$, and probability, $\theta_{m+1} = 1 - \sum_{j=1}^m \theta_j$, since they are determined by the others) so that

$$p(\boldsymbol{x} \mid n,\theta_1,\ldots,\theta_m) = \frac{n!}{\prod_{j=1}^m x_j!(n-\Sigma x_j)!} \prod_{j=1}^m \theta_j^{x_j}(1-\Sigma\theta_j)^{n-\Sigma x_j}.$$

Using (4.2), computation of the Fisher information matrix yields

$$I(\theta_1,\ldots,\theta_m) = \frac{n}{1-\Sigma\theta_j}\begin{bmatrix} \frac{1+\theta_1-\Sigma\theta_j}{\theta_1} & 1 & \ldots & 1 \\ 1 & \frac{1+\theta_2-\Sigma\theta_j}{\theta_2} & \ldots & 1 \\ \ldots & \ldots & \ldots & \ldots \\ 1 & 1 & \ldots & \frac{1+\theta_m-\Sigma\theta_j}{\theta_m} \end{bmatrix}$$

with

$$|I(\theta_1,\ldots,\theta_m)| = n^m \left[\left(1-\sum\nolimits_{j=1}^m \theta_j\right)\prod\nolimits_{j=1}^m \theta_j\right]^{-1}.$$

Thus, the Jeffreys-rule prior, proportional to $|I(\theta_1,\ldots,\theta_m)|^{1/2}$, is the proper Dirichlet prior

$$\pi^J(\theta_1,\ldots,\theta_m) \propto \left(1-\sum\nolimits_{j=1}^m \theta_j\right)^{-1/2} \prod_{j=1}^m \theta_j^{-1/2}, \tag{4.14}$$

i.e., the $\mathrm{Di}((\theta_1,\ldots,\theta_m) \mid (\frac{1}{2},\ldots,\frac{1}{2}))$ distribution. Multiplying this by the multinomial likelihood immediately yields that the corresponding posterior distribution is $\mathrm{Di}((\theta_1,\ldots,\theta_m) \mid (x_1+\frac{1}{2},\ldots,x_m+\frac{1}{2}))$.

Again, (4.14) does *not* have appropriate behavior for an objective prior. As a numerical illustration of this, consider the case where the sample size n is small relative to the number of classes $m+1$; thus we have a large sparse table. For instance, suppose $n = 3$ and $m = 1000$, with $x_{240} = 2$, $x_{876} = 1$, and all the other $x_i = 0$. The posterior means resulting from using the Jeffreys-rule prior can be shown to be

$$\mathrm{E}[\theta_i \mid \boldsymbol{x}] = \frac{x_i + 1/2}{\sum_{j=1}^{m} [x_j + 1/2]} = \frac{x_i + 1/2}{n + m/2} = \frac{x_i + 1/2}{503},$$

so that $\mathrm{E}[\theta_{240} \mid \boldsymbol{x}] = 2.5/503$, $\mathrm{E}[\theta_{876} \mid \boldsymbol{x}] = 1.5/503$, and the other cells have $\mathrm{E}[\theta_i \mid \boldsymbol{x}] = 0.5/503$. So, cells 240 and 876 only have total posterior probability of $4/503 = 0.008$, even though all 3 observations are in these cells.

The problem is that the Jeffreys-rule prior effectively added 1/2 to the 998 zero cells, making them more important than the cells with data! That the Jeffreys-rule prior can encode much more information than is contained in the data is hardly desirable for an objective analysis.

An alternative objective prior that is sometimes considered is the uniform prior on the simplex, but this is even worse than the Jeffreys-rule prior, adding 1 to each cell. The prior that adds 0 to each cell is the improper prior $\prod_{j=1}^{m} \theta_j^{-1}$, but this results in an improper posterior if any cell has a zero entry, a virtual certainty for very large tables.

Three priors will be developed that overcome these problems, the reference prior in Section 10.3.2 and two overall recommended priors in Chapter 12.

In this section, we have seen three important examples — the normal model with both parameters unknown, the Neyman-Scott problem and the multinomial model — where the multiparameter form of the Jeffreys-rule prior fails to provide appropriate objective posteriors. We actually know of no multiparameter example in which the Jeffreys-rule prior has been verified to be satisfactory. In higher dimensions, that prior always seems to be either 'too diffuse' as in the normal examples, or 'too concentrated' as in the multinomial example.

4.4 Discussion and Additional References

4.4.1 The Extensive Use of Jeffreys-Rule Priors

During the second half of the 20th century, the Jeffreys-rule prior has been extensively used in the literature, both in theoretical and in applied work. Examples of the relevant huge bibliography are Lindley (1965), Broeckx et al. (1974), Broeckx et al. (1977), Liu (1981), Liu & Hanssens (1981), Pericchi (1981) Achcar & Bolfarine (1986), Ibrahim & Laud (1991), Clarke & Wasserman (1993), Clarke & Barron (1994), Eric & Peter (1994), Poirier (1994), Slater

(1995, 1996), Chao & Phillips (1998), Siu & Kelly (1998), Kershaw et al. (1999), Oh & Kim (2000), Slater (2000), De Santis et al. (2001), Kershaw et al. (2001), Chao & Phillips (2002), Chen & Varshney (2002), Tardella (2002), Girón et al. (2003), Kaplan & Barnett (2003), Abul-Magd et al. (2004), Lee (2004), Paulo (2005), Abul-Magd et al. (2006), Bunouf & Lecoutre (2006), De Oliveira & Song (2008), Fonseca et al. (2008), Tuyl et al. (2008), Hamelryck (2009), Jung et al. (2009), Gubler & Oka (2010), Cabras et al. (2011), Lira & Grientschnig (2011), Alston et al. (2012), De Oliveira (2012), Harvey & Van der Merwe (2012), Kazianka & Pilz (2012), Ho (2012), Salazar et al. (2012), Wang & Sivaganesan (2013), Moala & Garcia (2013), Branco et al. (2013), Mohsin et al. (2014), Rubio & Liseo (2014), Rubio & Steel (2014), Villa & Walker (2014), Achcar et al. (2015), Vallejos & Steel (2015), Nosek & Nosková (2016), Ruli et al. (2016), Wang & Yang (2016), Dittrich et al. (2017), Feroze & Aslam (2017), Hattori et al. (2017), Li & Wang (2017), Sharma (2017), Coelho-Barros et al. (2018), Leisen et al. (2018), Long et al. (2018), Muré (2019), Nogarotto et al. (2020) and Yosboonruang & Niwitpong (2021).

Extensions, modifications and reinterpretations of Jeffreys-rule priors include Milne (1983), Rodriguez (1988), Kass (1989), Kass & Wasserman (1996), Natarajan (2001), de Cristofaro (2002, 2008), Harney (2003), Goyal (2005), Soyer (2009), Ho (2012), Shemyakin et al. (2012), Shemyakin (2014), Abe (2014), Ortega & Basulto (2016), Rainey (2016) and Fuglstad et al. (2019).

Many papers have been devoted to comparison of the performance of the Jeffrey-rule prior with other objective priors; some of these are Box & Tiao (1973), Bucephala & Gatsonis (1988), Nicolaou (1993), Bernardo (1997), Howson (1997), Silliman (1997), Krichevskiy (1998), Kuboki (1998), Mendoza & Gutiérrez-Peña (2000), Asakawa et al. (2001), Rosa & Gianola (2001*b*), Kim et al. (2002), Contreras-Cristán et al. (2003), Kardaun et al. (2003), Zhu & Lu (2004), Liseo (2005), Lee (2007), Berger & Sun (2006), Sun & Berger (2007*a*), Berger & Sun (2008), Ren et al. (2009), Berger, De Oliveira & Sansó (2001), Barger & Bunge (2010), Lecoutre et al. (2010), Liang & Sun (2012), Pilz et al. (2012), Ren et al. (2012), Ren (2013), Ren & Sun (2014), Cabras (2013), Kazianka (2013), Zaffalon & Miranda (2013), Ren et al. (2013), Ren & Sun (2013), Ferreira & Salazar (2014), Martin & Liu (2015), Ramos et al. (2015), Veronese & Melilli (2015), Rao & D'Cunha (2016), Ramos et al. (2017), Aitkin (2018), Abe (2018), Díaz (2018), Harvey & Van der Merwe (2018), Moala & Dey (2018), Veronese & Melilli (2018), Villa & Rubio (2018), Keefe et al. (2019), Kelly et al. (2019), van Zwet (2019), Yanuar et al. (2019), Leisen et al. (2020), Maneerat et al. (2020), Berger et al. (2020*b*), Song et al. (2020) and Ramos et al. (2021).

4.4.2 Alternative Derivations of the Jeffreys-Rule Prior

Perks (1947) used an argument based on frequentist matching (see Chapter 5) to propose an objective prior of the form

$$\pi(\theta) \propto s(\theta)^{-1}, \tag{4.15}$$

where $s(\theta)$ is the asymptotic standard deviation of the maximum likelihood estimate of θ. Under regularity conditions which ensure asymptotic normality, this is equivalent to the Jeffreys-rule prior, and can be applied even if those conditions fail. For instance, in the Poisson Example 4.1, the maximum likelihood estimate of θ is $\hat{\theta} = \overline{x}$, whose sampling distribution is asymptotically normal with mean θ and standard deviation $\theta^{1/2}/\sqrt{n}$; thus, using Perks argument, the suggested objective prior is, again $\pi(\theta) \propto \theta^{-1/2}$. Since Perks work was independent of Jeffreys (1946) and was published at about the same time, the one-dimensional rule (4.5) is sometimes referred to as the *Jeffreys-Perks rule*.

Lindley (1961) used information-theoretic ideas to provide an alternative justification of Jeffreys univariate rule. He argued that one could always replace a continuous range of θ by discrete values over a grid whose mesh size $\delta(\theta)$ described the precision of the measuring process, and that a possible objective prior would be one which assigned equal probability to all points of this grid. In the continuous case, this implies a prior of the form $\pi(\theta) \propto \delta(\theta)^{-1}$. To determine $\delta(\theta)$, Lindley showed that if θ and $\theta + \delta(\theta)$ were the only two possible values of the parameter, and they had equal prior probability, then the amount of information that one observation from $p(x \mid \theta)$ would be expected to provide about θ is $2\delta^2(\theta)I(\theta)$ where, again, $I(\theta)$ is Fisher's information function. This expected information would be independent of θ if, and only if, $\delta(\theta) \propto I(\theta)^{-1/2}$, thus defining an appropriate mesh. Arguing as before, this suggests Jeffreys-rule prior $\pi(\theta) \propto I(\theta)^{1/2}$.

Welch & Peers (1963) and Welch (1965)) discussed conditions under which there is formal mathematical equivalence between one-dimensional Bayesian credible regions and the corresponding frequentist confidence intervals. They showed that, under suitable regularity conditions, one-side intervals asymptotically coincide if the prior used for the Bayesian analysis is the Jeffreys-rule prior. This line of work was extensively developed in the 1990's, and will be discussed in Chapter 5.

Hartigan (1965) reported that the prior which minimizes the bias of the Bayes estimator $\tilde{\theta}$, that arises from a loss function $L(\tilde{\theta}, \theta)$, is given by

$$\pi(\theta) = I(\theta) \left[\frac{\partial^2}{(\partial\tilde{\theta})^2} L(\tilde{\theta}, \theta) \right]^{-1/2} \Bigg|_{\tilde{\theta}=\theta} ;$$

if, in particular, one uses the KL divergence

$$L(\tilde{\theta}, \theta) = \int_{\mathcal{X}} p(x \mid \theta) \log \frac{p(x \mid \theta)}{p(x \mid \tilde{\theta})} \, dx$$

as the loss function, this implies $\pi(\theta) \propto I(\theta)^{1/2}$ which, once again, is the Jeffreys-rule prior.

Yet another argument for use of the Jeffreys-rule prior in one parameter regular continuous models was provided by Box and Tiao (1973, Sec. 1.3) in their classic, influential monograph. They argued that, since the principle of insufficient reason invoked to suggest a constant prior is reasonable in location problems, one could use, as a conventional prior, that $\pi(\theta)$ which would imply a uniform prior

$$\pi(\psi) = \pi(\theta)\left|\frac{\partial\psi}{\partial\theta}\right|^{-1} \propto 1$$

for that function $\psi = \psi(\theta)$ of the original parameter such that $p(x \mid \psi)$ is, at least approximately, a location parameter family (*i.e.*, such that, for some functions f and g, the model is of the form $p(x \mid \theta) = g[f(x) - \psi(\theta)]$). Using standard asymptotic theory, they showed that, under suitable regularity conditions, this would happen if

$$\psi(\theta) = \int^{\theta} I(\theta)^{1/2}\, d\theta,$$

thus leading again to $\pi(\theta) \propto I(\theta)^{1/2}$, the Jeffreys-rule prior.

Good (1969) derived the Jeffreys-rule prior as the 'least favourable' initial distribution with respect to the logarithmic scoring rule, in the sense that it minimizes the expected score from reporting the true distribution. Thus, the Jeffreys-rule prior may technically be described, under suitable regularity conditions, as a minimax solution to the problem of scientific reporting when the utility function is the logarithmic score function. Kashyap (1971) provided a similar, more detailed argument; an axiom system is used to justify the use of an information measure as a payoff function and the Jeffreys-rule prior is shown to be the minimax solution in a two person zero sum game, where the statistician chooses the 'objective' prior and nature chooses the 'true' θ. Hartigan (1971) and Hartigan (1983, Ch. 5) define a similarity measure for events E and F to be $P(E \cap F)/P(E)P(F)$, and show that the Jeffreys-rule prior asymptotically ensures constant similarity for current and future observations.

4.4.3 Coherence under Reparameterization

Jeffreys rule achieved the goal of producing objective priors that are invariant to reparameterization, but his particular rule was rather *ad hoc*, in that there are many other such rules that would also exhibit the required invariance. Nevertheless, the requirement of invariance was a major step forward and one we certainly accept. But, if this is conceded, it follows that approaches which do not have this property should be regarded as unsatisfactory. Such approaches include the use of limiting forms of conjugate priors, as in Haldane (1948), Novick & Hall (1965), Novick (1969), DeGroot (1970, Ch. 10) and Piccinato (1973, 1978); a predictive version of the principle of insufficient reason, Geisser (1984); and different forms of information-theoretic arguments, such as those put forward by Zellner (1977, 1991), Geisser (1979) and Torgersen (1981).

Chapter 5

Frequentist Matching

This chapter addresses three important issues related to probability matching: matching priors, confidence distributions, and fiducial inference. This is meant only as an introduction to these subjects; enough so that their relationships with the objective Bayes approach can be discussed.

5.1 Frequentist Matching Priors

We have encountered situations (*cf.* Examples 1.17 and 2.2) where Bayesian credible sets are identical to frequentist confidence sets and have the same stated confidence or coverage. This is studied more formally in this section.

Without effective loss of generality, it is customary to consider one-sided credible sets $(-\infty, \theta_\alpha(\boldsymbol{x})]$, for a real parameter θ, where $\theta_\alpha(\boldsymbol{x})$ denotes the $\alpha-$posterior quantile of θ given $\boldsymbol{x}$, i.e.,

$$\Pr(\theta \le \theta_\alpha(\boldsymbol{x}) \,|\, \boldsymbol{x}) = \int_{-\infty}^{\theta_\alpha(\boldsymbol{x})} \pi(\theta \,|\, \boldsymbol{x})\, d\theta = \alpha, \tag{5.1}$$

for $\alpha \in (0, 1)$. Now, consider $C(\boldsymbol{x}) \equiv (-\infty, \theta_\alpha(\boldsymbol{x})]$ to be a frequentist confidence set, with $\boldsymbol{x}$ random and arising from $p(\boldsymbol{x} \,|\, \theta)$, for given θ. The frequentist coverage probability of this confidence set is clearly

$$\Pr(\theta \le \theta_\alpha(\boldsymbol{x}) \,|\, \theta) = \int_{\{\boldsymbol{x}:\theta_\alpha(\boldsymbol{x}) \ge \theta\}} p(\boldsymbol{x} \,|\, \theta)\, d\boldsymbol{x}\,. \tag{5.2}$$

Note that, in (5.1), θ is random while $\boldsymbol{x}$ is fixed but, in (5.2), $\boldsymbol{x}$ is random while θ is fixed. The idea is now to study how close the frequentist coverage probability, $\Pr(\theta \le \theta_\alpha(\boldsymbol{x}) \,|\, \theta)$, is to α, the stated Bayesian coverage.

5.1.1 Exact Frequentist Matching

Definition 5.1 *If the frequentist coverage (5.2), of the one-sided Bayesian credible intervals $(-\infty, \theta_\alpha(\boldsymbol{x})]$, is exactly equal to the Bayesian credible probability α, then the Bayesian credible intervals are said to be* exact frequentist matching *(or 'exact matching' for short).*

Example 5.1 *Normal mean with variance known, continued.* Consider again Example 4.2. Under the Jeffreys prior, $\pi(\mu) \propto 1$, it follows from (4.6) that $\pi(\mu \mid \boldsymbol{x}) = \mathrm{N}(\mu \mid \overline{x}, \sigma^2/n)$. In this case, the posterior α-quantile of μ given $\boldsymbol{x}$ is $\mu_\alpha(\boldsymbol{x}) = \overline{x} + Z_\alpha \sigma/\sqrt{n}$, where Z_α is the α-quantile of N(0, 1). Algebra yields that the frequentist coverage is

$$\Pr(\mu \leq \mu_\alpha(\boldsymbol{x}) \mid \mu) = \Pr\left(\frac{\sqrt{n}(\overline{x}-\mu)}{\sigma} \geq -Z_\alpha \;\middle|\; \mu\right) = \Pr\left(\frac{\sqrt{n}(\overline{x}-\mu)}{\sigma} \leq Z_\alpha \;\middle|\; \mu\right) = \alpha,$$

$\forall \mu \in \mathbb{R}$ and $\alpha \in (0,1)$. Consequently, the constant prior for μ is an exact matching prior for μ.

Having exact matching credible sets is actually a fairly common phenomenon. Indeed, this occurs for a wide class of models that are 'invariant' to certain transformations; see Chapter 6 for development. Here is another interesting situation where many priors are exact matching.

Exact Matching for the Bivariate Normal Problem. The bivariate normal distribution of (x_1, x_2) has mean (μ_1, μ_2) and covariance matrix

$$\boldsymbol{\Sigma} = \begin{pmatrix} \sigma_1^2 & \rho\sigma_1\sigma_2 \\ \rho\sigma_1\sigma_2 & \sigma_2^2 \end{pmatrix},$$

where ρ is the correlation between x_1 and x_2. The sufficient statistics, for a sample $\boldsymbol{x} = ((x_{11}, x_{21}), (x_{12}, x_{22}), \ldots, (x_{1n}, x_{2n}))$, are $\overline{\boldsymbol{x}} = (\overline{x}_1, \overline{x}_2)'$, where $\overline{x}_i = n^{-1}\sum_{j=1}^n x_{ij}$, and

$$\boldsymbol{S} = \sum_{i=1}^n (\boldsymbol{x}_i - \overline{\boldsymbol{x}})(\boldsymbol{x}_i - \overline{\boldsymbol{x}})' \equiv \begin{pmatrix} s_{11} & r\sqrt{s_{11}s_{22}} \\ r\sqrt{s_{11}s_{22}} & s_{22} \end{pmatrix},$$

where $s_{ij} = \sum_{k=1}^n (x_{ik} - \overline{x}_i)(x_{jk} - \overline{x}_j)$ and $r = s_{12}/\sqrt{s_{11}s_{22}}$ is the sample correlation.

For the bivariate normal distribution, there are many potential parameters of interest, beyond just $(\mu_1, \mu_2, \sigma_1, \sigma_2, \rho)$; some of the commonly considered parameters are

$$\begin{aligned}
&\eta_1 = \frac{1}{\sigma_1},\ \eta_2 = \frac{1}{\sigma_2\sqrt{1-\rho^2}},\ \eta_3 = -\frac{\rho}{\sigma_1\sqrt{1-\rho^2}},\ \theta_1 = \frac{\rho\sigma_2}{\sigma_1},\ \theta_2 = \sigma_2^2(1-\rho^2),\\
&\theta_3 = |\boldsymbol{\Sigma}| = \sigma_1^2\sigma_2^2(1-\rho^2),\ \theta_4 = \frac{\sigma_2\sqrt{1-\rho^2}}{\sigma_1},\ \theta_5 = \frac{\mu_1}{\sigma_1},\ \theta_6 = \sigma_1^2\sigma_2^2,\ \theta_7 = \frac{\sigma_2}{\sigma_1},\ \theta_8 = \frac{\mu_2}{\sigma_2},\\
&\theta_9 = \sigma_{12} = \rho\sigma_1\sigma_2,\ \theta_{10} = \sigma_1^2 + \sigma_2^2 - 2\rho\sigma_1\sigma_2,\ \theta_{11} = \boldsymbol{d}'\boldsymbol{\Sigma}\boldsymbol{d}\ (\boldsymbol{d}' = (d_1, d_2) \neq c(0,1)),\\
&\lambda_1 = ch_{\max}(\boldsymbol{\Sigma}),\ \lambda_2 = ch_{\min}(\boldsymbol{\Sigma})\ \ \text{(maximum and minimum eigenvalues)}. \qquad (5.3)
\end{aligned}$$

Some of these parameters are studied next, from the perspective of which have exact matching priors.

An Interesting Class of Priors. For $a, b > 0$, the priors for $(\mu_1, \mu_2, \sigma_1, \sigma_2, \rho)$ of the form

$$\pi_{ab}(\mu_1, \mu_2, \sigma_1, \sigma_2, \rho) = \frac{1}{\sigma_1^{3-a}\sigma_2^{2-b}(1-\rho^2)^{2-b/2}} \tag{5.4}$$

have essentially a closed form posterior, and include the following important special cases:

- $\pi^r = \pi_{12}(\mu_1, \mu_2, \sigma_1, \sigma_2, \rho) = \sigma_1^{-2}(1-\rho^2)^{-1}$; this is called the right-Haar prior and is developed in Lemma 6.1.
- $\pi^J = \pi_{10}$, the Jeffreys-rule prior (also expressable as $|\boldsymbol{\Sigma}|^{-2}d\boldsymbol{\Sigma}$).
- $\pi^{IJ} = \pi_{21}$, the independence-Jeffreys prior (also expressable as $(|\boldsymbol{\Sigma}|^{-3/2}d\boldsymbol{\Sigma})$.

Many such priors are exact matching for various parameters of interest arising in the bivariate normal model. Table 5.1, taken from Berger & Sun (2008) (in which proofs can be found), gives many such results. Note that no listed priors are exact matching for all the parameters. (Almost certainly there is no prior that is exact matching for all the parameters.)

Also listed in the table are what are called constructive random posterior representations for many parameters of interest. See Section 1.8.6 for an introduction to constructive random posteriors, and note that (1.21) is actually the constructive random posterior for μ_1 that would result from using the right-Haar prior. Of particular interest is the following constructive random posterior for ρ that arises from use of the right-Haar prior above.

Constructive random posterior for ρ (from Berger & Sun (2008)): Drawing independent $Z \sim N(0,1)$, χ^2_{n-1} and χ^2_{n-2} random variables, a draw from $\pi^r(\rho \mid \boldsymbol{x})$, the marginal posterior distribution of ρ, given $\boldsymbol{x}$ under the right-Haar prior, is

$$\rho = \frac{Y}{\sqrt{1+Y^2}}, \text{ where } Y = -\frac{Z}{\sqrt{\chi^2_{n-1}}} + \frac{\sqrt{\chi^2_{n-2}}}{\sqrt{\chi^2_{n-1}}}\frac{r}{\sqrt{1-r^2}}. \tag{5.5}$$

Note that one only needs r, the sample correlation coefficient, to draw samples from $\pi^r(\rho \mid \boldsymbol{x})$. Having the ability to trivially generate inexpensive random samples makes inferences concerning ρ particularly simple. And note that, since this prior is exact frequentist matching, inferences such as credible sets derived from the Bayesian side will also be exact frequentist confidence sets.

Table 5.1: Parameters with exact matching priors of the form π_{ab} in (5.4), and associated constructive random posteriors. Here Z^* is a standard normal random variable, and χ^{2*}_{n-1} and χ^{2*}_{n-2} are chi-squared random variables with the indicated degrees of freedom, all random variables being independent.

parameter	exact matching prior	constructive posterior
μ_1	π_{1b}, $\forall\, b$ (including π^J and π^r)	$\overline{x}_1 + \frac{Z^*}{\sqrt{\chi^{2*}_{n-1}}}\sqrt{\frac{s_{11}}{n}}$
μ_2	$\pi^J = \pi_{10}$	$\overline{x}_2 + \frac{Z^*}{\sqrt{\chi^{2*}_{n-1}}}\sqrt{\frac{s_{22}}{n}}$
$\boldsymbol{d}(\mu_1, \mu_2)'$, $\boldsymbol{d} \in \mathbb{R}^2$	$\pi^J = \pi_{10}$ and π^{r^*} (see Table 10.1)	$\boldsymbol{d}(\overline{x}_1, \overline{x}_2)' + \frac{Z^*}{\sqrt{\chi^{2*}_{n-1}}}\sqrt{\frac{\boldsymbol{dSd}'}{n}}$
σ_1	π_{1b}, $\forall\, b$ (including π^J and π^r)	$\sqrt{\frac{s_{11}}{\chi^{2*}_{n-1}}}$
$\boldsymbol{d\Sigma d}'$	$\pi^J = \pi_{10}$ and π^{r^*} (see Table 10.1)	$\sqrt{\frac{\boldsymbol{dSd}'}{\chi^{2*}_{n-1}}}$
$\frac{\mu_1}{\sigma_1}$	π_{1b}, $\forall\, b$ (including π^J and π^r)	$\frac{Z^*}{\sqrt{n}} + \frac{\overline{x}_1\sqrt{\chi^{2*}_{n-1}}}{\sqrt{s_{11}}}$
ρ	$\pi^r = \pi_{12}$	$\psi\left(-\frac{Z^*}{\sqrt{\chi^{2*}_{n-1}}} + \frac{\sqrt{\chi^{2*}_{n-2}}}{\sqrt{\chi^{2*}_{n-1}}}\frac{r}{\sqrt{1-r^2}}\right)$ $\psi(y) = y/\sqrt{1+y^2}$
$-\frac{\rho}{\sigma_1\sqrt{1-\rho^2}}$	π_{a2}, $\forall\, a$ (including π^r)	$\frac{Z^*}{\sqrt{s_{11}}} - \frac{\sqrt{\chi^{2*}_{n-2}}}{\sqrt{s_{11}}}\frac{r}{\sqrt{1-r^2}}$
$\frac{\rho\sigma_2}{\sigma_1}$	π_{a2}, $\forall\, a$ (including π^r)	$\frac{r\sqrt{s_{22}}}{\sqrt{s_{11}}} - \frac{Z^*}{\sqrt{\chi^{2*}_{n-2}}}\frac{\sqrt{1-r^2}\sqrt{s_{22}}}{\sqrt{s_{11}}}$
$\sigma_2^2(1-\rho^2)$	π_{a2}, $\forall\, a$ (including π^r)	$\frac{s_{22}(1-r^2)}{\sqrt{\chi^{2*}_{n-2}}}$
$\lvert\boldsymbol{\Sigma}\rvert \equiv \sigma_1^2\sigma_2^2(1-\rho^2)$	$\pi^r = \pi_{12}$ and $\pi^{IJ} = \pi_{21}$	$\frac{\lvert\boldsymbol{S}\rvert}{\chi^{2*}_{n-1}\chi^{2*}_{n-2}}$

5.1.2 Asymptotic Frequentist Matching

Introduction. The rest of this chapter is quite technical and can be skipped by those whose primary interest is in methodology. Asymptotic posterior distributions are utilized in the derivation of matching priors and can be dated back to Laplace (1812). Let $\boldsymbol{x} = (x_1, \cdots, x_n)$ be a simple random sample of size n from the model $\mathcal{M} = \{p(x \mid \boldsymbol{\theta}),\, x \in \mathcal{X},\, \boldsymbol{\theta} \in \boldsymbol{\Theta}\}$, where $\boldsymbol{\Theta} \subset \mathbb{R}^m$. Johnson (1970) was among the first authors to rigorously investigate pointwise expansions of the posterior distribution of $\sqrt{n}(\boldsymbol{\theta} - \hat{\boldsymbol{\theta}}_n)$, where $\hat{\boldsymbol{\theta}}_n$ is the maximum likelihood estimator of $\boldsymbol{\theta}$ based on $\boldsymbol{x}$. Ghosh et al. (1982) and Bickel et al. (1985) gave a uniform variant of Johnson's result for a one-parameter case.

A natural application of posterior expansions is to find approximate Bayesian credible intervals for θ. Let $\text{Lik}_n(\boldsymbol{\theta})$ denote the likelihood function and $l_n(\boldsymbol{\theta}) = \log(\text{Lik}_n(\boldsymbol{\theta}))$. Efron (1985) proved the asymptotic normality of the posterior distribution of $\Lambda_n = \{2[l_n(\boldsymbol{\theta}) - l_n(\hat{\boldsymbol{\theta}}_n)]\}^{1/2}\text{sign}(\boldsymbol{\theta} - \hat{\boldsymbol{\theta}}_n)$, the signed squared root of the log-likelihood ratio statistic, and recommended the use of Λ_n, rather than $\sqrt{n}(\boldsymbol{\theta} - \hat{\boldsymbol{\theta}}_n)$, to find a Bayesian credible interval, especially when $k = 1$. The use of the signed squared root goes back to Lawley (1956), Woodroofe (1978) and Barndorff-Nielsen (1986), all from a frequentist point of view. Bickel & Ghosh (1990) extended the result in Efron (1985) to the multiparameter case, and pointed out that the normal approximation to Λ_n is correct to a higher order than the normal approximation to $\sqrt{n}(\boldsymbol{\theta} - \hat{\boldsymbol{\theta}}_n)$.

A posterior expansion of Λ_n can also be used to characterize priors leading to posterior credible invervals (or regions) with approximate frequentist validity. Generalizing the setup in Section 5.1, suppose $\eta = g(\boldsymbol{\theta})$ is a smooth function of $\boldsymbol{\theta}$, let $\eta_\alpha(\boldsymbol{x})$ denote the $\alpha-$posterior quantile of η given $\boldsymbol{x}$, i.e.,

$$\Pr(\eta \leq \eta_\alpha(\boldsymbol{x}) \,|\, \boldsymbol{x}) = \alpha, \tag{5.6}$$

for $\alpha \in (0,1)$. The frequentist coverage probability of the one-sided credible interval $(-\infty, \eta_\alpha(\boldsymbol{x})]$ is

$$\Pr(\eta \leq \eta_\alpha(\boldsymbol{x}) \,|\, \boldsymbol{\theta}). \tag{5.7}$$

Definition 5.2 Matching Prior.

(a) *A prior* $\pi(\boldsymbol{\theta})$ *is an* i-th order matching *prior for* $\eta = g(\boldsymbol{\theta})$ *if*

$$P(\eta \leq \eta_\alpha(\boldsymbol{x}) \,|\, \boldsymbol{\theta}) = \alpha + o(n^{-i/2}), \quad \forall \boldsymbol{\theta} \in \boldsymbol{\Theta} \text{ and } \alpha \in (0,1). \tag{5.8}$$

(b) *A prior* $\pi(\boldsymbol{\theta})$ *is* exactly matching *for* $\eta = g(\boldsymbol{\theta})$ *if*

$$P(\eta \leq \eta_\alpha(\boldsymbol{x}_n) \,|\, \boldsymbol{\theta}) = \alpha, \quad \forall \boldsymbol{\theta} \in \boldsymbol{\Theta} \text{ and } \alpha \in (0,1). \tag{5.9}$$

These are the usual definitions in the literature, but typically being $o(n^{-1/2})$ occurs because the rate is $O(n^{-1})$, and being $o(n^{-1})$ occurs because the rate is $O(n^{-3/2})$ (or even $O(n^{-2})$ for symmetric distributions).

Example 5.2 *Uniform data on* $(0,\theta)$*, continued.* Consider again Example 4.6. Let $\boldsymbol{x} = (x_1, \ldots, x_n)$ be a random sample of size n from the $\mathrm{Un}(x \,|\, 0, \theta)$ distribution. Then $t = \max\{x_1, \ldots, x_n\}$ is a sufficient statistic for θ, with sampling distribution being the inverse Pareto density

$$p(t \,|\, \theta) = \mathrm{Ip}(t \,|\, n, \theta^{-1}) = n\,\frac{t^{n-1}}{\theta^n}\,, \quad 0 < t < \theta\,. \tag{5.10}$$

Consider the class of priors $\pi_a(\theta) \propto \theta^{-a}$, $a \in \mathbb{R}$. The posterior of θ given $\boldsymbol{x}$ (or t) under π_a is

$$\pi_a(\theta \,|\, \boldsymbol{x}) = \pi_a(\theta \,|\, t) = \begin{cases} (n+a-1)\, t^{n+a-1}\theta^{-(n+a-1)}, & \text{if } \theta \geq t \\ 0 & \text{otherwise.} \end{cases}$$

It is easy to see that the posterior $\alpha-$quantile of θ is

$$\theta_\alpha(\boldsymbol{x}) = t(1-\alpha)^{-1/(n+a-1)}.$$

The frequentist coverage probability of the $\alpha-$credible interval $(0, \theta_\alpha(\boldsymbol{x})]$ is

$$\begin{aligned}\Pr(0 < \theta \le \theta_\alpha(\boldsymbol{x}) \,|\, \theta) &= \Pr\left((1-\alpha)^{1/(n+a-1)} \le t/\theta \,|\, \theta\right)\\ &= \int_{\theta(1-\alpha)^{1/(n+a-1)}}^{\theta} n t^{n-1}\theta^{-n} dt = 1-(1-\alpha)^{n/(n+a-1)},\end{aligned}$$

where we use the sampling distribution of t, given in (5.10). When $a = 1$, this equals α, for any $\theta > 0$ and $\alpha \in (0,1)$; thus $\pi(\theta) = 1/\theta$ is an exact matching prior for θ.

If $a \neq 1$, it follows from a Taylor expansion that

$$1-(1-\alpha)^{n/(n+a-1)} - \alpha = (1-\alpha)\Big[\frac{(1-a)\alpha}{n+a-1} + O(n^{-2})\Big],$$

so π_a is a 1st order matching prior, since $\sqrt{n}[1-(1-\alpha)^{n/(n+a-1)} - \alpha] \to 0$; but not a 2nd order matching prior, since

$$n[1-(1-\alpha)^{n/(n+a-1)} - \alpha] \to \alpha(1-\alpha)(1-a) \neq 0 \text{ as } n \to \infty.$$

As mentioned earlier, note that the actual rate of the first order matching prior is $O(n^{-1})$.

The Single Parameter Case. When there is a single parameter, Welch & Peers (1963) proved that a one-sided posterior credible interval arising from use of the Jeffreys-rule prior (proportional to the square root of the Fisher information $I(\theta)$), has the desired frequentist coverage probability up to $O(n^{-1})$. In fact, they showed that a prior π is a first order matching prior if, and only if, the differential equation

$$\frac{\partial}{\partial\theta}\left(\frac{\pi(\theta)}{\sqrt{I(\theta)}}\right) = 0 \tag{5.11}$$

is satisfied. Consequently, under the regularity conditions where the Fisher information exists, the Jeffreys-rule prior is the only first order matching prior.

The Multiparameter Case. Peers (1965) was among first authors to study matching priors for the two parameter case, where one is a parameter of interest and the other is a nuisance parameter. Stein (1985) extended the results in Welch & Peers (1963) and Peers (1965) and introduced a method to find a prior that is matching for a parameter to order $O(n^{-1})$, in the presence of nuisance parameters Tibshirani (1989) generalized this method by using a one-to-one transformation of the parameter vector into a parameter of interest and a nuisance parameter vector orthogonal in the sense of Cox & Reid (1987). Datta & Ghosh (1995c) derived a matching prior for a smooth function $\eta = \eta(\boldsymbol{\theta})$ for $\boldsymbol{\theta} = (\theta_1, \cdots, \theta_m)$, their result being stated in the following theorem.

Theorem 5.1 *Let $I(\boldsymbol{\theta})$ be the Fisher information matrix for one observation, and define the gradient vector $\nabla\eta = \{\partial\eta/\partial\theta_1, \cdots, \partial\eta/\partial\theta_m\}'$. Also, define*

$$\boldsymbol{\xi}(\boldsymbol{\theta}) \equiv (\xi_1, \cdots, \xi_m)' = \frac{I(\boldsymbol{\theta})^{-1}\nabla\eta}{\sqrt{(\nabla\eta)' I(\boldsymbol{\theta})^{-1}\nabla\eta}}.$$

Then a first order matching prior $\pi(\boldsymbol{\theta})$ for η must satisfy

$$\sum_{i=1}^{m} \frac{\partial}{\partial\theta_i}[\pi(\boldsymbol{\theta})\boldsymbol{\xi}_i(\boldsymbol{\theta})] = 0. \tag{5.12}$$

A matching prior is not unique in the multiparameter case. Indeed, the class of solutions of (5.12) is often quite large, growing with the dimension m. If the interest is in all the coordinates of $\boldsymbol{\theta}$, note that there are m such matching equations, and one might hope that there is then a simultaneous solution to these m equations. Sometimes this is so, as shown in the following example, but not always (cf. Datta & Mukerjee (2004)).

Example 5.3 *Inference on a normal mean with unknown variance.* Consider again Example 4.8 and define $s_{n-1}^2 = \sum(x_i - \overline{x})^2/(n-1)$. Under the class of priors $\pi_a(\mu, \sigma^2) \propto 1/\sigma^a$, it follows from (4.11) that $\pi(\mu \,|\, \boldsymbol{x}, a) = \text{St}(\mu \,|\, \overline{x},\, s_n/\sqrt{n+a-3}, n+a-3)$. In this case, the posterior α quantile of μ given $\boldsymbol{x}$ is

$$\mu_\alpha(\boldsymbol{x}) = \overline{x} + t_\alpha(n+a-3)\frac{s_n}{\sqrt{n+a-3}},$$

where $t_\alpha(n+a-3)$ is the α-quantile of $\text{St}(\mu \,|\, 0, 1, n+a-3)$. Clearly,

$$\begin{aligned}
\Pr(\mu \le \mu_\alpha(\boldsymbol{x}) \,|\, \mu, \sigma^2) &= P\left(\frac{\overline{x}-\mu}{s_n} \ge -\frac{t_\alpha(n+a-3)}{\sqrt{n+a-3}} \,\Big|\, \mu, \sigma^2\right) \\
&= \Pr\left(\frac{\overline{x}-\mu}{s_n/\sqrt{n-1}} \ge -\frac{\sqrt{n-1}}{\sqrt{n+a-3}}\, t_\alpha(n+a-3) \,\Big|\, \mu, \sigma^2\right). \\
&= \Pr\left(\frac{\overline{x}-\mu}{s_n/\sqrt{n-1}} \le \frac{\sqrt{n-1}}{\sqrt{n+a-3}}\, t_\alpha(n+a-3) \,\Big|\, \mu, \sigma^2\right).
\end{aligned} \tag{5.13}$$

Since

$$t(\boldsymbol{x} \,|\, \mu, \sigma)) = \frac{\overline{x}-\mu}{s_n/\sqrt{n-1}} = \frac{\sqrt{n}\,(\overline{x}-\mu)}{s_{n-1}}$$

has sampling distribution $p(t \,|\, \mu, \sigma) = \text{St}(t \,|\, 0, 1, n-1)$, the right hand side of (5.13) equals α, for any $\mu \in \mathbb{R}$, $\sigma > 0$ and $\alpha \in (0, 1)$, if and only if $a = 2$. Consequently, the independence Jeffreys prior for (μ, σ^2) is an exact matching prior for μ, but the Jeffreys-rule prior is *not* exact matching.

It follows from (4.10) that the marginal posterior of σ^2 given $\boldsymbol{x}$, under the prior $\pi_a(\mu, \sigma^2) \propto 1/\sigma^a$, is

$$\pi_a(\sigma^2 \,|\, \boldsymbol{x}, a) \propto (\sigma^2)^{-[1+(n+a-3)/2]} \exp\left\{-\frac{n}{2\sigma^2}s_n^2\right\},$$

or $\sigma^2 \mid \boldsymbol{x} \sim \mathrm{Ig}((n+a-3)/2, ns_n^2/2)$. The posterior α-quantile of σ^2 given $\boldsymbol{x}$ can be shown to be

$$\sigma_\alpha^2(\boldsymbol{x}) = \frac{ns_n^2}{\chi_\alpha^2(n+a-3)},$$

where $\chi_\alpha^2(n+a-3)$ is the upper α critical value of the χ^2 distribution with $n+a-3$ degrees of freedom. Algebra yields

$$\Pr(\sigma^2 \le \sigma_\alpha^2(\boldsymbol{x}) \mid \mu, \sigma^2) = P\left(\frac{ns_n^2}{\sigma^2} \ge \chi_\alpha^2(n+a-3) \mid \mu, \sigma^2\right). \quad (5.14)$$

Since the sampling distribution of ns_n^2/σ^2 is $\chi^2(n-1)$, the right hand side of (5.14) equals α for any $\mu \in \mathbb{R}$, $\sigma > 0$ and $\alpha \in (0,1)$, if and only if $a = 2$. Here, again, the independent Jeffreys prior is an exact matching prior for σ^2, but the Jeffreys-rule prior is not.

Next, we use Theorem 5.1 to determine which priors would be first order matching for μ and/or σ^2. Recall that the Fisher information matrix of (μ, σ^2) is given in (4.8). To see which priors are matching priors for μ, let $\eta = \mu$ so that $\nabla\eta = (1,0)'$, and $\boldsymbol{\xi} \equiv (\sigma, 0)'$. Then (5.12) becomes

$$\sigma \frac{\partial \pi(\mu, \sigma^2)}{\partial \mu} = 0.$$

Any prior which is constant in μ is clearly a matching prior for μ, so both the Jeffreys-rule and independent Jeffreys priors are first order matching priors for μ.

When $\eta = \sigma^2$, $\nabla\eta = (0,1)'$, and $\boldsymbol{\xi} \equiv (0, \sqrt{2}\sigma^2)'$. Thus (5.12) becomes

$$\sqrt{2}\, \frac{\partial \sigma^2 \pi(\mu, \sigma^2)}{\partial \sigma^2} = 0.$$

Clearly the independent Jeffreys prior for (μ, σ^2) is a first order matching prior for σ^2, but the Jeffreys-rule prior ($a = 3$) is not. *Thus, contrary to the one-parameter case (under regularity), the Jeffreys-rule prior need not be first order matching in multiparameter problems.*

Mukerjee & Dey (1993) showed that a prior $\pi(\theta_1, \theta_2)$ is a second order matching prior for θ_1 with a nuisance parameter θ_2, if it satisfies (5.12) and an additional differential equation. Unfortunately, a second order matching prior may not exist. A comprehensive study of higher order matching priors can be found in Datta & Mukerjee (2004). See also Fraser & Reid (2002).

5.2 Confidence Distributions

The concept of 'confidence' was solidified by Neyman (1934, 1937), in his seminal papers on confidence intervals, clarifying the frequentist repetition properties of confidence. The idea has many antecedents in the literature (see Fraser (2011) for discussion). According to David and Edwards (2001, p. 191), the terminology 'confidence distribution' was first used in 1927 in a letter from E. S. Pearson to W. S. Gossett. Cox (1958) suggested that a confidence distribution "can either be defined directly, or can be introduced in terms of the set of all confidence intervals at different levels of probability."

In this section, only a one-dimensional θ will be considered. For data $\boldsymbol{x} \in \boldsymbol{\mathcal{X}}$ and $\alpha \in (0,1)$, let $\theta_\alpha(\boldsymbol{x})$ be the upper endpoint of a $100\alpha\%$ one-sided (frequentist) confidence interval for θ, *i.e.*,

$$\Pr(\theta_\alpha(\boldsymbol{x}) \geq \theta \,|\, \theta) = \alpha. \tag{5.15}$$

Here Pr is computed according to the sampling distribution of $\boldsymbol{x}$ given θ. Note that, in a Bayesian context, if $\theta_\alpha(\boldsymbol{x})$ is the posterior α-quantile, the left hand side of (5.15) is the frequentist coverage probability of the one-sided Bayesian credible interval (see (5.7)). Cox's (1958) formal definition of a confidence distribution follows.

Definition 5.3 Confidence distribution. *Suppose $\theta_\alpha(\boldsymbol{x})$, the upper endpoint of a $100\alpha\%$ one-sided confidence interval for θ, is monotonically increasing in $\alpha \in (0,1)$. Then $\alpha_{\boldsymbol{x}}(\theta)$, the inverse function of $\theta_\alpha(\boldsymbol{x})$* (i.e., *the value of α for which the specified θ is the upper endpoint of the confidence interval), is a* confidence distribution *for θ. The density*

$$\pi_x(\theta) = \frac{d}{d\theta}\,\alpha_{\boldsymbol{x}}(\theta) \tag{5.16}$$

is called the confidence density.

Example 5.4 *Confidence distributions for normal data.* Consider a random sample $\boldsymbol{x} = (x_1 \ldots, x_n)$ of size n from a normal distribution $\mathrm{N}(x \,|\, \mu, \sigma^2)$. When σ^2 is known, $\overline{x} \sim \mathrm{N}(\overline{x} \,|\, \mu, \sigma^2/n)$. A $100\,\alpha\%$ one-sided confidence interval for μ is $(-\infty, \overline{x} - Z_\alpha \sigma/\sqrt{n}]$, where Z_α is the upper α critical value of $\mathrm{N}(z \,|\, 0,1)$ (i.e., the $(1-\alpha)$-quantile of $\mathrm{N}(z \,|\, 0,1)$). So, if a specified value of μ is the upper confidence limit, then the corresponding α is that corresponding to solving for $-Z_\alpha$ (since $\alpha = \Phi(-Z_\alpha)$). Indeed, from $\mu = \overline{x} - Z_\alpha\sigma/\sqrt{n}$, we get $-Z_\alpha = \sqrt{n}(\mu - \overline{x})/\sigma$, and so the confidence distribution for μ is

$$\alpha_{\boldsymbol{x}}(\mu) = \Phi(\sqrt{n}(\mu - \overline{x})/\sigma)\,, \tag{5.17}$$

which is the cdf of $\mathrm{N}(\mu \,|\, \overline{x}, \sigma^2/n)$. The corresponding confidence density is, of course the specified normal density.

Note that a confidence distribution is typically not unique. In fact, since $\sqrt{n}(\overline{x} - \mu)/s_{n-1} \sim \mathrm{St}(\cdot \,|\, 0, 1, n-1)$, another $100\,\alpha\%$ one-sided confidence interval for μ, when σ^2 is known, is $(-\infty, \overline{x} - t_\alpha(n-1)s_{n-1}/\sqrt{n})$, where

$t_\alpha(n-1)$ is the upper α critical value of $\text{St}(\cdot\,|\,0,1,n-1)$. Using the same argument as above, it follows that another confidence distribution for μ is

$$\alpha_{\boldsymbol{x}}(\mu) = F_{t_{n-1}}(\sqrt{n}(\mu - \overline{x})/s_{n-1}), \tag{5.18}$$

where $F_{t_{n-1}}(\cdot)$ is the cumulative distribution function of $\text{St}(\cdot\,|\,0,1,n-1)$.

When σ^2 is unknown, $\alpha_{\boldsymbol{x}}(\mu)$ in (5.17) is no longer a confidence distribution for μ, since it contains the unknown parameter σ, but $\alpha_{\boldsymbol{x}}(\mu)$ in (5.18) is still a confidence distribution for μ.

It is tempting to interpret a confidence distribution (or density) as a posterior distribution (or density). For instance, given a confidence density it is tempting to, say, pick an interval $\theta \in [a, b]$, find the area under the density over that interval and state that this is the frequentist confidence level for that interval (analogous to the area under a posterior being the posterior probability of the interval).

This is not generally legitimate, however; frequentists work with procedures that are specified for all possible data, not observed data-dependent entities such as confidence distributions. So a frequentist needs to specify (in advance) how they would deal with the confidence distribution that would arise from any possible value of the data for a model. For instance, specifying "for every possible confidence distribution that could arise from the model, I will choose an equal-tailed 95% confidence interval" does specify a proper frequentist procedure, and the resulting inference is a valid frequentist inference.

On the other hand, waiting until one sees the confidence distribution for the observed data and choosing the type of confidence interval or confidence level based on intuition developed from this observed confidence distribution, would not be a valid frequentist procedure. In contrast, Bayesians can operate with the observed posterior distribution in any way they wish.

Example 5.5 Suppose two possible confidence densities (corresponding to different possible data) are those given in Figure 5.1. If these were posterior densities, a Bayesian might choose to report the one-sided 90% credible interval $(0, 0.013)$ for the posterior density in the left figure, and the equal-tailed 90% credible interval $(0.458, 0.502)$ for the posterior density in the right figure. A frequentist could not do so with confidence densities, however. It would be legitimate to always report one-sided intervals, or to always report equal-tailed intervals but choosing between the two, based on the data, would not be allowed. (One could always report both the one-sided and the equal-tailed intervals, but this starts to get complicated when one realizes all the intervals of possible interest to report.) A key advantage of objective Bayesian analysis here is thus the ability to tailor the report to the shape of the encountered posterior, in such a way as to maximize the communication of information.

One might think that the frequentist could adapt to this possible situation by, say, always taking the highest density 90% interval (*i.e.*, the

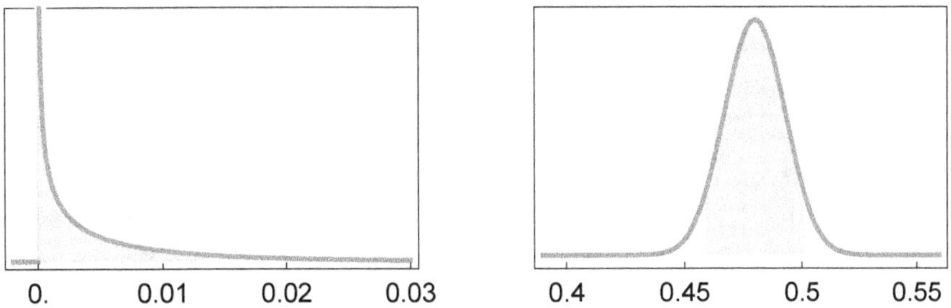

Figure 5.1: Two densities on $[0, 1]$ — either confidence densities or posterior densities — arising from different data. The shaded regions are 90% posterior credible intervals.

interval having area 0.9 under the density, formed from the points with highest density). This is a procedure that is well-defined in advance and so can be evaluated from a frequentist perspective, but the procedure does not necessarily have exact frequentist coverage probability of 0.9; only intervals defined in terms of pre-specified quantiles have guaranteed coverage of 0.9.

Here is a more general definition of a confidence distribution, following Schweder & Hjort (2002) and Singh *et al.* (2005, 2007).

Definition 5.4 *A function* $H(\boldsymbol{x}, \theta)$ *on* $\mathcal{X} \times \Theta \to [0, 1]$ *is called a* confidence distribution (CD) *for a parameter* θ*, if the following hold.*

(i) *For each given* $\boldsymbol{x} \in \mathcal{X}$, $H(\boldsymbol{x}, \theta)$ *is a cumulative distribution function on* Θ.

(ii) *For each fixed value of* θ, $H(\boldsymbol{x}, \theta)$, *as a function of the sample* $\boldsymbol{x}$, *has a uniform distribution on* $(0, 1)$.

A confidence distribution so defined can be shown to satisfy the frequentist properties discussed earlier. For an extensive discussion of the history, theory and practice of confidence distributions, see Schweder & Hjort (2016).

5.3 Fiducial Inference

5.3.1 Traditional Fiducial Inference

Fiducial inference was introduced in Fisher (1930) for the problem of inference concerning the correlation ρ from a bivariate normal distribution. The paper produces a fiducial distribution for ρ. We do not present the derivation here, because this is equivalent to the derivation of a confidence distribution, extensively discussed in the previous section. Of note is that Fisher's reasoning in this paper was effectively the same reasoning as with confidence distributions, based on a desire to be able to produce frequentist confidence procedures. Fisher showed that, if θ is one dimensional, t is a sufficient statistic for θ, and the cdf of t, $F(t \mid \theta)$, is monotonically decreasing in θ, then the fiducial density is

$$\pi^F(\theta) = -\frac{\partial}{\partial\theta} F(t \mid \theta) .$$

This is essentially equivalent to the definition of the confidence density in (5.16).

When dealing with more than one parameter, Fisher took a different path.

Example 5.6 *Normal fiducial distributions.* In Fisher (1935, 1939), inference in the normal model with unknown mean and variance was considered from a fiducial perspective. The sufficient statistics $\overline{x}$ and s_{n-1}^2 satisfy

$$\frac{\sqrt{n}\,(\overline{x} - \mu)}{s_{n-1}} = t_{n-1}, \qquad \frac{(n-1)\, s_{n-1}^2}{\sigma^2} = \chi_{n-1}^2 ,$$

where t_{n-1} is a random variable having a standard t-distribution with $n-1$ degrees of freedom and χ_{n-1}^2 is a random variable having a chi-square distribution with $n-1$ degrees of freedom. Naïve 'algebra' then yields

$$\mu = \overline{x} - \frac{s_{n-1}\, t_{n-1}}{\sqrt{n}}, \quad \sigma^2 = \frac{(n-1)\, s_{n-1}^2}{\chi_{n-1}^2} .$$

Interpreting these last equations literally implies that μ has a Student $\text{St}(\mu \mid \overline{x}, s_{n-1}/\sqrt{n}, n-1)$ distribution and that σ^2 has an inverse Chi-squared distribution $\chi^{-1}(\sigma^2 \mid s_{n-1}^2, n-1)$ (this last following from the observation that the inverse of a chi-square random variable has a χ^{-1} distribution).

While the results here are certainly the correct 'objective' answers (they are constructive posterior representations arising from use of the independence Jeffreys prior, and they are also a representation of the confidence distributions discussed in Example 5.3), the logic behind this derivation can be questioned. In particular, 'algebra' was put in quotations above because one cannot use simple algebra like this to shift randomness from $\overline{x}$ and s_{n-1}^2 to μ and σ^2; that is not how probability theory works. Indeed, Lindley (1958) showed (for scalar θ) that one could only justify the fiducial distribution through probability theory (*i.e.*, find a prior distribution which yields the fiducial distribution as a posterior) if the model could be transformed to a location model.

While Fisher was promoting this approach to fiducial inference, he was also coming into increasing conflict with Neyman's strict frequentist approach to statistics, and by the 1950's Fisher (1973) was no longer emphasizing the frequentist connection. Indeed, Fisher seemed closer to viewing the fiducial distribution as being like a posterior, but with no need to introduce a 'questionable' prior. See Zabell (1992) for a discussion of Fisher's evolving view of fiducial inference.

This approach to fiducial inference came under rather severe criticism, in addition to questioning the probability transference. The primary issue was the non-uniqueness of fiducial distributions in multi-parameter problems (discussed in the next section).

There were various subsequent efforts to 'fix' fiducial inference. *Pivotal inference* Barnard (1980) was one such approach, and *structural inference* Fraser (1979, 2004) was another; the latter was focused on models which have a 'group invariance structure' (see Chapter 6). While successful in eliminating most of the problems with fiducial inference, both approaches were limited in applicability, although they did cover most of the standard statistical models.

Fiducial and Bayesian Inference for the Bivariate Normal Problem. The bivariate normal distribution has a long history of being studied from different objective Bayesian and fiducial approaches. For μ_1, μ_2, σ_1, and σ_2, the exact frequentist matching posteriors given in Table 5.1 are also the fiducial distributions of the parameters, as derived in Fisher (1973). As mentioned earlier, the fiducial distribution for ρ was found in Fisher (1930), his first example of the fiducial argument. This is also the exact matching posterior in (5.5), arising from the right-Haar prior.

There is a long history of trying to match fiducial distributions with objective Bayesian posteriors, as above. The only (known) prior that simultaneously gives the four exact matching posterior/fiducial distributions for means and variances is the Jeffreys-rule prior. This prior is not exact matching for the correlation ρ, however, nor is it even first order matching. Indeed, it was shown in Berger & Sun (2008) that the Jeffreys-rule prior has quite poor frequentist matching properties for ρ in finite samples.

The attempts to find a posterior distribution for ρ that is also the fiducial distribution of ρ have an interesting history. Geisser & Cornfield (1963) studied the question of whether the fiducial distribution of ρ could be reproduced as an objective Bayesian posterior, and they concluded that this is most likely not possible. The strongest support for this view arose from Brillinger (1963), which used results from Lindley (1958) and a difficult analytic argument to show that there does not exist a prior $\pi(\rho)$ such that the fiducial density of ρ, $f(\rho \,|\, r)$ (which happens to depend on the data only through the sample correlation r), could be reproduced by a direct application of Bayes theorem with the density $p(r \mid \rho)$ of r. In other words, Brillinger proved that there is no prior $\pi(\rho)$ for which

$$f(\rho \,|\, r) \propto p(r \mid \rho)\pi(\rho).$$

Of course, we now know that the right-Haar prior results in a marginal distribution of ρ that is equal to the fiducial distribution, so the speculation that there was no compatible prior was incorrect; but it has to be a joint prior. That there is no $\pi(\rho)$ satisfying (5.19) is also of historical interest, however, because of what is called the marginalization paradox, discussed in Section 7.5.2. Indeed, we will return to this example therein.

5.3.2 Generalized Fiducial Inference

Recently, a revival of fiducial inference has been under way, under the name given to this section. The approach starts with a *structural equation* of the form $\boldsymbol{x} = G(\boldsymbol{\theta}, \boldsymbol{u})$, where $\boldsymbol{u}$ is a random variable having a completely known distribution. Any $p(\boldsymbol{x} \mid \boldsymbol{\theta})$ can be represented this way, although the representation is not necessarily unique; thus specifying a structural equation often means specifying more than just a probability density. One then considers

$$Q(\boldsymbol{x}, \boldsymbol{u}) = \{\boldsymbol{\theta} : \boldsymbol{x} = G(\boldsymbol{\theta}, \boldsymbol{u})\}.$$

This is the set of parameter values compatible with a given $\boldsymbol{x}$ and $\boldsymbol{u}$. The set can be empty and it can contain more than one value, in which case certain steps have to be taken (*cf.* Hannig, 2009), but we will focus on the common situation where $Q(\boldsymbol{x}, \boldsymbol{u})$ contains just one element, *i.e.*, it is essentially the inverse function of G. Then $\boldsymbol{\theta}$ can be identified with $Q(\boldsymbol{x}, \boldsymbol{u})$ and is assigned the distribution induced by $\boldsymbol{u}$; this is the generalized fiducial distribution.

Example 5.7 Consider again Example 5.6. A structural equation for the sufficient statistics $(\overline{x}, s_{n-1}^2)$ is

$$(\overline{x}, s_{n-1}^2) = G((\mu, \sigma^2), (z, \chi_{n-1}^2)) \equiv \left(\mu + \frac{\sigma}{\sqrt{n}}\, z,\; \frac{\sigma^2\, \chi_{n-1}^2}{(n-1)}\right),$$

with z being standard normal and χ_{n-1}^2 being chi-squared with $n-1$ degrees of freedom. (Thus $\boldsymbol{u} = (z, \chi_{n-1}^2)$.) Here the inverse function returns single values of the parameters, first

$$\sigma^2 = \frac{(n-1)\, s_{n-1}^2}{\chi_{n-1}^2},$$

and then

$$\mu = \overline{x} - \frac{\sigma}{\sqrt{n}}\, z = \overline{x} - \frac{s_{n-1}}{\sqrt{n}} \frac{\sqrt{n-1}\; z}{\sqrt{\chi_{n-1}^2}} = \overline{x} - \frac{s_{n-1}}{\sqrt{n}}\, t_{n-1},$$

so that

$$Q\{(\overline{x}, s_{n-1}^2), (z, \chi_{n-1}^2)\} = \left\{\left(\overline{x} - \frac{s_{n-1}}{\sqrt{n}} \frac{\sqrt{n-1}\; z}{\sqrt{\chi_{n-1}^2}},\; \frac{(n-1)\, s_{n-1}^2}{\chi_{n-1}^2}\right)\right\}$$

and $\boldsymbol{\theta} = (\mu, \sigma^2)$ is assigned the distribution implied by $\boldsymbol{u} = (z, \chi_{n-1}^2)$. This is, of course, the same inversion used in Example 5.6.

There are several sources of non-uniqueness in generalized fiducial theory. One has already been mentioned: there are often different structural models that can be used to represent the same $p(\boldsymbol{x} \mid \boldsymbol{\theta})$, and the different structural models can result in different generalized fiducial distributions. Sometimes one knows the way in which the data was generated, *i.e.*, one knows G, and then all is fine. If not, one just has to pick a G that seems reasonable, and note that others might get different answers with different choices of G.

A second source of non-uniqueness is more subtle. Note that the equation $\boldsymbol{x} = G(\boldsymbol{\theta}, \boldsymbol{u})$ is satisfied for the $\boldsymbol{\theta}$ and $\boldsymbol{u}$ used to generate $\boldsymbol{x}$. In other words, for this particular $\boldsymbol{u}$, the function $Q(\boldsymbol{x}, \boldsymbol{u}) \neq \emptyset$, the empty set. Thus we not only know the distribution of $\boldsymbol{u}$ (and hence of $Q(\boldsymbol{x}, \boldsymbol{u})$, assuming it contains only a single element), but we also know that the event $\{Q(\boldsymbol{x}, \boldsymbol{u}) \neq \emptyset\}$ has happened, and it is necessary to condition on this event to determine the fiducial distribution. But, if $\{Q(\boldsymbol{x}, \boldsymbol{u}) \neq \emptyset\}$ has probability 0, one is in a situation where Borel's paradox can arise, and there may well be different ways to implement the conditioning yielding different answers. Indeed, Example 10 in Hannig (2009) shows how this can result in a different (and considerably inferior) answer in the situation of Example 5.7. Generalized fiducial theory tackles this head on by proposing a choice of the limiting operation that leads to a particular type of conditioning on $\{Q(\boldsymbol{x}, \boldsymbol{u}) \neq \emptyset\}$; see Cui & Hannig (2019) for discussion.

Among the interesting results that have appeared in the generalized fiducial literature is a representation of the fiducial distribution as a posterior distribution corresponding to a data-dependent prior (*cf.* Hannig (2009)). There are also many results in the literature showing that generalized fiducial distributions usually have excellent frequentist properties, with performance comparable to the performance from optimal objective priors.

In conclusion, note that generalized fiducial theory has not really eliminated the problems that were raised concerning Fisher's fiducial inference; rather, the theory accepts that the problems are there, recommends what to do in practice to deal with them, and studies the behavior of the resulting methodology. This change in focus from 'the fiducial distribution is *the* correct prior-free distribution for $\boldsymbol{\theta}$' to 'here is an interesting methodology that seems to produce very promising procedures' is one of the main advances of generalized fiducial inference.

Closely related to generalized fiducial inference is the *Inferential Models* approach to inference, as developed in Martin & Liu (2015). To see the relationship, and also its relationship to confidence distributions, see Cui & Hannig (2022).

5.4 Discussion and Additional References

There is a huge literature concerning 'matching priors,' 'confidence distributions,' and 'fiducial inference.' An important subset of this literature is given in the bibliography; just search for the key words 'matching,' 'confidence,' or 'fiducial.'

Chapter 6

Invariance Priors

6.1 Introduction

The big advance of Jeffreys-rule priors was in providing inferences that were consistent no matter which parameterization was chosen for the model; this is often called 'invariance to parameterization.' Here we discuss an additional type of invariance, namely invariance to other transformations (of both the data and the parameters) that seem to leave the problem unchanged. There are many illustrations of this in the literature (Jeffreys 1961, Hartigan 1964, Fraser 1968, 1979, Villegas 1977*a*,*b*, 1981, Jaynes 2003), but the most extensively studied and reliable invariance theory is 'invariance to a group operation' on the model; here is an illustration.

Example 6.1 Consider a *scale parameter* density

$$p(x \mid \theta) = \frac{1}{\theta} f\left(\frac{x}{\theta}\right), \quad x > 0, \quad \theta > 0.$$

- Consider a change of scale for the problem (*e.g.*, measure in terms of meters, rather than feet) to $x^* = c\,x > 0$ and $\theta^* = c\,\theta > 0$.
- The density of x^* can be shown to be $p(x^* \mid \theta^*) = (1/\theta^*) f(x^*/\theta^*)$, which is of exactly the same mathematical form as the original density. Another density of exactly the same mathematical form is $(1/\sigma) f(y/\sigma)$.
- Objective priors are determined by the mathematical form of the density so it would seem that the objective prior for θ should be the same function as that for the objective prior for θ^* (or σ).

This logic will be seen to lead (when combined with 'invariance to parameterization') to what are called 'invariance priors.'

The development of invariance priors is achieved through the mathematical tools of *group theory*. In the above example, the transformation 'multiplication by positive constants' is called the *scale group* or *multiplication group.* We do not delve deeply into group theory in this chapter and, hence, only present invariance priors in simple cases (although we will mention the results of application of invariance theory in more complicated scenarios). More advanced examples are given in Section 14.3.2.

Curiously, the development of invariance priors initially mirrored the success and failure of the Jeffreys-rule prior. For one-parameter problems having a group invariance structure, the invariance prior is the Jeffreys-rule prior (also the reference prior and first order matching prior), so all is well. In multi-parameter problems, however, the natural invariance prior, which will be seen to be what is called the *left-Haar prior* [1], is problematical, as is the Jeffreys-rule prior in multi-parameter problems.

Unlike the situation with the Jeffreys-rule prior, however, a 'fix' was developed for the invariance prior in multi-parameter problems; instead of using the left-Haar prior, use what is called the *right-Haar prior.* This development was actually a frequentist development, finding the prior distribution in invariant problems that is exact frequentist matching. And the motivation was not at all Bayesian, but was computational, allowing the determination of, say, optimal frequentist confidence sets through much easier Bayesian computations than would be the direct frequentist computations. In any case, right-Haar priors can lay claim to be serious contenders for optimal objective priors in invariant problems.

6.2 Groups of Transformations

As in Example 6.1, the important concept will be that of transformations of the data model. In particular, consider transformations, to be denoted by g, of $\boldsymbol{\mathcal{X}}$ into itself. We will be concerned only with transformations that are one-to-one and onto. (A transformation g is one-to-one if $g(\boldsymbol{x}_1) = g(\boldsymbol{x}_2) \Rightarrow \boldsymbol{x}_1 = \boldsymbol{x}_2$, and it is onto if the range of g is all of $\boldsymbol{\mathcal{X}}$.) Attention will also be restricted to probability distributions that are continuous and live in $\mathbb{R}^p$ for dimension p.

If g and g^* are two transformations, it is necessary to consider the composition of g and g^*, which is the transformation, to be denoted $g^* \circ g$, defined by

$$g^* \circ g(\boldsymbol{x}) = g^*(g(\boldsymbol{x}))\,.$$

Definition 6.1 *A* group of transformations *of $\boldsymbol{\mathcal{X}}$, to be denoted $\mathcal{G}$, is a set of one-to-one and onto transformations of $\boldsymbol{\mathcal{X}}$ into itself, which satisfies the following two conditions:*

(i) *If $g \in \mathcal{G}$ and $g^* \in \mathcal{G}$, then $g^* \circ g \in \mathcal{G}$.*

(ii) *If $g \in \mathcal{G}$, then g^{-1}, the inverse transformation defined by the relation $g^{-1} \circ g(\boldsymbol{x}) = g^{-1}(g(\boldsymbol{x})) = \boldsymbol{x}$, is in $\mathcal{G}$.*

[1]The Haar measures, from which the Haar priors are derived, were introduced in the 1930's by the Hungarian mathematician Alfred Haar (*cf.* Szökefalvi-Nagy (1985)).

Define $e = g^{-1} \circ g$ *(for any* g*); this is the* identity transformation, *in that* $e(\boldsymbol{x}) = \boldsymbol{x}$.

For simplicity, we will restrict attention to groups that can be represented as sets in $\mathbb{R}^p$ for some dimension p.

Example 6.2 In Example 6.1, the group transformation is $g_c(x) = cx$, and $\mathcal{G} = \{g_c : c > 0\}$. Clearly the functions g_c are one-to-one and onto. Note that

$$g_{c^*} \circ g_c(\boldsymbol{x}) = g_{c^*}(g_c(\boldsymbol{x})) = g_{c^*}(c\,\boldsymbol{x}) = c^*\,c\,\boldsymbol{x} = g_{c^*c}(\boldsymbol{x}),$$

so that

$$g_{c^*} \circ g_c = g_{c^*c} \in \mathcal{G}\,.$$

It is also easy to see that $g_c^{-1} = g_{c^{-1}}$ and $e = g_1 \in \mathcal{G}$. Thus $\mathcal{G}$ is, indeed, a group of transformations of the model.

Example 6.3 Suppose the model is of the form $p(x \mid \mu) = f(x - \mu)$, where $x \in \mathcal{X} = \mathbb{R}^1$. Consider the transformation $g_c(x) = x + c$, with $\mathcal{G} = \{g_c : c \in \mathbb{R}^1\}$. Clearly the functions g_c are one-to-one and onto,

$$g_{c^*} \circ g_c(x) = x + c^* + c = g_{c^*+c}(x) \in \mathcal{G}\,,$$

$g_c^{-1} = g_{-c} \in \mathcal{G}$ and $e = g_0 \in \mathcal{G}$. Thus $\mathcal{G}$ is, indeed, a group of transformations of the model, here called the *location group of transformations.*

Example 6.4 Suppose the model is $p(x \mid \mu, \sigma) = (1/\sigma)f((x-\mu)/\sigma)$, where $x \in \mathcal{X} = \mathbb{R}^1$. Consider the transformations $g_{b,c}(x) = cx + b$, with $\mathcal{G} = \{g_{b,c} : b \in \mathbb{R}^1, c > 0\}$. Clearly the functions $g_{b,c}$ are one-to-one and onto and

$$\begin{aligned} g_{b^*,c^*} \circ g_{b,c}(x) &= g_{b^*,c^*}(cx + b) = c^*[cx + b] + b^* = c^*cx + [c^*b + b^*] \\ &= g_{[c^*b+b^*],c^*c}(x) \in \mathcal{G}\,. \end{aligned}$$

It is also easy to see that $g_{b,c}^{-1} = g_{-b/c,1/c} \in \mathcal{G}$ and $e = g_{0,1} \in \mathcal{G}$. Thus $\mathcal{G}$ is, indeed, a group of transformations of the model, here called the *location-scale group of transformations.*

6.3 Invariant Models

A statistical model is $\mathcal{M} = \{p(x \mid \boldsymbol{\theta}),\, x \in \mathcal{X},\, \boldsymbol{\theta} \in \boldsymbol{\Theta}\}$, where $\boldsymbol{\Theta} \subset \mathbb{R}^m$.

Definition 6.2 *The family of densities* $\mathcal{M}$ *is said to be* invariant under the group $\mathcal{G}$ *if, for every* $g \in \mathcal{G}$ *and* $\boldsymbol{\theta} \in \boldsymbol{\Theta}$*, there exists a unique* $\boldsymbol{\theta}^* \in \boldsymbol{\Theta}$ *such that* $\boldsymbol{x}^* = g(\boldsymbol{x})$ *has density* $p(\boldsymbol{x}^* \mid \boldsymbol{\theta}^*)$*. Then* $\boldsymbol{\theta}^*$ *will be denoted* $\bar{g}(\boldsymbol{\theta})$.

For a given g, the transformation $\boldsymbol{\theta} \to \bar{g}(\boldsymbol{\theta})$ is a transformation on $\boldsymbol{\Theta}$. It is easy to show that, if $\mathcal{M}$ is invariant under $\mathcal{G}$, then

$$\bar{\mathcal{G}} = \{\bar{g} : g \in \mathcal{G}\}$$

is a group of transformations on $\boldsymbol{\Theta}$. It is this group of transformations on $\boldsymbol{\Theta}$ that will be used to develop an objective prior for $\boldsymbol{\theta}$.

Example 6.5 In Example 6.2, the group transformation $g_c(x) = c\,x$ yields the transformed density for $x^* = c\,x$ (noting that $dx/dx^* = 1/c$)

$$p(x^* \,|\, \theta^*) = \frac{1}{c} \cdot \frac{1}{\theta} f\left(\frac{x^*/c}{\theta}\right) = \frac{1}{\theta^*} f\left(\frac{x^*}{\theta^*}\right),$$

with $\theta^* = c\,\theta = \bar{g}_c(\theta)$; so $\bar{\mathcal{G}} = \{\bar{g}_c : c > 0\}$. The group action on $\boldsymbol{\Theta}$ is thus the same as the group action on $\mathcal{X}$, and the scale model is clearly invariant under $\mathcal{G}$. Note that $\bar{g}_{c^*} \circ \bar{g}_c = \bar{g}_{c^* c}$.

Example 6.6 In Example 6.3, the group transformation $g_c(x) = x + c$ yields the transformed density for $x^* = x + c$,

$$p(x^* \,|\, \theta^*) = f(x^* - c - \theta) = f(x^* - \theta^*),$$

with $\theta^* = \theta + c = \bar{g}_c(\theta)$. The group action on $\boldsymbol{\Theta}$ is thus the same as the group action on $\mathcal{X}$, and the scale model is clearly invariant under $\mathcal{G}$. Note that $\bar{g}_{c^*} \circ \bar{g}_c = \bar{g}_{c^*+c}$.

Example 6.7 In Example 6.4, the unknown parameter is $\boldsymbol{\theta} = (\mu, \sigma)$. The group transformation $g_{b,c}(x) = cx + b$ yields the transformed density for $x^* = c\,x + b$ (noting that $x = (x^* - b)/c$ and $dx/dx^* = 1/c$)

$$\begin{aligned} p(x^* \,|\, \boldsymbol{\theta}^*) = \frac{1}{c} \cdot \frac{1}{\sigma} f\left(\frac{1}{\sigma}\left(\frac{(x^* - b)}{c} - \mu\right)\right) &= \frac{1}{c\sigma} f\left(\frac{x^* - b - c\mu}{c\sigma}\right) \\ &= \frac{1}{\sigma^*} f\left(\frac{x^* - \mu^*}{\sigma^*}\right), \end{aligned}$$

with $\boldsymbol{\theta}^* = (\mu^*, \sigma^*) = (c\,\mu + b, c\,\sigma) = \bar{g}_{b,c}(\boldsymbol{\theta})$ and $\bar{\mathcal{G}} = \{\bar{g}_{b,c} : b \in \mathbb{R}^1, c > 0\}$. The group action on $\boldsymbol{\Theta}$ is necessarily different than that on $\mathcal{X}$, because $\boldsymbol{\Theta}$ is two-dimensional and $\mathcal{X}$ is one-dimensional. Nevertheless, the conditions are satisfied for $\mathcal{M} = \{p(x \,|\, \boldsymbol{\theta}) = (1/\sigma) f((x - \mu)/\sigma) : \mu \in \mathbb{R}^1, \sigma > 0\}$ to be invariant under $\mathcal{G}$. Note that

$$\begin{aligned} \bar{g}_{b,c} \circ \bar{g}_{b^*,c^*}(\mu, \sigma) &= \bar{g}_{b,c}(c^* \mu + b^*, c^* \sigma) \\ &= (c[c^* \mu + b^*] + b, c\,c^* \sigma) = \bar{g}_{cb^*+b, c\,c^*}(\mu, \sigma) \end{aligned}$$

so that

$$\bar{g}_{b,c} \circ \bar{g}_{b^*,c^*} = \bar{g}_{c\,b^*+b, c\,c^*} \tag{6.1}$$

is the group action on $\bar{\mathcal{G}}$.

When $\bar{g}$ is indexed by variables, *e.g.*, $\bar{g}_{\boldsymbol{v}}$, we will sometimes write $\boldsymbol{v}$ instead of $\bar{g}_{\boldsymbol{v}}$. For instance, (6.1) could instead be written

$$(b, c) \circ (b^*, c^*) = (c\,b^* + b, c\,c^*);$$

indeed, we will actually be defining densities on $\boldsymbol{v}$, so this is a convenient notation.

6.4 Invariance Priors

If the group transformations, g, of a model result in models of exactly the same form, it would seem natural to require that an objective prior $\pi(\boldsymbol{\theta})$ should be invariant to the transformations, i.e.

$$\int_A \pi(\boldsymbol{\theta})\, d\boldsymbol{\theta} = \int_{\bar{g}\ \circ A} \pi(\boldsymbol{\theta})\, d\boldsymbol{\theta} \qquad \text{for any } \bar{g} \in \bar{\mathcal{G}} \text{ and set } A \subset \Theta\,, \tag{6.2}$$

where $\bar{g} \circ A = \{\bar{g} \circ \boldsymbol{\theta} : \boldsymbol{\theta} \in A\}$.

6.4.1 One-Dimensional Parameter

Fact 6.1 *Suppose $\bar{\mathcal{G}}$ is isomorphic to a continuous real parameter space. Then the density which satisfies (6.2) is the* Haar density *on $\bar{\mathcal{G}}$ (up to an arbitrary multiplicative constant if improper), given by*

$$\pi(\theta) = \frac{1}{J_\theta(e)}\,,$$

where $J_{\bar{g}}(e)$ is the Jacobian of the transformation $\theta \to \bar{g} \circ \theta$ and e is the identity element. (For formal development and proof, see Berger (1985, Chapter 7)).

Example 6.8 In Example 6.6, the group transformation on $\bar{\mathcal{G}}$ is $\bar{g}(\theta) = \theta + c$, which has Jacobian $J_c(\theta) = d(\theta + c)/d\theta = 1$. Hence the invariant prior is the Haar density

$$\pi(\theta) = \frac{1}{J_\theta(e)} = 1\,.$$

Example 6.9 In Example 6.5, the group transformation on $\bar{\mathcal{G}}$ is $\bar{g}(\theta) = c\,\theta$, which has Jacobian $J_c(\theta) = d(c\,\theta)/d\theta = c$. Hence the invariant prior is the Haar density (watch the indices and arguments carefully)

$$\pi(\theta) = \frac{1}{J_\theta(e)} = \frac{1}{\theta}\,.$$

Application *The table entry problem.* Consider tables of widely varying numbers; for instance, Table 2.3 in Abramowitz & Stegun (1964) is a table of physical constants (such as the speed of light), the entries of which vary from 1.05×10^{-34} (Planck's constant) to 6.02×10^{23} (Avogadro's constant). For any such table that has a large number of entries, determine the probability distribution of the first significant digit of the entries (*e.g.*, 1 and 6 for the above constants).

Answer: Digit i occurs with probability close to $\log(1+i^{-1})/\log 10$. (This is commonly called *Benford's Law*, from Benford (1938), although Simon Newcomb had stated it in 1881.)

"Proof:" The distribution of the first significant digit should be scale invariant since, in a sense, the scales of the numbers were originally arbitrary. (The speed of light could be measured in many different units.) In contrast, one would not expect the first significant digits to be location invariant; for instance, adding 1 to every number would change the first significant digit of every entry that is less than one to 1.

So suppose table numbers θ have density $\pi(\theta) = c/\theta$ over the range of numbers from 10^{-N} to 10^{N} with N large. Requiring $\pi(\theta)$ to be a proper density yields $c = 1/[2N\log(10)]$ and

$$\Pr(1^{st} \text{ significant digit} = i) = \sum_{n=-N}^{n=N-1} \int_{10^n i}^{10^n (i+1)} \frac{1}{2N\log(10)} \frac{1}{\theta}\, d\theta = \frac{\log(1+i^{-1})}{\log(10)} .$$

It can be shown that the Haar density is proportional to the Jeffreys-rule prior for one-dimensional parameters; we thus know that the invariance prior is good in one dimension, but we have not encountered anything new.

6.4.2 Higher Dimensional Parameters

Group theory yields the relevant invariant priors on groups $\bar{\mathcal{G}}$. First, however, we discuss the issue of transferring a distribution from $\bar{\mathcal{G}}$ to $\boldsymbol{\Theta}$.

Inducing priors on $\boldsymbol{\Theta}$ from $\bar{\mathcal{G}}$. Define the *image* of $\bar{\mathcal{G}}$ to be $\mathcal{I} = \{\bar{g}(e) : g \in \mathcal{G}\}$, where again e is the identity transformation. This space is the same as the parameter space $\boldsymbol{\Theta}$ but $\{\boldsymbol{\theta}\}$ and $\{\bar{g}(e)\}$ can be different representations of the space. Hence one might need to transfer a probability distribution on $\bar{g}$ to one on $\boldsymbol{\theta}$ through a change of variables. (In Berger (1985), it was essentially assumed that the representations used for the two spaces were the same, so the issue did not arise.)

Example 6.10 In Example 6.4, with $\boldsymbol{\theta} = (\mu, \sigma)$, $\bar{g}_{b,c}(\boldsymbol{\theta}) = (c\,\mu + b, c\,\sigma)$ and $e = (0, 1)$, the image of $\bar{\mathcal{G}}$ in $\boldsymbol{\Theta}$ is clearly $\mathcal{I}_1 = \{(b, c) : b \in \mathbb{R}^1, c > 0\}$. We can thus simply equate b with μ and c with σ and transfer any distribution on (b, c) directly to (μ, σ).

Now, however, consider the group operation on $\mathcal{X}$, $g_{b,c}(x) = c\,(x+b)$, yielding the transformed density for $x^* = c\,(x+b)$

$$p(x^* \,|\, \boldsymbol{\theta}^*) = \frac{1}{c} \cdot \frac{1}{\sigma} f\left(\frac{1}{\sigma}\left(\frac{x^*}{c} - b - \mu\right)\right) = \frac{1}{\sigma^*} f\left(\frac{x^* - \mu^*}{\sigma^*}\right),$$

with $\mu^* = c\,(\mu + b)$ and $\sigma^* = c\,\sigma$, so that $\bar{g}_{b,c}(\boldsymbol{\theta}) = (c\,(\mu+b), c\,\sigma)$. The model is thus also invariant under the new group operation, but now the image is $\mathcal{I}_2 = \{(c\,b, c) : b \in \mathbb{R}^1, c > 0\}$. Equating μ with $c\,b$ and σ with c is fine, but now any density $p(b,c)$ can be transferred to (μ, σ) only through a change of variables, i.e.

$$\pi(\mu,\sigma) = p\,(\mu/\sigma, \sigma)\,J(\mu,\sigma) = (1/\sigma)\,p\,(\mu/\sigma,\sigma)\;, \tag{6.3}$$

where $J(\mu,\sigma) = 1/\sigma$ is the Jacobian of the transformation.

In general, suppose $\bar{\mathcal{G}} = \{\bar{g}_{\boldsymbol{v}}, \boldsymbol{v} \in \mathcal{V}\}$, and we wish to transform from a density $p(\boldsymbol{v})$ on $\mathcal{V}$ to a density on $\boldsymbol{\Theta}$. Assuming the transformation is one-to-one, simply set $\boldsymbol{\theta} = \bar{g}_{\boldsymbol{v}}(e) \equiv f(\boldsymbol{v})$ and change variables, resulting in

$$\pi(\boldsymbol{\theta}) = p(f^{-1}(\boldsymbol{\theta}))J(\boldsymbol{\theta})\,, \tag{6.4}$$

where $J(\boldsymbol{\theta})$ is the Jacobian of the change of variables $\boldsymbol{\theta} = f(\boldsymbol{v})$.

The Left-Haar Density on $\bar{\mathcal{G}}$. There is a unique density (up to a multiplicative constant), $h^l(\cdot)$, called the *left invariant Haar density* (left-Haar for short) that satisfies

$$\int_A h^l(\bar{g}^*)\,d\bar{g}^* = \int_{\bar{g}\,\circ A} h^l(\bar{g}^*)\,d\bar{g}^* \qquad \text{for any } \bar{g} \text{ and set } A \subset \bar{\mathcal{G}}\,, \tag{6.5}$$

where $\bar{g} \circ A = \{\bar{g} \circ \bar{g}^* : \bar{g}^* \in A\}$. See Halmos (1950) for a proof. It can be shown (*cf.* Berger (1985)), that the left-Haar density is given by

$$h^l(\bar{g}) = \frac{1}{J^l_{\bar{g}}(e)}\,, \tag{6.6}$$

where $J^l_{\bar{g}}(\bar{g}^*)$ is the Jacobian of the transformation $\bar{g}^* \to \bar{g} \circ \bar{g}^*$.

Example 6.11 In Example 6.10, the Jacobian of the group transformation $\bar{g}_{b^*,c^*} \to \bar{g}_{b,c} \circ \bar{g}_{b^*,c^*}$, or (see (6.1)) $(b,c) \circ (b^*,c^*) = (c\,b^* + b, c\,c^*)$, is

$$J^l_{\bar{g}_{b,c}}(\bar{g}_{b^*,c^*}) = \left|\begin{pmatrix} c & 0 \\ 0 & c \end{pmatrix}\right| = c^2\,,$$

so that $h^l(\bar{g}_{b,c}) = h^l(b,c) = c^{-2}$. Since (b,c) can be identified with (μ,σ), transferring this to θ yields the left-Haar prior $\pi^l(\mu,\sigma) = \sigma^{-2}$.

If, instead, we had considered the group operation $g_{b,c}(x) = c\,(x+b)$, the induced group operation on $\bar{\mathcal{G}}$ can be shown to be $\bar{g}_{b,c}(\mu,\sigma) = (c\,[\mu+b], c\,\sigma)$, so that

$$\begin{aligned}\bar{g}_{b,c} \circ \bar{g}_{b^*,c^*}(\mu,\sigma) &= \bar{g}_{b,c}(c^*[\mu+b^*], c^*\sigma) = (c(c^*[\mu+b^*]+b), c\,c^*\sigma) \\ &= (c\,c^*\,[\mu+b^*+b/c^*]\,, c\,c^*\sigma) = \bar{g}_{b^*+b/c^*, c\,c^*}(\mu,\sigma)\end{aligned}$$

is the group action on $\bar{\mathcal{G}}$. The Jacobian of this action, or equivalently of $(b,c)\circ(b^*,c^*) = (b^*+b/c^*, c, c^*)$, is

$$J^l_{\bar{g}_{b,c}}(\bar{g}_{b^*,c^*}) = \left|\begin{pmatrix} 1 & 0 \\ -b/c^{*2} & c \end{pmatrix}\right| = c\,,$$

so that $h^l(\bar{g}_{b,c}) = h^l(b,c) = c^{-1}$. Recall, however, that $\bar{\mathcal{G}}$ and $\boldsymbol{\Theta}$ are parameterized differently here, so that the transformation (6.3) is needed to convert the left-Haar density on $\bar{\mathcal{G}}$ to that on $\boldsymbol{\Theta}$, resulting in

$$\pi^l(\mu,\sigma) = (1/\sigma)h^l\,(\mu/\sigma,\sigma) = \sigma^{-2}\,.$$

Note that the two different group operations resulted in the same left invariant prior on $\boldsymbol{\Theta}$.

Unfortunately, $\pi^l(\mu,\sigma) = \sigma^{-2}$ in the above example is the Jeffreys-rule prior and this is not the correct objective prior; the independence Jeffreys-rule prior (also the reference prior) $\pi(\mu,\sigma) = \sigma^{-1}$ is the correct objective prior. Thus it seems like we have simply encountered the same problem as with the Jeffreys-rule prior; the left invariant prior works great in one-dimension, but fails in higher dimensions. Luckily there is a fix with invariance theory, to which we now turn.

The Right-Haar Density on $\bar{\mathcal{G}}$. There is a unique density (up to a multiplicative constant), $h^r(\cdot)$, called the *right invariant Haar density* (right-Haar for short) that satisfies

$$\int_A h^r(\bar{g}^*)\,d\bar{g}^* = \int_{A\circ\bar{g}} h^r(\bar{g}^*)\,d\bar{g}^* \qquad \text{for any } \bar{g} \text{ and set } A\subset\bar{\mathcal{G}}\,, \tag{6.7}$$

where $A\circ\bar{g} = \{\bar{g}^*\circ\bar{g} : \bar{g}^*\in A\}$; see Halmos (1950). It can be shown (*cf.* Berger (1985)), that the right Haar density is given by

$$h^r(\bar{g}) = \frac{1}{J^r_{\bar{g}}(e)}\,, \tag{6.8}$$

where $J^r_{\bar{g}}(\bar{g}^*)$ is the Jacobian of the transformation $\bar{g}^* \to \bar{g}^*\circ\bar{g}$.

Example 6.12 As in Example 6.11, the Jacobian of the group transformation $\bar{g}_{b^*,c^*} \to \bar{g}_{b^*,c^*}\circ\bar{g}_{b,c}$, or $(b^*,c^*)\circ(b,c) = (c^*b+b^*, c^*c)$, is

$$J^l_{\bar{g}_{b,c}}(\bar{g}_{b^*,c^*}) = \left|\begin{pmatrix} 1 & 0 \\ b & c \end{pmatrix}\right| = c\,,$$

so that $h^r(\bar{g}_{b,c}) = h^r(b,c) = c^{-1}$. Since (b,c) can be identified with (μ,σ), transferring this to $\boldsymbol{\Theta}$ yields the right Haar prior $\pi^r(\mu,\sigma) = \sigma^{-1}$. The same prior results from use of the other group operation.

So the right-Haar prior has succeeded in producing the correct objective prior for the location-scale case. In the next section we will see another compelling property of the right-Haar prior.

6.4.3 The Right-Haar Prior is Exact Matching

Under modest technical conditions, the right-Haar prior is exact matching so that, say, resulting Bayesian 90% posterior credible sets will also define exact 90% frequentist confidence sets. This is not only a strong justification for use of the right-Haar prior but also provides frequentists with a powerful tool for deriving confidence sets, since we will see that Bayesian derivations are typically much simpler than direct frequentist derivations.

To begin, a few more concepts from invariance theory are needed. We will restrict consideration to iid data $\boldsymbol{x} = (x_1, x_2, \ldots, x_n)$ for simplicity, although the exact matching result holds in much greater generality. The group action g on $\mathcal{X}^n$ is then $g(\boldsymbol{x}) \equiv (g(x_1), g(x_2), \ldots, g(x_n))$.

Definition 6.3 $\boldsymbol{s}(\boldsymbol{x})$ *is a* maximal invariant statistic *if*

(i) *it is invariant, i.e.,* $\boldsymbol{s}(\boldsymbol{x}) = \boldsymbol{s}(g(\boldsymbol{x}))$ *for all* $g \in \mathcal{G}$; *and*

(ii) *any other invariant statistic is a function of* $\boldsymbol{s}(\boldsymbol{x})$.

Example 6.13 In the location-scale Example 6.12, a maximal invariant statistic is given by

$$\boldsymbol{s}(\boldsymbol{x}) = \left(\frac{x_1 - \bar{x}}{s}, \frac{x_2 - \bar{x}}{s}, \ldots, \frac{x_{n-1} - \bar{x}}{s}\right), \tag{6.9}$$

where $\bar{x}$ is the sample mean and $s^2 = \sum_i (x_i - \bar{x})^2$. It is easy to see that replacing each x_i by $g_{b,c}(x_i) = cx_i + b$ results in the same statistic, so $\boldsymbol{s}$ is invariant. Proving that any other invariant statistic is a function of $\boldsymbol{s}$ is fairly easy, once one notices that (say) $(x_1 - x_2)/(x_3 - x_4)$ is another invariant statistic and equals $(s_1 - s_2)/(s_3 - s_4)$.

A maximal invariant statistic is generally not unique. For instance,

$$\left(\frac{x_1 - x_n}{s}, \frac{x_2 - x_n}{s}, \ldots, \frac{x_{n-1} - x_n}{s}\right)$$

is another maximal invariant statistic. However, different maximal invariant statistics will result in equivalent inferences.

There is no particular role for maximal invariant statistics in Bayesian theory, but they are crucial in frequentist statistics Their primary important property is that they are *ancillary* statistics, meaning that their distribution does not depend on $\boldsymbol{\theta}$. In the above example, for instance, the $(x_i - \overline{x})$ and s clearly do not depend on μ, and taking their ratio removes their dependence on σ.

It is well-accepted that frequentists should perform inferences conditional on the value of ancillary statistics and, especially, maximal invariant statistics. One justification is to note that an ancillary statistic carries no information about $\boldsymbol{\theta}$ and hence one should condition upon it. More importantly, an ancillary statistic can be thought of as reflecting the strength of the evidence in the data (as does $s = |x_1 - x_2|$ in Example 2.2). Thus, conditioning provides inferences reflecting the strength of evidence in the actual data, not just an average over all possible strengths of evidence.

Invariance theory can be applied to almost any statistical inference — estimation, testing, confidence procedures, decision analysis, etc. — as long as the desired inference or decision itself has a suitable invariance structure; see, *e.g.*, Berger (1985) for development. Here we discuss only one such example — that invoving confidence procedures — because the primary goal of this book is development of objective priors, and the frequentist matching properties of priors is an important part of this development. And, once one has an appropriate objective prior, any desired statistical inference or decision analysis is straightforward in principle.

Fact 6.2 *In the setting of this chapter and under certain technical conditions, the right-Haar prior is exact frequentist matching, conditional on a maximal invariant statistic* $\boldsymbol{s}$*, for any parameter* ξ *for which the Bayesian credible sets* $C(\boldsymbol{x})$ *are invariant, i.e., for which* $C(g(\boldsymbol{x})) = \tilde{g} \circ C(\boldsymbol{x}) = \{\tilde{g}(\xi) : \xi \in C(\boldsymbol{x})\}$*, where* $\tilde{g}$ *is the group action induced on* ξ *by* $\bar{g}$ *(the group action on* Θ*.)*

Example 6.14 In the location-scale Example 6.13, where $p(x \mid \mu, \sigma) = (1/\sigma) f((x - \mu)/\sigma)$, consider one-sided posterior confidence intervals for $\xi_\alpha = \xi_\alpha(\mu, \sigma) = \mu + q_\alpha \sigma$, where $\int_{-\infty}^{q_\alpha} f(x)dx = \alpha$; it is easy to see that ξ_α is the α-quantile of the location-scale density. Using the right-Haar prior $\pi^r(\mu, \sigma) = \sigma^{-1}$, we will construct the one-sided Bayesian credible set for ξ_α based on data $\boldsymbol{x} = (x_1, \ldots, x_n)$, namely the interval $C(\boldsymbol{x}) = (-\infty, q_\alpha(\boldsymbol{x}))$, where $q_\alpha(\boldsymbol{x})$ is the α posterior quantile for ξ_α, defined by

$$P(\xi_\alpha < q_\alpha(\boldsymbol{x}) \mid \boldsymbol{x}) = \int_{-\infty}^{q_\alpha(\boldsymbol{x})} \pi^r(\xi_\alpha \mid \boldsymbol{x})\, d\xi_\alpha = \alpha\,,$$

with $\pi^r(\xi_\alpha \mid \boldsymbol{x})$ being the posterior distribution of ξ_α arising from the right-Haar prior.

To apply Fact 6.2, note first that

$$\tilde{g}_{b,c}(\xi_\alpha(\mu, \sigma)) = \xi_\alpha(\bar{g}_{b,c}(\mu, \sigma)) = c\,\mu + b + q_\alpha\, c\sigma = c\,\xi_\alpha(\mu, \sigma) + b$$

is the group action on the parameter of interest, induced from $\bar{g}$. It can also be shown that $q_\alpha(g(\boldsymbol{x})) = \tilde{g}(q_\alpha(\boldsymbol{x}))$, from which it is immediate that the credible sets are invariant. Fact 6.2 thus implies that the one-sided Bayesian credible interval is also a one-sided frequentist confidence interval for ξ_α, conditional on the maximal invariant statistic $\boldsymbol{s}$ in (6.9), with exact coverage of α.

If, say, an equal-tailed $100(1-\alpha)\%$ confidence interval is desired (*i.e.*, a confidence interval where the probability of missing on the left and right both equal $\alpha/2$), one simply finds the corresponding equal-tailed Bayesian credible interval.

As a specific example, suppose the x_i are iid from the Cauchy density $\text{Ca}(x \mid \mu, \sigma)$, and it is desired to find the 90% frequentist confidence set for the third quartile of the distribution, $\xi = \mu + \sigma$, conditional on the maximal invariant statistic $\boldsymbol{s}$. Here is a simple way to construct the confidence interval, using posterior simulation.

The posterior distribution of μ and σ, using the right-Haar prior, is

$$\pi(\mu, \sigma \mid \boldsymbol{x}) \propto \frac{1}{\sigma} \prod_{i=1}^{n} \frac{1}{\sigma\left[1 + (x_i - \mu)^2/\sigma^2\right]} \,.$$

(i) Draw samples (μ_j, σ_j), $j = 1, \ldots, m$ (m large), from this posterior, using *e.g.*, a Metropolis-Hastings sampler (see Section 1.8).

(ii) Compute the corresponding $\xi_j = \mu_j + \sigma_j$, $j = 1, \ldots, m$.

(iii) Order the ξ_j and remove the lower 5% and upper 5%. The span of the remaining samples forms the desired confidence interval.

Note the simplicity of the analysis when done on the Bayesian side. There is no need to even find a maximal invariant statistic $\boldsymbol{s}$, much less figure out how to develop a confidence interval for the very complicated distribution $p(\boldsymbol{x} \mid \mu, \sigma, \boldsymbol{s})$.

Presenting a formal version of Fact 6.2 is beyond the scope of this book. A very simple version (without conditioning on $\boldsymbol{s}$) can be found in Berger (1985). Other versions can be found in (Kiefer 1957, 1966, Stein 1965, Hora & Buehler 1966, Zidek 1969, Bondar & Milnes 1981, Chang & Villegas 1986). The whole class of such theorems is sometimes called the Hunt-Stein theorem, which was developed in Hunt & Stein (1946) for testing.

6.4.4 The Triangular Group

We finish with an example that will be needed elsewhere in the book, and also an example that illustrates that the 'technical conditions' referred to in Fact 6.2 are important and not always satisfied.

The multivariate generalization of the location-scale model is a density

$$p(\boldsymbol{x} \mid \boldsymbol{\mu}, \Sigma) = |\Sigma|^{-1/2} f(\Sigma^{-1/2}(\boldsymbol{x} - \boldsymbol{\mu}))\,,$$

where $\boldsymbol{x}$ and $\boldsymbol{\mu}$ are p-dimensional vectors and Σ is a $p \times p$ positive definite matrix, with $\boldsymbol{\mu}$ and Σ being unknown. A possible generalization of the location-scale group transformation is

$$g_{\boldsymbol{b},\boldsymbol{C}}(\boldsymbol{x}) = \boldsymbol{C}\boldsymbol{x} + \boldsymbol{b}, \quad \boldsymbol{b} \in \mathbb{R}^p \text{ and } \boldsymbol{C} \in \{\text{nonsingular } p \times p \text{ matrices}\}\,. \tag{6.10}$$

$\bar{\mathcal{G}}$, the group of induced transformations on $(\boldsymbol{\mu}, \Sigma)$, can be seen to be defined by

$$\bar{g}_{\boldsymbol{b},\boldsymbol{C}} = (\boldsymbol{C}\boldsymbol{\mu} + \boldsymbol{b}, \boldsymbol{C}\Sigma\boldsymbol{C}')\,,$$

and the right-Haar density (also the left-Haar density) on $\bar{\mathcal{G}}$ can be shown to be

$$h^r(\boldsymbol{b}, \boldsymbol{C}) = \frac{1}{|\boldsymbol{C}|^p}\, d\boldsymbol{b}\, d\boldsymbol{C}\,.$$

The induced prior on $(\boldsymbol{\mu}, \Sigma)$ (following (6.4)) is

$$\pi^r(\boldsymbol{\mu}, \Sigma) = \frac{1}{|\Sigma|^{(p+1)/2}}\, d\boldsymbol{\mu}\, d\Sigma\,,$$

which can be recognized as the independence Jeffreys prior (see Sun & Berger (2007*b*)) for the parameters.

While the independence Jeffreys prior is fine for $p = 1$, it is seriously suboptimal for larger p, as discussed in Sun & Berger (2007*b*). The problem here is the 'technical conditions' referred to in Fact 6.2. The most important of these conditions is that the group $\mathcal{G}$ must be *amenable* (see Bondar & Milnes (1981) for definition and discussion in this context); very crudely, this means that the group cannot be 'too big' and the group formed from all nonsingular matrices is too big.

A smaller group that acts on $\boldsymbol{x}$ and is amenable is the group formed from the *lower triangular matrices*, defined as

$$\boldsymbol{\Omega} = \{p \times p \text{matrices with zeroes above the diagonal and positive diagonal elements}\}\,.$$

Thus we define a new group action

$$g_{\boldsymbol{b},\boldsymbol{C}}(\boldsymbol{x}) = \boldsymbol{C}\boldsymbol{x} + \boldsymbol{b}, \quad \boldsymbol{b} \in \mathbb{R}^p \text{ and } \boldsymbol{C} \in \boldsymbol{\Omega}\,.$$

$\bar{\mathcal{G}}$, the group of induced transformations on $(\boldsymbol{\mu}, \Sigma)$, can be seen to be defined by

$$\bar{g}_{\boldsymbol{b},\boldsymbol{C}} = (\boldsymbol{C}\boldsymbol{\mu} + \boldsymbol{b}, \boldsymbol{C}\Sigma\boldsymbol{C}')\,. \tag{6.11}$$

The right-Haar density on $\bar{\mathcal{G}}$ for this smaller group can be shown to be

$$h^r(\boldsymbol{b}, \boldsymbol{C}) = \prod_{i=1}^{p} c_{ii}^{-(p+1-i)}\, d\boldsymbol{b}\, d\boldsymbol{C}\,, \tag{6.12}$$

where the c_{ii} are the diagonal elements of $\boldsymbol{C}$. The induced prior on $(\boldsymbol{\mu}, \Sigma)$ (following (6.4)) can be computed, but it does not have a simple form. Hence,

Bayesian computations with this prior are typically either done in terms of the variables $(\boldsymbol{\mu}, \boldsymbol{C})$ directly or other parameterizations, as illustrated in the next section. In any case, $\bar{\mathcal{G}}$ here is an amenable group and so the resulting posterior will be exact matching for parameters for which the posterior credible sets are invariant.

An interesting complication here is that one can, alternatively, work with the group of upper triangular matrices. This will lead to a different right-Haar prior on $(\boldsymbol{\mu}, \boldsymbol{\Sigma})$ and this one will also be exact matching for parameters which have invariant posterior credible sets (which can differ from the invariant parameters for the lower triangular group action). This complication is further discussed in Section 12.3.

The Right-Haar Prior in the Bivariate Normal Problem.

Lemma 6.1 *The right-Haar prior for $(\mu_1, \mu_2, \sigma_1, \sigma_2, \rho)$, induced by the right-Haar measure in (6.12) is*

$$\pi^r(\mu_1, \mu_2, \sigma_1, \sigma_2, \rho) = \frac{1}{\sigma_1^2(1-\rho^2)}\,. \tag{6.13}$$

Proof. Using (6.11), the image (6.4) becomes

$$\bar{g}_{\boldsymbol{b},\boldsymbol{C}}(\boldsymbol{e}) = (\boldsymbol{C}\mathbf{0} + \mathbf{b}, \mathbf{CIC}') = (\mathbf{b}, \mathbf{CC}')\,.$$

Thus we transform from $(\boldsymbol{b}, \boldsymbol{C})$ to $(\boldsymbol{\mu} = \boldsymbol{b}, \boldsymbol{\Sigma} = \boldsymbol{CC}')$. Since $\boldsymbol{b}$ had a constant density under the right-Haar prior, so does $\boldsymbol{\mu}$. Thus we focus on the transformation from $\boldsymbol{C}$ to $\boldsymbol{\Sigma} = \boldsymbol{CC}'$ or, equivalently, the transformation from (c_{11}, c_{21}, c_{22}) (the nonzero entries of $\boldsymbol{C}$) to $(\sigma_1, \sigma_2, \rho)$. Computing the Jacobian of the transformation results in

$$dc_{11}\, dc_{21}\, dc_{22} = \frac{\sigma_2}{(1-\rho^2)^{1/2}}\, d\sigma_1\, d\sigma_2\, d\rho\,.$$

Finally, noting that $c_{11} = \sigma_1$ and $c_{22} = \sigma_2(1-\rho^2)^{1/2}$, the right-Haar density in (6.12) transforms to

$$h^r(\boldsymbol{b}, \boldsymbol{C})\, dc_{11}\, dc_{21}\, dc_{22} = \frac{1}{c_{11}^2 c_{22}}\, dc_{11}\, dc_{21}\, dc_{22} = \frac{1}{\sigma_1^2(1-\rho^2)}\, d\sigma_1\, d\sigma_2\, d\rho\,.$$

□

Note the curious asymmetry between the two variances in the prior. This has a significant effect on the quality of inferences using this prior. Indeed, the prior is exact frequentist matching for μ_1, σ_1, and ρ. This follows from Fact 6.2, because it is easy to show that any credible sets for these parameters are invariant, in the sense defined in the Fact. On the other hand, credible sets for μ_2 and σ_2 are not invariant, and so the Fact cannot be applied to them. Indeed, the right-Haar prior is not frequentist matching for these two parameters. However, the other 'natural' right-Haar prior, $\pi(\mu_1, \mu_2, \sigma_1, \sigma_2, \rho) = [\sigma_2^2(1-\rho^2)]^{-1}$, is exact matching for μ_2 and σ_2 (but not for μ_1 and σ_1). This other right-Haar prior is derived from the group operation with upper triangular matrices.

6.5 Discussion and Additional References

Good general references about invariance in statistics and invariant priors include Eaton (1989), Wijsman (1990), and George & McCulloch (1993). Many other articles concerning invariance are given in the appendix; just search under the key word 'invariance.

A number of authors, including Villegas and Fraser (see the references at the beginning of this chapter) argued strongly for versions of the left-Haar prior, based on first principles. Indeed, (6.5) is a considerably more natural condition than (6.7). But the fact that the right-Haar prior is usually exact matching, together with the demonstrated possible poor performance of the left-Haar prior, strongly suggest that the right-Haar prior be chosen to be the objective prior in invariant situations.

Hartigan (1964) raises an interesting argument in terms of principled use of the right-Haar prior. He observes that there is nothing wrong with (6.2), as long as the resulting priors are proper distributions (which will only rarely happen in invariant situations – essentially $\bar{\mathcal{G}}$ would have to be a compact set). But, when the invariance priors are improper, then the priors for $\boldsymbol{\theta}$ and $\bar{g}(\boldsymbol{\theta})$ are really the same if they are multiples of each other, so that (6.2) should instead be

$$\int_A \pi(\boldsymbol{\theta})\, d\boldsymbol{\theta} = c \int_{\bar{g}\ \circ A} \pi(\boldsymbol{\theta})\, d\boldsymbol{\theta} \qquad \text{for any } g\,, \tag{6.14}$$

for a positive constant c, which could depend on $\bar{g}$. Only requiring this restriction, which Hartigan calls *relative invariance*, allows many other priors to be labelled 'invariance priors.' In particular, the right-Haar prior can be shown to satisfy this restriction.

Chapter 7

Evaluating Objective Priors

The only objective prior that 'looks' objective is the constant prior, but we saw that this is not logically sound because one cannot simultaneously be constant in different parameterizations of a problem. Thus, intuition is not a satisfactory guide in evaluating objective priors. This section discusses the criteria that are commonly used to evaluate objective priors, and which will serve as the guides for our objective prior recommendations.

We will, to some extent, 'look ahead' in this chapter, allowing evaluation not just of the objective priors already discussed (*e.g.*, constant, Jeffreys, matching and invariance priors) but also allowing for consideration of reference priors, our preferred method for deriving an objective prior and discussed in Part III of the book.

7.1 Propriety of Posteriors

Any improper prior carries the potential danger of yielding an improper posterior. An improper posterior is useless for most purposes (although see Lindqvist & Taraldsen (2018) for an opposing view) and obviates the mathematical justification for using improper priors (Theorem 2.1): thus we view posterior propriety as a fundamentally needed property of an objective prior.

When the parameter space is unbounded, the likelihood function $p(\boldsymbol{x} \mid \boldsymbol{\theta})$ (considered as a function of $\boldsymbol{\theta}$) will typically decay to zero in an integrable way for large $\boldsymbol{\theta}$, at least for sufficiently large sample sizes. Then reasonable improper priors will usually have proper posteriors. One example where this was not the case was the mixture situation in Example 2.19. Here is another example.

Example 7.1 *Coefficient of variation.* Suppose $\boldsymbol{x} = (x_1, \ldots, x_n)$ is a random sample from $\mathrm{N}(x \mid \mu, \sigma^2)$, with both μ and σ^2 unknown, and we are interested in the coefficient of variation, $\theta = \sigma/\mu$. Consider priors of the form $\pi(\theta)\pi(\sigma^2)$. Posterior propriety then requires

$$\int_{-\infty}^{\infty}\int_{0}^{\infty}\frac{1}{\sigma^n}\exp\Big\{-\frac{1}{2}\sum_{i=1}^{n}\Big(\frac{x_i}{\sigma}-\frac{1}{\theta}\Big)^2\Big\}\,\pi(\sigma^2)\,d\sigma^2\;\pi(\theta)\,d\theta<\infty\,.$$

Consider first the constant prior. As $|\theta|\to\infty$, it is clear the inner integral above converges to

$$\int_{0}^{\infty}\frac{1}{\sigma^n}\exp\left\{-\frac{1}{2}\sum_{i=1}^{n}\Big(\frac{x_i}{\sigma}\Big)^2\right\}\,\pi(\sigma^2)\,d\sigma^2\,,$$

which is a nonzero constant in θ. Hence the outer integral is necessarily infinite if $\pi(\theta)$ is constant.

The Jeffreys-rule prior here turns out to be σ/θ^2, so that posterior propriety would require

$$\int_{-\infty}^{\infty}\int_{0}^{\infty}\frac{1}{\sigma^{(n-1)}}\exp\Big\{-\frac{1}{2}\sum_{i=1}^{n}\Big(\frac{x_i}{\sigma}-\frac{1}{\theta}\Big)^2\Big\}\,d\sigma^2\frac{1}{\theta^2}\,d\theta<\infty\,.$$

It can be checked that this is, indeed, finite, since the prior $1/\theta^2$ is integrable as $|\theta|\to\infty$. Thus the Jeffreys-rule prior rather miraculously becomes proper at infinity to deal with a likelihood function that does not go to zero at infinity.

The above example is one illustration of a problem in the Gleser & Hwang (1987) class of problems, where a marginal likelihood function does not go to zero at infinity. This class of problems is very interesting from a foundational perspective, not only because care must be taken in choice of an objective prior, but also because any problem in this class has the property that any frequentist confidence set must yield infinite sets with positive probability. Indeed, for some data in the above example, a frequentist must say "my 95% confidence set for θ is $(-\infty,\infty)$," a rather unsettling statement. (It is not unsettling that the confidence set might be $(-\infty,\infty)$ — that could just be a reflection that one has learned essentially nothing about θ from the data — but it is unsettling to ascribe only 95% confidence to the entire real line.)

A Bayesian using the Jeffreys-rule prior in the above example would always have finite 95% credible sets, but the frequentist coverage of the Bayesian credible sets could be poor. Because of this, finding definitive solutions for the Gleser-Hwang class of problems remains elusive.

If one could always easily recognize when the likelihood function is not integrable for large $\boldsymbol{\theta}$, the problem would be considerably lessened (although the question of what to do would not be resolved), but determining integrability can be quite difficult. Here is an example.

Example 7.2 *Spatial statistics.* A commonly analyzed spatial model for a real random process $z(\boldsymbol{s})$, observed at points $\boldsymbol{s}_i \in \mathbb{R}^k, i = 1, \ldots, n$, is

$$z(\boldsymbol{s}) = \mu + \epsilon(\boldsymbol{s}),$$

where μ is the unknown mean level of the process and ϵ is a Gaussian process with zero mean and covariance function

$$\text{Cov}[\epsilon(\boldsymbol{s}), \epsilon(\boldsymbol{u})] = \sigma^2 \exp\big(- \|\boldsymbol{s} - \boldsymbol{u}\|^2/\theta \big),$$

with σ^2 and $\theta > 0$ also being unknown. Natural objective priors are of the form

$$\pi(\mu, \sigma, \theta) = \pi(\mu, \sigma)\, \pi(\theta) = \pi(\theta)\, \sigma^{-1}. \tag{7.1}$$

(Given θ, (μ, σ) is a location-scale parameter, exactly as in Definition 4.1, and the almost universally recommended prior for a location-scale parameter is $\pi(\mu, \sigma) = \sigma^{-1}$.)

The key issue is how to choose $\pi(\theta)$ in (7.1), and it was common, for many years, to choose the simple $\pi(\theta) = 1$. This was shown to result in an improper posterior distribution by a quite lengthy technical argument in Lemma 1 of Berger, De Oliveira & Sansó (2001). It was also shown therein that the Jeffreys-rule prior can result in an improper posterior distribution for this problem, one of the few examples (other than the mixture model) where this happens. In Part III of the book we discuss the reference prior approach (our preferred method of developing an objective prior), and the reference prior does yield a proper posterior here.

The problem of improper posteriors with objective priors is worsened by the common use today of Markov chain Monte Carlo (MCMC) methods of computation, since they can make it difficult to identify improper posterior distributions. (See Example 2.11 for an illustration.) Indeed, in the above spatial example, the constant prior for θ was used in practice for many years, without anyone realizing that the MCMC was being done with an improper posterior. It was only after Victor De Oliveira (private communication) noticed, in an application, that the posterior would not stabilize if he kept on running the MCMC, that the impropriety was discovered. At best, the conclusions from all the Bayesian spatial analyses run during that time period are questionable. Also recall, from Example 2.11, that replacing improper priors by vague proper priors does not solve the problem; it only makes 'effective impropriety' harder to identify.

7.2 Overly Diffuse Priors and Inconsistency

In spite of the above discussion, constant priors (and, even more so, Jeffreys-rule priors) do not usually result in posterior impropriety. Considerably more serious in practice is that, if the parameter is high-dimensional, these priors can often be 'too diffuse' to yield reasonable answers. Here are two examples indicating the problem.

Example 7.3 *Stein's paradox involving a multivariate mean (Stein, 1959).* Let $(x_1, \ldots x_m)$ be independent normal random variables, $\mathrm{N}(x_i \mid \mu_i, 1)$. The constant prior yields the natural posterior $\mathrm{N}(\mu_i \mid x_i, 1)$ for each individual μ_i. We focus on using this constant prior for inference concerning

$$\theta = \frac{1}{m}|\boldsymbol{\mu}|^2 = \frac{1}{m}\sum\nolimits_{i=1}^{m} \mu_i^2 .$$

Note that the resulting posterior mean of θ is

$$\mathrm{E}[\theta \mid x_1, \ldots x_m] = \frac{1}{m}\sum\nolimits_{i=1}^{m} \mathrm{E}[\mu_i^2 \mid x_i] = \frac{1}{m}\sum\nolimits_{i=1}^{m} (x_i^2 + 1) = 1 + \frac{1}{m}\sum\nolimits_{i=1}^{m} x_i^2 .$$

The law of large numbers says that this posterior mean will converge to its frequentist expectation as $m \to \infty$, which is

$$\begin{aligned} \mathrm{E}\left[1 + \frac{1}{m}\sum\nolimits_{i=1}^{m} X_i^2 \;\middle|\; \mu_1, \ldots, \mu_m\right] &= 1 + \frac{1}{m}\sum\nolimits_{i=1}^{m} \mathrm{E}[X_i^2 \mid \mu_i] \\ &= 1 + \frac{1}{m}\sum\nolimits_{i=1}^{m} (1 + \mu_i^2) = 2 + \theta . \end{aligned}$$

Thus the posterior mean is inconsistent, *i.e.*, does not converge to θ. The problem arises because the constant prior on a large number of normal means is 'too diffuse.' (This is also suggested from the large frequentist literature on shrinkage estimation.) It will later be shown (see Example 10.7) that use of the reference prior yields a consistent estimator.

For further discussion of the consistency of Bayes estimates see Stein (1959), Freedman & Diaconis (1983), Dawid & Stone (1972, 1973), Cox & Hinkley (1974), Stone (1976), Lane & Sudderth (1984), Diaconis & Freedman (1986*b*), and Diaconis & Freedman (1986*a*).

Example 7.4 *Covariance matrix.* Consider again Example 2.8. We saw that common priors for a covariance matrix, including Inverse Wishart priors, the Jeffreys-rule prior, and the constant prior, have the undesirable property that they force the eigenvalues of the covariance matrix away from each other, *i.e.*, they are 'too diffuse.' In contrast, the reference prior (Yang & Berger 1994)

$$\pi(\boldsymbol{D}, \boldsymbol{O}) = |\boldsymbol{D}|^{-1}\, d\boldsymbol{D}\, d\boldsymbol{O}, \tag{7.2}$$

does not force the eigenvalues apart and gives much better statistical inferences. (See Chang & Eaves (1990) and Consonni & Veronese (2003) for different reference priors for a covariance matrix.)

As an illustration, consider, estimation of $\boldsymbol{\Sigma}$, under the loss function

$$L(\hat{\boldsymbol{\Sigma}},\ \boldsymbol{\Sigma}) = \mathrm{tr}(\hat{\boldsymbol{\Sigma}}\boldsymbol{\Sigma}^{-1}) - \log|\hat{\boldsymbol{\Sigma}}\boldsymbol{\Sigma}^{-1}| - k, \tag{7.3}$$

where $\hat{\boldsymbol{\Sigma}}$ denotes an arbitrary estimator. This loss was advocated by Stein (1956), and is the most commonly used loss function for covariance matrix estimation.

For this loss, the Bayes estimate of $\boldsymbol{\Sigma}$ can be shown (see Yang and Berger, 1994) to be $\hat{\boldsymbol{\Sigma}} = (\mathrm{E}[\boldsymbol{\Sigma}^{-1} \mid \boldsymbol{S}])^{-1}$. Now consider, as data, the following matrix $\boldsymbol{S}/n$, where $\boldsymbol{S}$ is generated from a Wishart distribution with 10 degrees of freedom and scale matrix $\boldsymbol{\Sigma} = \mathrm{diag}(5, 4, 3, 2, 1)$:

$$\boldsymbol{S}/n = \begin{pmatrix} 1.925 & 1.618 & 0.132 & -1.101 & 0.264 \\ 1.618 & 8.437 & 1.638 & -0.880 & -0.983 \\ 0.132 & 1.638 & 2.147 & -0.439 & -0.646 \\ -1.101 & -0.880 & -0.439 & 1.331 & -0.035 \\ 0.264 & -0.983 & -0.646 & -0.035 & 1.280 \end{pmatrix}.$$

The corresponding Bayes estimate, under the loss in (7.3), is

$$\hat{\boldsymbol{\Sigma}} = \begin{pmatrix} 1.726 & 0.946 & 0.077 & -0.712 & 0.207 \\ 0.946 & 5.536 & 0.917 & -0.483 & -0.567 \\ 0.077 & 0.917 & 1.889 & -0.282 & -0.385 \\ -0.712 & -0.483 & -0.282 & 1.366 & -0.022 \\ 0.207 & -0.567 & -0.385 & -0.022 & 1.412 \end{pmatrix}.$$

The actual loss of this estimate is $L(\hat{\boldsymbol{\Sigma}}, \boldsymbol{\Sigma}) = 1.270$ (computable since we know $\boldsymbol{\Sigma}$). In contrast, use of the common Jeffreys-rule prior (2.2) results in the Bayes estimate $\boldsymbol{S}/n$ given above (which is also the classical unbiased estimate of $\boldsymbol{\Sigma}$ and the mle), and $L(\boldsymbol{S}/n, \boldsymbol{\Sigma}) = 2.267$, significantly greater than the loss of the reference prior Bayes estimate.

This particular data set is not an isolated example; in Yang and Berger (1994) it is shown that use of the reference prior often results in improvements on the order of 50% over $\boldsymbol{S}/n$.

Note that use of the constant prior here would result in an estimator that is substantially worse than even that arising from the Jeffreys-rule prior. See Section 12.6 for further discussion.

7.3 Overly Concentrated Objective Priors

While objective priors being overly diffuse is a rather common occurrence, the opposite can also happen — objective priors can be too concentrated. We saw two examples in Chapter 4: that of Neyman-Scott (Example 4.9), where the Jeffreys-rule prior seemed to be much too concentrated about 0, and the multinomial model (Example 4.10), where the Jeffreys-rule prior (and the constant prior) could be much more influential than the actual data.

7.4 Importance of the Quantity of Interest

It would be nice if a single prior distribution could always be used for inference in a model. For instance, with the $\mathrm{N}(x \mid \mu, \sigma^2)$ model, the independence Jeffreys prior $\pi(\mu, \sigma^2) = 1/\sigma^2$ is often used for all inferences (but see Example 10.5). Part of the appeal of both the constant prior and the Jeffreys-rule prior is this generality of use.

Unfortunately, optimal objective priors will typically depend on the inference being undertaken. For instance, in Example 7.3 (the Stein example), the constant prior would be satisfactory for inference concerning, say, μ_1, but was horrible for inference concerning θ.

Peers (1965) and Welch (1965) were the first to clearly notice this; in their theory of matching priors, each parameter in a model could have a different matching prior. In contrast, the Jeffreys-rule prior is the same for all parameters in a model, which suggests one of the reasons why it seems to fail for multi-parameter models.

Invariance priors are also the same for any parameter of a model, although they can be tailored to be exact matching for certain parameters. Here is an example.

Example 7.5 *Bivariate normal distribution (continued).* In Section 5.1.1 the bivariate normal distribution was extensively discussed. This distribution has mean and covariance matrix

$$\boldsymbol{\mu} = \begin{pmatrix} \mu_1 \\ \mu_2 \end{pmatrix}, \qquad \boldsymbol{\Sigma} = \begin{pmatrix} \sigma_1^2 & \rho\sigma_1\sigma_2 \\ \rho\sigma_1\sigma_2 & \sigma_2^2 \end{pmatrix},$$

where ρ is the correlation. This distribution is invariant with respect to various group transformations, two of which are called the lower triangular matrix group and the upper triangular matrix group (see Section 6.4.4); these lead to the right-Haar priors

$$\pi_1^r(\mu_1, \mu_2, \sigma_1, \sigma_2, \rho) = \frac{1}{\sigma_1^2(1-\rho^2)} \quad \text{and} \quad \pi_2^r(\mu_1, \mu_2, \sigma_1, \sigma_2, \rho) = \frac{1}{\sigma_2^2(1-\rho^2)},$$

respectively. In Sun & Berger (2007*b*), it is shown that π_i yields optimal objective posterior distributions for μ_i, σ_i and ρ (for instance, they are exact matching), but performs quite poorly for the other mean and standard deviation (for which posterior credible sets are not invariant to the group transformation). Thus, two different invariance priors are needed to do a good job of inference for all five parameters.

While, foundationally, it seems necessary to focus separately on each parameter of interest, this can be cumbersome in practice. Thus, in Chapter 12, we also consider overall priors that can be used with reasonable assurance for any parameter of a model.

7.5 Coherence

Coherence in the Bayesian world refers to the property that inferences should behave in logical ways. There are several types of coherence relevant to evaluation of objective priors.

7.5.1 Invariance to Parameterization

This criterion has already been discussed in the evolution from the constant prior (which is not invariant) to the Jeffreys-rule prior (which is invariant). We view this as another mandatory criterion for an objective prior. It is not necessarily practically important, as we saw in Example 3.2, but a methodology that does not have this property does not seem tenable from a foundational perspective. For further discussion of this issue, see Jeffreys (1946), Hartigan (1964), Jaynes (1968), Box & Tiao (1973, Sec. 1.3), Villegas (1977*a*, 1990), Dawid (1983*b*), Yang (1995), Datta & Ghosh (1995*a*) and Datta & Ghosh (1996).

7.5.2 Marginalization Paradoxes

The *marginalization paradox* is often described as an inconsistency in objective Bayesian inference. It was first stated by Dawid and Stone (1972) and further analyzed by Dawid, Stone and Zidek (1973) and Dawid (1980), and its implications have remained polemic (*cf.* Jaynes, 2003, and references therein). The essence of the paradox is best introduced with a simple example.

Example 7.6 *Standardized normal mean.* Let $\boldsymbol{x} = (x_1, \ldots, x_n)$ be a random sample from a normal $\mathrm{N}(x \mid \mu, \sigma^2)$ distribution, and suppose that one is interested in making objective inferences about its standardized mean $\theta = \mu/\sigma$. The conventionally recommended prior in this model is the independence Jeffreys prior $\pi(\mu, \sigma) = \sigma^{-1}$ and, indeed, this leads to excellent objective posteriors for inferences about μ (or any one-to-one function of μ) and σ (or any one-to-one function of σ); this is not the case, however, if the quantity of interest is $\theta = \mu/\sigma$, for this leads to the first published example of a marginalization paradox. Dawid and Stone (1972) showed that, using the prior $\pi(\mu, \sigma) = \sigma^{-1}$, the posterior distribution of θ, $\pi(\theta \mid \boldsymbol{x})$, only depends on the data through the statistic

$$t = \frac{\sum_{i=1}^{n} x_i}{(\sum_{i=1}^{n} x_i^2)^{1/2}},$$

so that we can write the posterior as a density of the form $\pi(\theta \mid t)$. They also showed that the sampling distribution of t only depends on θ; indeed,

$$\begin{aligned} p(t \mid \mu, \sigma) &= p(t \mid \theta) \\ &\propto e^{-n\theta^2/2}[1 - t^2/n]^{(n-3)/2} \int_0^\infty \omega^{n-1} e^{-\omega^2/2 + t\theta\omega} d\omega \,. \end{aligned}$$

It seems natural, therefore, to expect to be able to reproduce the posterior distribution of θ using Bayes theorem with $p(t \mid \theta)$, *i.e.*, to expect to be able to find a prior distribution $\pi(\theta)$ such that $\pi(\theta \mid t) \propto p(t \mid \theta)\pi(\theta)$.

Dawid and Stone (1972) showed, however, that this is *not* possible. Thus imagine you are reading a paper, in which only t and n are reported because they were the only quantities that were involved in the posterior distribution. You know the sampling distribution of t, $p(t \mid \theta)$, but find that you cannot reproduce the conclusion in the paper, since no prior will yield the posterior in the paper with $p(t \mid \theta)$ as the likelihood function.

The reference prior approach to this problem will be discussed later (Example 10.5) and shown to yield the prior

$$\pi^R(\theta, \sigma) = (2 + \theta^2)^{-1/2}\sigma^{-1},$$

and the reference posterior

$$\pi^R(\theta \mid \boldsymbol{x}) = \pi^R(\theta \mid t) \propto (2 + \theta^2)^{-1/2}\left[e^{-n\theta^2/2}\int_0^\infty \omega^{n-1}e^{-\omega^2/2+t\theta\omega}d\omega\right],$$

which also depends only on t. But now, since the factor within the brackets is proportional to $p(t \mid \theta)$, there is a prior for which $\pi^R(\theta \mid t) \propto p(t \mid \theta)\pi(\theta)$, namely $\pi(\theta) = (2 + \theta^2)^{-1/2}$, and the apparent inconsistency disappears.

Definition 7.1 Marginalization paradox. *Let $p(\boldsymbol{z} \mid \boldsymbol{\theta})$ be a probability model, $\pi(\boldsymbol{\theta})$ be a (possibly improper) prior, and $\pi(\boldsymbol{\theta} \mid \boldsymbol{z})$ be the corresponding posterior distribution. Suppose that $\boldsymbol{z} = (\boldsymbol{x}, \boldsymbol{y})$ and $\boldsymbol{\theta} = (\boldsymbol{\phi}, \boldsymbol{\omega})$, with the marginal density $p(\boldsymbol{x} \mid \boldsymbol{\theta}) = p(\boldsymbol{x} \mid \boldsymbol{\phi})$ depending on $\boldsymbol{\theta}$ only through $\boldsymbol{\phi}$ and the marginal posterior $\pi(\boldsymbol{\phi} \mid \boldsymbol{z}) = \pi(\boldsymbol{\phi} \mid \boldsymbol{x})$ depending on $\boldsymbol{z}$ only through $\boldsymbol{x}$. Intuition would then suggest that $\pi(\boldsymbol{\phi} \mid \boldsymbol{x}) \propto p(\boldsymbol{x} \mid \boldsymbol{\phi})\pi(\boldsymbol{\phi})$ for some prior $\pi(\boldsymbol{\phi})$, so that the marginalized quantities would satisfy Bayes theorem. A marginalization paradox occurs if this is* not *the case.*

By standard probability theory, the marginalization paradox cannot occur if only proper priors are used, *although we are not aware of any examples with countably additive proper priors that satisfy the marginal conditions of the formal paradox.* (See Sudderth (1980) for finitely additive examples.) Some might argue that the italicized fact weakens the motivation for avoiding the paradox, but our view is that we want objective Bayesian analysis to satisfy such natural intuitions, and so this can be a useful way to discriminate amongst objective priors.

In this regard, we know of no example in which use of the reference prior results in a marginalization paradox, yielding further support to that approach to objective Bayesian analysis (in contrast to the other objective Bayes approaches). We conjecture that this is because the reference prior (when improper) is defined in terms of the limit, using expected KL convergence, of proper priors on compact sets. Intuitively, the limit should then inherit the good properties of proper priors, such as the (potential) avoidance of the marginalization paradox. This also suggests that priors which violate the marginalization paradox cannot be derived as limits of proper priors under expected KL convergence. See Dawid & Stone (1973) and Wallstrom (2007) for support for this conjecture.

Here is another historically interesting scenario involving a marginalization paradox.

Example 7.7 *Normal coefficient of correlation.* Consider, again, Example 7.5. Let $(\boldsymbol{x}, \boldsymbol{y}) = ((x_1, y_1), \ldots, (x_n, y_n))$ be a random sample from a bivariate normal distribution, and suppose that one is interested in making objective inferences about its coefficient of correlation ρ. It can be shown that, if the prior is of the form

$$\pi(\mu_1, \mu_2, \sigma_1, \sigma_2, \rho) = \pi(\rho)\, \sigma_1^{-a} \sigma_2^{-a}\,,$$

then the marginal posterior distribution of ρ only depends on the sample coefficient of correlation r (see Section 5.1.1 for definition), and is given by

$$\pi(\rho \,|\, \boldsymbol{x}, \boldsymbol{y}, a) = \frac{\pi(\rho)(1-\rho^2)^{(n+2a-3)/2}}{(1-\rho r)^{(n+2a-5)/2}} F\left(\tfrac{1}{2}, \tfrac{1}{2}, n+a-3/2, (1+\rho r)/2\right)\,,$$

where F is the hypergeometric function. On the other hand, the sampling distribution of r only depends on ρ and is given by

$$p(r \,|\, \rho) = \frac{(1-\rho^2)^{(n-1)/2}(1-r^2)^{(n-4)/2}}{(1-\rho r)^{(n-3)/2}} F\left(\tfrac{1}{2}, \tfrac{1}{2}, n-1/2, (1+\rho r)/2\right)\,.$$

Comparison of $\pi(\rho \,|\, \boldsymbol{x}, \boldsymbol{y}, a)$ and $p(r \,|\, \rho)$ shows that the marginalization paradox is avoided if, and only if, $a = 1$, *i.e.*, if the joint prior is of the form $\pi(\mu_1, \mu_2, \sigma_1, \sigma_2, \rho) = \pi(\rho)\, \sigma_1^{-1} \sigma_2^{-1}$.

Section 10.3.3 discusses reference priors for this situation, the most important being $\pi(\mu_1, \mu_2, \sigma_1, \sigma_2, \rho) = (1-\rho^2)^{-1}\, \sigma_1^{-1} \sigma_2^{-1}$; this clearly avoids the marginalization paradox. The constant prior, the Jeffreys-rule prior, the fiducial density and the right-Haar priors in Example 7.5 all exhibit a marginalization paradox.

7.5.3 Compatibility with the Likelihood Principle

The Likelihood Principle was discussed in Section 1.2.2 and is compelling to many. Use of a constant prior is clearly compatible with the Likelihood Principle, since the posterior distribution is then just proportional to $p(\boldsymbol{x} \,|\, \boldsymbol{\theta})$ for the observed data, and the posterior distribution conveys all the evidence about $\boldsymbol{\theta}$ to a Bayesian. Other objective priors, such as the Jeffreys-rule prior, are not formally compatible with the Likelihood Principle, however. The Jeffreys-rule prior, for instance, depends on the (expected) Fisher information matrix for $\boldsymbol{\theta}$, and this is computed as a frequentist average over all possible data in the model and thus does not depend only on $p(\boldsymbol{x} \mid \boldsymbol{\theta})$. Likewise, invariance priors depend on the entire model's invariance structure, not the invariance structure of $p(\boldsymbol{x} \mid \boldsymbol{\theta})$ for the actual observation. Finally, derivation of our preferred reference priors also depends on the entire model.

This is an unusual situation in that, once the prior distribution is determined, objective Bayesian analysis is, of course, fully compatible with the Likelihood Principle. One defense that can be given for the prior constructions that violate the Likelihood Principle is to note that the 'universe' for objective Bayesian analysis is a given model, and an objective prior can be thought of as one that can routinely (and repeatedly) be used for data arising from that model (think, *e.g.*, of the objective prior as being the default prior in a computer package that can be used to analyze any data from the given model). The objective prior then necessarily depends on the model.

This would be a quite satisfactory explanation if a version of the Likelihood Principle could be formulated and, say, derived from versions of the sufficiency and conditionality principles, in such a way that the objective prior would be allowed to depend on the model, but only $p(\boldsymbol{x} \mid \boldsymbol{\theta})$ for the observed data could be used with this prior to obtain the evidence for $\boldsymbol{\theta}$. Such a principle and derivation have, alas, remained elusive.

7.6 Frequentist Validation

We saw, in Chapter 5 and Chapter 6, that frequentist tools and concepts could be useful in deriving objective priors, through the matching or the invariance approaches. Here we consider the use of frequentist tools to evaluate objective priors, not to derive them. In particular, we view it to be positive if an objective prior leads to inferences that have attractive frequentist properties. We do not view this to be absolutely necessary, but it is a positive. In this regard, the constant prior and the Jeffreys-rule prior have been seen to often have quite bad frequentist properties, so we will primarily focus on comparison of matching priors, invariance priors and reference priors in this section.

7.6.1 Matching

The 'frequentist matching' concept was described in terms of the frequentist coverage probabilities of confidence procedures derived as posterior credible sets arising from an objective prior. There are several ways of establishing matching.

- Proving exact matching (frequentist coverage exactly equals Bayesian coverage), usually by invariance arguments.
- Asymptotically (as the sample size n grows) demonstrating matching at least to order $O(n^{-1})$ (essentially any prior is matching to order $O(n^{-1/2})$, so that is rather meaningless), and to higher orders if possible.

- Doing small sample numerical studies, usually by simulation, to study the frequentist coverage of posterior credible sets; Table 2.1 was such an example.

Small sample numerical studies are especially important, because inference for small sample sizes is one of the real strengths of Bayesian analysis; using objective priors with good matching properties is perhaps the foremost way of guaranteeing good small sample frequentist inference.

Small samples are also useful for distinguishing between objective priors. For instance, it is not the case that all 'first order matching priors' are created equal. Even though they are asymptotically the same in terms of matching, it is very common for certain first order priors to be considerably better for small sample sizes than others. For instance, first order matching reference priors typically outperform first order matching priors derived directly from the differential equations in Theorem 5.1. (But reference priors are not guaranteed to be first order matching in general.)

7.6.2 Decision-Theoretic Evaluation and Admissbility

In Section 7.2, the inadequacy of certain priors was judged, not by frequentist coverage of posterior credible regions, but by inadequacy of parameter estimates that arose from the priors. So the expected loss from using objective priors in estimation is another major criterion in the frequentist evaluation of priors. Often, as with the matching criterion, this is simply evaluated in small samples by simulation, comparing the expected loss of the Bayes estimates of a parameter for different priors.

A sophisticated version of this arises from the frequentist concept of admissibility. Suppose we consider estimation of $\boldsymbol{\theta}$ by an estimator $\boldsymbol{\delta}(\boldsymbol{x})$ under squared error loss $||\boldsymbol{\delta}(\boldsymbol{x}) - \boldsymbol{\theta}||^2$. The frequentist risk of the estimator is $R(\boldsymbol{\delta}, \boldsymbol{\theta}) = \mathrm{E}[||\boldsymbol{\delta}(\boldsymbol{x}) - \boldsymbol{\theta}||^2 \mid \boldsymbol{\theta}]$, the expectation being over the sampling distribution of $\boldsymbol{x}$ given $\boldsymbol{\theta}$.

Definition 7.1 *An estimator $\boldsymbol{\delta}(\boldsymbol{x})$ of $\boldsymbol{\theta}$ is* inadmissible, *with respect to squared error loss, if there exists another estimator $\boldsymbol{\delta}^*(\boldsymbol{x})$ such that $R(\boldsymbol{\delta}^*, \boldsymbol{\theta}) \leq R(\boldsymbol{\delta}, \boldsymbol{\theta})$, with strict inequality for some $\boldsymbol{\theta}$. If no such $\boldsymbol{\delta}^*(\boldsymbol{x})$ exists, then $\boldsymbol{\delta}(\boldsymbol{x})$ is* admissible.

The motivation for considering admissibility is that priors that are too diffuse result in inadmissible estimators, while priors that are concentrated enough result in admissible estimators. Priors that are 'on the boundary of admissibility' are sensible choices for objective priors, being as diffuse as possible without resulting in inadmissible procedures. This criterion will be heavily exploited in Section 12.6 to develop objective priors for general normal hierarchical models.

7.7 Computational Issues

There are two main computational issues involved with objective priors: deriving the prior, and utilizing the prior in performing Bayesian computations. We will occasionally recommend objective priors that are not quite optimal (*i.e.*, are not reference priors), because of computational considerations. One example is Section 12.6, where we present the recommended priors for general hierarchical normal models, and another is for sequential problems, where we recommend, in Section 14.4.4, a much easier-to-use objective prior.

7.7.1 Deriving the Objective Prior

The simplest objective prior — a constant prior — has no cost associated with its derivation. Significantly more complicated is the Jeffreys-rule prior, which requires determination of the expected Fisher information matrix. Matching priors require (in more than one-dimension) the solving of a system of differential equations.

Invariance priors require determination of the right-Haar prior corresponding to the group transformation. Reference priors (in multi-parameter problems) are often the most difficult objective priors to derive. Note that these derivations are generally considered part of the theoretical background of statistics; they are done by a researcher, who then records the result for general practical use. (This is not always the case; there are been derivations of challenging reference priors by practitioners — *cf.* Demortier et al. (2010), Grientschnig & Lira (2011), Casadei (2012) and Grientschnig (2015)).

It is possible to approach the derivation of an objective prior numerically. For instance, one could numerically compute the Jeffreys-rule prior at a particular value of $\boldsymbol{\theta}$ by simulating the expected value of the appropriate function of numerical derivatives of the log-likelihood function; indeed, one could easily develop a generic computer program to do so. This has not been done because it would be fairly intensive computationally, as it would need to be done thousands or millions of times within, say, an MCMC computation. We will return to this issue when discussing derivation of the reference prior.

7.7.2 Utilizing the Objective Prior in Computations

The constant prior and the Jeffreys-rule prior are often easy to use in performing Bayesian computations, since they often end up being conjugate for the model. This may, or may not, be true of other objective priors. When not conjugate, special computational algorithms may need to be developed to utilize the prior. Example 7.4 is one such situation, where the Jeffreys-rule prior is of conjugate form, but the (much superior) reference prior is not, and requires a special MCMC algorithm (described in Section 12.6).

Sometimes it is expensive to evaluate the objective prior, even when its form is known. In Example 7.2 for instance, evaluation of the reference prior requires multiple operations (including inversions) of potentially high dimensional matrices, and the cost can be prohibitive. Various approximations to the objective prior may then be needed.

7.8 Discussion and Additional References

7.8.1 Summary of Comparisons of Methods

Table 7.1 summarizes the examples and discussion from this chapter in regards to the 9 evaluation measures considered and the 5 main approaches to development of an objective prior.

Table 7.1: Performance ('no' is bad and 'yes' is good) of the five objective Bayesian methods with respect to the nine evaluation criteria. More details: [1]but is a proper posterior most of the time; [2]fine for invariant parameters; [3]but compatible with the Likelihood Principle within the 'universe' of the model; [4]examples of marginalization paradoxes exist; [5]no known examples of marginalization paradoxes; [6]for invariant parameters; [7]when the prior is conjugate.

Evaluation Criteria	Objective Bayes Priors				
	Constant	Jeffreys	Matching	Invariance	Reference
Post. Propriety	worst[1]	good	good	good	best
Diffuseness	worst	worst	good	good	best
Par. of Interest	no	no	yes	maybe[2]	yes
Par. Invariance	no	yes	yes	yes	yes
Likelihood Prin.	yes	no[3]	no[3]	no[3]	no[3]
Marginalization	no[4]	no[4]	no[4]	no[4]	yes[5]
Frequentist	worst	worst	good	good[6]	good
Derivation	it's 1	average	worse	worse	worst
Ease of MCMC	best	okay[7]	worse	okay[7]	worst

The conclusion from this table would seem to be that, unless computation (difficulty of derivation and difficulty of implementation through MCMC) is of primary importance, the reference prior approach to objective Bayesian analysis is best overall. Again, however, note that both the derivation of reference priors and development of MCMC algorithms for their implementation are available in the literature for many situations, so that simply using reference priors can be straightforward. See Eaton (1982) for another perspective.

7.8.2 Other Objective Bayesian Approaches

The appeal of objective Bayesian analysis has resulted in many attempts to create an objective Bayesian methodology over the last hundred years (rather arbitrarily setting the start to be Egon Pearson's effort in Pearson (1925)). Here we mention a few of the most prominent efforts that have not yet been discussed.

Lindley (1965) utilized a variety of intuitive arguments to derive objective priors for specific problems. The objective priors that were recommended were quite sensible; indeed, they were usually equal to the reference priors derived later. Of course, Lindley later abandoned objective Bayesian analysis in favor of subjective Bayesian analysis.

Section 13.7.1 discusses the maximum entropy approach to development of objective priors subject to constraints. This was quite popular in the physical sciences for a period of time, but has fallen out of favor. Zellner (1971, 1977) proposed an approach related to maximum entropy that he called 'maximal data information prior densities.' This approach had the disadvantage of not being invariant to reparameterization, which was the original problem with always using the constant prior.

Box & Tiao (1973) was based on the 'data-translated' likelihood approach to development of objective priors. In this approach, one transforms the data and parameter to achieve a location density (often only approximately), and then uses a constant prior in this transformed problem. (See Section 4.4.2 for the application of this approach to derive Jeffreys prior.) This approach produces reasonable priors when applicable (often the reference priors), but can require complications such as conditioning on ancillary statistics (*cf.* Kass (1990)) to achieve needed generality.

Other versions this approach include Hartigan (1996) and Fraser et al. (2010), which considers two approaches to development of objective priors, one based on extending the location approximation argument and one based on matching higher order approximations. Interestingly, these priors depend on the parameter of interest and, in general, are data-dependent, based on local approximations to the likelihood. The priors resulting from these approaches seem to be excellent (and, indeed, seem to be the only ones that generally achieve higher-order frequentist matching), although some might find the potential data dependence of the priors to be curious.

Approaches from computer science, based on the idea of finding the prior which can be coded with minimal code length, have been popular; see Rissanen (1988) and Barron et al. (1998), for examples.

An interesting topological approach to developing objective priors is given in Ghosal et al. (1997).

Another approach to developing objective priors is given in Villa & Walker (2014), Villa & Rubio (2018), Leisen et al. (2020). The proposal is to produce objective priors starting from a so-called self-information loss combined with utilization of the KullbackLeibler divergence between models.

Part III

Reference Analysis

Chapter 8

Introduction to Part III

8.1 Introduction

Table 7.1 summarized the comparison of the most prominent approaches to deriving objective priors, in regards to the desiderata discussed in that chapter. The conclusion was that *reference analysis* appears to be the only available approach that is optimal or nearly optimal according to each of the desiderata, with the possible exception of computation. Thus Part III of the book is devoted to an in-depth look at reference analysis.

This chapter introduces the key concepts and ideas behind reference analysis. There are three major components:

1. Tailoring the reference prior to the quantify of interest, a necessity that has been illustrated through many examples.

2. Defining an objective prior as a prior resulting in a maximal difference in information between the posterior and the prior because, then, the data is much more important than the prior. Note that because the posterior depends on the model, so will this difference in information (and, hence, the reference prior).

3. Applying reference analysis iteratively to a sequence of parameters, in order to eliminate the problems with multi-parameter problems that were seen with, say, the Jeffreys-rule prior.

These will be introduced heuristically in the next three sections of the chapter. The last section of the chapter illustrates these key ideas for two examples with discrete parameter spaces. Having a discrete parameter space greatly simplifies the calculations, and so is useful pedagogically.

Chapter 9 considers the case where θ is one-dimensional, and only information-based ideas are necessary to define a reference prior. An explicit expression for the reference prior will be given, an expression that holds in essentially complete generality (e.g., there is no need for the problem to have regular asymptotics).

Chapter 10 considers the multi-parameter case, where there is a parameter of interest and nuisance parameters, and finding the reference prior requires an iterative application of the information-based ideas. These two chapters provide the essence of the reference prior approach.

Chapter 11 considers the situation of discrete parameter spaces (*e.g.*, the parameter space of a hypergeometric distribution), where the reference prior method – if applied directly – yields a constant prior, which may not be optimal. Several method of dealing with such parameter spaces are discussed and illustrated.

Chapter 12 addresses the major pragmatic problem introduced by Point 1 above, that the reference prior can differ when one has different parameters of interest. It is obviously convenient to have a single objective prior distribution for a model and this chapter addresses how to develop such a prior – suggesting a trade-off between optimal objective priors for the different parameters.

Chapter 13 discusses the common issue where partial prior information is available. For instance, there may be an acknowledged scientific or historic (data-based) prior for the parameter of interest $\boldsymbol{\theta}$, or accepted priors for the nuisance parameters (*e.g.*, in high-energy physics). This chapter discusses how the reference prior method adapts to these situations.

Chapter 14 begins the discussion of the huge variety of situations in which reference priors have been derived. It is not possible for us to review all of the specific problems in which reference priors have been derived, but this chapter provides some of the major classes of problems where this has been done.

8.2 Importance of the Quantity of Interest

Section 7.4 discussed the need for objective priors to depend on the parameter of interest; further illustrations were given in Section 7.5.2. Note that recognition of this was a crucial point in the development of Bayesian statistics, for more reasons than just putting objective Bayesian analysis on the correct track. Indeed, some Bayesians decided that, if the objective Bayesian approach could not yield a single prior for a model, then it should be rejected in favor of the subjective Bayesian approach. Of course, others strived to adapt the objective Bayesian approach to meet this challenge.

To focus on this issue, the model with parameter vector $\boldsymbol{\omega}$ can be written (without loss of generality) in terms of parameters of interest $\boldsymbol{\theta} = \boldsymbol{\theta}(\boldsymbol{\omega})$ (typically one-dimensional) and a vector of nuisance parameters $\boldsymbol{\lambda} = \boldsymbol{\lambda}(\boldsymbol{\omega})$ such that $\{\boldsymbol{\theta}, \boldsymbol{\lambda}\}$ is a one-to-one reparameterization of $\boldsymbol{\omega}$; the model thus becomes

$$\mathcal{M} \equiv \{p(\boldsymbol{x} \,|\, \boldsymbol{\theta}, \boldsymbol{\lambda}),\ \boldsymbol{x} \in \boldsymbol{\mathcal{X}},\ \boldsymbol{\theta} \in \boldsymbol{\Theta}, \boldsymbol{\lambda} \in \boldsymbol{\Lambda}\}. \tag{8.1}$$

Later we will not insist that the parameter space be the product $\boldsymbol{\Theta} \times \boldsymbol{\Lambda}$, but it is a useful assumption for the pedagogical introduction.

8.3 Maximizing Missing Information

KL divergence, given in Definition 1.5, provides an excellent measure of the difference between two probability distributions. Thus $\kappa\{\pi(\boldsymbol{\theta}) \mid \pi(\boldsymbol{\theta} \mid \boldsymbol{x})\}$ can be viewed as the difference in the information about $\boldsymbol{\theta}$ provided by the posterior distribution from that in the prior distribution. If we view the prior distribution as reflecting the original information about $\boldsymbol{\theta}$, then $\kappa\{\pi(\boldsymbol{\theta}) \mid \pi(\boldsymbol{\theta} \mid \boldsymbol{x})\}$ can be viewed as the information provided by the data $\boldsymbol{x}$. The average information about $\boldsymbol{\theta}$ (averaging over $\boldsymbol{x}$) is then defined as follows.

Definition 8.1 Expected information in the observation. *The information to be expected from one observation from model $\mathcal{M} \equiv \{p(\boldsymbol{x} \mid \boldsymbol{\theta}), \boldsymbol{x} \in \boldsymbol{\mathcal{X}}, \boldsymbol{\theta} \in \Theta\}$, when the proper prior for $\boldsymbol{\theta}$ is $\pi(\boldsymbol{\theta})$, is*

$$
\begin{aligned}
I\{\pi \mid \mathcal{M}\} &= E\left[\kappa\{\pi(\boldsymbol{\theta}) \mid \pi(\boldsymbol{\theta} \mid \boldsymbol{x})\}\right] \\
&= \int\int_{\boldsymbol{\mathcal{X}}\times\Theta} p(\boldsymbol{x} \mid \boldsymbol{\theta})\pi(\boldsymbol{\theta}) \log \frac{\pi(\boldsymbol{\theta} \mid \boldsymbol{x})}{\pi(\boldsymbol{\theta})}\, d\boldsymbol{x}\, d\boldsymbol{\theta}\,. \qquad (8.2)
\end{aligned}
$$

The amount of information, $I\{\pi \mid \mathcal{M}\}$, to be expected from observing $\boldsymbol{x}$ from $\mathcal{M}$ depends on the prior $\pi(\boldsymbol{\theta})$; the sharper the prior the smaller the amount of information to be expected from the data. (In the extreme, a point mass prior provides all the information that is possible, so the data can add no information.) Thus the prior that maximizes $I\{\pi \mid \mathcal{M}\}$ can claim to be that prior which maximizes the amount of information provided by the data, and is a natural candidate for an objective prior.

Unfortunately, the resulting prior turns out to be problematical, in that it is very difficult to compute and it will typically turn out to be a discrete prior, even when $\boldsymbol{\theta}$ is a continuous parameter. Furthermore, the discrete points of support will often depend on the sample size. These properties of the prior essentially preclude it from being useful in practice. See Berger et al. (1989) for discussion.

An alternative, proposed in Bernardo (1979*b*), is to apply this idea asymptotically, considering k independent replications of $\mathcal{M}$, resulting in the sequence of realizations $\boldsymbol{x}^{(k)} = \{\boldsymbol{x}_1, \ldots, \boldsymbol{x}_k\}$ and the ensuing replicated model $\mathcal{M}^k$. Then $I\{\pi \mid \mathcal{M}^k\}$ is the amount of information in $\boldsymbol{x}^{(k)}$. As $k \to \infty$, the posterior distribution will typically converge to a point mass, providing complete information about $\boldsymbol{\theta}$. Thus, intuitively, $\lim_{k\to\infty} I\{\pi \mid \mathcal{M}^k\}$ provides a measure of the missing information about θ associated to the prior $\pi(\theta)$. Maximizing this missing information is thus another natural way to define an objective prior and is the basis of the *reference prior* approach that we will be discussing. For various reasons, some technical and one substantive (discussed in the next section), we do not work directly with $\lim_{k\to\infty} I\{\pi \mid \mathcal{M}^k\}$, but the exact approach that will be taken captures this idea.

8.4 Determining the Reference Prior Through Iteration

Applying the idea of maximizing missing information, discussed in the previous section, works wonderfully if θ is one-dimensional but fails – if utilized directly – when $\boldsymbol{\theta}$ is multi-dimensional; indeed, the result will often be the multivariate Jeffreys-rule prior, which was seen in Section 4.3 to be highly problematical. The third key component of the reference prior approach is thus to apply the idea of maximizing missing information sequentially, one parameter at a time.

To see the idea, suppose θ is a one-dimensional parameter of interest and that we are told that the 'correct' prior for the nuisance parameter is $\pi(\boldsymbol{\lambda}\,|\,\theta)$ (the prior being allowed to depend on θ and assumed here to be proper). Then $\boldsymbol{\lambda}$ could be integrated out of the model, obtaining the marginal model

$$\mathcal{M}^* \equiv \{p(\boldsymbol{x}\,|\,\theta),\ \boldsymbol{x} \in \boldsymbol{\mathcal{X}},\ \theta \in \Theta,\},\quad p(\boldsymbol{x}\,|\,\theta) = \int_{\boldsymbol{\Lambda}} p(\boldsymbol{x}\,|\,\theta, \boldsymbol{\lambda})\,\pi(\boldsymbol{\lambda}\,|\,\theta)\,d\boldsymbol{\lambda}. \tag{8.3}$$

As this marginal model only depends on the one-dimensional parameter θ, the method of maximizing missing information can be applied to obtain the (marginal) reference prior $\pi^*(\theta)$. The final overall prior would then simply be $\pi(\theta, \boldsymbol{\lambda}) = \pi(\boldsymbol{\lambda}\,|\,\theta)\pi^*(\theta)$.

If no such conditional prior is known, one may work with model (8.1) conditionally on θ to obtain the conditional reference prior $\pi(\boldsymbol{\lambda}\,|\,\theta)$, as that which maximizes the missing information about $\boldsymbol{\lambda}$ given θ; use this in (8.3) to obtain the marginal model $\mathcal{M}^*$; maximize the missing information about θ in this model to obtain the marginal reference prior $\pi^*(\theta)$; and use both to define the required *joint* reference prior as $\pi(\theta, \boldsymbol{\lambda}) = \pi(\boldsymbol{\lambda}\,|\,\theta)\,\pi^*(\theta)$.

Unfortunately, the conditional reference prior $\pi(\boldsymbol{\lambda}\,|\,\theta)$ is typically not proper, in which case $p(\boldsymbol{x}\,|\,\theta)$ is not proper and so cannot be used to determine $\pi^*(\theta)$. To overcome this difficulty we operate, as in Section 2.3.3, on compact subsets of the parameter space which increase to the full space, taking appropriate limits of the constrained reference priors to define the reference prior on the full space.

8.5 Discrete Parameter Space Illustrations

To illustrate the three components of the reference prior method, we start with two discrete parameter examples, where the parameter space is $\Theta = \{\theta_1, \theta_2, \ldots, \theta_m\}$ the θ_i being distinct points in the sense that the KL divergence between $p(\boldsymbol{x}\,|\,\theta_i)$ and $p(\boldsymbol{x}\,|\,\theta_j)$ is nonzero for any two different parameter values. Discrete parameter spaces are simple in that maximizing missing information becomes the same as choosing what is called a maximum entropy distribution. Indeed, for k replications of the experiment, $\boldsymbol{x}^{(k)}$, the missing information is

$$\lim_{k\to\infty} I\{\pi\,|\,\mathcal{M}^k\} = \sum_{i=1}^{m} \lim_{k\to\infty} \int_{\boldsymbol{\mathcal{X}}^{(k)}} p(\boldsymbol{x}^{(k)} \mid \theta_i)\pi(\theta_i) \log \frac{\pi(\theta_i \mid \boldsymbol{x}^{(k)})}{\pi(\theta_i)}\, d\boldsymbol{x}^{(k)}$$

$$= - \sum_{i=1}^{m} \pi(\theta_i) \log \pi(\theta_i) \,, \tag{8.4}$$

the *entropy* of $\pi(\cdot)$. Equation 8.4 follows because, for a discrete parameter space, $\lim_{k \to \infty} \pi(\theta_i \mid \boldsymbol{x}^{(k)}) = 1$, for $\boldsymbol{x}^{(k)}$ arising from $p(\boldsymbol{x}^{(k)} \mid \theta_i)$. To maximize the missing information, one thus wants the *maximum entropy* distribution on Θ, which is well-known to be the uniform density $\pi(\theta_i) = 1/m$.

Example 8.1 *Medical diagnosis.* Let $\boldsymbol{x}$ be a vector which contains the results of a blood analysis designed to provide information on whether or not a person suffers from hepatitis, and suppose that, from laboratory tests, the sampling distribution of $\boldsymbol{x}$ has been determined for healthy people (ω_0) and for m different types of hepatitis patients $\{\omega_1, \ldots, \omega_m\}$. Suppose that it is desired to obtain an objective posterior probability of whether or not a patient has *some* type of hepatitis.

The model can be parametrized in terms of the quantity of interest $\theta = 1$ if the patient is healthy and $\theta = 0$ if the patient has hepatitis, and a nuisance parameter λ (the actual type of hepatitis if not healthy), so that the model becomes

$$p(\boldsymbol{x} \mid \theta, \lambda) = \begin{cases} p(\boldsymbol{x} \mid \omega_0), & \text{if } \theta = 1 \\ p(\boldsymbol{x} \mid \omega_j), & \text{if } \theta = 0 \text{ and } \lambda = j, \ j = 1, \ldots, m. \end{cases}$$

The conditional reference prior for λ given $\theta = 0$ (there is no distribution for λ given $\theta = 1$) will be the uniform distribution $\pi(\lambda_j \mid \theta = 0) = 1/m$, as discussed above. In determining the marginal reference prior for θ, there is no need to marginalize over λ, since θ is in the discrete parameter space $\Theta = \{0, 1\}$ and, hence, the marginal reference prior must be $\pi(\theta = 1) = \pi(\theta = 0) = 1/2$. The joint reference prior $\pi(\lambda \mid \theta)\pi(\theta)$ is most easily written by transforming back to the ω_j, and is

$$\pi(\omega) = \begin{cases} 1/2, & \text{if } \omega = \omega_0 \\ 1/2m, & \text{if } \omega = \omega_j, \ j = 1, \ldots, m. \end{cases}$$

This yields the objective posterior probability that the patient is healthy

$$\pi(\omega_0 \mid \boldsymbol{x}) = \frac{\frac{1}{2} p(\boldsymbol{x} \mid \omega_0)}{\frac{1}{2} p(\boldsymbol{x} \mid \omega_0) + \sum_{i=1}^{m} \frac{1}{2m} p(\boldsymbol{x} \mid \omega_i)} = \frac{p(\boldsymbol{x} \mid \omega_0)}{p(\boldsymbol{x} \mid \omega_0) + \frac{1}{m} \sum_{i=1}^{m} p(\boldsymbol{x} \mid \omega_i)} \,, \tag{8.5}$$

which is roughly independent of the number m of different hepatitis types being considered (as makes sense).

The key point here is that, if one was not focusing on the parameter of interest (being healthy or having hepatitis), one might just use the direct reference prior for the parameter space $\{\omega_0, \omega_1, \ldots, \omega_m\}$, which would be the uniform prior $\pi(\omega_i) = 1/(m+1)$. Then the objective posterior probability that the patient is healthy, noticing that all the prior probabilities

$1/(m+1)$ cancel, is

$$\pi(\omega_0 \mid \boldsymbol{x}) = \frac{p(\boldsymbol{x} \mid \omega_0)}{p(\boldsymbol{x} \mid \omega_0) + \sum_{i=1}^{m} p(\boldsymbol{x} \mid \omega_i)},$$

and the term $\sum_{i=1}^{m} p(\boldsymbol{x} \mid \omega_i)$ is not downweighted by $1/m$ as it was in (8.5). So, here, as more types of hepatitis are identified, the posterior probability of being healthy decreases, which makes no sense.

Of course, the more common approach to this problem would have been to view it as a hypothesis testing problem, testing $H_0 : \omega = \omega_0$ versus $H_1 : \omega \neq \omega_0$, and objective Bayesian testing would have also arrived at the correct answer (8.5). Still, the example indicates the importance of focusing on the question (parameter) of interest.

Example 8.2 *Induction.* Consider a finite population of (known) size N, all of whose elements may or may not have a specified property. A random sample of size n is taken without replacement, and all the elements in the sample turn out to have that property. Scientific interest often centers on the probability that all the N elements in the population have the property under consideration (often called 'natural induction').

As an example, if all the $n = 225$ randomly chosen turtles in an isolated volcanic island are found to show a particular difference with respect to those in the mainland, zoologists would tend to believe that all the turtles in the island share that property and would like to determine the probability that this is so. (This was a question posed to one of the authors by zoologists at the Galápagos Islands in the eighties.)

Formally, if r and R respectively denote the number of elements in the sample and in the population, which have the property under study, the statistical model is

$$\mathcal{M} \equiv \Big\{ p(r \mid n, R, N), \quad r \in \{0, \ldots, n\}, \quad R \in \{0, \ldots, N\} \Big\},$$

where R is the unknown parameter, and

$$p(r \mid n, R, N) = \binom{R}{r} \binom{N-R}{n-r} \Big/ \binom{N}{n} \tag{8.6}$$

is the relevant hypergeometric distribution. Once the prior, $p(R)$, for the unknown number R of elements in the population which have the property, has been determined, Bayes theorem yields the desired posterior probability (noting that, necessarily, $R \geq n$) as

$$p(R = N \mid r = n, N) = \frac{p(r = n \mid n, R = N, N)\, p(R = N)}{\sum_{R=n}^{N} p(r = n \mid n, R, N)\, p(R)}. \tag{8.7}$$

If the parameter of interest were R itself, a possible objective prior would be the maximum entropy uniform prior over its range as discussed above, namely $p(R) = (N+1)^{-1}$ (but see Section 11.4 for an improvement). Using (8.7) and the identity

$$\sum_{R=n}^{N} p(r = n \,|\, n, R, N) = \frac{N+1}{n+1}\,, \tag{8.8}$$

this results in the posterior probability

$$p(R = N \,|\, r = n, N) = \frac{n+1}{N+1}\,,$$

which will be very small when (as is usually the case) the sampling fraction n/N is small. This is a solution that most scientists would not accept when n is relatively large (e.g., if 225 randomly selected turtles on the island all have a certain property, most zoologists would conclude that all turtles on the island probably have the property). If the modified reference prior for the hypergeometric model (11.6) is used, the results are a bit better, but qualitatively similar.

Again, however, the quantity of interest here is not the value of R but whether or not $R = N$, and a reference prior is desired which maximizes the missing information about this specific question. Rewrite the unknown parameter as $R = (\theta, \lambda)$, where $\theta = 1$ if $R = N$ and $\theta = 0$ otherwise, and $\lambda = R$; thus we have explicitly parameterized the quantity of interest θ.

Clearly $\pi(\lambda \,|\, \theta = 1)$ is a point mass at $\lambda = N$ and $\pi(\lambda \,|\, \theta = 0)$ is the conditional reference prior for the discrete space $\{\lambda \in \{0, 1, 2, \ldots, N-1\}\}$, which, as argued above, is the uniform distribution $\pi(\lambda \,|\, \theta = 0) = 1/N$. And, as in the previous example, the marginal reference prior for θ will be $\pi(\theta = 0) = \pi(\theta = 1) = 1/2$.

Recall that the observation is $r = n$. Bayes theorem yields the posterior probability of $\theta = 1$ (noting that $\lambda = R \geq n$ when $\theta = 0$ and, from (8.8), that $\sum_{\lambda=n}^{N-1} p(n \,|\, \theta = 0, \lambda, N) = (N-n)/(n+1)$)

$$\begin{aligned}
\pi(\theta = 1 \,|\, r = n, N) &= \frac{p(n \,|\, \theta = 1, N)\pi(1)}{p(n \,|\, \theta = 1, N)\pi(1) + \sum_{\lambda=n}^{N-1} p(n \,|\, \theta = 0, \lambda, N)\pi(0, \lambda)} \\
&= \frac{1 \times \frac{1}{2}}{1 \times \frac{1}{2} + \frac{1}{2N} \sum_{\lambda=n}^{N-1} p(n \,|\, \theta = 0, \lambda, N)} \\
&= \frac{1 \times \frac{1}{2}}{1 \times \frac{1}{2} + \frac{1}{2N} \frac{N-n}{n+1}} = \left[1 + \frac{1}{n+1}\left(1 - \frac{n}{N}\right)\right]^{-1} \\
&\approx \frac{n+1}{n+2}\,.
\end{aligned} \tag{8.9}$$

So, while (8.8) is unrealistically small when N is large relative to n, (8.9) behaves very sensibly, being close to 1 when n is large, regardless of N.

For the turtles example, this result yields

$$p(\theta = 1 \,|\, r = n = 225, N) = p(R = N \,|\, r = n = 225,\, N) \approx 0.995$$

for all large N. This was viewed as a very reasonable result by the zoologists at the Galápagos Islands, and they were happy with the approximation because they then did not have to try to determine N.

Again, this result could have been obtained by the more common approach of performing objective Bayesian testing of $H_0 : \theta = 1$ (or $H_0 : R = N$) versus $H_1 : \theta = 0$ (or $H_0 : R < N$). But it is instructive to see that, even if not properly formulated as a testing problem, focusing on the parameter of interest and using the reference prior approach arrives at the correct objective Bayesian answer.

For further discussion of this problem (with important applications in philosophy of science, physical sciences and reliability), see Jeffreys (1961, pp. 128–132), Geisser (1984), Bernardo (1985*a*) and Singpurwalla & Wilson (2004).

8.6 Discussion and Additional References

8.6.1 History of the Reference Prior Approach

The original formulation of reference priors in Bernardo (1979*a*) was largely informal. In continuous one parameter problems, heuristic arguments were given to justify an explicit expression in terms of the expectation under sampling of the logarithm of the asymptotic posterior density, which reduced to Jeffreys prior under asymptotic posterior normality. In multiparameter problems it was argued that one should not maximize the joint missing information but proceed sequentially, thus avoiding known problems such as marginalization paradoxes. Berger and Bernardo (1992*a*) gave more precise definitions of this sequential reference process, but they restricted consideration to continuous multiparameter problems under asymptotic posterior normality. Ghosal & Samanta (1997) and Ghosal (1997) provided explicit results for reference priors in some types of nonregular models. More recently, Berger, Bernardo and Sun (2009*a*) provided a general, formal definition of reference priors, and an explicit algorithm to compute them which may also be used for numerical derivation.

Lindley (1956) was the first to introduce Shannon's information measures into Bayesian statistics. For additional references on information, see Barnard (1951), Stone (1959), Ibragimov & Khasminskii (1973), Basu (1975), Bernardo (1979*a*), Akaike (1983), Dawid (1984), Rissanen (1986), Haussler & Opper (1997), de Waal & Groenewald (1989), Clarke & Barron (1990, 1994), Barron (1999), and references therein.

Chapter 9

Models with One Continuous Parameter

9.1 Formal Definition of the Reference Prior

Let $\boldsymbol{x}$ be the result of an experiment which consists of one observation from model $\mathcal{M} = \{p(\boldsymbol{x} \mid \theta), \boldsymbol{x} \in \boldsymbol{\mathcal{X}}, \theta \in \Theta \subset \mathrm{I\!R}\}$. Note that $\boldsymbol{x}$ represents the *complete* set of available data; often, but not necessarily, $\boldsymbol{x}$ will be a random sample $\boldsymbol{x} = (x_1, \ldots, x_n)$ from some simpler model.

The reference prior will be defined in terms of the asymptotic behavior of the model. The asymptotics we consider is with respect to an increasing number of replications of the data. Thus let $\boldsymbol{x}^{(k)} = \{\boldsymbol{x}_1, \ldots, \boldsymbol{x}_k\}$ be the result of k conditionally independent replications of the original experiment, so that

$$p(\boldsymbol{x}^{(k)} \mid \theta) = \prod_{j=1}^{k} p(\boldsymbol{x}_j \mid \theta), \quad \boldsymbol{x}^{(k)} \in \boldsymbol{\mathcal{X}}^k, \quad \theta \in \Theta \subset \mathrm{I\!R},$$

with $\mathcal{M}^k$ denoting the corresponding model.

As in (8.2), the amount of information about θ, which $\mathcal{M}^k$ is expected to provide, under the proper prior $\pi(\theta)$, is

$$\mathcal{I}\{\pi \mid \mathcal{M}^k\} = \int_{\Theta} \pi(\theta) \int_{\boldsymbol{\mathcal{X}}^k} p(\boldsymbol{x}^{(k)} \mid \theta) \log \left[\frac{\pi(\theta \mid \boldsymbol{x}^{(k)})}{\pi(\theta)} \right] d\boldsymbol{x}^{(k)} \, d\theta \,. \tag{9.1}$$

As discussed in Section 8.3, perfect knowledge about θ will be approached as k grows, so that, as $k \to \infty$, $\mathcal{I}\{\pi \mid \mathcal{M}^k\}$ will approach the missing information about θ which corresponds to the prior $\pi(\theta)$. That prior which maximizes this missing information, the *reference prior*, is a natural objective prior.

The formal definition of a reference prior is as follows, taken from Berger et al. (2009*a*). In the definition, it is presumed that one has a class $\mathcal{P}$ of candidate priors and that one wants the most objective prior in this class. Usually, $\mathcal{P}$ will be taken to be the class of all reasonable priors, proper and improper. That improper priors cannot be used directly in (9.1) will be handled by operating on an increasing sequence of compact subsets of the parameter space.

Definition 9.1 ***(Reference Prior).*** *Let $\mathcal{M} \equiv \{p(\boldsymbol{x} \mid \theta),\, \boldsymbol{x} \in \boldsymbol{\mathcal{X}},\, \theta \in \Theta \in \mathbb{R}\}$ be a model with one continuous parameter θ, and let $\mathcal{P}$ be a class of prior functions for θ for which $\int_\Theta p(\boldsymbol{x} \mid \theta)\, p(\theta) d\theta < \infty$. A function $\pi(\theta) = \pi(\theta \mid \mathcal{M}, \mathcal{P})$ is the* reference prior *for model $\mathcal{M}$ given $\mathcal{P}$ if*

(i) *it is permissible (see Definition 2.4) and*

(ii) *for any compact set $\Theta_0 \in \Theta$, any $p \in \mathcal{P}$, and any k, $I\{\pi_0 \mid \mathcal{M}^k\}$ and $\mathcal{I}\{p_0 \mid \mathcal{M}^k\}$ are finite, where π_0 and p_0 are, respectively, the renormalized restrictions of $\pi(\theta)$ and $p(\theta)$ to the compact set Θ_0, and*

$$\lim_{k \to \infty} \{\mathcal{I}\{\pi_0 \mid \mathcal{M}^k\} - \mathcal{I}\{p_0 \mid \mathcal{M}^k\}\} \geq 0\,. \tag{9.2}$$

The restriction of the definition to a compact set typically ensures the existence of the missing information for given k. Although the definition is posed for any compact set, the key idea is that it needs to hold for a sequence of compact sets that grows to Θ so that we have a sequence of models that converges to $\mathcal{M}$. That the missing information will diverge for large k is handled by insisting that the missing information for the reference prior be larger, as $k \to \infty$, than the missing information for any other candidate $p(\theta)$.

In simple examples, it may possible to verify that a particular prior function is a reference prior by direct use of Definition 9.1, but an explicit expression for the reference prior will be provided in Section 9.4.

9.2 Properties of Reference Priors

Important properties of reference priors can immediately be deduced from their definition.

Proposition 9.1 Independence of sample size. *If the data consists of a random sample from model $\mathcal{M} = \{p(\boldsymbol{x} \mid \theta), \boldsymbol{x} \in \boldsymbol{\mathcal{X}},\, \theta \in \Theta\}$, then the reference prior is the same for any sample size n.*

Proof. k independent replicates of a random sample $(\boldsymbol{x}_1, \ldots, \boldsymbol{x}_n)$ is equivalent to a random sample of size kn, and $k \to \infty$ is equivalent to $kn \to \infty$. □

Note that if observations are dependent, as in time series, spatial, or hierarchical models, the reference prior may well depend on the sample size; see, respectively, Berger & Yang (1994), Berger, De Oliveira & Sansó (2001), and (1.9).

Proposition 9.2 Compatibility with sufficient statistics. *Consider the model $\mathcal{M} = \{p(\boldsymbol{x} \mid \theta),\, \boldsymbol{x} \in \boldsymbol{\mathcal{X}},\, \theta \in \Theta\}$ with sufficient statistic $\boldsymbol{t} = \boldsymbol{t}(\boldsymbol{x}) \in \boldsymbol{\mathcal{T}}$, and let*

$\mathcal{M}_t = \{p(\boldsymbol{t}\,|\,\theta), \boldsymbol{t} \in \boldsymbol{\mathcal{T}},\, \theta \in \Theta\}$ *be the corresponding model in terms of* $\boldsymbol{t}$. *Then,* $\pi(\theta\,|\,\mathcal{M},\mathcal{P}) = \pi(\theta\,|\,\mathcal{M}_t,\mathcal{P})$.

Proof. This follows because expected information is invariant under such transformations, so that, for all k, $\mathcal{I}\{q\,|\,\mathcal{M}^k\} = \mathcal{I}\{q\,|\,\mathcal{M}_t^k\}$. □

Recall that obtaining invariance under reparameterization was the first crucial step in modern objective Bayesian theory.

Proposition 9.3 Invariance under reparameterization. *Consider the model* $\mathcal{M}_1 = \{p(\boldsymbol{x}\,|\,\theta),\, \boldsymbol{x} \in \boldsymbol{\mathcal{X}},\, \theta \in \Theta\}$, *let* $\phi(\theta)$ *be an invertible transformation of* θ, *and let* $\mathcal{M}_2$ *be the model parametrized in terms of* ϕ. *Then,* $\pi(\phi\,|\,\mathcal{M}_2,\mathcal{P})$ *is the prior density induced from* $\pi(\theta\,|\,\mathcal{M}_1,\mathcal{P})$ *by the appropriate change of variables.*

Proof. This follows from the fact that the expected information is invariant under one-to-one reparameterization, so that, for all k, $I\{q_1\,|\,\mathcal{M}_1^k\} = I\{q_2\,|\,\mathcal{M}_2^k\}$, where $q_2(\phi) = q_1(\theta) \times |\partial\theta/\partial\phi|$. □

Often, instead of operating with the entire parameter space Θ, one has constraints on the parameter (e.g., a normal mean is known to be positive), leading to the model having parameter space $\Theta_0 \subset \Theta$.

Proposition 9.4 *If* θ *is restricted to a subset* Θ_0 *of* Θ, *the reference prior is simply that for the unconstrained problem, restricted to* Θ_0.

Proof. Since any compact subset of Θ_0 is also a compact subset of Θ, the proof is immediate from Definition 9.1. □

This is an attractive property, in that one does not have to derive the reference prior for each possible constrained parameter space.

9.3 Existence of the Expected Information

The definition of a reference prior requires that $\mathcal{I}\{\pi_0\,|\,\mathcal{M}^k\}$ and $\mathcal{I}\{p_0\,|\,\mathcal{M}^k\}$ be finite. Here are some conditions under which this is so, taken from Berger et al. (2009*a*) and in which the proofs of the results can be found. This is a technical section that can be skipped by those interested primarily in methodology.

Definition 9.2 *Consider the model* $\mathcal{M} \equiv \{p(\boldsymbol{x}\,|\,\theta),\, \boldsymbol{x} \in \boldsymbol{\mathcal{X}},\, \theta \in \Theta \in \mathbb{R}\}$. *The class* $\mathcal{P}_s$ *of* standard prior functions *for model* $\mathcal{M}$ *is the set of strictly positive and continuous prior functions on* Θ *which have proper formal posterior distributions so that,*

$$\forall\theta \in \Theta,\ \forall p \in \mathcal{P}_s,\quad p(\theta) > 0 \quad and \quad \forall \boldsymbol{x} \in \boldsymbol{\mathcal{X}},\ \int_\Theta p(\boldsymbol{x}\,|\,\theta)\,p(\theta)d\theta < \infty.$$

This will be the class of priors that we typically use to define the reference prior. The motivation for requiring a standard prior to be positive and continuous on Θ is that any prior not satisfying these conditions would be difficult to accept as reasonable objective prior.

Definition 9.3 Standard models. *Let* $\mathcal{M} \equiv \{p(\boldsymbol{x} \mid \theta), \boldsymbol{x} \in \boldsymbol{\mathcal{X}}, \theta \in \Theta \subset \mathbb{R}\}$ *be a model with one continuous parameter. The model* $\mathcal{M}$ *is said to be* standard *if, for any prior function* $p(\theta) \in \mathcal{P}_s$ *and any compact set* Θ_0,

$$\mathcal{I}\{p_0 \mid \mathcal{M}^k\} < \infty\,, \tag{9.3}$$

where $p_0(\theta)$ *is the proper prior obtained by restricting* $p(\theta)$ *to* Θ_0.

There are a variety of conditions under which (9.3) is satisfied. It is most useful, for ease in application, to present them in terms of a sufficient statistic $\boldsymbol{t}_k = \boldsymbol{t}_k(\boldsymbol{x}_1, \ldots, \boldsymbol{x}_k) \in \boldsymbol{\mathcal{T}}_k$ for $\mathcal{M}^k$ ($\boldsymbol{t}_k$ could be just the observation vectors themselves). The first result is useful when all $p(\boldsymbol{t}_k \mid \theta)$, for $\theta \in \Theta_0$, have the same support.

Proposition 9.5 *For* $p(\theta) \in \mathcal{P}_s$ *and any compact set* Θ_0, *(9.3) is satisfied if, for any* θ_1 *and* θ_2 *in* Θ_0, $\kappa\{p(\boldsymbol{t}_k \mid \theta_1) \mid p(\boldsymbol{t}_k \mid \theta_2)\} < \infty$.

When the $p(\boldsymbol{t}_k \mid \theta)$ can have different supports for $\theta \in \Theta_0$, the following result can be useful to verify (9.3).

Proposition 9.6 *For* $p(\theta) \in \mathcal{P}_s$ *and any compact set* Θ_0, *the two conditions*

(i) $\mathcal{H}[p(\boldsymbol{t}_k \mid \theta)] \equiv -\int_{\boldsymbol{\mathcal{T}}_k} p(\boldsymbol{t}_k \mid \theta) \log[p(\boldsymbol{t}_k \mid \theta)]\, d\boldsymbol{t}_k$ *is bounded below over* Θ_0;

(ii) $\int_{\boldsymbol{\mathcal{T}}_k} p_0(\boldsymbol{t}_k) \log[p_0(\boldsymbol{t}_k)]\, d\boldsymbol{t}_k > -\infty$, *where* $p_0(\boldsymbol{t}_k) = \int_{\Theta_0} p(\boldsymbol{t}_k \mid \theta)\, d\theta \Big/ \int_{\Theta_0} d\theta$;

are jointly sufficient for (9.3) to be satisfied.

9.4 An Explicit Expression for the Reference Prior

The following theorem, from Berger et al. (2009*a*) in which the proof can be found, provides an explicit expression for the reference prior, under mild conditions. Recall that $\boldsymbol{x}$ refers to the entire vector of observations from the model, $\boldsymbol{x}^{(k)} = \{\boldsymbol{x}_1, \ldots, \boldsymbol{x}_k\}$ refers to a vector of (artificial) independent replicates of these vector observations from the model, and $\boldsymbol{t}_k = \boldsymbol{t}_k(\boldsymbol{x}^{(k)}) \in \boldsymbol{\mathcal{T}}_k$ is a sufficient statistic for the replicated observations.

Proposition 9.7 Explicit form of the reference prior.

Consider a standard model $\mathcal{M} \equiv \{p(\boldsymbol{x} \,|\, \theta), \boldsymbol{x} \in \boldsymbol{\mathcal{X}}, \theta \in \Theta \subset \mathrm{I\!R}\}$, and let $\mathcal{P}_s$ be the corresponding class of standard priors. Let $\pi^*(\theta)$ be a continuous strictly positive function such that the corresponding formal posterior

$$\pi^*(\theta \,|\, \boldsymbol{t}_k) = \frac{p(\boldsymbol{t}_k \,|\, \theta)\, \pi^*(\theta)}{\int_\Theta p(\boldsymbol{t}_k \,|\, \theta)\, \pi^*(\theta)\, d\theta} \tag{9.4}$$

is proper and asymptotically consistent (see Section 1.7) and, for any interior point $\theta_0 \in \Theta$, define

$$f_k(\theta) = \exp\left\{\int_{\boldsymbol{\mathcal{T}}_k} p(\boldsymbol{t}_k \,|\, \theta)\, \log\left[\pi^*(\theta \,|\, \boldsymbol{t}_k)\right] d\boldsymbol{t}_k\right\} \quad \textit{and} \tag{9.5}$$

$$f(\theta) = \lim_{k\to\infty} \frac{f_k(\theta)}{f_k(\theta_0)}\ . \tag{9.6}$$

Then $\pi(\theta \,|\, \mathcal{M}, \mathcal{P}_s) = f(\theta)$ is a reference prior *for model $\mathcal{M}$ and prior class $\mathcal{P}_s$ if (i) it is a permissible prior function, and*

(ii) $f_k(\theta)$ is continuous and, for any fixed θ and sufficiently large k, $\{f_k^0(\theta)/f_k^0(\theta_0)\}$ is either monotonic in k or is bounded above by some $h(\theta)$ which is integrable on any compact set.

The choice of π^* is essentially arbitrary and can be chosen for computational convenience, and the choice of θ_0 is immaterial. As opposed to Definition 9.1, note that no compact set is mentioned in the theorem; indeed, (9.6) satisfies Definition 9.1 simultaneously for all compact subsets of Θ.

Example 9.1 *Exponential data.* Suppose x is one observation from the exponential distribution $p(x \,|\, \theta) = \theta e^{-\theta x}$, $\theta > 0$. The i.i.d. replication data is the vector $\boldsymbol{x}^{(k)} = \{x_1, \ldots, x_k\}$ and using (for mathematical convenience) $\pi^*(\theta) = \theta^{-1}$, the corresponding posterior distribution is $\pi^*(\theta \,|\, t_k) = \mathrm{Ga}(\theta \,|\, k, t_k)$, where $t_k = \sum_{j=1}^k x_j$ is a sufficient statistic. The sampling distribution of t_k is $p(t_k \,|\, \theta) = \mathrm{Ga}(t_k \,|\, k, \theta)$ and the required functions are

$$f_k(\theta) = \exp\left[\int_0^\infty \mathrm{Ga}(t_k \,|\, k, \theta)\, \log\left(\mathrm{Ga}(\theta \,|\, k, t_k)\right) dt_k\right] = c_k\, \frac{1}{\theta}\ ,$$

where c_k is a constant. It follows that, for all k (not just in the limit), $f_k^0(\theta)/f_k^0(\theta_0) = \theta_0/\theta$. Since $f(\theta) = \theta^{-1}$ is a permissible prior for this problem (see Example 2.20), the reference prior is $\pi^*(\theta \,|\, \mathcal{M}, \mathcal{P}_s) = \theta^{-1}$.

Example 9.2 *(Uniform data, Example 2.21, continued.)* Suppose x is one observation from a uniform distribution $\{p(x \,|\, \theta) = \theta^{-1}, 0 < x < \theta, \theta > 0\}$. Here the sampling space is $\mathcal{X} = [0, \theta]$, which depends on the parameter θ, so that Fisher information is not defined and, hence, there is no Jeffreys-rule prior.

The i.i.d. replication data is the vector $\boldsymbol{x}^{(k)} = \{x_1, \ldots, x_k\}$ and using (for mathematical convenience) $\pi^*(\theta) = \theta^{-1}$, the corresponding posterior distribution is the Pareto density.

$$\pi^*(\theta \mid t_k) = \mathrm{Pa}(\theta \mid k, t_k) = k\,\frac{t_k^k}{\theta^{k+1}}\,, \quad \theta > t_k, \tag{9.7}$$

where $t_k = \max\{x_1, \ldots, x_k\}$ is a sufficient statistic.

The sampling distribution of t_k is the inverse Pareto density

$$p(t_k \mid \theta) = \mathrm{Ip}(t_k \mid k, \theta^{-1}) = k\,\frac{t^{k-1}}{\theta^k}\,, \quad 0 < t_k < \theta. \tag{9.8}$$

Computation yields

$$f_k(\theta) = \exp\left[\int_0^\theta \mathrm{Ip}(t_k \mid k, \theta)\, \log\left(\mathrm{Pa}(\theta \mid k, t_k)\right) dt_k\right] = \frac{c(k)}{\theta}\,, \tag{9.9}$$

where $c(k) = k\,e^{-(1+1/k)}$. It follows that, for all k (not just in the limit), $f_k^0(\theta)/f_k^0(\theta_0) = \theta_0/\theta$. Since $\pi(\theta) = \theta^{-1}$ is a permissible prior for this problem (see Example 2.21), it is the reference prior.

Proposition 9.8 (Location models and scale models) *Consider a* loca-tion *model $\mathcal{M}_1$, so that for some function f_1, $\mathcal{M}_1 \equiv \{f_1(x-\mu),\ x \in \mathbb{R},\ \mu \in \mathbb{R}\}$, and let $\mathcal{P}_1$ be the corresponding class of standard prior functions; then the reference prior for μ is of the form $\pi(\mu \mid \mathcal{M}_1, \mathcal{P}_1) = c$, assuming it is permissible. Moreover, if $\mathcal{M}_2$ is a* scale *model, $\mathcal{M}_2 \equiv \{\sigma^{-1} f_2(|x|/\sigma),\ x \in \mathbb{R},\ \sigma > 0\}$, and $\mathcal{P}_2$ is its class of standard prior functions, a reference prior for σ is of the form $\pi(\sigma \mid \mathcal{M}_2, \mathcal{P}_1) = c\,\sigma^{-1}$, assuming it is permissible. (Permissibility is established under mild conditions in Corollary 2.1 and the later discussion.)*

Proof. The explicit form of the reference prior could be used to prove this, but it is interesting to proceed by a difference route. Let $\pi(\mu)$ be the reference prior which corresponds to model $\mathcal{M}_1$; the changes $y = x+\alpha$ and $\theta = \mu+\alpha$ produce $\{f_1(y-\theta),\ y \in \mathcal{Y},\ \theta \in \mathbb{R}\}$, which is again model $\mathcal{M}_1$. Hence, using Proposition 9.3, $\pi(\mu) = \pi(\mu+\alpha)$ for all α and, therefore, $\pi(\mu)$ must be constant. The changes of variables $y = \log x$ and $\phi = \log \sigma$ transform the scale model $\mathcal{M}_2$ into a location model; hence, $\pi(\phi) = c$ and, therefore, $\pi(\sigma) \propto \sigma^{-1}$. □

9.5 When There is a Known Asymptotic Distribution

Let $\hat{\theta}_k = \hat{\theta}_k(\boldsymbol{x}^{(k)})$ be the maximum likelihood estimate (mle) of θ in $\mathcal{M}^k$ (any consistent, asymptotically efficient estimate would do). There is often a known asymptotic distribution of $\hat{\theta}_k$ and it is typically of the form (possibly after a change of variables)

$$p(\hat{\theta}_k \mid \theta) = c_k(\theta) g(c_k(\theta)[\hat{\theta}_k - \theta]), \tag{9.10}$$

for some fixed density $g(\cdot)$ and series of constants $c_k(\theta)$. The corresponding asymptotic posterior distribution of θ (for any reasonable prior) is then

$$\pi(\theta \mid \hat{\theta}_k) = c_k(\hat{\theta}_k) g(c_k(\hat{\theta}_k)[\hat{\theta}_k - \theta]). \tag{9.11}$$

Example 9.3 *The Regular Case:* This is the situation when asymptotic normality holds, so that

$$p(\hat{\theta}_k \mid \theta) = \frac{\sqrt{kI(\theta)}}{\sqrt{2\pi}} \exp\left\{-\frac{1}{2} kI(\theta)(\hat{\theta}_k - \theta)^2\right\},$$

where $I(\theta)$ is the single observation Fisher information, as defined in (4.5). This is of the form (9.10) with $g(\cdot)$ being the standard normal density and $c_k(\theta) = \sqrt{kI(\theta)}$. The asymptotic posterior distribution of θ is the same, with $I(\theta)$ replaced by $I(\hat{\theta}_k)$, the observed information. (For needed technical conditions on the behavior of the mle for this to hold, see Le Cam (1986), Schervish (1995) and references therein.)

Example 9.4 *The Irregular Uniform Case:* In Example 9.2, $\hat{\theta}_k = t_k$. Transforming to $\eta = \log\theta$ and $\hat{\eta}_k = \log t_k$ one obtains, from (9.8),

$$p(\hat{\eta}_k \mid \eta) = k \exp\{k(\hat{\eta}_k - \eta)\}, \quad \text{for} \quad \hat{\eta}_k < \eta.$$

This is of the form (9.10) with $g(\cdot)$ being the standard exponential density (on $(-\infty, 0)$) and $c_k(\theta) = k$. This is also the asymptotic posterior distribution of η.

Proposition 9.9 ***(Case of known univariate asymptotic posterior).*** *Let $\mathcal{M} \equiv \{p(\boldsymbol{x} \mid \theta),\ \boldsymbol{x} \in \boldsymbol{\mathcal{X}},\ \theta \in \Theta \in \mathbb{R}\}$ be a standard model and let $\mathcal{P}_s$ be the corresponding class of standard prior functions. Let $\hat{\theta}_k = \hat{\theta}_k(\boldsymbol{x}^{(k)})$ be the maximum likelihood estimate of θ in $\mathcal{M}^k$ (or any consistent, asymptotically efficient estimate), having limiting distribution of the form (9.10) as $k \to \infty$. Then, the reference prior is*

$$\pi(\theta \mid \mathcal{M}, \mathcal{P}_s) = \lim_{k\to\infty} \frac{c_k(\theta)}{c_k(\theta_0)} = \lim_{k\to\infty} \frac{p(\theta \mid \theta)}{p(\theta_0 \mid \theta_0)}, \tag{9.12}$$

provided this is a permissible prior.

Proof. This follows from Proposition 9.7. We only provide here a heuristic argument. A formal proof in the regular case (asymptotic normality) can be obtained from the results in Clarke & Barron (1994), with minor modifications.

Using (9.10) and (9.11), (9.5) becomes, for large k,

$$
\begin{aligned}
f_k(\theta) &= \exp\left\{\int_{\Theta} p(\hat{\theta}_k \,|\, \theta) \log\left[\pi^*(\theta \,|\, \hat{\theta}_k)\right] d\hat{\theta}_k\right\} && (9.13)\\
&\cong \exp\left\{\int_{\Theta} c_k(\theta) f(c_k(\theta)[\hat{\theta}_k - \theta]) \log\left[c_k(\hat{\theta}_k) f(c_k(\hat{\theta}_k)[\hat{\theta}_k - \theta])\right] d\hat{\theta}_k\right\}\\
&= \exp\left\{\int_{\Theta} c_k(\theta) f(c_k(\theta)[\hat{\theta}_k - \theta]) \log[c_k(\hat{\theta}_k)]\, d\hat{\theta}_k\right\}\\
&\quad \times \exp\left\{\int_{\Theta} c_k(\theta) f(c_k(\theta)[\hat{\theta}_k - \theta]) \log[f(c_k(\theta)[\hat{\theta}_k - \theta])]\, d\hat{\theta}_k\right\}. && (9.14)
\end{aligned}
$$

Since $c_k(\theta) f(c_k(\theta)[\hat{\theta}_k - \theta])$ concentrates about θ as k grows,

$$
\exp\left\{\int_{\Theta} c_k(\theta) f(c_k(\theta)[\hat{\theta}_k - \theta]) \log[c_k(\hat{\theta}_k)]\, d\hat{\theta}_k\right\} \cong \exp\{\log[c_k(\theta)]\} = c_k(\theta) \qquad (9.15)
$$

and

$$
\begin{aligned}
&\exp\left\{\int_{\Theta} c_k(\theta) f(c_k(\theta)[\hat{\theta}_k - \theta]) \log[f(c_k(\hat{\theta}_k))[\hat{\theta}_k - \theta])]\, d\hat{\theta}_k\right\}\\
&\cong \exp\left\{\int_{\Theta} c_k(\theta) f(c_k(\theta)[\hat{\theta}_k - \theta]) \log[f(c_k(\theta))[\hat{\theta}_k - \theta])]\, d\hat{\theta}_k\right\}\\
&= \exp\left\{\int f(y) \log[f(y)] dy\right\} = C \text{ (making the transformation } y = c_k(\theta)[\hat{\theta}_k - \theta]).
\end{aligned}
$$

Combining this with (9.15) in (9.14), and inserting into (9.6) yields the first equality in (9.12). The second equality is just algebra, using (9.10). □

Corollary 9.9.1 ***(Jeffreys univariate rule, regular case).***
Let $\mathcal{M} \equiv \{p(\boldsymbol{x} \,|\, \theta),\, \boldsymbol{x} \in \boldsymbol{\mathcal{X}},\, \theta \in \Theta \in \mathbb{R}\}$ be a standard model, let $\mathcal{P}_s$ be the corresponding class of standard prior functions, and let $\hat{\theta}_k = \hat{\theta}_k(\boldsymbol{x}^{(k)})$ be the mle of θ and have a sampling distribution that is asymptotically $N(\hat{\theta} \,|\, \theta, [kI(\theta)]^{-1})$, where $I(\theta)$ is the Fisher information of a single observation, as defined in (4.5). The the reference prior is the Jeffreys-rule prior

$$
\pi(\theta \,|\, \mathcal{M}, \mathcal{P}_s) \propto \sqrt{I(\theta)}\,,
$$

provided this is a permissible function.

Proof. This is immediate from Proposition 9.9, since $c_k(\theta) = \sqrt{kI(\theta)}$. □

Note that this is the same as *Perks Rule*, discussed in Subsection 4.4.2, which stated that, if the sampling distribution of $\hat{\theta}_k(\boldsymbol{x}^{(k)})$ is asymptotically normal with variance $s^2(\theta)/k$, then one should use the prior $\pi(\theta) \propto s(\theta)^{-1}$. The argument of Perks (1947) was different, however, based on a probability matching criterion.

9.6 Irregular Continuous Problems

The situation is more involved when the support of the sampling space depends on the unknown parameter, although Proposition 9.9 can sometimes be applied.

Example 9.5 *The Irregular Uniform Case continued:* In Example 9.4, we saw that an asymptotic distribution existed with $c_k(\eta) = k$. Using Proposition 9.9, it follows that the reference prior for η is $\pi(\eta) \propto 1$, assuming it is permissible (which was verified earlier). Transforming back to $\theta = e^\eta$ yields the reference prior for θ, $\pi(\theta) = 1/\theta$.

Here is a much more general result for non-regular problems.

Proposition 9.10 (Ghosal and Samantha class, Ghosal & Samanta (1997)). *Let $\mathcal{M} \equiv \{p(\boldsymbol{x} \mid \theta),\ \boldsymbol{x} \in \boldsymbol{\mathcal{X}}(\theta),\ \theta \in \Theta \in \mathbb{R}\}$ be a model whose support depends on θ, and let $\mathcal{P}_s$ be the corresponding class of standard prior functions. If $\boldsymbol{\mathcal{X}}(\theta)$ is either monotonically decreasing or monotonically increasing in θ, the reference prior is*

$$\pi(\theta \mid \mathcal{M}, \mathcal{P}_s) \propto \left| \int_{\boldsymbol{\mathcal{X}}(\theta)} p(x \mid \theta) \frac{\partial}{\partial \theta} \log[p(x \mid \theta)]\, dx \right|,$$

provided this is a permissible function.

Proof. While this can be deduced from Proposition 9.7, a direct proof is given in Ghosal & Samanta (1997). □

Proposition 9.7 can be used to derive reference priors in non-regular cases that are even more complicated.

Example 9.6 *(Uniform data, support unknown at both extremes.)* Consider the uniform model $\mathcal{M} = \{p(x \mid \theta), x \in \mathcal{X}(\theta), \theta \in \Theta\}$ with

$$p(x \mid \theta) = \mathrm{Un}(x \mid a_1(\theta), a_2(\theta)) = \frac{1}{a_2(\theta) - a_1(\theta)}, \quad a_1(\theta) < x < a_2(\theta),$$

with support on $\mathcal{X}(\theta) = (a_1(\theta), a_2(\theta))$, where $\theta > \theta_0$, and $0 < a_1(\theta) < a_2(\theta)$ are both strictly monotonic increasing functions on $\Theta = (\theta_0, \infty)$, with derivatives satisfying $0 < a_1'(\theta) < a_2'(\theta)$. This is not a regular model, and does not belong to the class of non-regular models covered by Proposition 9.10.

Using Proposition 9.7, the corresponding reference prior is

$$\pi(\theta \mid \mathcal{M}, \mathcal{P}_s) = \frac{a_2'(\theta) - a_1'(\theta)}{a_2(\theta) - a_1(\theta)} \exp\left\{b_1 + \frac{1}{b_1 - b_2}\left[b_1 \psi\Big(\frac{1}{b_1}\Big) - b_2 \psi\Big(\frac{1}{b_2}\Big)\right]\right\},$$

$$b_j \equiv b_j(\theta) = \frac{a_2'(\theta) - a_1'(\theta)}{a_j'(\theta)}, \quad j = 1, 2, \tag{9.16}$$

where $\psi(z)$ is the digamma function $\psi(z) = \frac{d}{dz}\log(\Gamma(z))$, $z > 0$. For a proof, see Appendix G of Berger et al. (2009*a*).

Uniform distribution on (θ, θ^2), $\theta > 1$. This is a special case of the above result with $\theta_0 = 1$, $a_1(\theta) = \theta$, and $a_2(\theta) = \theta^2$. Then $b_1 = 2\theta - 1$ and $b_2 = (2\theta - 1)/(2\theta)$. It is easy to show that $b_2^{-1} = b_1^{-1} + 1$. For the digamma function, $\psi(z+1) = \psi(z) + 1/z$ for $z > 0$, so that $\psi(1/b_1) = \psi(1/b_2) - b_1$. The reference prior thus becomes

$$\begin{aligned}
\pi(\theta) &= \frac{2\theta - 1}{\theta(\theta - 1)} \exp\Big\{b_1 + \frac{1}{b_1 - b_2}\Big[b_1\psi\Big(\frac{1}{b_2}\Big) - b_1^2 - b_2\psi\Big(\frac{1}{b_2}\Big)\Big]\Big\} \\
&= \frac{2\theta - 1}{\theta(\theta - 1)} \exp\Big\{-\frac{b_1 b_2}{b_1 - b_2} + \psi\Big(\frac{1}{b_2}\Big)\Big]\Big\}. \\
&\propto \frac{2\theta - 1}{\theta(\theta - 1)} \exp\Big\{\psi\Big(\frac{2\theta}{2\theta - 1}\Big)\Big\},
\end{aligned} \tag{9.17}$$

the last equation following from the identity $b_1 b_2/(b_1 - b_2) = b_2^{-1} - b_1^{-1} = 1$. This can be numerically shown to be nearly proportional to $1/(\theta - 1)$.

Sareen (2003) makes an interesting observation on the relationship between Jeffreys priors and reference priors in non-regular situations.

9.7 Numerical Reference Priors

Analytical derivation of reference priors can be technically challenging in non-regular models. However, Proposition 9.7 can also be used to obtain a numerical approximation to the reference prior, providing random observations can be efficiently simulated from the model.

The proposed algorithm requires a numerical evaluation of Equation (9.5). Moderate values of k (to simulate the asymptotic posterior) are typically sufficient to obtain a good approximation to the reference prior $\pi(\theta \mid \mathcal{M}, \mathcal{P}_0)$ (up to an irrelevant proportionality constant). Here is an appropriate pseudo code.

Step 0. Starting values. Choose a moderate value for k, choose an arbitrary positive function $\pi^*(\theta)$, say $\pi^*(\theta) = 1$, and choose the number, m, of samples $\boldsymbol{x}^k$ that will be used to simualate the reference prior.

Step 1. For any θ value at which the reference prior is to be computed and for $j = 1, \ldots, m$,

(i) simulate a random sample $(\boldsymbol{x}_{1j}, \ldots, \boldsymbol{x}_{kj})$ of size k from $p(\boldsymbol{x} \mid \theta)$;

(ii) numerically compute the integral $c_j = \int_\Theta \prod_{i=1}^k p(\boldsymbol{x}_{ij} \mid \theta)\, \pi^*(\theta)\, d\theta$ (easy, as it is only one-dimensional);

(iii) evaluate $r_j(\theta) = \log[\prod_{i=1}^{k} p(\boldsymbol{x}_{ij} \,|\, \theta)\, \pi^*(\theta)/c_j\,]$.

Step 2. Compute $\pi(\theta) = \exp[\,m^{-1} \sum_{j=1}^{m} r_j(\theta)\,]$ and store the pair $\{\theta, \pi(\theta)\}$.

Sometimes $\pi^*(\theta)$ can be chosen so that there is an analytical form for the approximate asymptotic posterior,

$$\pi^*(\theta \,|\, \boldsymbol{x}_{1j}, \ldots, \boldsymbol{x}_{kj}) \propto \prod_{i=1}^{k} p(\boldsymbol{x}_{ij} \,|\, \theta)\, \pi^*(\theta);$$

in that case, the values of the $r_j(\theta) = \log[\,\pi^*(\theta \,|\, \boldsymbol{x}_{1j}, \ldots, \boldsymbol{x}_{kj})\,]$ are immediately obtained, and the numerical algorithm reduces to one Monte Carlo integration for each desired pair $\{\theta, \pi(\theta)\}$.

Example 9.7 *Exponential data, continued.* Figure 9.1 represents the exact reference prior for the exponential model $\pi(\theta) = \theta^{-1}$ (continuous line) and the reference prior numerically calculated with the algorithm above for nine θ values, ranging from e^{-3} to e^3, uniformly log-spaced and rescaled to have $\pi(1) = 1$; $m = 500$ samples of $k = 25$ observations were used to compute each of the nine $\{\theta_i, \pi(\theta_i)\}$ points.

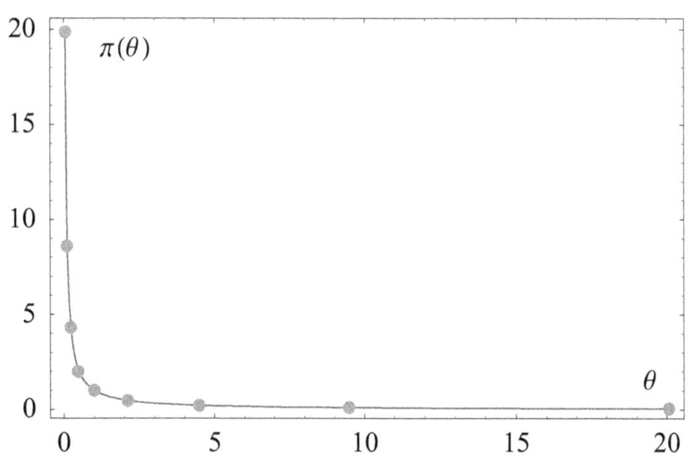

Figure 9.1: Numerical reference prior for the exponential model.

If required, a continuous approximation to $\pi(\theta)$ can easily be obtained from the computed points using standard interpolation techniques. For an alternative, MCMC based, numerical computation method for determining reference priors, see Lafferty & Wasserman (2001).

9.8 Discussion and Additional References

There is a huge literature concerning the development and use of reference priors. An important subset of this literature is given in the bibliography; just search for the key word 'reference.'

Chapter 10

Multiple Continuous Parameters

10.1 The Challenge of Many Parameters

In this chapter, models with many parameters are considered and reference priors are developed through an iterative application of the procedure for one-parameter models, together with appropriate model marginalization. Given a multiparameter model

$$\mathcal{M} \equiv \{p(\boldsymbol{x} \mid \boldsymbol{\theta}), \boldsymbol{x} \in \mathcal{X}, \boldsymbol{\theta} = (\theta_1, \ldots, \theta_m) \in \boldsymbol{\Theta} \subset \mathbb{R}^m\},$$

suppose that the parameters have been labeled so that θ_1 is the parameter of interest and the remaining θ_i are nuisance parameters that have also been placed in a specified order (typically most important to least important). The reference prior $\pi(\boldsymbol{\theta} \mid \mathcal{M}, \mathcal{P}_s)$, for this specific ordering, will be constructed from a series of *conditional reference priors*, as

$$\pi(\theta_m \mid \theta_1, \ldots, \theta_{m-1}) \times \pi(\theta_{m-1} \mid \theta_1, \ldots, \theta_{m-2}) \times \ldots \times \pi(\theta_2 \mid \theta_1) \times \pi(\theta_1), \quad (10.1)$$

starting at the left and moving right. Not only does this typically depend on the choice of the parameter of interest, θ_1, but it can depend on the ordering of the nuisance parameters.

It its tempting to try to define a global objective prior by directly maximizing the missing information about the vector $\boldsymbol{\theta}$, as this would yield a unique objective prior for each model, a prior which is invariant under reparameterization. However, under regularity conditions, this leads to the Jeffreys-rule prior, which we have seen is often demonstrably inadequate for multi-parameter problems. Sequentially computing the conditional reference priors overcomes this problem.

Sometimes the parameter of interest and nuisance parameters are transformations of the original parameters. For instance, in a normal model with mean μ and standard deviation σ, the parameter of interest might be $\theta_1 = \mu/\sigma$ and the nuisance parameter might be $\theta_2 = \sigma$. In such situations, we will determine the reference prior $\pi(\theta_1, \theta_2)$, and do a change of variables transformation to obtain the reference prior for the original parameters.

Section 10.2 considers the case of just two parameters – a parameter of interest and a single nuisance parameter – to illustrate the key steps in the sequential method. The general case is consider in Section 10.3. Because the notation becomes involved, we will henceforth drop the conditioning on $\mathcal{M}$ and $\mathcal{P}_s$ in the notation for the reference prior.

Sections 10.2.1 and 10.2.2, while providing the formal definitions and underlying motivations for reference priors with multiple parameters, are often not easy to use to actually develop reference priors.

In the regular case of asymptotic normality, however, quite specific and comparatively simple algorithms are available for implementing the formal definitions; these are given in Section 10.2.3 (for the two parameter case) and Section 10.3 (for the general case) and do not require reading of the previous material in the chapter.

10.2 Iterative Algorithm for One Nuisance Parameter

To simplify notation in this section and to highlight the different treatments of the parameter of interest and the nuisance parameter, θ will be used to denote the parameter of interest and λ will be the nuisance parameter. Thus the model is $\mathcal{M} \equiv \{p(\boldsymbol{x} \,|\, \theta, \lambda), \boldsymbol{x} \in \boldsymbol{\mathcal{X}}, (\theta, \lambda) \in \boldsymbol{\Upsilon}\}$.

For a given θ, let $\Lambda(\theta) = \{\lambda : (\theta, \lambda) \in \boldsymbol{\Upsilon}\}$ denote the projection of $\boldsymbol{\Upsilon}$ onto the λ axis. Often $\boldsymbol{\Upsilon}$ is of the form $\Theta \times \Lambda$, in which case θ and λ are said to be *variation independent*, and then $\Lambda(\theta)$ clearly just equals Λ.

10.2.1 Definition of the Reference Prior and Examples

Here are the steps of the sequential reference prior algorithm.

First one finds the conditional reference prior, $\pi(\lambda \,|\, \theta)$, for λ, given θ; then one finds the marginal reference prior $\pi(\theta)$; so the overall reference prior is $\pi(\theta, \lambda) = \pi(\lambda \,|\, \theta)\pi(\theta)$, providing it is permissible. Two cases have to be considered, that in which $\pi(\lambda \,|\, \theta)$ is proper and that in which it is not.

Step 1. Find the Conditional Reference Prior of λ Given θ

Conditional on θ, $p(\boldsymbol{x}\,|\,\theta,\lambda)$ depends only on the one-dimensional λ, so the algorithm in Proposition 9.7 can be used to obtain the reference prior $\pi(\lambda\,|\,\theta)$ as follows, initiating the algorithm with any prior $\pi^*(\lambda) > 0$ (a constant will do):

$$f_k(\lambda\,|\,\theta) = \exp\left\{\int p(\boldsymbol{x}^{(k)}\,|\,\lambda,\theta)\,\log\left[\pi^*(\lambda\,|\,\boldsymbol{x}^{(k)},\theta)\right]d\boldsymbol{x}^{(k)}\right\}, \tag{10.2}$$

$$f(\lambda\,|\,\theta) = \lim_{k\to\infty}\frac{f_k(\lambda\,|\,\theta)}{f_k(\lambda_0\,|\,\theta)}, \tag{10.3}$$

where λ_0 is any interior point in $\Lambda(\theta)$. Recall, from Corollary 9.9.1, that, under asymptotic normality of λ given θ,

$$f(\lambda\,|\,\theta) = \sqrt{I(\lambda\,|\,\theta)},\quad I(\lambda\,|\,\theta) = -\int_{\mathcal{X}} p(\boldsymbol{x}\,|\,\theta,\lambda)\,\frac{\partial^2\log p(\boldsymbol{x}\,|\,\theta,\lambda)}{\partial\lambda^2}\,d\boldsymbol{x}\,, \tag{10.4}$$

being the Fisher information of λ given θ.

Case 1. If $c(\theta) = \int_{\Lambda(\theta)} f(\lambda\,|\,\theta)\,d\lambda < \infty$, the conditional reference prior is

$$\pi(\lambda\,|\,\theta) = \frac{1}{c(\theta)}f(\lambda\,|\,\theta)\,. \tag{10.5}$$

Case 2. If $c(\theta)$ is infinite, constrain (θ,λ) to compact subsets $\{\boldsymbol{\Upsilon}_l,\ l=1,2,\ldots\}$ of $\boldsymbol{\Upsilon}$, such that $\lim_{l\to\infty}\boldsymbol{\Upsilon}_l = \boldsymbol{\Upsilon}$, with associated projections $\{\Lambda_l(\theta), l=1,2,\ldots\}$; the constrained conditional reference prior is then

$$\pi_l(\lambda\,|\,\theta) = \frac{1}{c_l(\theta)}f(\lambda\,|\,\theta)1_{\Lambda_l(\theta)}(\lambda),\quad c_l(\theta) = \int_{\Lambda_l(\theta)} f(\lambda\,|\,\theta)\,d\lambda\,. \tag{10.6}$$

Step 2. Form the Marginal Model

Case 1. If $c(\theta)$ is finite, form the marginal model for $\boldsymbol{x}$

$$p(\boldsymbol{x}\,|\,\theta) = \int_{\Lambda(\theta)} p(\boldsymbol{x}\,|\,\theta,\lambda)\,\pi(\lambda\,|\,\theta)\,d\lambda\,. \tag{10.7}$$

Case 2. If If $c(\theta)$ is infinite, form the marginal model for $\boldsymbol{x}$

$$p_l(\boldsymbol{x}\,|\,\theta) = \int_{\Lambda_l(\theta)} p(\boldsymbol{x}\,|\,\theta,\lambda)\,\pi_l(\lambda\,|\,\theta)\,d\lambda\,. \tag{10.8}$$

Step 3. Find the Marginal Reference Prior for θ

Case 1. If $c(\theta) < \infty$, the marginal reference prior for θ in the marginal model is

$$\pi(\theta) = \lim_{k\to\infty}\frac{f_k(\theta)}{f_k(\theta_0)},\quad f_k(\theta) = \exp\left\{\int p(\boldsymbol{x}^{(k)}\,|\,\theta)\,\log\left[\pi^*(\theta\,|\,\boldsymbol{x}^{(k)})\right]d\boldsymbol{x}^{(k)}\right\}. \tag{10.9}$$

Case 2. If $c(\theta)$ is infinite, the constrained marginal reference prior for θ in the marginal model p_l is

$$\pi_l(\theta) = \lim_{k\to\infty} \frac{f_{(l,k)}(\theta)}{f_{(l,k)}(\theta_0)}, \; f_{(l,k)}(\theta) = \exp\left\{\int p_l(\boldsymbol{x}^{(k)}\,|\,\theta)\,\log\left[\pi_l^*(\theta\,|\,\boldsymbol{x}^{(k)})\right] d\boldsymbol{x}^{(k)}\right\}. \quad (10.10)$$

Again using Corollary 9.9.1, if this marginal model has regular asymptotics, these can be determined from the Fisher information in the marginal model.

Step 4. Determine the Joint Reference Prior

Case 1. If $c(\theta) < \infty$, the joint reference prior is simply $\pi(\theta,\lambda) = \pi(\lambda\,|\,\theta)\pi(\theta)$. Note that only the marginal likelihood function, $p(\boldsymbol{x}\,|\,\theta)$, and marginal reference prior, $\pi(\theta)$, are needed for inference about the parameter of interest θ, since $\pi(\theta\,|\,\boldsymbol{x}) \propto p(\boldsymbol{x}\,|\,\theta)\pi(\theta)$.

Case 2. If $c(\theta)$ is infinite, the joint reference prior should be computed as

$$\pi(\theta,\lambda) = \lim_{l\to\infty} \frac{\pi_l(\lambda\,|\,\theta)\pi_l(\theta)}{\pi_l(\lambda_0\,|\,\theta_0)\pi_l(\theta_0)}, \quad (10.11)$$

for any interior point (θ_0, λ_0).

Example 10.1 *Ratio of multinomial parameters.* We start with an example in which the conditional reference prior is proper (Case 1 above). Let data $\boldsymbol{x} = (x_1, x_2, x_3)$ consist of the result from n trinomial observations (hence, $x_3 = n - x_1 - x_2$), with parameters ω_1, ω_2 and $\omega_3 = 1 - \omega_1 - \omega_2$, so that, for $i = 1, 2$, $0 < \omega_i < 1$, and $\omega_1 + \omega_2 < 1$. The trinomial density is then

$$p(x_1, x_2\,|\,n, \omega_1, \omega_2) = c(x_1, x_2, n)\,\omega_1^{x_1}\omega_2^{x_2}(1 - \omega_1 - \omega_2)^{n-x_1-x_2},$$

where $c(x_1, x_2, n) = (n!)/(x_1!\,x_2!\,(n - x_1 - x_2)!)$.

Suppose that the quantity of interest is the *ratio* $\theta = \omega_1/\omega_2$ of the first two parameters. Reparametrization in terms of θ and (say) $\lambda = \omega_2$ yields

$$p(x_1, x_2\,|\,n, \theta, \lambda) = c(x_1, x_2, n)\,\theta^{x_1}\,\lambda^{x_1+x_2}\,[1 - \lambda(1+\theta)]^{n-x_1-x_2},$$

for $\theta > 0$ and $0 < \lambda < (1+\theta)^{-1}$.

Step 1. Conditional on θ, this is a model with one regular continuous parameter λ, and the conditional Fisher information can be computed to be $I(\lambda\,|\,\theta) = n(1+\theta)/[\lambda(1 - \lambda(1+\theta))]$. From (10.4), it follows that

$$f(\lambda\,|\,\theta) = \sqrt{I(\lambda\,|\,\theta)} \propto \lambda^{-1/2}[1 - \lambda(1+\theta)]^{-1/2}, \quad \lambda \in [0, (1+\theta)^{-1}],$$

the last step because, given θ, any multiplicative function of θ can be ignored. Computation yields (where π below is the constant, not a prior)

$$c(\theta) = \int_0^{(1+\theta)^{-1}} \lambda^{-1/2}[1 - \lambda(1+\theta)]^{-1/2}\,d\lambda = \frac{\pi}{(1+\theta)^{1/2}},$$

which is finite, so that the conditional reference prior of λ, given θ, is

$$\pi(\lambda \mid \theta) = \pi^{-1}(1+\theta)^{1/2}\lambda^{-1/2}[1-\lambda(1+\theta)]^{-1/2},\ \lambda \in [0,(1+\theta)^{-1}]\,. \quad (10.12)$$

Step 2. The marginal model for θ is then

$$\begin{aligned} p(x_1, x_2 \mid n, \theta) &= \int_0^{(1+\theta)^{-1}} p(x_1, x_2 \mid n, \theta, \lambda)\, \pi(\lambda \mid \theta)\, d\lambda \\ &= \frac{\Gamma(x_1 + x_2 + \frac{1}{2})\,\Gamma(n - x_1 - x_2 + \frac{1}{2})}{x_1!\, x_2!\, (n - x_1 - x_2)!} \, \frac{\theta^{x_1}}{(1+\theta)^{x_1+x_2}}\,. \end{aligned}$$

Step 3. The Fisher information for this marginal model can be computed to be $I(\theta) = n/[2\theta(1+\theta)^2]$, so the marginal reference prior of the parameter of interest is $\pi(\theta) \propto I(\theta)^{1/2} \propto \theta^{-1/2}(1+\theta)^{-1}$.

Step 4. The joint reference prior is $\pi(\theta, \lambda) = \pi(\lambda \mid \theta)\pi(\theta)$, given by

$$\pi(\theta, \lambda) \propto [\theta\lambda]^{-1/2}[1-\lambda(1+\theta)]^{-1/2}(1+\theta)^{-1/2}, \lambda \in [0,(1+\theta)^{-1}]\,. \quad (10.13)$$

If only inference about the parameter of interest θ is desired, it follows directly from $\pi(\theta \mid x_1, x_2) \propto p(x_1, x_2 \mid n, \theta)\, \pi(\theta)$ that

$$\pi(\theta \mid x_1, x_2) = \frac{\Gamma(x_1 + x_2 + 1)}{\Gamma(x_1 + \frac{1}{2})\,\Gamma(x_2 + \frac{1}{2})} \, \frac{\theta^{x_1 - 1/2}}{(1+\theta)^{x_1+x_2+1}}\,, \quad \theta > 0. \quad (10.14)$$

Notice that $\pi(\theta \mid x_1, x_2)$ does not depend on n. To draw conclusions about the value of $\theta = \omega_1/\omega_2$, only the numbers x_1 and x_2 observed in the first two classes matter; the data $\{55, 45, 100\}$ carries precisely the same information about the *ratio* ω_1/ω_2 as does the data $\{55, 45, 10000\}$.

That this is appealing is apparent from considering the example of an electoral survey of size n, which yields x_1 voters for party A and x_2 voters for party B. The reference posterior distribution of the ratio θ, of the proportion of A voters to B voters in the population, only depends on the respective number of voters in the sample, x_1 and x_2, whatever the size and political intentions of the other $n - x_1 - x_2$ citizens in the sample. In particular, the reference posterior probability that party A gets better results than party B is $\Pr[\theta > 1 \mid x_1, x_2] = \int_1^{\infty} \pi(\theta \mid x_1, x_2)\, d\theta$. As one would expect, this is equal to 1/2 if, and only if, $x_1 = x_2$; one-dimensional numerical integration (or use of the incomplete beta function) is required to compute other values. For instance, $\Pr[\theta > 1 \mid x_1 = 55, x_2 = 45] = 0.841$ and $\Pr[\theta > 1 \mid x_1 = 550, x_2 = 450] = 0.999$.

Example 10.2 *Location-Scale Models.* Location-scale models are of the form $p(x \mid \mu, \sigma) = \sigma^{-1} g((x - \mu)/\sigma)$ for some density g. Consider, first, that μ is the parameter of interest, with σ being the nuisance parameter. Step 1 of the reference prior algorithm yields $f(\sigma \mid \mu) \propto 1/\sigma$ since, given μ, $p(x \mid \mu, \sigma)$ is a scale density, and we can apply Proposition 9.8.

Since $1/\sigma$ is not integrable, we choose an increasing sequence of compact subsets of the full parameter space, $\boldsymbol{\Upsilon}_l = [-l\sigma, l\sigma] \times [l^{-1}, l]$; the choice of $[-l\sigma, l\sigma]$ as the domain for μ is because σ provides a scale for μ. Clearly $\Lambda_l = [l^{-1}, l]$ (not depending on μ). Then the normalizing constant for $1/\sigma$ is $c_l = \int_{l^{-1}}^{l} \sigma^{-1}\, d\sigma = 2\log l$, and (10.8) becomes

$$p_l(x \mid \mu) = \int_{l^{-1}}^{l} \frac{1}{2[\log l]\sigma^2} g\left(\frac{x-\mu}{\sigma}\right) d\sigma = \int_{(x-\mu)/l}^{(x-\mu)l} \frac{1}{2[\log l]|x-\mu|} g\,(y)\; dy\,,$$

the last step following from the change of variables $y = (x-\mu)/\sigma$. This is of the form $h(x-\mu)$ and so is a location density, for which we know the marginal reference prior is $\pi(\mu) = 1$; hence we do not have to formally solve Step 3. Finally, taking the limit in Step 4 yields, as the joint reference prior, $\pi(\mu, \sigma) = 1/\sigma$.

If σ is the parameter of interest, we note that, conditional on σ, the nuisance parameter μ is a location parameter and so has conditional reference prior $\pi(\mu \mid \sigma) \propto 1$. Clearly $\Lambda_l(\sigma) = [-l\sigma, l\sigma]$, so that the normalizing constant is $c_l = 1/[2l\sigma]$. Changing variables to $y = \mu/\sigma$ yields

$$p_l(x \mid \sigma) = \int_{-l\sigma}^{l\sigma} \frac{1}{2l\,\sigma^2}\, g\left(\frac{x-\mu}{\sigma}\right) d\mu = \int_{-l}^{l} \frac{1}{2l\,\sigma}\, g\left(\frac{x}{\sigma} - y\right) dy\,.$$

This is recognizable as a scale density of the form $\frac{1}{\sigma}h(\frac{x}{\sigma})$, so the marginal reference prior for σ is $1/\sigma$ and the joint reference prior, taking the limit in l, is $\pi(\sigma, \mu) = 1/\sigma$. It is interesting in this example that $1/\sigma$ is the reference prior regardless of whether μ or σ is the parameter of interest.

See also Fernández & Steel (1999b).

A Missing Link. Step 1 and Step 3 of the reference prior algorithm both maximize missing information to determine the conditional and marginal reference priors, respectively. These are then multiplied together (with appropriate limiting arguments over compact sets, in the case of impropriety) to form the joint reference prior. But it is not known if this procedure satisfies any global property related to maximizing missing information.

Assuming $\boldsymbol{\Upsilon} = \Theta \times \Lambda$ for simplicity, the joint (θ, λ) missing information is

$$I_{\theta,\lambda}\{\pi(\cdot,\cdot) \mid \mathcal{M}^k\} = \int_{\Theta\times\Lambda} \int_{\boldsymbol{\mathcal{X}}^k} p(\boldsymbol{x}^{(k)} \mid \theta, \lambda)\; \pi(\theta, \lambda) \log\left[\frac{\pi(\theta, \lambda \mid \boldsymbol{x}^{(k)})}{\pi(\theta, \lambda)}\right] d\boldsymbol{x}^{(k)}\, d\theta\, d\lambda\,.$$

Unfortunately, maximizing this (as $k \to \infty$) will just result in the Jeffreys-rule prior, which is (often) suboptimal. One can decompose this as

$$I_{\theta,\lambda} = \int_{\Theta\times\Lambda}\int_{\mathcal{X}^k} p(\boldsymbol{x}^{(k)}\,|\,\theta,\lambda)\,\pi(\theta,\lambda)\log\left[\frac{\pi(\theta\,|\,\boldsymbol{x}^{(k)})}{\pi(\theta)}\right] d\boldsymbol{x}^{(k)}\,d\theta\,d\lambda$$
$$+\int_{\Theta\times\Lambda}\int_{\mathcal{X}^k} p(\boldsymbol{x}^{(k)}\,|\,\theta,\lambda)\,\pi(\theta,\lambda)\log\left[\frac{\pi(\lambda\,|\,\theta,\boldsymbol{x}^{(k)})}{\pi(\lambda\,|\,\theta)}\right] d\boldsymbol{x}^{(k)}\,d\theta\,d\lambda\,, \quad (10.15)$$

and write the second integral as

$$\int_{\Theta}\pi(\theta)\int_{\Lambda}\int_{\mathcal{X}^k} p(\boldsymbol{x}^{(k)}\,|\,\theta,\lambda)\,\pi(\lambda\,|\,\theta)\log\left[\frac{\pi(\lambda\,|\,\theta,\boldsymbol{x}^{(k)})}{\pi(\lambda\,|\,\theta)}\right] d\boldsymbol{x}^{(k)}\,d\lambda\,d\theta\,.$$

The two inner integrals here, over $\boldsymbol{x}^{(k)}$ and λ, define the missing information for λ, given θ, and this is maximized as in Step 1 of the sequential reference prior algorithm.

At this point, the sequential algorithm effectively ignores the second term of (10.15), and focuses on maximizing the missing information about θ in the first term, using the conditional reference prior found from the second term to integrate out λ. This is not unreasonable; θ is the parameter of interest, so the primary focus should be in maximizing the missing information about θ, i.e., maximizing the first term. The second term is being used only as a technical device to determine the needed $\pi(\lambda\,|\,\theta)$.

It would be nice, however, to have a clean criterion that is being utilized. Trying to maximize only the first term in (10.15), over choice of the joint prior $\pi(\theta,\lambda)$, does not work ($\pi(\lambda\,|\,\theta)$ can go crazy); and including the second term in the maximization leads to the Jeffreys-rule prior. Thus, while the sequential reference prior algorithm seems perfectly reasonable (and there is no obvious alternative), there is no criterion which it can claim to be optimizing; there is a missing link. Thus the procedure is probably best characterized as the most natural algorithm for determining an objective prior that has been devised – and one with a highly impressive track record – but, still, it is just an algorithm.

10.2.2 The Asymptotic Reference Prior

Although Step 3 in the previous section is mathematically well-defined, it can be difficult to work with the marginal model $p(\boldsymbol{x}\,|\,\theta)$ (or the constrained marginal models $p_l(\boldsymbol{x}\,|\,\theta)$) because they are mixture models (mixing over λ) and it is well known that it can be difficult to determine the Fisher information, $I(\theta)$, for such models. Another way to say this is that working, for a large number of replications k, with

$$p(\boldsymbol{x}^{(k)}\,|\,\theta) = \prod_{i=1}^{k} p(\boldsymbol{x}_i\,|\,\theta) = \prod_{i=1}^{k}\int_{\Lambda(\theta)} p(\boldsymbol{x}_i\,|\,\theta,\lambda)\pi(\lambda\,|\,\theta)\,d\lambda$$

can be difficult.

An alternative, that is frequently used in reference prior theory is to instead work, for large k, with

$$p^A(\boldsymbol{x}^{(k)} \mid \theta) \equiv \int_{\Lambda(\theta)} p(\boldsymbol{x}^{(k)} \mid \theta, \lambda)\pi(\lambda \mid \theta)\, d\lambda$$
$$= \int_{\Lambda(\theta)} \prod_{i=1}^{k} p(\boldsymbol{x}_i \mid \theta, \lambda)\pi(\lambda \mid \theta)\, d\lambda\,. \tag{10.16}$$

The reason is that the asymptotic distribution of $p^A(\boldsymbol{x}^{(k)} \mid \theta)$ is usually much more accessible than that of $p(\boldsymbol{x}^{(k)} \mid \theta)$, because the asymptotics of $p(\boldsymbol{x}^{(k)} \mid \theta, \lambda)$ are often easy; for instance in the regular case, $p(\boldsymbol{x}^{(k)} \mid \theta, \lambda)$ can, for large k, be replaced by its asymptotic normal distribution $\mathrm{N}_2((\hat{\theta}, \hat{\lambda}) \mid (\theta, \lambda), [kI(\theta, \lambda)]^{-1})$, and dealing with this expression is often straightforward.

Definition 10.1 Asymptotic Reference Prior. *This reference prior is defined by following Steps 1 and 4 in Section 10.2.1, removing Step 2, and, in Step 3, replacing $p(\boldsymbol{x}^{(k)} \mid \theta)$ (or its constrained version) by $p^A(\boldsymbol{x}^{(k)} \mid \theta)$ from (10.16) (or its constrained version); the posteriors π^* (or π_l^*) are also computed with respect to $p^A(\boldsymbol{x}^{(k)} \mid \theta)$ instead of $p(\boldsymbol{x}^{(k)} \mid \theta)$. Denote the result from this version of Step 3 as $\pi^A(\theta)$ (or $\pi_l^A(\theta)$).*

$\pi^A(\theta)$ is not interpretable as an approximation to $\pi(\theta)$; it is simply a different way to define missing information. Also, in some models, this is the only natural type of asymptotics that is available. In time series and sequential models, for instance, a sequence $(x_1, x_2, \ldots)$ of observations is considered and, if the sequence is nominally infinite, it makes no sense to consider replications of the infinite experiment. One could say that the experiment will be truncated at, say, N observations, and then replicate that experiment, but this is unnatural in that it would not be following the natural asymptotics of the situation.

In any case, it is of interest to compare this asymptotic reference prior with that derived in the previous section. This can be done in the situation of Example 10.1, as it was possible to carry out the base reference prior algorithm exactly. The asymptotic reference prior is computed in the next section, and turns out to be the same. Indeed, in virtually all examples that we have analyzed, the two are equal. An exception appears to be Example 10.7.

For the rest of the chapter (and, indeed, almost all of Part III of the book), we will be dealing with the regular situation of asymptotic normality, and so we will almost exclusively use the asymptotic reference prior algorithm. We will, therefore, not complicate notation by using the π^A notation, but will instead mention when a reference prior is of the original base reference prior type.

10.2.3 Reference Priors under Asymptotic Bi-Normality

The model is still $\mathcal{M} \equiv \{p(\boldsymbol{x} \mid \theta, \lambda), \boldsymbol{x} \in \boldsymbol{\mathcal{X}}, (\theta, \lambda) \in \boldsymbol{\Upsilon}\}$, with $\Lambda(\theta) = \{\lambda : (\theta, \lambda) \in \boldsymbol{\Upsilon}\}$ denoting the projection of $\boldsymbol{\Upsilon}$ onto the λ axis. As usual, we consider replications $\boldsymbol{x}^{(k)} = \{\boldsymbol{x}_1, \ldots, \boldsymbol{x}_k\}$, and assume we have asymptotic normality (of, say, the mle $(\hat{\theta}, \hat{\lambda})$) as $k \to \infty$.

Proposition 10.1 (Reference priors under asymptotic bi-normality) *The k replications result in model $\mathcal{M} \equiv \{p(\boldsymbol{x}^{(k)} \mid \theta, \lambda), \boldsymbol{x}^{(k)} \in \boldsymbol{\mathcal{X}}^k, (\theta, \lambda) \in \Upsilon\}$. The Fisher information matrix for this model is $kI(\theta, \lambda)$, where $I(\theta, \lambda)$ is the Fisher information matrix for a single $\boldsymbol{x}$. Define $V(\theta, \lambda) = I(\theta, \lambda)^{-1}$ and*

$$I(\theta, \lambda) = \begin{pmatrix} I_{11}(\theta, \lambda) & I_{12}(\theta, \lambda) \\ I_{21}(\theta, \lambda) & I_{22}(\theta, \lambda) \end{pmatrix}, \quad V(\theta, \lambda) = \begin{pmatrix} V_{11}(\theta, \lambda) & V_{12}(\theta, \lambda) \\ V_{21}(\theta, \lambda) & V_{22}(\theta, \lambda) \end{pmatrix}.$$

Under asymptotic normality, the reference prior when θ is the parameter of interest is defined as follows:

Case 1. *If $c(\theta) = \int_{\Lambda(\theta)} \sqrt{I_{22}(\theta, \lambda)}\, d\lambda < \infty$, let $\pi(\lambda \mid \theta) = \sqrt{I_{22}(\theta, \lambda)}/c(\theta)$ and*

$$\pi(\theta) \propto \exp\left\{ \int_{\Lambda(\theta)} \pi(\lambda \mid \theta) \log[V_{11}^{-1/2}(\theta, \lambda)]\, d\lambda \right\}. \tag{10.17}$$

Then the reference prior is $\pi(\theta, \lambda) \propto \pi(\lambda \mid \theta)\pi(\theta)$, provided it is permissible.

Case 2. *If $c(\theta)$ is infinite, constrain (θ, λ) to compact subsets $\{\boldsymbol{\Upsilon}_l,\ l = 1, 2, \ldots\}$ of $\boldsymbol{\Upsilon}$, such that $\lim_{l \to \infty} \boldsymbol{\Upsilon}_l = \boldsymbol{\Upsilon}$, with associated projections $\{\Lambda_l(\theta), l = 1, 2, \ldots\}$ and define*

$$\pi_l(\lambda \mid \theta) = \frac{1}{c_l(\theta)} \sqrt{I_{22}(\theta, \lambda)} 1_{\Lambda_l(\theta)}(\lambda), c_l(\theta) = \int_{\Lambda_l(\theta)} \sqrt{I_{22}(\theta, \lambda)}\, d\lambda \tag{10.18}$$

$$\pi_l(\theta) \propto \exp\left\{ \int_{\Lambda_l(\theta)} \pi_l(\lambda \mid \theta) \log[V_{11}^{-1/2}(\theta, \lambda)]\, d\lambda \right\},$$

$$\pi(\theta, \lambda) = \lim_{l \to \infty} \frac{\pi_l(\lambda \mid \theta)\pi_l(\theta)}{\pi_l(\lambda_0 \mid \theta_0)\pi_l(\theta_0)}, \tag{10.19}$$

for any interior point (θ_0, λ_0). Then $\pi(\theta, \lambda)$ is the reference prior, provided it is permissible.

Proof. We only prove this for Case 1; the proof for Case 2 is identical, modulo the limiting argument over compact sets that is justified by permissibility.

The derivation of $\pi(\lambda \mid \theta)$ was Step 1 in Section 10.2.1. Since we are using the alternate reference prior definition,

Step 2 therein was skipped and Step 3 was applied with $p(\boldsymbol{x}^{(k)} \mid \theta) = \int_{\Lambda(\theta)} p(\boldsymbol{x}^{(k)} \mid \theta, \lambda)\pi(\lambda \mid \theta)\, d\lambda$, requiring determination of

$$\begin{aligned} f_k(\theta) &= \exp\left\{ \int p(\boldsymbol{x}^{(k)} \mid \theta) \log\left[\pi^*(\theta \mid \boldsymbol{x}^{(k)})\right] d\boldsymbol{x}^{(k)} \right\} \\ &= \exp\left\{ \int_{\Lambda(\theta)} \pi(\lambda \mid \theta) \int p(\boldsymbol{x}^{(k)} \mid \theta, \lambda) \log\left[\pi^*(\theta \mid \boldsymbol{x}^{(k)})\right] d\boldsymbol{x}^{(k)}\, d\lambda \right\}. \end{aligned}$$

Proceeding heuristically, because of asymptotic normality of both the mle $(\hat{\theta}, \hat{\lambda})$ and the posterior, and noting that $\frac{1}{k}V(\hat{\theta}, \hat{\lambda})$ is the asympotic covariance matrix of the posterior (see *e.g.*, Bernardo & Smith (1994, Sec. 5.3) and references therein for needed regularity conditions), the inner integral above, $I = \int p(\boldsymbol{x}^{(k)} \mid \theta, \lambda) \log\left[\pi^*(\theta \mid \boldsymbol{x}^{(k)})\right] d\boldsymbol{x}^{(k)}$, becomes

$$
\begin{aligned}
I &\approx \int p(\hat{\theta}, \hat{\lambda} \mid \theta, \lambda) \log\left[\mathrm{N}(\theta \mid \hat{\theta}, k^{-1}V_{11}(\hat{\theta}, \hat{\lambda})\right] d\hat{\theta}\, d\hat{\lambda} \\
&= \int p(\hat{\theta}, \hat{\lambda} \mid \theta, \lambda) \left[\log\left(\frac{k}{2\pi V_{11}(\hat{\theta}, \hat{\lambda})}\right)^{1/2} - \frac{k(\theta-\hat{\theta})^2}{2V_{11}(\hat{\theta}, \hat{\lambda})}\right] d\hat{\theta}\, d\hat{\lambda} \\
&\approx \left[\log\left(\frac{k}{2\pi V_{11}(\theta, \lambda)}\right)^{1/2} - \frac{1}{2}\right],
\end{aligned}
$$

the first term because $p(\hat{\theta}, \hat{\lambda} \mid \theta, \lambda)$ concentrates to a point mass at (θ, λ) and the second term because $\int p(\hat{\theta}, \hat{\lambda} \mid \theta, \lambda)(\theta - \hat{\theta})^2\, d\hat{\theta}\, d\hat{\lambda}$ is the asymptotic variance of $\hat{\theta}$, which is $V_{11}(\hat{\theta}, \hat{\lambda})/k$.

Thus, defining $c_k = \frac{1}{2}[\log(k/2\pi) - 1]$,

$$
\begin{aligned}
f_k(\theta) &\approx \exp\left\{\int_{\Lambda(\theta)} \pi(\lambda \mid \theta)(c_k + \log[V_{11}^{-1/2}(\theta, \lambda)])\, d\lambda\right\} \\
&= e^{c_k} \exp\left\{\int_{\Lambda(\theta)} \pi(\lambda \mid \theta) \log[V_{11}^{-1/2}(\theta, \lambda)]\, d\lambda\right\}.
\end{aligned}
$$

The marginal reference prior for θ is defined as $\pi(\theta) = \lim_{k\to\infty} f_k(\theta)/f_k(\theta_0)$, from which (10.17) is immediate. □

Example 10.3 *Ratio of multinomial parameters (continued).* Here we determine the (asymptotic) reference prior, in the situation of Example 10.1, using the above proposition. Computation yields

$$
I(\theta, \lambda) = \frac{n}{[1 - \lambda(1+\theta)]} \begin{pmatrix} \frac{\lambda(1-\lambda)}{\theta} & 1 \\ 1 & \frac{(1+\theta)}{\lambda} \end{pmatrix}, \quad V(\theta, \lambda) = \frac{\theta}{n} \begin{pmatrix} \frac{(1+\theta)}{\lambda} & -1 \\ -1 & \frac{\lambda(1-\lambda)}{\theta} \end{pmatrix}.
$$

Determination of the conditional reference prior $\pi(\lambda \mid \theta)$, in the Proposition, proceeds exactly as in Example 10.1, yielding the result in (10.12); note that this is a Case 1 situation. Finally, (10.17) becomes (π in the denominators below is the constant, not a prior)

$$
\pi(\theta) \propto \exp\left\{\int_0^{(1+\theta)^{-1}} \frac{(1+\theta)^{1/2}}{\pi(\lambda[1 - \lambda(1+\theta)^2])^{1/2}} \log\left(\frac{n\lambda}{\theta(1+\theta)}\right)^{1/2} d\lambda\right\}.
$$

Changing variables from λ to $y = \lambda(1+\theta)$ in the integral above results in

$$
\log\left(\frac{n}{\theta(1+\theta)^2}\right)^{1/2} + \int_0^1 \frac{\log\sqrt{y}}{\pi(y[1-y])^{1/2}}\, dy\,.
$$

The integral above is finite, and so is some constant C; thus exponentiating yields $\pi(\theta) \propto \theta^{-1/2}(1+\theta)^{-1}$ which, when multiplied by $\pi(\lambda \mid \theta)$, yields the same joint reference prior derived in (10.13).

Here is a Case 2 example, where a limiting operation over increasing compact sets is required.

Example 10.4 *Product of positive normal means.* Let the data consist of two independent random samples $\boldsymbol{x} = (x_1, \ldots, x_n)$ and $\boldsymbol{y} = (y_1, \ldots, y_m)$ from the $N(x \mid \mu_1, 1)$ and $N(y \mid \mu_2, 1)$ distributions, with $\mu_1 > 0$ and $\mu_2 > 0$. The model is thus

$$p(\boldsymbol{x}, \boldsymbol{y} \mid \mu_1, \mu_2) = \prod_{i=1}^{n} N(x_i \mid \mu_1, 1) \prod_{j=1}^{m} N(y_j \mid \mu_2, 1), \quad \mu_1 > 0,\ \mu_2 > 0\,.$$

Suppose that the quantity of interest is the product of the means, $\mu_1\mu_2$, a frequent situation in physics and engineering (*e.g.*, when trying to precisely measure the area of a table, based on measurements of its length and width).

Note first that it suffices to consider the standardized sufficient statistics $x = \sqrt{n}\bar{x} \sim \mathrm{N}(x \mid \mu_2^* \equiv \sqrt{n}\mu_1, 1)$ and $y = \sqrt{m}\bar{y} \sim \mathrm{N}(y \mid \mu_2^* \equiv \sqrt{m}\mu_2, 1)$. Since $\mu_1\mu_2 = \mu_1^*\mu_2^*/\sqrt{nm}$, it suffices to find the reference prior for $\theta = \mu_1^*\mu_2^*$ and transform back. For the nuisance parameter, it is convenient to choose $\lambda = [\mu_2^*/\mu_1^*]^{1/2}$.

Making the one-to-one transformation to (θ, λ), the Fisher information matrix $I(\theta, \lambda)$ (corresponding to one observation pair (x, y)) and its inverse, $V(\theta, \lambda)$, can be computed to be

$$I = \begin{pmatrix} \frac{(\lambda^2+\lambda^{-2})}{4\theta} & \frac{\lambda(1-\lambda^{-4})}{2} \\ \frac{\lambda(1-\lambda^{-4})}{2} & \theta(1+\lambda^{-4}) \end{pmatrix}, \quad V = \begin{pmatrix} \theta(\lambda^2+\lambda^{-2}) & \frac{(\lambda^{-1}-\lambda^{3})}{2} \\ \frac{(\lambda^{-1}-\lambda^{3})}{2} & \frac{(1+\lambda^4)}{4\theta} \end{pmatrix}.$$

To apply Proposition 10.1, we first consider $\sqrt{I_{22}(\theta, \lambda)} = \sqrt{\theta}(1+\lambda^{-4})^{1/2}$. Unfortunately, this does not have a finite integral over λ, so we need to consider a sequence of increasing compact sets of the parameter space.

Choosing compact subsets in the (θ, λ) parameter space seems natural, but recall that the motivation for using a sequence of compact subsets (besides the technical need) is to have a sequence of compact sets reflecting prior knowledge that the relevant parameter space is large, but the actual size is unknown (*cf.* Section 2.3). This rationale makes much more sense in the original (μ_1^*, μ_2^*) parameterization, so the increasing sequence of subsets that we choose is $\Upsilon_l = \{(\mu_1^*, \mu_2^*);\ 0 < \mu_1^* < l,\ 0 < \mu_2^* < l\}$. These project, into the parameter space of λ, given θ, as $\Lambda_l(\theta) = \{\lambda;\ \theta\, l^{-2} < \lambda < l^2\, \theta^{-1}\}$.

Unfortunately, $c_l(\theta) = \int_{\Lambda_l(\theta)} \sqrt{\theta}(1+\lambda^{-4})^{1/2}\, d\lambda$, the normalizing constant in (10.18), is not available in closed form, and neither is the integral in

$$\pi_l(\theta) \propto \exp\left\{ \int_{\Lambda_l(\theta)} c_l^{-1}(\theta)\sqrt{\theta}(1+\lambda^{-4})^{1/2}\, \log[(\theta[\lambda^2+\lambda^{-2}])^{-1/2}]\, d\lambda \right\}.$$

However, the limit in (10.19) can be evaluated (see Berger & Bernardo (1989)) and results in the joint reference prior, when θ is the parameter of interest, $\pi(\theta, \lambda) \propto [\theta(1+\lambda^{-4})]^{1/2}$. In the (μ_1^*, μ_2^*) parametrization, this corresponds to $\pi(\mu_1^*, \mu_2^*) \propto ([\mu_1^*]^2 + [\mu_2^*]^2)^{1/2}$. Finally, transforming back to (μ_1, μ_2) gives the reference prior, when $\mu_1\mu_2$ is the parameter of interest,

$$\pi(\mu_1, \mu_2) \propto (n\mu_1^2 + m\mu_2^2)^{1/2}, \tag{10.20}$$

which depends on the sample sizes through the ratio m/n. Notice that the reference prior (10.20) is very different from the constant prior, $\pi(\mu_1, \mu_2) = 1$, which is used for objective inferences about either μ_1 or μ_2.

Berger & Bernardo (1989) demonstrated that the reference prior here has much better finite sample frequentist matching properties (in terms of confidence sets for $\mu_1\mu_2$) than the constant prior and Datta & Ghosh (1995*e*) showed that it is asymptotic first order matching. Indeed, this prior was originally suggested by Stein (1985), for this reason. Efron (1986) used this particular problem as an example in which traditional objective Bayesian theories (e.g., always use a constant prior) encounters difficulties. For the reference priors associated to generalizations of the first order product of normal means problem, see Sun & Ye (1995, 1999).

An interesting feature of the above example is that, while $c_l(\theta)$ and $\pi_l(\theta)$ are not available in closed form, the limit in (10.19) is available in closed form. The same was true in Example 10.2 for the completely general reference prior algorithm; while a closed form expression for the marginal model $p(\boldsymbol{x} \mid \theta)$ was not available, it could be recognized as being a location (or scale) density so that the reference prior was known. This phenomenon seems to hold quite generally; the reference prior can often be determined in closed form (up to a normalizing constant), even if the relevant integrals defining missing information cannot be evaluated in closed form. But use of ϵ,δ mathematical analysis is often required, identifying the dominant regions (as $l \to \infty$) in the relevant integrals.

Corollary 10.1.1 (Reference priors under factorization).
Under the conditions of Proposition 10.1, if

(i) *θ and λ are variation independent, so that $\Lambda = \Lambda(\theta)$ does not depend on the parameter of interest θ, and*

(ii) *both $I_{22}(\theta, \lambda)$ and $V_{11}(\theta, \lambda)$ factorize, so that*

$$V_{11}^{-1/2}(\theta, \lambda) \propto f_1(\theta)\, g_1(\lambda), \quad I_{22}^{1/2}(\theta, \lambda) \propto f_2(\lambda)\, g_2(\theta), \tag{10.21}$$

then the joint reference prior is $\pi(\theta, \lambda) = f_1(\theta)\, f_2(\lambda)$, even if the conditional reference prior, $\pi(\lambda \mid \theta) \propto f_2(\lambda)$, is improper.

Proof. The conditional reference prior is $\pi(\lambda \mid \theta) \propto I_{22}^{1/2}(\theta, \lambda) \propto f_2(\lambda) g_2(\theta)$. First, if $\int_\Lambda f_2(\lambda)\, d\lambda < \infty$, normalizing yields $\pi(\lambda \mid \theta) = c_1\, f_2(\lambda)$, which does not depend on θ. If, furthermore, $V_{11}^{-1/2}(\theta, \lambda) = f_1(\theta) g_1(\lambda)$ and noting that Λ here does not depend on θ, (10.17) reduces to

$$\pi(\theta) \propto \exp\left\{ \int_\Lambda c_1\, f_2(\lambda) \log[f_1(\theta) g_1(\lambda)]\, d\lambda \right\} = c_2\, f_1(\theta)$$

and, hence, the reference prior is $\pi(\theta, \lambda) = \pi(\lambda \mid \theta)\, \pi(\theta) = c\, f_1(\theta)\, f_2(\lambda)$. If $\int_\Lambda f_2(\lambda)\, d\lambda = \infty$, operating on an increasing sequence of compact sets and taking the limit yields the same result. □

The Jeffreys-rule prior, under the factorization in (10.21), can be computed to be $\pi^J(\theta, \lambda) \propto f_1(\theta)\, g_1(\lambda) f_2(\lambda) g_2(\theta)$, which is quite different from the reference prior.

Example 10.5 *Inference on normal parameters.* Let data $\boldsymbol{x} = (x_1, \ldots, x_n)$ consist of a random sample of fixed size n from a normal distribution $\mathrm{N}(x \mid \mu, \sigma^2)$. For a single observation, the information matrix $I(\mu, \sigma)$ and its inverse matrix are, respectively,

$$I(\mu, \sigma) = \begin{pmatrix} \sigma^{-2} & 0 \\ 0 & 2\sigma^{-2} \end{pmatrix}, \quad V(\mu, \sigma) = I^{-1}(\mu, \sigma) = \begin{pmatrix} \sigma^2 & 0 \\ 0 & \frac{1}{2}\sigma^2 \end{pmatrix}.$$

Hence, $I_{22}^{1/2}(\mu, \sigma) = \sqrt{2}\, \sigma^{-1} = f_2(\mu)\, g_2(\sigma)$, with $g_2(\sigma) = \sigma^{-1}$, so that $\pi(\sigma \mid \mu) = \sigma^{-1}$. Similarly, $V_{11}^{-1/2}(\mu, \sigma) = \sigma^{-1} = f_1(\mu)\, g_1(\sigma)$, with $f_1(\mu) = 1$, and thus $\pi(\mu) = 1$. Therefore, using Corollary 10.1.1, the reference prior when μ is the parameter of interest is $\pi(\mu, \sigma) = \pi(\sigma \mid \mu)\, \pi(\mu) = \sigma^{-1}$ for all $n \geq 2$. For $n = 1$, the posterior distribution is not proper, so that $\pi(\mu, \sigma) = \sigma^{-1}$ is not a permissible prior and a reference prior does not exist.

A nearly identical argument shows that the reference prior, when σ is the parameter of interest, is also σ^{-1}. Of course, this also follows from Example 10.2, because the normal distribution is a location-scale model.

Next consider the case where the parameter of interest is *not* the mean μ or the standard deviation σ, but the *standardized* mean $\theta = \mu/\sigma$ (or, equivalently, its reciprocal, the coefficient of variation σ/μ). The Fisher information matrix, in terms of the parameters θ and σ (we keep σ as the nuisance parameter rather than renaming it λ), can be computed as $I(\theta, \sigma) = J^t\, I(\mu, \sigma)\, J$, where $J = (\partial(\mu, \sigma)/\partial(\theta, \sigma))$ is the Jacobian of the inverse transformation. The result is

$$I(\theta, \sigma) = \begin{pmatrix} 1 & \theta\sigma^{-1} \\ \theta\sigma^{-1} & \sigma^{-2}(2 + \theta^2) \end{pmatrix}, \quad V(\theta, \sigma) = \begin{pmatrix} 1 + \frac{1}{2}\theta^2 & -\frac{1}{2}\theta\sigma \\ -\frac{1}{2}\theta\sigma & \frac{1}{2}\sigma^2 \end{pmatrix}.$$

Thus, $I_{22}^{1/2}(\theta, \sigma) = \sigma^{-1}(2 + \theta^2)^{1/2}$ and $V_{11}^{-1/2}(\theta, \sigma) = (1 + \frac{1}{2}\theta^2)^{-1/2}$. Hence, again using Corollary 10.1.1, $\pi(\theta, \sigma) = (1 + \frac{1}{2}\theta^2)^{-1/2} \sigma^{-1}$ $(n \geq 2)$. Transforming back to the original parameterization, the reference prior when θ is the parameter of interest becomes $\pi(\mu, \sigma) = (1 + \frac{1}{2}(\mu/\sigma)^2)^{-1/2}\sigma^{-2}$, which is very different from $\pi(\mu, \sigma) = \sigma^{-1}$, the reference prior when either μ or σ was the parameter of interest.

The resulting reference posterior density for θ, given data $\boldsymbol{x} = (x_1, \ldots, x_n)$, can be shown to be

$$\pi(\theta \,|\, \boldsymbol{x}) \propto (1 + \tfrac{1}{2}\theta^2)^{-1/2}\, p(t \,|\, \theta)\,, \tag{10.22}$$

where $t = (\sum x_j)/(\sum x_j^2)^{1/2}$ is a one-dimensional statistic whose sampling distribution, $p(t \,|\, \mu, \sigma) = p(t \,|\, \theta)$, only depends on θ. Thus, the reference prior algorithm is seen here to avoid the marginalization paradox (see Section 7.5.2).

10.3 Multiple Nuisance Parameters

Section 10.3.1 presents the recommended reference prior when there are multiple nuisance parameters and under asymptotic normality. This recommended prior is called the *one-at-a-time* reference prior, because each nuisance parameter is treated separately in the reference prior algorithm iterations.

Sections 10.3.2, 10.3.3, and 10.3.4 give an in-depth look at application to the multinomial distribution, the bivariate normal distribution and prediction, respectively, to point out key issues that arise in reference prior development.

Section 10.3.1 necessarily has quite involved notation. A much simpler (but potentially less effective) reference prior method for the situation of multiple nuisance parameters is the two-group reference prior discussed in Section 10.3.5. This is part of the more general reference prior theory, developed in Berger & Bernardo (1992*a*) and Berger & Bernardo (1992*b*), where the parameter vector is divided into a series of parameter groupings, to which the reference prior algorithm is sequentially applied. We delay discussion of this more general reference prior theory until Chapter 14.

10.3.1 Recommended Reference Prior under Asymptotic Normality

This section deals with the situation where $\boldsymbol{\theta} = (\theta_1, \ldots, \theta_m)$, with θ_1 being the parameter of interest and the ordering of parameters specified, as discussed in Section 10.1. Note that the ordering of the nuisance parameters may matter, although it is often irrelevant. The model is $\mathcal{M} \equiv \{p(\boldsymbol{x} \,|\, \boldsymbol{\theta}), \boldsymbol{x} \in \boldsymbol{\mathcal{X}}, \boldsymbol{\theta} \in \boldsymbol{\Theta}\}$. Throughout we use the notational conventions that $\prod_{i-1}^{i} = 1$ and $\sum_{i-1}^{i} = 0$.

Derivation of the Recommended Reference Prior

Proposition 10.2 (Reference priors under asymptotic normality.) *The k replications result in the model* $\mathcal{M} \equiv \{p(\boldsymbol{x}^{(k)} \,|\, \boldsymbol{\theta}), \boldsymbol{x}^{(k)} \in \boldsymbol{\mathcal{X}}^k, \boldsymbol{\theta} \in \boldsymbol{\Theta}\}$. *The* *corresponding Fisher information matrix is $kI(\boldsymbol{\theta})$, where $I(\boldsymbol{\theta})$ is the Fisher information matrix for a single $\boldsymbol{x}$. Define $I_0(\boldsymbol{\theta}) = 1$ and, for $j = 1, \ldots, m$,*

$$h_j(\boldsymbol{\theta}) = \frac{|I_{m-j+1}(\boldsymbol{\theta})|}{|I_{m-j}(\boldsymbol{\theta})|}, \; I_i(\boldsymbol{\theta}) \text{ being the lower right } i \times i \text{ corner of } I(\boldsymbol{\theta}). \tag{10.23}$$

Also define $\boldsymbol{\theta}_j = (\theta_1, \ldots, \theta_j)$ and $\boldsymbol{\theta}^j = (\theta_j, \ldots, \theta_m)$ and suppose that, given $\boldsymbol{\theta}_j$ for $j \leq m-1$, the domain of $\boldsymbol{\theta}^{j+1}$ is $\boldsymbol{\Theta}^{j+1}(\boldsymbol{\theta}_j)$. (Typically, $\boldsymbol{\Theta} = \Theta_1 \times \Theta_2 \times \cdots \times \Theta_m$, in which case $\boldsymbol{\Theta}^{j+1}(\boldsymbol{\theta}_j) = \Theta_{j+1} \times \cdots \times \Theta_m$.)

Under asymptotic normality, the reference prior, for the ordered parameterization $(\theta_1, \ldots, \theta_m)$ and providing it is permissible, is

$$\pi(\boldsymbol{\theta}) = \pi(\theta_m \mid \boldsymbol{\theta}_{m-1}) \times \pi(\theta_{m-1} \mid \boldsymbol{\theta}_{m-2}) \times \cdots \times \pi(\theta_2 \mid \theta_1) \times \pi(\theta_1),$$

where $\pi(\theta_m \mid \boldsymbol{\theta}_{m-1}) \propto h_m^{1/2}(\boldsymbol{\theta}) = I_{mm}^{1/2}(\boldsymbol{\theta})$ and, for $j = 1, \ldots, m-1$ (with $\pi(\theta_1 \mid \boldsymbol{\theta}_0)$ just being the marginal reference prior $\pi(\theta_1)$),

$$\pi(\theta_j \mid \boldsymbol{\theta}_{j-1}) \propto \exp\left\{ \int_{\boldsymbol{\Theta}^{j+1}(\boldsymbol{\theta}_j)} \prod_{i=j+1}^{m} \pi(\theta_i \mid \boldsymbol{\theta}_{i-1}) \times \log[h_j^{1/2}(\boldsymbol{\theta})] \, d\boldsymbol{\theta}^{j+1} \right\}, \tag{10.24}$$

provided $\pi(\theta_j \mid \boldsymbol{\theta}_{j-1})$ is proper (integrating to 1) for all $j \geq 2$; this requires normalization of $h_m^{1/2}(\boldsymbol{\theta})$ and the righthand side of the above expression for $2 \leq j \leq m-1$. This defines the reference prior even if $\pi(\theta_1)$ is improper.

Next, suppose that the $\{\pi(\theta_j \mid \boldsymbol{\theta}_{j-1}),\ j = r+1, \ldots, m\}$ are proper but $\pi(\theta_r \mid \boldsymbol{\theta}_{r-1})$ is not, for some $r \geq 2$. Letting $\boldsymbol{\Upsilon}$ denote the domain of $\boldsymbol{\theta}_r$, it is then necessary to constrain $\boldsymbol{\theta}_r$ to compact subsets $\{\boldsymbol{\Upsilon}_l,\ l = 1, 2, \ldots\}$ of $\boldsymbol{\Upsilon}$, such that $\lim_{l \to \infty} \boldsymbol{\Upsilon}_l = \boldsymbol{\Upsilon}$. For $j \leq r$, denote the implied domains of $\boldsymbol{\theta}_j$ by $\boldsymbol{\Upsilon}_l^j$. Then the constrained conditional reference priors are, for $j = 2, \ldots, r$ (with $\pi_l(\theta_1 \mid \boldsymbol{\theta}_0)$ just being the marginal reference prior $\pi_l(\theta_1)$)

$$\begin{aligned} \pi_l(\theta_j \mid \boldsymbol{\theta}_{j-1}) \propto 1_{\boldsymbol{\Upsilon}_l^j}(\boldsymbol{\theta}_j) \exp\Bigg\{ \int_{\boldsymbol{\Theta}^{j+1}(\boldsymbol{\theta}_j)} & \prod_{i=j+1}^{r} \pi_l(\theta_i \mid \boldsymbol{\theta}_{i-1}) \\ & \times \prod_{i=r+1}^{m} \pi(\theta_i \mid \boldsymbol{\theta}_{i-1}) \times \log[h_j^{1/2}(\boldsymbol{\theta})] \, d\boldsymbol{\theta}^{j+1} \Bigg\}, \end{aligned} \tag{10.25}$$

which again must be normalized before iterating. If $r = m$,

$$\pi_l(\theta_m \mid \boldsymbol{\theta}_{m-1}) = \frac{h_m^{1/2}(\boldsymbol{\theta}) 1_{\boldsymbol{\Upsilon}_l^m}(\boldsymbol{\theta})}{\int_{\boldsymbol{\Upsilon}_l^m} h_m^{1/2}(\boldsymbol{\theta}) \, d\theta_m}.$$

Finally, assuming it is permissible, the joint reference prior is given by

$$\pi(\boldsymbol{\theta}) = \prod_{j=r+1}^{m} \pi(\theta_j \mid \boldsymbol{\theta}_{j-1}) \times \lim_{l \to \infty} \frac{\prod_{j=1}^{r} \pi_l(\theta_j \mid \boldsymbol{\theta}_{j-1})}{\prod_{j=1}^{r} \pi_l(\theta_{j0} \mid \theta_{10}, \ldots, \theta_{[j-1]0})},$$

for any interior point $(\theta_{10}, \ldots, \theta_{r0})$.

Proof. The proof is just iteration of the proof for the binormal model – modulo the algebra discussed below – and is, hence, omitted. □

It is common (as in the binormal situation) to develop the reference prior in terms of the asymptotic covariance matrix which, under regularity conditions, is proportional to $V(\boldsymbol{\theta}) = I(\boldsymbol{\theta})^{-1}$. Indeed, (10.23) could have been replaced by defining $V_0(\boldsymbol{\theta}) = 1$ and, for $j = 1, \ldots, m$,

$$h_j(\boldsymbol{\theta}) = \frac{|V_{j-1}(\boldsymbol{\theta})|}{|V_j(\boldsymbol{\theta})|}, \quad V_i(\boldsymbol{\theta}) \text{ being the upper left } i \times i \text{ corner of } V(\boldsymbol{\theta}). \quad (10.26)$$

What is needed at the jth iteration of the reference prior algorithm is the Fisher information about θ_j that would result from the marginal model involving only $(\theta_1, \ldots, \theta_j)$. In the asymptotic setting, the marginal model has asymptotic covariance matrix $V_j(\boldsymbol{\theta})$ and the information about θ_j in this model is the lower right corner entry of $V_j^{-1}(\boldsymbol{\theta})$; using Cramer's rule, this is given by (10.26). The alternate expression in (10.23) (utilized in Lemma 1 of Datta & Ghosh (1996)) arises from the usual formula for the determinant of a block matrix, with $V_j(\boldsymbol{\theta})$ being the upper left block of $V(\boldsymbol{\theta})$ and $I_{j+1}(\boldsymbol{\theta})$ being the lower right block of $I(\boldsymbol{\theta}) = V^{-1}(\boldsymbol{\theta})$.

Sometimes, only $I(\boldsymbol{\theta})$ or $V(\boldsymbol{\theta})$ is known, so having both (10.23) and (10.26) available to determine the $h_j(\boldsymbol{\theta})$ is helpful. And one can mix and match when both are available; for instance, here are the simplest expressions for the h_j (dropping the argument $\boldsymbol{\theta}$) when $m = 3$ and $m = 4$:

$$m = 3: \qquad h_1 = \tfrac{1}{V_{11}}, \quad h_2 = I_{22} - \tfrac{I_{23}^2}{I_{33}}, \quad h_3 = I_{33}\,;$$

$$m = 4: \quad h_1 = \tfrac{1}{V_{11}}, \quad h_2 = \tfrac{V_{11}}{[V_{11}V_{22} - V_{12}^2]}, \quad h_3 = I_{33} - \tfrac{I_{34}^2}{I_{44}}, \quad h_4 = I_{44}\,.$$

The following result, which is a slight generalization of Theorem 1 in Datta & Ghosh (1995*e*), provides a very useful computational simplification.

Corollary 10.2.1 *(***Reference priors under factorization.***) If, moreover,* (i) $\boldsymbol{\Theta} = \Theta_1 \times \Theta_2 \times \cdots \times \Theta_m$, i.e., *the θ_i are variation independent, and* (ii) *the functions $h_j(\boldsymbol{\theta})$ factorize, so that,*

$$h_j^{1/2}(\boldsymbol{\theta}) = f_j(\theta_j)\, g_j(\theta_1, \ldots, \theta_{j-1}, \theta_{j+1}, \ldots, \theta_m), \quad j = 1, \ldots, m,$$

then the reference prior is $\pi(\boldsymbol{\theta}) = \prod_{j=1}^m f_j(\theta_j)$, even if the conditional reference priors are improper.

Proof. Assuming that all conditional reference priors are proper (the proof in the improper case is analogous), the first step of the reference prior algorithm determines

$$\pi(\theta_m \mid \boldsymbol{\theta}_{m-1}) = \frac{h_m^{1/2}(\boldsymbol{\theta})}{\int_{\Theta_m} h_m^{1/2}(\boldsymbol{\theta})\, d\theta_m} = \frac{f_m(\theta_m)\, g_m(\theta_1, \ldots, \theta_{m-1})}{\int_{\Theta_m} f_m(\theta_m)\, g_m(\theta_1, \ldots, \theta_{m-1})\, d\theta_m} = c_m f_m(\theta_m)$$

for some constant c_m. Next, note that

$$\begin{aligned}
&\int_{\Theta_m} c_m f_m(\theta_m) \log[h_{m-1}^{1/2}(\boldsymbol{\theta})]\, d\theta_m \\
&= \int_{\Theta_m} c_m f_m(\theta_m) \left(\log[f_{m-1}(\theta_{m-1})] + \log[g_{m-1}(\boldsymbol{\theta}_{m-2}, \theta_m)]\right) d\theta_m \\
&= \log[f_{m-1}(\theta_{m-1})] + q(\boldsymbol{\theta}_{m-2}),
\end{aligned}$$

for some function $q(\cdot)$. Thus (10.25) becomes, for $j = m-1$,

$$\pi(\theta_{m-1} \,|\, \boldsymbol{\theta}_{m-2}) = \frac{\exp\left\{ \log[f_{m-1}(\theta_{m-1})] + q(\boldsymbol{\theta}_{m-2}) \right\}}{\int_{\Theta_{m-1}} \exp\left\{ \log[f_{m-1}(\theta_{m-1})] + q(\boldsymbol{\theta}_{m-2}) \right\} d\theta_{m-1}} = c_{m-1} f_{m-1}(\theta_{m-1})$$

for some constant c_{m-1}. Continuing through the iterations of the reference prior algorithm yields the result. □

Example 10.6 *Multivariate normal data.* Let the data be $\boldsymbol{x} = (\boldsymbol{y}_1, \ldots, \boldsymbol{y}_n)$, $n \geq 2$, where the $\boldsymbol{y}_i = (y_{i1}, \ldots y_{im})$ are a random sample from the m-variate normal distribution with mean $\boldsymbol{\mu} = (\mu_1, \ldots, \mu_m)$ and covariance matrix $\sigma^2 \mathbf{I}_m$, $m \geq 1$, so that the single observation Fisher information matrix is

$$I(\boldsymbol{\mu}, \sigma) = \begin{pmatrix} \frac{1}{\sigma^2} \mathbf{I}_m & 0 \\ 0 & \frac{2m}{\sigma^2} \end{pmatrix}.$$

It follows from Corollary 10.2.1 that the reference prior for $(\mu_1, \ldots, \mu_m, \sigma)$ is $\pi(\mu_1, \ldots, \mu_m, \sigma) \propto \sigma^{-1}$, and also that the result does not depend on the order in which the parametrization is taken. The corresponding reference posterior for any of the μ_j's is easily shown to be the Student t density

$$\pi(\mu_j \,|\, \boldsymbol{y}_1, \ldots, \boldsymbol{y}_n) = \mathrm{St}\left(\mu_j \,\middle|\, \overline{y}_j,\, s_n/\sqrt{(n-1)},\, m(n-1)\right)$$

with $n\overline{y}_j = \sum_{i=1}^n y_{ij}$ and $nm\, s_n^2 = \sum_{j=1}^m \sum_{i=1}^n (y_{ij} - \overline{y}_j)^2$. Similarly, the reference posterior of σ^2 is the inverted Gamma density

$$\pi(\sigma^2 \,|\, \boldsymbol{y}_1, \ldots, \boldsymbol{y}_n) = \mathrm{IGa}(\sigma^2 \,|\, m(n-1)/2,\, n\, m\, s_n^2/2).$$

When $m = 1$, these results reduce to those obtained in Example 10.5.

Example 10.7 *Stein's paradox (continued):* Let $\boldsymbol{x} = (x_1, \ldots, x_m)$ be an observation from the m-variate normal distribution $\mathrm{N}_m(\boldsymbol{x} \,|\, \boldsymbol{\mu}, \boldsymbol{I}_m)$, with mean $\boldsymbol{\mu} = (\mu_1, \ldots, \mu_m)$ and identity covariance matrix. Suppose that the quantity of interest is $\theta = \sum_{j=1}^m \mu_j^2/m$. This was discussed in Example 7.3, where it was shown that use of the uniform prior, $\pi(\boldsymbol{\mu}) = 1$, results in a seriously inconsistent posterior distribution for θ, as $m \to \infty$.

The reference prior is found most easily by using the parameterization $(\theta, \boldsymbol{\lambda})$, where $\boldsymbol{\lambda}$ is the unit directional vector of $\boldsymbol{\mu}$ in $\mathbb{R}^m$; thus $\boldsymbol{\mu} = \sqrt{m\theta}\boldsymbol{\lambda}$. The reference prior in this parameterization was found, following the above algorithm, by a lengthy argument in Ferrándiz (1985) (see Datta & Ghosh (1995*e*) for a simpler derivation); Berger et al. (1998) give generalizations of the argument. This reference prior is $\pi(\theta, \boldsymbol{\lambda}) \propto 1/\sqrt{\theta}$. Transforming back to $\boldsymbol{\mu}$ results in the reference prior, when θ is the parameter of interest, $\pi(\boldsymbol{\mu}) = (\sum_{j=1}^m \mu_j^2)^{-(m-1)/2}$, the prior also suggested by Stein (1985) and Tibshirani (1989) from a frequentist matching perspective (see also Datta & Ghosh (1995*c*)). This prior results in fully consistent Bayes estimates and results in excellent frequentist performance of Bayes credible sets.

This example provides an interesting scenario in which to compare the basic reference prior algorithm in Section 10.2.1 with the asymptotic reference prior algorithm in Section 10.2.2 (under which $\pi(\boldsymbol{\mu}) = (\sum_{j=1}^m \mu_j^2)^{-(m-1)/2}$ was derived). The reason is that $\boldsymbol{\lambda}$ is variation independent of θ and is a vector on the unit ball in $\mathbb{R}^m$. Thus, given θ, the reference prior for $\boldsymbol{\lambda}$ will simply be the (proper) uniform prior, $\pi(\boldsymbol{\lambda}) \propto 1$, on the unit ball in $\mathbb{R}^m$. This means that Step 2 in Section 10.2.1 can be carried out, resulting in the marginal model $p(\boldsymbol{x} \mid \theta) = \int p(\boldsymbol{x} \mid \theta, \boldsymbol{\lambda})\pi(\boldsymbol{\lambda})\, d\boldsymbol{\lambda}$.

It can be shown that $p(\boldsymbol{x} \mid \theta)$ only depends on $t = |\boldsymbol{x}|$, so that one can instead consider the transformed density $p(t \mid \theta)$, which happens to be the noncentral chi-squared density, $\chi^2(t \,|\, m, m\theta)$. The marginal reference prior for θ, using the original reference prior algorithm in Section 10.2.1, would thus be the Jeffreys-rule prior, $\pi^J(\theta)$, for θ from the noncentral chi-square distribution, with the overall reference prior being $\pi(\theta, \boldsymbol{\lambda}) \propto \pi^J(\theta)$.

To our knowledge, $\pi^J(\theta)$ has never been computed; indeed, the Fisher information, $I(\theta)$, for the noncentral chi-square distribution does not seem to be available in closed form. So it seems that we have a situation in which the two reference prior approaches result in different answers. It would be interesting to compare $\pi^J(\theta)$ here with $\pi(\theta) \propto 1/\sqrt{\theta}$ above to see how significant the difference is. This example also reinforces that the asymptotic reference prior approach can be much easier to implement than the original reference prior approach.

Invariance to Parameterization. In Section 7.5.1, we reaffirmed the importance of invariance to parameterization, the initial driving force behind modern objective Bayesian theory. Here we consider the extent to which the one-at-a-time reference prior theory satisfies this criterion. The discussion will be based on the following proposition, which is a special case of the remarkable Theorem 2.1 in Datta & Ghosh (1996). Note that this proposition is also often very useful in calculating the reference prior, as will be illustrated in the next section.

Proposition 10.3 *A one-to-one transformation from* $\boldsymbol{\theta}$ *to* $\boldsymbol{\eta} = (\eta_1, \ldots, \eta_m)$, *where* $\eta_i(\boldsymbol{\theta}) = g_i(\theta_1, \ldots, \theta_i)$, *does not change the reference prior, in the sense* *that the reference prior in either parameterization will be the transform of the other. Indeed, if $\pi(\boldsymbol{\eta})$ is the reference prior in the $\boldsymbol{\eta}$ parameterization, then the reference prior in the $\boldsymbol{\theta}$ parameterization is*

$$\pi(\boldsymbol{\theta}) = \pi(g_1(\boldsymbol{\theta}), \ldots, g_m(\boldsymbol{\theta})) \times \left| \prod_{i=1}^{m} \frac{\partial}{\partial \theta_i} g_i(\theta_1, \ldots, \theta_i) \right| . \tag{10.27}$$

Consider, first, the case where $m = 2$. Choosing the parameter of interest is necessary, but choice of the nuisance parameter is rather arbitrary. Proposition 10.3 shows that the choice does not matter; any choice will lead to the same reference prior. Hence we have invariance to parameterization, modulo the need to choose the parameter of interest.

When $m \geq 3$, however, the situation is not as clean. Different choices of the nuisance parameters – or different orderings of them – could lead to different reference priors. We did suggest that nuisance parameters be chosen in terms of descending importance, which is admittedly a vague concept. But Proposition 10.3 reinforces this notion, in that reparameterizing by allowing a given nuisance parameter to be a function of 'more important' parameters will not change the reference prior. This lends significant support to the invariance of reference priors to parameterization.

In applications, it is rare for the choice of nuisance parameters to actually matter, especially for the marginal posterior distribution of the parameter of interest; the joint reference prior can change, but the marginal posterior is usually the same. This also addresses the parameterization issue because the initial criticisms of, say, inverse probability centered on the fact that use of the constant prior in different parameterizations could result in different final answers.

10.3.2 Multinomial Distribution

Introduction. A very interesting example of the general iterated procedure to derive reference priors is provided by the multinomial distribution. Recall that the multinomial probalility function for $m + 1$ cells is given by

$$\text{Mu}(x_1, \ldots, x_m \,|\, \theta_1, \ldots, \theta_m, n) = \frac{n!}{\prod_{j=1}^{m+1} x_j!} \prod_{j=1}^{m+1} \theta_j^{x_j},$$

where x_j is the number of observations in cell j, n is the total number of observations, and θ_j is the probability of cell j, with $0 \leq \theta_1 + \cdots + \theta_m \leq 1$, $x_{m+1} = n - \sum_{j=1}^{m} x_j$, and $\theta_{m+1} = 1 - \sum_{j=1}^{m} \theta_j$.

In Example 4.10, the Jeffreys-rule prior, $\pi^J(\boldsymbol{\theta})$, for this problem was derived and shown to have problematical behavior, as was the behavior of the uniform prior, $\pi^U(\boldsymbol{\theta})$. In Berger & Bernardo (1992a), the reference prior for $\boldsymbol{\theta}$ was derived, and is given (when θ_1 is the parameter of interest), by

$$\pi^R(\theta_1,\ldots,\theta_m) = \pi^{-m}\prod\nolimits_{j=1}^{m}\{\theta_j^{-1/2}(1-\delta_j)^{-1/2}\}, \text{ with } \delta_j = \sum_{i=1}^{j}\theta_i. \quad (10.28)$$

This is a proper prior (as are the Jeffreys-rule and uniform priors).

Hazard Rates and Derivation of the Multinomial Reference Prior. The derivation of the reference prior in Berger & Bernardo (1992*a*) followed the algorithm in Proposition 10.2, but a much simpler development was suggested in He (2011), using a parameterization for the multinomial model in terms of the corresponding hazard rates, as given in He (2003). Indeed, letting $\theta_{m+1} = 1-\theta_1-\ldots-\theta_m$, define the hazard rates

$$\eta_j = \frac{\theta_j}{\theta_j+\cdots+\theta_{m+1}}, \quad \text{for} \quad j=1,\cdots,m, \quad \text{and} \quad \eta_{m+1}=1.$$

The inverse transformation is $\theta_1 = \eta_1$ and $\theta_j = \eta_j\prod_{i=1}^{j-1}(1-\eta_i)$ for $j \geq 2$.

Notice that $\boldsymbol{\eta} = (\eta_1,\ldots,\eta_m)$ is a one-to-one transformation of $(\theta_1,\cdots,\theta_m)$ that satisfies the condition of Corollary 10.3, so the reference prior found in this new parameterization will be the transformation of the reference prior in the original parameterization (and vice versa).

The ranges for the hazard rates are $0 < \eta_i < 1$ and the multinomial model expressed in terms of the hazard rates is

$$p(\boldsymbol{x}\,|\,\boldsymbol{\eta}) \propto \prod\nolimits_{j=1}^{m}\eta_j^{x_j}(1-\eta_j)^{x_{j+1}+\cdots+x_m+x_{m+1}}.$$

The corresponding single-observation Fisher information matrix is diagonal, with diagonal entries, for $j=1,\cdots,m$,

$$I_{jj}(\boldsymbol{\eta}) = [\eta_j(1-\eta_j)]^{-1}\prod_{i=1}^{j-1}(1-\eta_i).$$

This satisfies the conditions of Corollary 10.2.1, with $f_j(\eta_j) = [\eta_j(1-\eta_j)]^{-1/2}$ and $g(\eta_1,\ldots,\eta_{j-1},\eta_{j+1},\ldots,\eta_m) = \prod_{i=1}^{j-1}(1-\eta_i)^{1/2}$. It follows from the Corollary that the joint reference prior for $\boldsymbol{\eta}$, with η_1 the parameter of interest, is

$$\pi(\boldsymbol{\eta}) = \prod_{j=1}^{m}[\eta_j(1-\eta_j)]^{-1/2}. \quad (10.29)$$

Noting that $\eta_1 = \theta_1$ and $\eta_i = \theta_i/(1-\sum_{j=1}^{i-1}\theta_j)$ for $i \geq 2$, we obtain

$$\prod_{i=1}^{m}\frac{\partial}{\partial\theta_i}\eta_i(\boldsymbol{\theta}) = \frac{1}{\prod_{i=1}^{m}(1-\sum_{j=1}^{i-1}\theta_j)} = \frac{1}{\prod_{i=1}^{m}\sum_{j=i}^{m+1}\theta_j}.$$

Using this and (10.29) in (10.27), results in the reference prior for $\boldsymbol{\theta}$ given in (10.28). Of course, as inference about hazard rates is often desired, having the reference prior for hazard rates is also important.

Marginalization of the Reference Prior and Posterior. The marginal probability function of $(x_1, \ldots, x_l)$ is also multinomial, with cell probabilities $(\theta_1, \ldots, \theta_l)$ and sample size n, so that all other observations are lumped together into the new $(l+1)$st cell, which is the union of the $(l+1)$st through $(m+1)$st cells in the original multinomial. It is of considerable interest to see whether or not the reference prior 'marginalizes' consistently, in the sense that the reference prior for the 'collapsed' $(l+1)$-cell multinomial is the same as that obtained by finding the marginal distribution $(\theta_1, \ldots, \theta_l)$ from the original reference prior. The answer is positive, as shown in Berger & Bernardo (1992*a*).

Typically, of course, one will be interested in the marginal posteriors, rather than the marginal priors. These are immediate from the marginal priors, however; simply multiply by the likelihood from the corresponding marginal multinomial distribution. Of course, this could also have been obtained by calculating the marginal density of the joint reference posterior.

That the reference posterior marginalizes consistently is important, because of the following well-known problem with the Jeffreys posterior. Suppose we begin analysis with the multinomial model with $(\theta_1, \ldots, \theta_{m_1})$. The Jeffreys-rule prior for this model is $\pi^J(\theta_1, \ldots, \theta_{m_1}) \propto \left(\prod_{i=1}^{m_1} \theta_i^{-1/2}\right)(1-\delta_{m_1})^{-1/2}$, having posterior means

$$\mathrm{E}[\theta_j \mid x_1, \ldots, x_{m_1}, n] = \frac{x_j + 1/2}{n + (m_1+1)/2} \,.$$

After completing this analysis, suppose someone points out that the $(n - x_1 - \ldots x_{m_1})$ observations in the (m_1+1)st cell could have been further subdivided into $m_2 + 1$ new categories, with corresponding counts $x_{m_1+1}, \ldots, x_{m_1+m_2}$; say one discovers a new classification scheme into m_2 categories for elements in this cell (*e.g.*, the original multinomial model was cells for different reptiles, but then a turtle expert pointed out that there are m_2 distinct types of turtles, so the cell for turtles could be replaced by m_2 cells reflecting the different turtle types).

One now has the option of adding $x_{m_1+1}, \ldots, x_{m_1+m_2}$ and $\theta_{m_1+1}, \ldots, \theta_{m_1+m_2}$ to the multinomial model. If one did so and used the Jeffreys-rule prior, which would be $\pi^J(\theta_1, \ldots, \theta_{m_1+m_2}) \propto \left(\prod_{i=1}^{m_1+m_2} \theta_i^{-1/2}\right)(1-\delta_{m_1+m_2})^{-1/2}$, the posterior means of the θ_j would now be

$$\mathrm{E}^J[\theta_j \mid x_1, \ldots, x_{m_1+m_2}, n] = \frac{x_j + 1/2}{n + (m_1+m_2+1)/2} \,.$$

The creation of new cells can thus have a pronounced effect on posterior beliefs about pre-existing cells (*e.g.*, inference about θ_1 – the probability of, say, snakes – would be affected by reclassifying turtles into different types, even though the total number of turtles is unchanged). This strikes many as highly problematical.

The reference prior is essentially immune to this difficulty, since the marginal prior and posterior are the same no matter how many additional groups, or cells, are added. More specifically, the marginal reference posterior of θ_1 (recall that the cells have been ordered so that θ_1 is the parameter of interest), arising from the larger multinomial model and reference prior, is shown in Berger & Bernardo (1992*a*) to be $\mathrm{Be}(\theta_1 \mid x_1 + \frac{1}{2}, n - x_1 + \frac{1}{2})$, which is the same as the posterior arising from the marginal distribution of x_1 ($\mathrm{Bi}(x_1 \mid \theta_1, n)$) and using the marginal reference prior for θ_1 ($\mathrm{Be}(\theta_1 \mid \frac{1}{2}, \frac{1}{2})$). The estimate of θ_1,

$$\mathrm{E}^R[\theta_1 \mid x_1, \ldots, x_m, n] = \frac{x_1 + \frac{1}{2}}{1 + n},$$

is, thus, not affected by the number of cells.

Discussion. The multinomial example dramatically demonstrates how reference priors can 'decouple' groups of coordinates. Thus the inferences obtained on $(\theta_1, \ldots, \theta_l)$ will depend only on $(x_1, \ldots, x_l)$ and n, and not on what happens in other cells, or how many other cells there are. This is a natural property when, indeed, $(\theta_1, \ldots, \theta_l)$ is of interest and the other parameters are nuisance parameters. The desirability of this property can, however, be questioned. It requires an asymmetric treatment of the θ_i and in problems where there is a small number of fixed 'indistinguishable' cells, such asymmetry may be unappealing. In the situation of classification of reptiles, for instance, dividing the turtles up into differing types, while not doing so for the other reptiles, would seem unnatural. Note, however, that if all reptiles were divided up into their differing subtypes and the numbers of subtypes differed significantly between reptiles, there could still be a strong effect on marginalization.

To dramatize the difference, compare the Jeffreys-rule and the reference prior means of θ_j which are, respectively, $(1+m)^{-1}$ and 2^{-j}. Thus, the Jeffreys prior treats the θ_j equally, while the reference prior gives exponentially decreasing mass to the θ_j as j increases. This situation clearly demonstrates the impossibility of unambiguously defining 'noninformative'. Initially it seems reasonable to insist that a noninformative prior for a multinomial problem be exchangeable, and also to require that it have the marginalization property; but these requirements are completely incompatible. Through consideration of a variety of examples, we have concluded that the marginalization property is typically more important, and hence that the reference prior is typically more attractive, but some flexibility is clearly required.

Indeed, this is a situation where subjective considerations should not be ignored, if it is possible to involve them. Thus, in this multinomial problem, there have been many developments (Fienberg & Holland (1972, 1973), Good & Crook (1974), Leonard (1977), Sono (1983), Tuyl et al. (2009), He (2011)) concerning how to appropriately handle multinomial data, when there is additional knowledge.

As a final comment, note that, even if a single overall exchangeable prior is required, we do not recommend the Jeffreys-rule prior. Preferred alternatives are discussed in Chapter 12.

10.3.3 Bivariate Normal Distribution

In sections 5.1.1, 6.4.4 and 7.4, the bivariate normal distribution, having means, standard deviations, and correlation parameter $(\mu_1, \mu_2, \sigma_1, \sigma_2, \rho)$, was discussed. A number of possible quantities of interest were highlighted in (5.3) and were studied from the viewpoints of (i) availability of invariance priors; and (ii) the availability of exact frequentist matching priors. Here we discuss the existence of reference priors for at least some of the quantities of interest. (Between the original five parameters and the 16 quantities of interest in (5.3), we do not attempt to present reference priors for everything!)

There are two main interests in the study of this example (beyond its practical importance). The first is that it enables an in-depth comparison of reference priors and invariance priors, which are often viewed as the 'gold standards' for objective priors. We restrict this comparison to one situation (in the next section), that of inference concerning ρ. The second motivation for the section is to illustrate the problem of there being potentially a plethora of reference priors, and discuss how to deal with this.

The Lindley/Bayarri Reference Prior for ρ. The following prior was suggested by Lindley (1965), as a natural prior for the bivariate normal problem, and was studied extensively in Bayarri (1981), where it was shown to be a reference prior when ρ is the parameter of interest, for any ordering of the nuisance parameters μ_1, μ_2, σ_1 and σ_2 in the sequential reference prior algorithm:

$$\pi^{LB}(\mu_1, \mu_2, \sigma_1, \sigma_2, \rho) = \frac{1}{\sigma_1 \sigma_2 (1 - \rho^2)}.$$

It is of interest to compare this prior with the right-Haar prior that was discussed in Section 6.4.4, namely $\pi^r(\mu_1, \mu_2, \sigma_1, \sigma_2, \rho) = \sigma_1^{-2}(1 - \rho^2)^{-1}$. One appealing feature of π^{LB} is that it treats the standard deviations symmetrically, as opposed to the right-Haar prior. Also, as illustrated in Section 7.5.2, it does not suffer a marginalization paradox, as did the right Haar prior.

Turning to frequentist matching, the right-Haar prior has the advantage, in that it is exact frequentist matching. However, as shown in Berger & Sun (2008), π^{LB} is first order frequentist matching and has excellent small sample matching properties for ρ. Furthermore, it was shown in Berger & Sun (2008) that π^{LB} has considerably better overall matching properties than the right-Haar prior, in the sense that the small sample frequentist coverage of resulting credible sets for almost any quantity of interest is fine, while that for the right-Haar prior can be bad for some quantities of interest. This is what leads us in Chapter 12 to recommend π^{LB} as the prior to use for the bivariate normal problem, if only one overall prior can be entertained.

Another evaluation criteria is ease of computation with the prior. The right-Haar prior cannot be beat in this regard, as it has the constructive random posterior given in (5.5) from which to simulate. However, π^{LB} is also easy to handle computationally, through the following rejection algorithm (see Section 1.8.4 for discussion and notation). Analytic expressions for the posteriors and moments are also available in Ly et al. (2018).

Rejection Algorithm to generate iid samples from the reference prior posterior

$$\pi^{LB}(\mu_1, \mu_2, \sigma_1, \sigma_2, \rho \,|\, \boldsymbol{x}) = C_1 \frac{1}{\sigma_1\sigma_2(1-\rho^2)} p(\boldsymbol{x} \,|\, \mu_1, \mu_2, \sigma_1, \sigma_2, \rho)\,,$$

C_1 a constant. A convenient proposal function here is

$$q(\theta) = \sigma_1^{-1}\sigma_2^{-1}(1-\rho^2)^{-3/2} p(\boldsymbol{x} \,|\, \mu_1, \mu_2, \sigma_1, \sigma_2, \rho)\,,$$

which, when normalized by a constant C_2, is the posterior distribution arising from the independence Jeffreys prior $\pi^{IJ}(\mu_1, \mu_2, \sigma_1, \sigma_2, \rho) = \sigma_1^{-1}\sigma_2^{-1}(1-\rho^2)^{-3/2}$. Since

$$\sup_{(\mu_1,\mu_2,\sigma_1,\sigma_2,\rho)} \frac{\sigma_1^{-1}\sigma_2^{-1}(1-\rho^2)^{-1} p(\boldsymbol{x} \,|\, \mu_1, \mu_2, \sigma_1, \sigma_2, \rho)}{\sigma_1^{-1}\sigma_2^{-1}(1-\rho^2)^{-3/2} p(\boldsymbol{x} \,|\, \mu_1, \mu_2, \sigma_1, \sigma_2, \rho)} = \sup_{\rho} \sqrt{1-\rho^2} = 1\,,$$

the rejection algorithm in Section 1.8.4 can be used with $M = 1$, and is as follows:

Step 1. Draw $\boldsymbol{\Sigma}$ from the independence Jeffreys prior posterior,

$$\boldsymbol{\Sigma} = \begin{pmatrix} \sigma_1^2 & \rho\sigma_1\sigma_2 \\ \rho\sigma_1\sigma_2 & \sigma_2^2 \end{pmatrix} \sim \text{Inverse Wishart}(\cdot \,|\, \boldsymbol{S}^{-1}, n-1).$$

Step 2. Simulate $u \sim \text{Un}(0,1)$. If $u \le \sqrt{1-\rho^2}$, report $\boldsymbol{\Sigma}$. Otherwise go back to Step 1 and keep iterating until a proposed $\boldsymbol{\Sigma}$ is accepted.

Step 3. Simulate $\boldsymbol{u} = (\mu_1, \mu_2) \sim N_2(\cdot \,|\, \overline{x}, n^{-1}\boldsymbol{\Sigma})$ for the accepted $\boldsymbol{\Sigma}$.

The acceptance rate of this algorithm is $\sqrt{1-\rho^2}$. Only if ρ^2 is near one will this be problematical.

Conclusion in the comparison of π^{LB} and π^r: If the focus is only on inference concerning ρ, there is no clear conclusion; both π^{LB} and the right-Haar prior are remarkable objective priors for ρ; the right-Haar prior suffers a marginalization paradox, but is exact frequentist matching. One's choice will depend on the weights one gives to the criteria discussed above. Of course, trying both is a sensible sensitivity step.

The Plethora of Reference Priors for Other Quantities of Interest. Table 10.1 gives the reference priors for some of the most important quantities of interest. (For reference priors for other quantities, see Sun & Berger (2007*b*); note that some of the reference priors given here were first developed in Datta & Ghosh (1995*a*).) Recall that the reference prior depends on the quantity of interest and (possibly) on the ordering of the nuisance parameters. Thus, the second column in the table indicates the quantity of interest (first element in the ordering) and the subsequent ordering of the nuisance parameters that leads to the reference prior. Note that different quantities of interest and different nuisance parameter orderings can lead to the same reference prior.

Table 10.1: Reference priors for the bivariate normal model, for various parameters of interest and nuisance parameter orderings. Here $\theta_1 = \rho\sigma_2/\sigma_1$, $\theta_2 = \sigma_2^2(1-\rho^2)$, $\theta_7 = \sigma_2/\sigma_1$, $\theta_8 = \sqrt{1-\rho^2}\sigma_2/\sigma_1$, $\eta_1 = 1/\sigma_1$, $\eta_2 = 1/[\sigma_2\sqrt{1-\rho^2}]$, $\eta_3 = -\rho/[\sigma_1\sqrt{1-\rho^2}]$, $\lambda_1 \geq \lambda_2$ are the eigenvalues of $\boldsymbol{\Sigma}$, ϑ is the angle defining the orthogonal matrix that diagonalizes $\boldsymbol{\Sigma}$, $\tilde{\mu}_1 = \boldsymbol{d}(\mu_1,\mu_2)'$, $(\tilde{\sigma}_1)^2 = \theta_7$, $\tilde{\rho} = \boldsymbol{d}\,\boldsymbol{\Sigma}(0,1)'/(\sigma_1\sqrt{\theta_7})$, $\tilde{\theta}_2 = \sigma_2^2[1-(\tilde{\rho})^2]$, and $\tilde{\theta}_1 = \tilde{\rho}\sigma_2/\tilde{\sigma}_1$.

prior $\pi(\mu_1,\mu_2,\sigma_1,\sigma_2,\rho)$	reference prior for parameter ordering
$\pi^J \propto \frac{1}{\sigma_1^2\sigma_2^2(1-\rho^2)^2}$	no ordering, all grouped together
$\pi^{IJ} \propto \frac{1}{\sigma_1\sigma_2(1-\rho^2)^{3/2}}$	two group prior: $\{(\mu_1,\mu_2),(\sigma_1,\sigma_2,\rho)\}$
$\pi^{LB} \propto \frac{1}{\sigma_1\sigma_2(1-\rho^2)}$	$\{\mu_1,\mu_2,\rho,\sigma_1,\sigma_2\}$ or $\{\rho,\sigma_1,\sigma_2,\mu_1,\mu_2\}$
$\pi^r \propto \frac{1}{\sigma_1^2(1-\rho^2)}$	$\{\sigma_1,\theta_2,\theta_1\}$ or $\{\lvert\boldsymbol{\Sigma}\rvert,\theta_8,\theta_1\}$, μ_i anywhere
$\pi^\sigma \propto \frac{\sqrt{1+\rho^2}}{\sigma_1\sigma_2(1-\rho^2)}$	$\{\mu_1,\mu_2,\sigma_1,\sigma_2,\rho\}$ or $\{\sigma_1,\sigma_2,\rho,\mu_1,\mu_2\}$
$\pi^{\sigma*} \propto \frac{1}{\sigma_1\sigma_2(1-\rho^2)\sqrt{2-\rho^2}}$	$\{\mu_1,\mu_2,\sigma_1,\rho,\sigma_2\}$ or $\{\sigma_1,\rho,\sigma_2,\mu_1,\mu_2\}$ or $\{\mu_1,\mu_2,\eta_1,\eta_3,\eta_2\}$ or $\{\eta_1,\eta_3,\eta_2,\mu_1,\mu_2\}$
$\pi^\eta \propto \frac{1}{\sigma_1^2\sigma_2(1-\rho^2)^{3/2}}$	$\{\mu_1,\mu_2,\eta_1,\eta_2,\eta_3\}$ or $\{\eta_1,\eta_2,\eta_3,\mu_1,\mu_2\}$
$\pi^\lambda \propto \frac{\left[\left(\frac{\sigma_1}{\sigma_2}-\frac{\sigma_2}{\sigma_1}\right)^2+4\rho^2\right]^{-1/2}}{\sigma_1\sigma_2(1-\rho^2)}$	$\{\mu_1,\mu_2,\lambda_1,\lambda_2,\vartheta\}$ or $\{\lambda_1,\lambda_2,\vartheta,\mu_1,\mu_2\}$
$\tilde{\pi}^r \propto \frac{1}{(\tilde{\sigma}_1)^2[1-(\tilde{\rho})^2]}d\tilde{\mu}_1 d\mu_2 d\tilde{\sigma}_1 d\sigma_2 d\tilde{\rho}$	$\{\boldsymbol{d}(\mu_1,\mu_2)',\mu_2,\theta_7,\tilde{\theta}_2,\tilde{\theta}_1\}$

The first four priors have already been discussed. The Jeffreys-rule prior arises when one maximizes missing information globally and, hence, does not reflect any parameter ordering. The independence Jeffreys prior is an example of a two-group reference prior, Step 1 of the reference prior algorithm being applied simultaneously to the group of parameters (σ_1,σ_2,ρ), and Step 3 being applied simultaneously to the group (μ_1,μ_2) of parameters of interest. The key fact about π^{LB} is that it arises as a reference prior only when ρ is placed before the standard deviations in the ordering. The right-Haar prior, π^r, is curious in that it arises as a reference prior only if nonstandard parameterizations involving θ_1, θ_2, θ_8 and $\lvert\boldsymbol{\Sigma}\rvert$ are employed.

The reference priors π^σ and $\pi^{\sigma*}$ arise primarily when ρ appears later in the parameter ordering than at least one of the standard deviations. These are serious priors, shown in Berger & Sun (2008) to have quite good performance.

π^η is another reference prior that arises with nonstandard parameterizations; it's main claim to fame is that it is exact frequentist matching for μ_1/σ_1, with a constructive random posterior, as shown in Table 5.1.

The reference prior π^λ is actually not new; this was presented for a general covariance matrix in (7.2), but with a completely different parameterization. $\tilde{\pi}^r$ is also not really new; it was mentioned in Example 7.5 that there is more than one right-Haar prior for the bivariate normal problem. Actually, there are an infinite number, corresponding to all the possible parameters of interest indicated in the table ($\boldsymbol{d}$ is just a fixed vector).

Highly efficient and simple computational algorithms are given in Sun & Berger (2007*b*) and Berger & Sun (2008) for all of the posteriors resulting from these priors. These papers also extensively discussed the frequentist matching properties of the reference priors (see also Datta & Ghosh (1995*c*)), including small sample frequentist matching. Taking all the criteria into account Table 10.2 provides our final recommendations concerning the bivariate normal distribution.

Table 10.2: Recommendations of objective priors for various quantities of interest in the bivariate normal model: $\square$ indicates that, for this quantity of interest, the posterior will *not* be exact frequentist matching. (For μ_2 and parameters with σ_1 replaced by σ_2, use the right-Haar prior with the variances interchanged.)

Use this prior	For these quantities of interest
π^{LB}	$\boxed{\rho}$, $\boxed{\frac{\sigma_1}{\sigma_2}}$, $\boxed{\textbf{general use}}$
π^r	$\mu_1,\ \sigma_1,\ \frac{\mu_1}{\sigma_1},\ \lvert\Sigma\rvert,\ \frac{\rho\sigma_2}{\sigma_1},\ \sigma_2^2(1-\rho^2),\ \eta_3 = \frac{-\rho}{\sigma_1\sqrt{1-\rho^2}},\ \rho$
$\tilde{\pi}^r$	$\boldsymbol{d}\,(\mu_1,\mu_2)',\ \boldsymbol{d}\,\boldsymbol{\Sigma}\,\boldsymbol{d}'$
π^λ	$\boxed{\lambda_1 \text{ or } \lambda_2}$
π^σ	$\boxed{\sigma_{12} = \rho\sigma_1\sigma_2}$

10.3.4 Prediction

Prediction problems are not specifically covered by the methods described above. The difficulty is that the quantity of interest is not a parameter of the model and, furthermore, typically depends on all parameters of the model. For instance, if one wants to predict the value of y based on $\boldsymbol{x}$ when $(y, \boldsymbol{x})$ has density $p(y, \boldsymbol{x} \mid \boldsymbol{\theta})$, the unknown of interest is y, whose distribution depends on all of $\boldsymbol{\theta}$, and one needs a prior for $\boldsymbol{\theta}$, not a prior for y.

A possible solution is to define the quantity of interest to be a conditional estimate of the quantity to be predicted. For instance, in predicting y in the situation described above, the quantity of interest could be defined to be $\phi(\boldsymbol{\theta}) = \text{Med}[y \mid \boldsymbol{\theta}]$, the median of the sampling distribution of y. This is often a better choice than $\text{E}[y \mid \boldsymbol{\theta}]$, since $\text{Med}[y \mid \boldsymbol{\theta}]$ always exists and has attractive invariance properties.

Example 10.8 *Normal prediction* (continuation of Example 1.20). The data $\boldsymbol{x} = (x_1, \ldots, x_n)$ is a random sample from the $\text{N}(x \mid \mu, \sigma^2)$ distribution, and suppose that one is interested in predicting a new, future observation x from $\text{N}(x \mid \mu, \sigma^2)$. Using the argument described above, the quantity of interest could be defined to be $\phi(\mu, \sigma) = \text{Med}[x \mid \mu, \sigma] = \mu$ and hence (see Example 10.5) the appropriate reference prior would be $\pi(\mu, \sigma) = \sigma^{-1}$ $(n \geq 2)$. The corresponding posterior is $\pi(\mu, \sigma \mid \boldsymbol{x}) \propto \prod_{j=1}^{n} \text{N}(x_j \mid \mu, \sigma^2)\, \sigma^{-1}$ and the posterior predictive is the $\text{St}(x \mid \overline{x},\ s[(n+1)/(n-1)]^{1/2},\ n-1)$ distribution, as shown in (1.5).

This posterior predictive distribution is also exact frequentist matching. As an amusing example when $n = 2$, the reference posterior predictive probability that a third observation lies between the first two is

$$\Pr[x_{(1)} < x < x_{(2)} \mid x_1, x_2] = \int_{x_{(1)}}^{x_{(2)}} \pi(x \mid x_1, x_2)\, dx = \frac{1}{3}\,,$$

where $x_{(1)} = \min\{x_1, x_2\}$, and $x_{(2)} = \max\{x_1, x_2\}$. This is consistent with the fact that, for all μ and σ, the frequentist coverage of the corresponding region of $\mathbb{R}^3$, namely $C = \{(x_1, x_2, x_3) :\ x_{(1)} < x_3 < x_{(2)}\}$ is precisely

$$\int\!\!\int\!\!\int_C \prod_{i=1}^{3} \text{N}(x_j \mid \mu, \sigma^2)\, dx_1\, dx_2\, dx_3 = \frac{1}{3}\,\cdot \tag{10.30}$$

Indeed, this result holds for any continuous exchangeable distribution for (x_1, x_2, x_3).

In sharp contrast, if the Jeffreys-rule prior, $\pi^J(\mu, \sigma) \propto \sigma^{-2}$, were used, the posterior predictive would have been the $\text{St}(x \mid \overline{x},\ s[(n+1)/n]^{1/2},\ n)$ distribution. For this predictive distribution, the probability that a third observation lies between the first two can be computed to be 1/2, rather than 1/3, a less than attractive result, in view of (10.30).

Example 10.9 *Simple linear regression* (continuation of Example 1.4.2). Recall that the goal in this example is to predict y, which will arise from the model $p(y \mid \boldsymbol{v}, \boldsymbol{\beta}, \sigma^2) = \text{N}(y \mid \boldsymbol{v}\boldsymbol{\beta}', \sigma^2)$, where $\boldsymbol{v} = (1, v_1, v_2, \ldots, v_m)$ is the given covariate vector at which prediction is desired, and $\boldsymbol{\beta} = (\beta_0, \beta_1, \ldots \beta_m)$ and σ are unknown. The quantity of interest could be taken to be $\phi(\boldsymbol{\beta}, \sigma) = \text{Med}[y \mid \boldsymbol{\beta}, \sigma] = \boldsymbol{v}\boldsymbol{\beta}'$. Choosing the nuisance parameters to be $\beta_1, \ldots, \beta_m$, one can compute that the Fisher information matrix, for a sample of size n (see Example 1.4.2 for the corresponding likelihood), is of the form

$$I(\phi, \beta_1, \ldots, \beta_m, \sigma) = \begin{pmatrix} \frac{1}{\sigma}\boldsymbol{H} & 0 \\ 0 & \frac{2n}{\sigma^2} \end{pmatrix},$$

where $\boldsymbol{H}$ is specific matrix that is irrelevant to the following argument. Note that the determinant of the lower $i \times i$ corner of I is $|I_i| \propto \sigma^{-(i+1)}$, so $h_m = I_{mm} \propto 1/\sigma^2$ and, for $j \leq m$, $h_j = |I_{m+1-j+1}/|I_{m+1-j}| \propto 1/\sigma$. Application of Corollary 10.2.1 yields the joint reference prior $\pi(\phi, \beta_1, \dots, \beta_m, \sigma) = 1/\sigma$ which, upon transforming back to $(\beta_0, \beta_1, \dots, \beta_m, \sigma)$ is also just $1/\sigma$, the prior used to complete the predictive analysis in Example 1.4.2.

10.3.5 Grouped Reference Priors

We have only considered the situation in which the reference prior is determined through one-at-a-time parameter iteration. One can also do grouped reference priors, where the iterations involve groups of parameters. For instance one might consider the grouping $(\{\theta_1\}, \{\theta_2, \dots, \theta_m\})$. With $\boldsymbol{\theta}^2 = (\theta_2, \dots, \theta_m)$, the idea would be to first find the conditional reference prior $\pi(\boldsymbol{\theta}^2 \,|\, \theta_1)$, use this to find the marginal probability model depending only on θ_1, and find the marginal reference prior for θ_1 from this marginal model; the product $\pi(\boldsymbol{\theta}) = \pi(\boldsymbol{\theta}^2 \,|\, \theta_1)\pi(\theta_1)$ would then be the two-group reference prior.

Our experience is that using one-at-a-time parameter iteration is superior, and should be done if feasible. Sometimes, however, it is technically too difficult to implement the one-at-a time method, and then one might be forced to develop a grouped reference prior, the notion being that any parameter iteration is better than none. In particular, one might hope that utilizing the grouping $(\{\theta_1\}, \{\theta_2, \dots, \theta_m\})$ at least separates the parameter of interest from the nuisance parameters in the iteration and this might be the crucial separation. Also, modulo choosing the quantity (parameter) of interest, Theorem 2.1 in Datta & Ghosh (1996) shows that the two-group reference prior is invariant to parameterization, in the sense that the choice of the parameterization of the nuisance parameters does not matter.

Proposition 10.4 (Two group reference prior under asymptotic normality.) *Using the same notation as in Proposition 10.2, define*

$$\pi(\boldsymbol{\theta}^2 \,|\, \theta_1) = c(\theta_1)|I_{m-1}(\boldsymbol{\theta})|^{1/2} \quad \textit{if} \quad [c(\theta_1)]^{-1} = \int_{\boldsymbol{\Theta}^2(\theta_1)} |I_{m-1}(\boldsymbol{\theta})|^{1/2}\, d\boldsymbol{\theta}^2 < \infty\,,$$

$$\pi(\theta_1) \propto \exp\left\{ \int_{\boldsymbol{\Theta}^2(\theta_1)} \pi(\boldsymbol{\theta}^2 \,|\, \theta_1) \log[h_1^{1/2}(\boldsymbol{\theta})]\, d\boldsymbol{\theta}^2 \right\}.$$

Then $\pi(\boldsymbol{\theta}) = \pi(\boldsymbol{\theta}^2 \,|\, \theta_1)\pi(\theta_1)$ is the two-group reference prior, providing it is permissible.

If $[c(\theta_1)]^{-1}$ is not finite, let $\{\boldsymbol{\Theta}_l,\ l = 1, 2, \dots\}$ be compact subsets of $\boldsymbol{\Theta}$, such that $\lim_{l\to\infty} \boldsymbol{\Theta}_l = \boldsymbol{\Theta}$, and define

$$\pi_l(\boldsymbol{\theta}^2 \,|\, \theta_1) = c_l(\theta_1)|I_{m-1}(\boldsymbol{\theta})|^{1/2}, \quad [c_l(\theta_1)]^{-1} = \int_{\boldsymbol{\Theta}_l^2(\theta_1)} |I_{m-1}(\boldsymbol{\theta})|^{1/2}\, d\boldsymbol{\theta}^2\,,$$

$$\pi_l(\theta_1) \propto \exp\left\{ \int_{\boldsymbol{\Theta}_l^2(\theta_1)} \pi_l(\boldsymbol{\theta}^2 \,|\, \theta_1) \log[h_1^{1/2}(\boldsymbol{\theta})]\, d\boldsymbol{\theta}^2 \right\}.$$

Then, for any interior point $(\theta_{10}, \boldsymbol{\theta}_0^2)$,

$$\pi(\boldsymbol{\theta}) = \lim_{l\to\infty} \frac{\pi_l(\boldsymbol{\theta}^2 \mid \theta_1)\pi_l(\theta_1)}{\pi_l(\boldsymbol{\theta}_0^2 \mid \theta_{10})\pi_l(\theta_{10})}$$

is the two-group reference prior, providing it is permissible.

Proof. $\pi(\boldsymbol{\theta}^2 \mid \theta_1)$ is just the Jeffreys-rule prior, given θ_1, which is the square root of the determinant of the lower $(m-1) \times (m-1)$ corner of $I(\boldsymbol{\theta})$. The rest of the proof is standard. □

Example 10.10 *Neyman-Scott problem continued:* In Example 4.9, the parameter of interest is σ^2 and the grouped nuisance parameter would be $\boldsymbol{\mu} = (\mu_1, \ldots, \mu_m)$. The Fisher information matrix is given in (4.12), from which it follows that $I_m(\boldsymbol{\theta}) = \frac{2}{\sigma^2}\boldsymbol{I}_m$ (note that $\boldsymbol{\theta}$ has dimension $m+1$ here). It follows that the conditional reference prior is $\pi(\boldsymbol{\mu} \mid \sigma) \propto 1$. Also, the marginal reference prior for σ^2 is $\pi(\sigma^2) = 1/\sigma^2$, so the two-group reference prior is $\pi(\sigma^2, \boldsymbol{\mu}) \propto 1/\sigma^2$, the correct answer, *i.e.*, the one-at-a-time reference prior. (Strictly we would need to operate on compact sets and take the limit, but we know the result from previous such computations.)

To show that the two-group reference prior can fail badly, consider this same model but now assume that μ_1 is the parameter of interest, with the grouped nuisance parameter being $\boldsymbol{\theta}^2 = (\mu_2, \ldots, \mu_m, \sigma^2)$. The conditional reference prior for $\boldsymbol{\theta}^2$ will be the Jeffreys-rule prior for $\boldsymbol{\theta}^2$, which is the square root of the determinant of the lower $m\times m$ matrix of $I(\mu_1, \ldots, \mu_m, \sigma^2)$, namely $\pi(\boldsymbol{\theta}^2 \mid \mu_1) \propto [\sigma^2]^{-(m+1)/2}$. The marginal reference prior for μ_1 will be constant, leading to the two-group reference prior, when μ_1 is the parameter of interest, $\pi(\mu_1, \ldots, \mu_m, \sigma^2) \propto [\sigma^2]^{-(m+1)/2}$.

An easy computation, using (4.13), shows that the marginal posterior density of μ_1 is then $\pi(\mu_1 \mid \boldsymbol{x}) \propto [1 + 2(\mu_1 - \bar{x}_1)^2/S^2]^{-m}$, which is recognizable as the $\mathrm{St}(\mu_1 \mid \bar{x}_1, S^2/[4m-2], 2m-1)$ density. The correct answer, with the one-at-a-time reference prior, yields the $\mathrm{St}(\mu_1 \mid \bar{x}_1, S^2/[2m], m)$ density, so the two-group reference prior fails by (i) having $2m-1$ degrees of freedom instead of m and (ii) having scale parameter $S/\sqrt{4m-2}$ instead of $S/\sqrt{2m}$, a particularly egregious error.

Example 10.11 *Multinomial problem (continued).* For the multinomial problem discussed in Section 10.3.2, the expression for any possible grouped reference prior is given in Berger & Bernardo (1992*a*). For the two-group case mentioned above, the result is

$$\pi(\theta_1, \ldots, \theta_m) \propto \left(\prod_{i=1}^{m} \theta_i^{-1/2}\right) (1-\theta_1)^{-(m-1)/2} \left(1 - \sum_{i=1}^{m} \theta_i\right)^{-1/2} .$$

This is clearly quite different from the one-at-a-time reference prior given in (10.28). Interestingly, however, a straightforward computation shows that the resulting posterior distribution of the parameter of interest, θ_1, is exactly the same as that arising from (10.28). Thus this two-group reference prior will produce all the desirable marginalization properties discussed in Section 10.3.2.

The above examples show that the two-group reference prior is sometimes fine (indeed, can give the same result as the one-at-a-time reference prior, if only inference on the parameter of interest, θ, is desired), but can also fail badly. It would be nice to know when it is fine and when it is not, but currently no guidelines are available.

10.4 Discussion and Additional References

10.4.1 Reverse Reference Priors

Suppose we have a three parameter model, with location-scale parameters (μ, σ) and a third parameter η. For instance, the model could be a t-distribution, where η is the unknown degrees of freedom. The usual parameter ordering, for application of the reference prior algorithm when μ is the parameter of interest, would be $\{\mu, \sigma, \eta\}$. But it can be hard to carry out the analysis if the conditional reference prior $\pi(\eta \mid \mu, \sigma)$ (the first step in the iteration) is difficult to work with.

Consider, instead, the parameter ordering $\{\eta, \mu, \sigma\}$. Now one starts by finding the conditional reference priors for σ and then μ, given η. But, given η, the problem is a location-scale problem so we know that the conditional reference prior of (μ, σ) is $\pi(\mu, \sigma \mid \eta) = 1/\sigma$. Finding the final marginal reference prior for η might then be relatively simple.

Such priors, suggested by J.K. Ghosh (*cf.* Ghosh & Mukerjee (1992*b*)), are called *reverse reference priors*, and have been considered in Berger & Bernardo (1992*a*), Datta & Ghosh (1995*c*,*a*), Datta (1996), Sun (1997), Berger et al. (1998), Sun & Berger (1998), Sun et al. (1998), Scricciolo (1999), van der Linde (2000), Datta et al. (2002), Gutiérrez-Peña & Rueda (2003), Datta & Sweeting (2005), Solari et al. (2008), Moala et al. (2009), Liseo & Macaro (2013), among others. It has been observed that reverse reference priors often are very satisfactory objective priors, even if not optimal. It appears to be the case that doing the iterative one-at-a-time computation of a reference prior is more important than using the natural parameter ordering.

Chapter 11

Discrete Parameter Problems

11.1 Introduction

The 'obvious' objective prior for a discrete parameter is the constant prior, as discussed in Section 8.5. If the parameter space is finite and has no structure, the constant prior certainly seems reasonable and is what we recommend.

If the discrete parameter space has structure, however, and/or is infinite (*e.g.*, the positive integers) a non-constant objective prior can be more natural, as the following two examples show.

Example 11.1 Let x have the binomial density $\text{Bi}(x \mid n, p)$, where the sample size n is the unknown quantity of interest and the success probability p is known. The parameter space is thus the set $\mathcal{N} = \{1, 2, \ldots\}$ of positive integers.

This situation is particularly interesting when p is also unknown since it is well known from the literature (see, e.g., Kahn (1987)) that use of a constant prior density for n is then bad, resulting in an improper posterior distribution for standard objective priors for p (such as the Jeffreys-rule prior or uniform prior). Thus we need to use the structure of the binomial model to develop a sensible objective prior for n.

Example 11.2 Consider the hypergeometric distribution for a population of size N (known). The unknown parameter is R, the number of items in the population having a specified property; R can take values $\{0, 1, \ldots, N\}$. The data is r, the number of items with the specified property out of a random sample of size n, taken without replacement.

The parameter space is finite here, but still reflects an obvious structure through the hypergeometric distribution. For instance, when N is large, it is well known that the hypergeometric distribution is essentially equivalent to the binomial distribution with parameter $p = R/N$. An objective prior for R should thus be compatible with an objective prior for p. If the uniform prior for p is believed to be appropriate, having a uniform prior on R would be natural. But we have argued for use of the non-uniform Jeffreys-rule prior $\pi(p) \propto p^{-1/2}(1-p)^{-1/2}$, and so the objective prior on R should be compatible with this in some sense.

There have been a number of proposals for objective priors in specific discrete parameter problems, and we will refer to these proposals when considering such problems. There have also been a few general proposals. Jeffreys (1961) simply suggests using $1/\theta$ for any unbounded positive parameter θ, including discrete parameters. This choice has the appeal of having the largest polynomial decay that retains impropriety (although the same could be said for $1/[c+\theta]$ with any fixed c). Rissanen (1983) proposed a proper prior for positive integers that is the vaguest proper prior in terms of tail behavior. Barger & Bunge (2008) propose using the formal Jeffreys-rule method (or reference prior method), based on the concept of an information matrix for a discrete parameter that was developed in Lindsay & Roeder (1987).

Motivated by the hypergeometric example, we approach structured discrete parameter problems by embedding the original model into a model with a continuous parameter (and possibly continuous data). In this continuous parameter problem, we can then apply the ordinary reference prior theory, discretizing the resulting continuous reference prior (if necessary) to return to the original problem. The discussion herein is taken from Berger et al. (2012), which developed the embedding approach.

The particular discrete problems that are considered in this chapter are the *discrete uniform distribution* (in Sections 11.2.1 and 11.2.4), a *population-size model* (in Section 11.3), the *hypergeometric distribution* (in Section 11.4), the *binomial-beta distribution* (in Section 11.5), and the *binomial distribution* (in Section 11.6). Finally, Section 11.7 looks at the quite different discrete problem of assigning prior probabilities to different models. The main goal of each section is to illustrate one of the embedding approaches described in the next section.

11.2 Possible Embeddings

We have found four different embedding approaches to be useful in dealing with discrete, structured parameter spaces. Here we let θ denote the unknown discrete parameter, taking values in a countable discrete space Θ; in the later specific examples we will revert to using standard notation for the example (e.g., n for the binomial problem and R for the hypergeometric problem).

11.2.1 Approach 1: Just Assume θ is Continuous

The simplest possibility is to treat θ as a continuous variable in the original model $p(x \mid \theta)$. Typically, however, $C(\theta) = \int p(x \mid \theta)\, dx$ will not equal 1 for $\theta \notin \Theta$, so the reference analysis would need to be performed with the density $C(\theta)^{-1}p(x \mid \theta)$; this can be difficult computationally and it is hard to argue that it is an appropriate embedding. Sometimes, however, one can also treat x as continuous and analyze the resulting completely continuous problem without having an additional normalizing factor.

Example 11.3 *Discrete Uniform Distribution.* Suppose that x is uniform on the positive integers $\{1, \ldots, \theta\}$, having discrete density $p(x \mid \theta) = 1/\theta$. Treating both θ and x as continuous variables on $(0, \infty)$, we have that x is Uniform on $(0, \theta)$ with density $p(x \mid \theta) = 1/\theta$; no additional normalizing factor is introduced. The reference prior for this continuous problem is $\pi(\theta) = 1/\theta$ and so this is the recommended prior for the original discrete problem.

11.2.2 Approach 2: Introducing a Continuous Hierarchical Hyperparameter

In some problems, it is natural to introduce a hierarchical model for the unknown discrete parameter, a hierarchical model with a continuous hyperparameter that can be analyzed with usual reference prior methods. For instance, in the hypergeometric problem, Jeffreys (1946, 1961) postulated that the unknown R arises as a $\mathrm{Bi}(R \mid N, p)$ random variable with p an unknown continuous parameter. The problem can then be reduced to finding the reference prior $\pi^R(p)$ for p, which will determine the objective prior for R via $\pi^*(R) = \int \mathrm{Bi}(R \mid N, p)\pi^R(p)dp$ (if the integral exists). This program is carried out for the hypergeometric distribution in Section 11.4.

When such hierarchical modeling is possible and natural, the case for the ensuing objective prior for the discrete parameter is compelling. Unfortunately, natural hierarchical models are often not available. And, if available, there may be more than one, resulting in different possible objective priors. This seems unavoidable; different 'hierarchical origins' for the discrete parameter model will naturally result in different objective priors.

11.2.3 Approach 3: Applying Reference Prior Theory with a Consistent Estimator

This approach uses the standard reference prior theory, based on replication of the experiment under study. One considers $\boldsymbol{x}^{(k)} = \{\boldsymbol{x}_1, \ldots, \boldsymbol{x}_k\}$, where the $\boldsymbol{x}_i$ are k independent replications of the data $\boldsymbol{x}$ from the model under consideration. (Recall that $\boldsymbol{x}$ will, itself, typically be a vector of observations, with the replication considered being imaginary replication of the entire original experiment and the number of replications, k, going to ∞.) Approach 3 proceeds by

(i) choosing a consistent linear (or some other simple) estimate $\hat{\theta}_k = \hat{\theta}_k(\boldsymbol{x}^{(k)})$ of θ;

(ii) finding the asymptotic sampling distribution of $\hat{\theta}_k$ as $k \to \infty$;

(iii) pretending that θ is continuous in this asymptotic distribution and finding the corresponding reference prior.

Curiously, one cannot use efficient estimators in this process, since their convergence to the true discrete parameter is typically too fast. Indeed, we will suppose that, as $k \to \infty$ and for some series of constants c_k (typically $c_k = \sqrt{k}$), $c_k(\hat{\theta}_k - \theta)$ has a limiting normal distribution with mean 0 and variance $s^2(\theta)$. In this limiting normal distribution, one now assumes that θ is continuous, and hence the desired objective prior is (see 4.15) $\pi^*(\theta) \propto s(\theta)^{-1}$.

That this approach requires use of inefficient estimators is philosophically problematical and practically ambiguous, since use of different inefficient estimators can result in different reference priors. Hence the best use of this approach is to suggest possible objective priors, which are then validated by other criteria.

For discrete models that have *linear difference score* (see Lindsay & Roeder (1987) for definition) one can define an analogue of the expected Fisher information and apply the Jeffreys-rule formula or the reference prior formula to obtain an objective prior; this was proposed in Barger & Bunge (2008). The only example we consider here that has a linear difference score is the binomial problem with p known, considered in Section 11.6. For this problem and, indeed, any problem with a linear difference score, the objective prior obtained from the Barger & Bunge (2008) approach will be the same as that obtained using our Approach 3 with a linear estimator. Barger & Bunge (2008) thus provides an additional justification for Approach 3 with linear estimators, in problems with a linear difference score.

11.2.4 Approach 4: Using Parameter-Based Asymptotics

Following Lindsay & Roeder (1987) and Sweeting (1992), this approach uses a formal limiting operation in θ to create a continuous problem.

Example 11.4 *Example 11.3 continued.* For instance, in the discrete uniform example, $\frac{x}{\theta}$ has a uniform distribution on the discrete set $\{\frac{1}{\theta}, \ldots, \frac{\theta-1}{\theta}, 1\}$. As θ gets large, it seems reasonable to replace the discrete values on the unit interval by the unit interval itself, so we end up with the limiting distribution of $\frac{x}{\theta}$ being Uniform on (0,1), or x being Uniform on $(0, \theta)$. Pretending that x and θ are continuous in this problem results in $\pi^*(\theta) = 1/\theta$, as before.

The generic steps of this approach are as follows:

(i) Let $\theta \to \infty$ (or ensure this by forcing some other parameter to ∞).

(ii) In the limiting asymptotic distribution of x (or some related random variable), let θ be continuous.

(iii) Do the reference prior replications over k with this asymptotic distribution, to define the objective prior.

This approach is well-defined, as it is based on the full (or an asymptotically efficient) posterior, but it only gives an objective prior for 'large θ.' Hence, this is perhaps best used in conjunction with one of the other approaches, to help validate the answer from the other approach.

11.2.5 Preference among Embeddings

In the remainder of the chapter, we illustrate the application of the above four approaches to development of an objective prior to particular discrete parameter problems. When Approach 1 (without renormalization) or Approach 2 (when there is a single natural hierarchical model) are applicable, we view the resulting prior as being the natural objective prior, and look no further. For problems where neither Approach 1 or 2 applies, we turn to Approaches 3 and 4 to propose an objective prior. Since these latter approaches do not typically result in unique objective priors, we study the properties of the possible priors (using frequentist validation) to make a recommendation.

11.3 Estimating a Population Size

11.3.1 Background of the Model

Consider an experiment with Type II censoring (cf. Lawless (1982), Section 1.4), in which there is a sample of N units whose lifetimes follow an exponential distribution with mean $1/\lambda$. The experiment is stopped when R units have failed; let $t_1 \leq \cdots \leq t_R$ denote the resulting failure times. Here $N \geq R$ and λ are both unknown, while R is given.

The problem of estimating N has many applications. Starr (1974) and Kramer & Starr (1990) desired to estimate the number of fish in a lake, assuming that the time to catch any particular fish is exponentially distributed with mean $1/\lambda$. Goudie & Goldie (1981) considered a linear death process, where the data consisted only of the lifetimes of those who had died, assumed to be iid exponentially distributed with mean $1/\lambda$; of interest was the initial population size N. This model can also arise in software reliability, where N is the unknown number of bugs and the lengths of time to discover the first R bugs are assumed to be independently exponential with mean $1/\lambda$; see Basu & Ebrahimi (2001).

The density of $(t_1, \cdots, t_R)$ is

$$p(t_1, \cdots, t_R \mid N, \lambda) = \frac{N!}{(N-R)!}\lambda^R \exp\Big\{-\lambda\Big[t_1 + t_2 + \cdots + t_R + (N-R)t_R\Big]\Big\}.$$

The pair $(V, W) = ([t_1 + \cdots + t_R)]/t_R, t_R)$ is a minimal sufficient statistic for (N, λ) and has density (Goudie & Goldie (1981), formula (4))

$$p(V, W \mid N, \lambda) = \frac{R}{(R-2)!}\binom{N}{R}\lambda^R W^{R-1} \exp\Big\{-\lambda(V + N - R)W\Big\}\, g_R(V),$$

for $1 < V < R, W > 0$, where, letting $[V]$ denote the integer part of V,

$$g_R(V) = \sum_{i=1}^{[V]} (-1)^{i-1}\binom{R-1}{i-1}(V-i)^{R-2}, \quad 1 < V < R.$$

Goudie & Goldie (1981) also show that this joint density can be written as

$$p(W \mid N, \lambda, V)\, p(V \mid N),$$

where $p(W \mid N, \lambda, V)$ is the gamma density $\mathrm{Ga}(W \mid R-1, \lambda(V+N-R))$ and

$$p(V \mid N) = \frac{1}{(R-2)!}\frac{N!}{(N-R)!}\frac{1}{(V+N-R)^R}\, g_R(V), \quad 1 < V < R. \tag{11.1}$$

Following reference prior logic, one would first find the reference prior for the nuisance parameter λ, conditional on the parameter of interest N. Since $p(W \mid N, \lambda, V)$ is a gamma distribution for which λ is a scale parameter, the reference prior is $\pi^R(\lambda) = 1/\lambda$. It remains to find the objective prior for N, based on $p(V \mid N)$ in (11.1).

Note that this problem is difficult to address by standard methods. For instance, Goudie & Goldie (1981) showed that there is no unbiased estimate of N and that moment estimates and maximum likelihood estimates do not exist with moderate probability. Also, note that the likelihood $p(V \mid N)$ in (11.1) tends to one as $N \to \infty$, so that the posterior would not exist if either a constant prior or a prior proportional to $1/N$ (two common objective choices for integer N) were used.

11.3.2 Using Approach 1 to Determine an Objective Prior

We can directly apply Approach 1 here, embedding the discrete parameter space for N, $\{R, R+1, \ldots\}$ into the continuous space $(R-0.5, \infty)$, since, for each $N \geq R-0.5$, $p(V \mid N)$ is still a probability density for V with the same normalization (the factorials being replaced by the corresponding Gamma functions for non-integer N). The reference prior for N in this continuous problem is simply the Jeffreys-rule prior

$$\pi(N \mid R) \propto \sqrt{I_R(N)}, \quad N \geq R - 0.5,$$

where $I_R(N)$ is the Fisher information of N in the continuous parameter problem, shown in Berger et al. (2012) to be (for the needed integer values of N)

$$I_R(N) = \sum_{j=0}^{R-1} \frac{1}{(N-j)^2} - \frac{R\,N!}{(R-2)!(N-R)!} J_{R,N}\,, \tag{11.2}$$

with

$$J_{R,N} = \sum_{i=1}^{R-1} \frac{(-1)^{i-1}}{(N-R+i)^3} \left(\frac{R-i}{N}\right)^{R-1} \binom{R-1}{i-1} \left[\frac{1}{R-1} - \frac{2(R-i)}{RN} + \frac{(R-i)^2}{(R+1)N^2}\right].$$

Things simplify if we consider the reparameterization $\theta = N - R + 1$, since the range of θ is then the positive integers $\mathcal{N} = \{1, 2, \cdots\}$. The corresponding objective prior for θ is then

$$\pi^*(\theta \mid R) \propto \sqrt{I_R(\theta + R - 1)}, \quad \theta \in \mathcal{N}.$$

For $R = 2$, this becomes

$$\pi^*(\theta) = \frac{1}{\theta(\theta+1)}\,, \tag{11.3}$$

and it is shown in Berger et al. (2012) that this a good approximation to $\pi^*(\theta \mid R)$ for all R. Hence, for simplicity, one might just recommend (11.3) as the objective prior for the problem. Note that this is a proper prior, necessary since the likelihood is constant as $N \to \infty$; this is thus another example of the rather remarkable tendency of the reference or Jeffreys-rule prior to yield a proper posterior for challenging likelihoods.

11.4 The Hypergeometric Distribution

Consider the hypergeometric distribution,

$$p(r \mid n, R, N) = \frac{\binom{R}{r}\binom{N-R}{n-r}}{\binom{N}{n}}, \text{ for } r \in \{0, \cdots, R\}, \tag{11.4}$$

where R is unknown and $R \in \{0, 1, \cdots, N\}$.

11.4.1 Using Approach 2 to Determine an Objective Prior

As already mentioned, a natural hierarchical model for the unknown R is to assume that it is $\text{Bi}(R \mid N, p)$, with unknown p. The problem then reduces to finding the reference prior for the continuous hyperparameter p, and using it to induce the prior distribution on R. Thus Approach 2 can be directly implemented.

The appropriate reference prior for a hyperparameter in a hierarchical setting is not the reference prior for the top level of the hierarchy (here $\text{Bi}(R \mid N, p)$) but that found by first marginalizing out the lower level parameters having specified distributions. (See Example 12.5 for further discussion.) Here, that would mean marginalizing over $\text{Bi}(R \mid N, p)$, resulting in

$$\begin{aligned} \Pr(r \mid n, N, p) &= \sum_{R=0}^{N} \Pr(r \mid n, R, N)\, \text{Bi}(R \mid N, p) \\ &= \sum_{R=0}^{N} \frac{\binom{R}{r}\binom{N-R}{n-r}}{\binom{N}{n}} \binom{N}{R} p^R (1-p)^{N-R} \\ &= \binom{n}{r} p^r (1-p)^{n-r}, \end{aligned} \tag{11.5}$$

which is the binomial model $\text{Bi}(r \mid n, p)$. The reference prior for the binomial model is known to be the corresponding Jeffreys-rule prior, which is the Beta distribution $\pi^R(p) = \text{Be}(p \mid \frac{1}{2}, \frac{1}{2})$ and, therefore, integrating out p in the $\text{Bi}(R \mid N, p)$ hierarchical prior for R yields the induced objective prior for R in the hypergeometric model

$$\begin{aligned} \pi^R(R \mid N) &= \int_0^1 \text{Bi}(R \mid N, p)\, \text{Be}(p \mid \tfrac{1}{2}, \tfrac{1}{2})\, dp \\ &= \frac{1}{\pi} \frac{\Gamma(R+\frac{1}{2})\,\Gamma(N-R+\frac{1}{2})}{\Gamma(R+1)\,\Gamma(N-R+1)}, \end{aligned} \tag{11.6}$$

for $R \in \{0, 1, \ldots, N\}$, as suggested in Jeffreys (1946, 1961). (Curiously, the argument of Jeffreys here was the 'incorrect' argument that the Jeffreys-rule prior for p from the hierarchical model $\text{Bi}(R \mid N, p)$ is $\text{Be}(p \mid \frac{1}{2}, \frac{1}{2})$ but, since this happens to be the same as the prior from the marginal model (11.5), he obtained the correct objective prior for R.) Note that, by construction, this is a proper prior. As shown in Berger et al. (2012), this prior becomes very close to the the continuous reference prior $\pi^R(p) = \text{Be}(p \mid \frac{1}{2}, \frac{1}{2})$ for the binomial probability model as N grows.

11.4.2 Laplace's Rule of Succession

A related curiosity is that of determining the probability that the next event in a randomly selected sequence of events (*e.g.*, the sun rising each morning) will occur (+), given that all previous n events occurred. The classical answer, known as Laplace's rule of succession (Laplace (1774)), is

$$\Pr(+ \mid \text{previous n +'s}) = \frac{n+1}{n+2}\,. \tag{11.7}$$

When the population is known to be finite of size N, the same rule was derived by Broad (1918). If the analysis is carried out using the reference prior $\pi^R(R \mid N)$

in (11.6), the resulting 'reference prior' law of succession is (see Berger et al. (2009*b*))

$$\Pr^R(+\,|\,N, \text{previous n +'s}) = \frac{n+1/2}{n+1}\,. \tag{11.8}$$

11.5 The Binomial-Beta Distribution

11.5.1 The Model and Likelihood

Consider the binomial-beta distribution formed as the marginal distribution from

$$x\,|\,(n,p) \sim \text{Bi}(x\,|\,n,p) \text{ and } p \sim \text{Be}(p\,|\,a,b),$$

where (a, b) are known positive constants. The marginal density of x is

$$\begin{aligned} p(x\,|\,n) &= \frac{n!}{x!(n-x)!}\frac{\Gamma(a+b)}{\Gamma(a)\Gamma(b)}\int_0^1 p^{x+a-1}(1-p)^{n-x+b-1}dp \\ &= \frac{n!}{x!(n-x)!}\frac{\Gamma(a+b)}{\Gamma(a)\Gamma(b)}\frac{\Gamma(x+a)\Gamma(n-x+b)}{\Gamma(n+a+b)}, \end{aligned} \tag{11.9}$$

for $x = 0, 1, \cdots, n$. For fixed x, the tail (when n is large) of the marginal likelihood function for n is

$$p(x\,|\,n) \propto \frac{n!}{(n-x)!}\frac{\Gamma(n-x+b)}{\Gamma(n+a+b)} \approx \frac{1}{n^{a+b-1}(n-x)^{1-b}} \approx \frac{1}{n^a}. \tag{11.10}$$

Note that, for $a \le 1$, this is not integrable so that, for instance, a constant prior for n would not yield a proper posterior.

To find the objective prior for n from this discrete distribution, application of Approach 1 would result in a non-constant normalization factor, and there is no natural hierarchical embedding for Approach 2. Hence we consider Approach 3.

11.5.2 Using Approach 3 to Determine an Objective Prior

Let $\{x_1, \cdots, x_k\}$ be k independent replications from (11.9). The mean and variance of the x_i are easily found to be

$$\text{E}(x_i\,|\,n) = \text{E}(np\,|\,n) = n\frac{a}{a+b}\,, \quad \text{Var}(x_i\,|\,n) = \frac{n(n+a+b)}{(a+b)^2(a+b+1)}\,.$$

A simple estimate of n is the linear estimate

$$\hat{n} = \frac{a+b}{a}\bar{x} = \frac{a+b}{a\,k}\sum_{j=1}^{k} x_j.$$

Note that this is not an efficient estimate but, as mentioned earlier, efficient estimates cannot be used in direct reference prior asymptotics for discrete distributions. This estimator has mean and variance

$$\mathrm{E}(\hat{n}\,|\,n) = n \quad \text{and} \quad \mathrm{Var}(\hat{n}\,|\,n) = \frac{n(n+a+b)}{a^2(a+b+1)\,k}\,.$$

It follows from the central limit theorem that

$$p(\hat{n}\,|\,n) \approx N\left(\hat{n}\;\middle|\; n, \frac{n(n+a+b)}{a^2(a+b+1)k}\right).$$

We extend the asymptotic distribution to a continuous parameter model by simply pretending that n is a continuous parameter. The reference prior for n in this extended model is the Jeffreys-rule prior,

$$\pi_1(n) \propto \left(\frac{n(n+a+b)}{a^2(a+b+1)k}\right)^{-\frac{1}{2}} \propto \frac{1}{\sqrt{n(n+a+b)}}. \tag{11.11}$$

11.5.3 Comparison with $1/n$

It is interesting to compare $\pi_1(n)$ with the generic recommendation of Jeffreys (1961) to use $\pi_2(n) \propto 1/n$ for a positive unbounded integer.

To further study this, we utilize the common device of studying the frequentist coverage of credible sets arising from the priors. For $\boldsymbol{x} = (x_1, \ldots, x_m)$ from (11.9) and given n, we construct the one-sided credible sets that have posterior probability exceeding a desired threshold $1-\alpha$, namely

$$C_i(\boldsymbol{x}) = \{1, \ldots, n_i^*\}, \quad n_i^* = \inf_{n^* \in \mathcal{N}} \Pr[n \le n^* \,|\, \boldsymbol{x}, i] \ge 1-\alpha\,,$$

corresponding to the two priors π_i, $i = 1, 2$. We then consider the frequentist coverage of these credible sets, namely $\Pr(C_i(\boldsymbol{X}) \ni n\,|\,n)$, with the goal being to have frequentist coverage close to the stated posterior probability. Because of the roughness caused by discreteness, we choose the target not to be $1-\alpha$ itself, but rather the frequentist expectation (given n) of the posterior coverage; this will be the actual average frequentist coverage in repeated use of the posterior credible sets.

A simple implementation of this idea, for given true parameter value n and sample size m, is to generate, say, 25,000 sample m-vectors, $\boldsymbol{x}_{(j)}$, with each coordinate being drawn independently from (11.9), compute $C_i(\boldsymbol{x}_{(j)})$ for each sample, and approximate the frequentist coverage and the average posterior coverage (APC) as, respectively,

$$\text{Coverage} \cong \frac{\#C_i(\boldsymbol{x}_{(j)}) \text{ that contain } n}{25{,}000},$$

$$\text{APC} \cong \frac{1}{25{,}000} \sum_{j=1}^{25{,}000} \Pr(n \in C_i(\boldsymbol{x}_{(j)})\,|\,\boldsymbol{x}_{(j)})\,.$$

Table 11.1: Average posterior coverage and frequentist coverage of one-sided 50% and 90% posterior credible intervals arising from the priors π_1 and π_2, based on samples of size $m = 1$ from the Binomial-Beta distribution with true $n = 10$ and various values of $a = b$.

		50% Credible Sets		90% Credible Sets	
$a = b$	Prior	APC	Coverage	APC	Coverage
5	π_1	0.549	0.579	0.910	0.872
	π_2	0.502	0.408	0.908	0.872
20	π_1	0.559	0.608	0.914	0.927
	π_2	0.501	0.390	0.916	0.927
50	π_1	0.562	0.618	0.914	0.937
	π_2	0.500	0.385	0.917	0.813

From Table 11.1 (the entries coming ftom Berger et al. (2012)), it is clear that π_2 can significantly overstate the coverage probability of its credible sets, its APC's often exceeding the actual coverage by at least 0.10. In contrast, π_1 is much more accurate in terms of stated credibilities approximately equalling actual coverage. Also, π_2 significantly degrades as $a = b$ grows larger, while the performance of π_1 remains stable as $a = b$ grows. The performance of π_1 is actually rather remarkable, considering that the study is only for a sample of size $m = 1$. In conclusion, the comparison clearly suggests preferring $\pi_1(n) = 1/\sqrt{n(n+a+b)}$ as the objective prior for the Binomial-Beta distribution.

11.6 The Binomial Sample Size

Assume that $x \sim \mathrm{Bi}(x \mid n, p)$. Estimating n has been a challenging problem for 60 years, with literature dating from Haldane (1941). Recent articles (which have many other references) include Berger et al. (1999) and DasGupta & Rubin (2005). Basu & Ebrahimi (2001) develop objective priors for a related problem in software reliability. We are interested in objective priors for n, first for the case of known p and then when p is unknown.

11.6.1 Known p

To find an objective prior for n from this discrete distribution, application of Approach 1 would again result in a non-constant normalization factor. Raftery (1988*b*), for the case of unknown p (see also Moreno & Girón (1998)), studied a version of Approach 2, using a hierarchical model, but it does not result in a closed-form answer. We thus turn to Approach 3.

Using Approach 3 to Determine an Objective Prior. Let $x_1, \cdots, x_k$ be k independent replications from $\mathrm{Bi}(x \mid n, p)$. A linear estimate of n is

$$\hat{n} = \frac{\bar{x}_k}{p} = \frac{1}{p\,k} \sum_{j=1}^{k} x_j\,,$$

for which

$$\mathrm{E}(\hat{n} \mid n, p) = n, \text{ and } \mathrm{Var}(\hat{n} \mid n) = \frac{n(1-p)}{k\,p}.$$

It follows from the central limit theorem that

$$p(\hat{n} \mid n, p) \approx N\left(\hat{n} \,\middle|\, n, \sqrt{\frac{n(1-p)}{k\,p}}\right). \tag{11.12}$$

Extending this to the model continuous in n, the Jeffreys-rule prior is

$$\pi_1(n \mid p) \propto \frac{1}{\sqrt{n}}\,. \tag{11.13}$$

As mentioned in Section 11.2.3, this is (necessarily) the same result derived in Barger & Bunge (2008), since we utilized a linear estimator and the problem can be shown to have a linear difference score.

Comparison with $\pi_2(n) = n^{-1}$ and $\pi_3(n) = 1$: We compare the performance of the suggested $\pi_1(n) = 1/\sqrt{n}$ with π_2 and π_3, two other priors that have been considered for this problem, using the method of comparison discussed in Section 11.5.3. Note that all three priors yield a proper posterior.

Table 11.2 is a representative sample of the many simulation results that were examined in Berger et al. (2012). The following conclusions can be reached from this (and many other) simulation results:

(i) The prior $\pi_2(n) = 1/n$ results in systematic overstatement of the actual coverage, and by a large amount; hence it can be eliminated from consideration.

(ii) The prior $\pi_3(n) = 1$ results in systematic understatement of the actual coverage by a significant amount – up to 11% in the table; hence this prior can also be eliminated from consideration.

(iii) $\pi_1(n) = 1/\sqrt{n}$ does reasonably well in the simulation, with APC's differing from actual coverage by no more than 6%.

In Sections 7.2 and 7.3, we discussed the issue that some priors are too concentrated, some are too diffuse, and some are just right; this example perfectly illustrates the issue. The recommended prior for the binomial distribution with known p is thus $\pi_1(n) = 1/\sqrt{n}$.

Table 11.2: Average posterior coverage and frequentist coverage of one-sided 50% credible intervals, for the priors $\pi_1(n) = 1/\sqrt{n}$, $\pi_2(n) = 1/n$ and $\pi_1(n) = 1$, based on 20,000 simulations of sample size m from the $\text{Bi}(x \mid 10, p)$ distribution.

(m, p)	Prior	APC	Coverage
$(5, 0.5)$	π_1	0.639	0.638
	π_2	0.637	0.615
	π_3	0.635	0.662
$(5, 0.1)$	π_1	0.545	0.578
	π_2	0.544	0.432
	π_3	0.539	0.598
$(50, 0.01)$	π_1	0.527	0.555
	π_2	0.529	0.410
	π_3	0.523	0.573

11.6.2 Unknown p

With p unknown as well as n, suppose there are $m \geq 2$ independent observations $x_1, \cdots, x_m$ from the $\text{Bi}(x \mid n, p)$ distribution. (It actually works to take $m = 1$ in the following but using one observation, when there are two unknown parameters, would be rather strange.) The likelihood function of (n, p), based on $(x_1, \cdots, x_m)$, is

$$p(x_1, \cdots, x_m \mid n, p) = \prod_{j=1}^{m} \binom{n}{x_j} p^s (1-p)^{mn-s}, \tag{11.14}$$

where $s = \sum_{j=1}^{m} x_j$. Because the Jeffreys-rule or reference prior for p given n is $\text{Be}(p \mid 1/2, 1/2)$, the marginal likelihood function of n is thus

$$p(x_1, \cdots, x_m \mid n) = \frac{1}{\pi} \prod_{j=1}^{m} \binom{n}{x_j} \frac{\Gamma(s+1/2)\Gamma(mn-s+1/2)}{\Gamma(mn+1)}. \tag{11.15}$$

We, therefore, need to find a reasonable objective prior for n for this discrete parameter model.

Using Approach 4 to Determine an Objective Prior. For fixed $p \in (0, 1)$ and as $n \to \infty$, the x_i/n are asymptotically independent and normally distributed with mean p and variance $p(1-p)/n$. Thus, for any $y_i \in (0, 1)$ and as $n \to \infty$,

$$P\Big(\frac{x_1}{n} \leq y_1, \ldots, \frac{x_m}{n} \leq y_m\Big) = \int_0^1 P\Big(\frac{x_1}{n} \leq y_1, \ldots, \frac{x_m}{n} \leq y_m \;\Big|\; p\Big) \text{Be}(p \mid \tfrac{1}{2}, \tfrac{1}{2})\, dp$$

$$\to \int_0^1 \prod_{i=1}^{m} 1_{(0, y_i)}(p) \text{Be}(p \mid \tfrac{1}{2}, \tfrac{1}{2})\, dp = \int_0^{\min\{y_1, \cdots, y_m\}} \text{Be}(p \mid \tfrac{1}{2}, \tfrac{1}{2})\, dp.$$

The limiting distribution of $(x_1/n, \cdots, x_m/n)$ does not depend on n, and $n > 0$ is a scale parameter of the limiting distribution of $(x_1, \cdots, x_m)$. The prior for n is thus the usual scale reference prior $\pi_1(n) \propto 1/n$.

The recommended objective prior for the binomial problem with both p and n unknown is thus

$$\pi_1(p, n) \propto \frac{1}{n} \times \text{Be}(p \mid \tfrac{1}{2}, \tfrac{1}{2}). \tag{11.16}$$

It is easy to see that this yields a proper posterior. Although Jeffreys never explicitly studied this problem, π_1 is presumably the prior he would have used, since he recommended the $\text{Be}(p \mid \frac{1}{2}, \frac{1}{2})$ prior for p and recommended $1/n$ for an infinite positive variable.

Comparison of Two Objective Priors When p is Unknown. Raftery (1988*b*) proposed the variant $\pi_2(p, n) = 1/n$ (constant in p), and it is of interest to compare this with the choice above. A partial comparison is given in Berger et al. (1999), where it is shown that

$$\frac{\pi_2(n \mid x_1, \ldots, x_m)}{\pi_1(n \mid x_1, \ldots, x_m)} \approx \frac{\pi\sqrt{n - s + 0.3}}{n+1},$$

so that the Raftery posterior has a sharper tail as n grows. In Berger et al. (2012) a number of numerical comparisons are given, but the results are inconclusive: $\pi_1(p, n)$ seems better for more extreme values of p, and $\pi_2(p, n)$ performs better for non-extreme p. For compatibility with reference prior theory, we recommend use of π_1, but use of π_2 is certainly not unreasonable.

11.7 Prior Model Probabilities

Suppose we have m potential models $\mathcal{M}_i$, $i = 1, \ldots, m$, for the data, and consider the problem of determining objective prior model probabilities, $\pi(\mathcal{M}_i)$.

11.7.1 Finite Number of Models with No Structure

If the models being considered have no structure, we would simply use the usual reference prior $\pi(\mathcal{M}_i) = 1/m$. This would occur for instance if there were just two models, or if the models had no relationships to each other. For instance, in selecting from among three different location-scale models – normal, Cauchy, and exponential – with only the location and scale parameters of each model being unknown, there is no structure that could lead to any objective assignment of prior probabilities other than $\pi(\mathcal{M}_i) = 1/3$.

11.7.2 Variable Selection

Consider the variable selection problem where there are m variables (or graphical nodes or links) that can make up a model. Let $\boldsymbol{\gamma}$ be the vector of zeroes and ones that indicates which variables are in the model, and denote the resulting model as $\mathcal{M}_{\boldsymbol{\gamma}}$.

> **Example 11.5** *Linear Regression, Example 1.4.2 continued.* Often one wants to keep only some of the covariates in the regression model; for instance one might use only the intercept and the first and third covariates, resulting in the model
>
> $$y = \beta_0 + v_1\beta_1 + v_3\beta_3 + \varepsilon\,.$$
>
> It is common to always include the intercept term, so including the first and third covariates would correspond to having $\boldsymbol{\gamma} = (1, 0, 1, 0, \ldots, 0)$.

Note that there are 2^m possible models so that, if structure were ignored, the reference prior would assign each model probability $\pi(\mathcal{M}_{\boldsymbol{\gamma}}) = 2^{-m}$. It is being increasingly realized, however (cf. Ley & Steel (2009) and Scott & Berger (2010)) that this assignment of prior model probabilities has serious deficiencies (as first recognized by Jeffreys (1961)), such as not allowing for a multiplicity correction based on the number of variables being considered. Hence the use of objective priors based on structure is appealing.

Structure 1. One obvious structure here is that each variable can be in or out of the model. One can thus implement the hierarchical Approach 2, and assume that each variable is in the model with unknown probability p. Then the probability of any model with r variables is $\pi(\mathcal{M}_{\boldsymbol{\gamma}} \mid p) = p^r(1-p)^{m-r}$. Formal application of Approach 2 would now require marginalizing over $\boldsymbol{\gamma}$, and then finding the reference (Jeffreys-rule) prior for p in the ensuing marginal model.

Unfortunately, this computation is extremely difficult and problem dependent, and cannot be carried out in closed form. A rather ad hoc way forward is to, instead, simply use the usual reference prior for p in a $\mathrm{Bi}(r \mid m, p)$ model, which is $\mathrm{Be}(p \mid \frac{1}{2}, \frac{1}{2})$. Utilizing this yields objective model prior probabilities

$$\pi(\mathcal{M}_{\boldsymbol{\gamma}}) = \int_0^1 p^r(1-p)^{m-r}\mathrm{Be}(p \mid \tfrac{1}{2}, \tfrac{1}{2})\,dp = \frac{\Gamma(r+\frac{1}{2})\Gamma(m-r+\frac{1}{2})}{\pi\Gamma(m+1)}\,. \tag{11.17}$$

Structure 2. Suppose we were instead to structure the problem (following Jeffreys (1961)) by saying that model size r is the parameter of interest and that, given r, there is no model structure. One might then assign the $m+1$ possible values of r equal prior probabilities, i.e. $\Pr(r) = (m+1)^{-1}$ and assign all models of a given size r equal prior probability $1/\binom{m}{r}$. This leads to the prior probability assignment

$$\pi(\mathcal{M}_{\boldsymbol{\gamma}}) = \frac{1}{(m+1)\binom{m}{r}}\,. \tag{11.18}$$

This argument is somewhat less satisfying than that for Structure 1, in that the assignment of equal prior probabilities to each model size r seems rather arbitrary. In contrast, the main assumption going into Structure 1 was an exchangeability assumption. Interestingly, however, (11.18) can also be derived as in (11.17) if, instead, the uniform prior is used for p; the choice of Jeffreys is thus reasonable from either perspective.

Chapter 12

Overall Objective Priors

12.1 Introduction

12.1.1 The Problem

In single parameter problems, the reference prior (usually the Jeffreys-rule prior) is uniquely defined but, in multiparameter models, we have seen that the reference prior often changes depending on the quantity of interest. There are however situations where one is *simultaneously* interested in all the parameters of the model or, at least in several of them; in such situations having a single overall prior for use is appealing. In prediction or decision analysis for instance, all of the parameters of the model are typically relevant and often none are of major interest individually.

Another situation in which having an overall prior would be beneficial is when there is a non-standard quantity of interest (*e.g.*, a non-standard function of the model parameters), and formal derivation of the reference prior is not tenable. Computation can also be a consideration, since having to separately do Bayesian computations with different reference priors for each parameter can be challenging. Finally, when dealing with non-specialists it may be best to just present them with one overall objective prior, rather than attempting to explain the reasons for preferring different priors for different quantities of interest.

To proceed, assume that $\boldsymbol{\theta} = (\theta_1, \ldots, \theta_m)$ is is the set of $m > 1$ parameters that are of interest. The goal is to find a prior, $\pi^o(\boldsymbol{\theta})$, whose corresponding marginal posteriors, $\{\pi^o(\theta_i \mid \boldsymbol{x}), i = 1, \ldots m\}$, are sensible from an objective Bayesian (typically reference prior) perspective; for, instance, it would be good if these marginal posteriors are very close to the reference prior marginal posteriors $\{\pi^R_{\theta_i}(\theta_i \mid \boldsymbol{x}), i = 1, \ldots m\}$, the subscript θ_i indicating that this is the marginal reference posterior when θ_i is the parameter of interest.

Example 12.1 *Multinomial data.* Suppose $\boldsymbol{x} = (x_1, \ldots, x_m)$ is multinomial $\mathrm{Mu}(\boldsymbol{x} \mid n; \theta_1, \ldots, \theta_m)$, with $\sum_{i=1}^m x_i = n$, and $\sum_{i=1}^m \theta_i = 1$. In

Section 10.3.2, it was shown that the reference prior, when the parameter of interest is θ_i, is a different prior for each θ_i, and that the reference prior for θ_i results in a Beta reference marginal posterior $\pi^R_{\theta_i}(\theta_i \mid \boldsymbol{x}) = \text{Be}(\theta_i \mid x_i + \frac{1}{2}, n - x_i + \frac{1}{2})$. One possible goal would be to identify a single *overall* objective prior, $\pi^o(\boldsymbol{\theta})$ whose marginal posteriors are close to each of these reference marginal posteriors, in some average sense.

We will discuss various possible approaches to the problem in this chapter. Most of the approaches and examples considered in this chapter come from either Berger et al. (2015) or Berger et al. (2020*b*).

12.1.2 Background

It is useful to begin by recalling earlier efforts at obtaining an overall objective prior. One example is given in Berger & Sun (2008), relating to the five parameter bivariate normal model discussed in Example 7.5. All the possible quantities of interest that had previously been considered were studied, with one of the goals being to find a good overall prior, defined as one which yielded reasonable frequentist coverage properties when used for at least the most important quantities of interest. The conclusion was that the prior

$$\pi^o(\mu_1, \mu_2, \sigma_1, \sigma_2, \rho) = \sigma_1^{-1}\sigma_2^{-1}(1 - \rho^2)^{-1}, \tag{12.1}$$

where the μ_i are the means, the σ_i are the standard deviations, and ρ is the correlation in the bivariate normal model, was a good choice for an overall prior.

Invariance-Based Priors. If $p(\boldsymbol{x} \mid \boldsymbol{\theta})$ has a group invariance structure, then the reference prior for all parameters that define the invariance structure will typically be the right-Haar prior and this is a natural candidate for an overall prior. For instance, if the sampling model is $\text{N}(x_i \mid \mu, \sigma)$, the right-Haar prior is $\pi^r(\mu, \sigma) = \sigma^{-1}$, which is also the reference prior for either μ or σ (and for a number of functions of them) and is thus a natural overall prior.

Such a nice situation does not always obtain, however. For instance, in Example 7.5 it was shown that there is more than one possible right-Haar prior for the bivariate normal example, and each works well for some parameters but not for others. The situation can be even worse if the right-Haar prior is used for other parameters that can be considered. In the multi-normal means Example 7.3, for instance, the right-Haar prior was just the constant prior and worked well for each individual mean μ_i but was terrible as a prior if the parameter of interest was $\theta = m^{-1}|\boldsymbol{\mu}|^2 = m^{-1}\sum_{i=1}^m \mu_i^2$, yielding an inconsistent estimate of θ as the number of means m grows.

Jeffreys-Rule Prior. The Jeffreys-rule prior is the same for all parameters in a model and is invariant under reparameterization and is, hence, an obvious candidate for an overall prior. But it is frequently a bad prior, as first discussed in Section 4.3. Additional examples of bad behavior of the Jeffreys-rule prior are the multi-normal means problem just mentioned (since the Jeffreys-rule prior is the constant prior for the means) and the Jeffreys-rule prior for a covariance matrix, discussed in Example 7.4.

For the bivariate normal example, the Jeffreys-rule prior is $\sigma_1^{-2}\sigma_2^{-2}(1-\rho^2)^{-2}$; this yields reference marginal posteriors for the means and standard deviations (and is, hence, great for these parameters) but, as shown in Berger & Sun (2008), results in quite inferior marginal posteriors for ρ and various derived parameters.

In Example 4.10, the Jeffreys-rule prior for the multinomial distribution was shown to result in highly questionable inferences from an objective viewpoint. For instance, some data was given there for which the posterior mean for θ_{240} was 0.005, even though 2/3 of the observations were in that cell; this hardly seems objective. This example will be carefully examined in this chapter.

There have been efforts to improve upon the Jeffreys-rule prior, such as consideration of the 'independence Jeffreys prior,' but such prescriptions have been rather *ad hoc* and have not led to a general alternative definition. We actually know of *no* multiparameter example in which we would recommend the Jeffreys-rule prior. In higher dimensions, the prior always seems to be either 'too diffuse,' as in the multinormal means example, or 'too concentrated,' as in the multinomial example.

12.1.3 Outline of the Chapter

Section 12.2 considers the situation where the reference prior for each parameter of interest is the same. This is clearly a great situation, but one must remember that the common reference prior might not be good if another parameter of interest arises.

Section 12.3 discusses the *prior averaging approach* to construction of an overall prior, which is to take, say, the various reference priors corresponding to different quantities of interest and average them in some way. Although this seems natural, we will see that it can be problematical.

Section 12.4 discusses the *reference distance approach*, which is to formally try to minimize the distances from the overall prior marginal posteriors to the reference prior marginal posteriors (for the quantities of interest). The approach is illustrated with the multinomial example.

Section 12.5 discusses the *hierarchical reference approach*, which is to utilize hierarchical Bayesian analysis to construct an overall prior. This approach is also illustrated with the multinomial example.

Section 12.6 gives another example of the the hierarchical approach, but applied to general normal hierarchical models. The use of these models is ubiquitous in applied Bayesian statistics and having overall priors available for their analysis is crucial. We present and discuss the recommended overall objective priors from Berger et al. (2020*b*), for means and covariance matrices that occur at any level of a hierarchical model. Computational implementation of the overall priors is also discussed.

12.2 When the Reference Prior is Common

If one is able to find a single joint prior $\pi^o(\boldsymbol{\theta})$ whose corresponding marginal posteriors are precisely equal to the reference prior posteriors for each of the θ_i's, so that, for all $\boldsymbol{x} \in \boldsymbol{\mathcal{X}}$,

$$\pi^o(\theta_i \,|\, \boldsymbol{x}) = \pi_{\theta_i}(\theta_i \,|\, \boldsymbol{x}), \quad i = 1, \ldots, m,$$

then it is natural to argue that this should be an appropriate solution to the problem. The $\mathrm{N}(x_i \mid \mu, \sigma)$ model has already been mentioned as an example, where the reference priors when μ or σ are the parameters of interest are the same. (Their common reference prior is also excellent if μ and σ are jointly of interest, *e.g.*, if a joint credible set for them is sought.)

Other examples where this holds are Example 13.6, the examples in Section 14.1, and Example 14.1.

12.3 Prior Averaging Approach

Starting with a collection of (say) reference priors $\{\pi_{\theta_i}(\boldsymbol{\theta}), i = 1, \ldots, m\}$ for differing parameters or quantities of interest, it seems natural to use an average of the priors as the overall prior.

Two natural averages to consider are the arithmetic mean

$$\pi^A(\boldsymbol{\theta}) = \frac{1}{m} \sum\nolimits_{i=1}^{m} \pi_{\theta_i}(\boldsymbol{\theta}),$$

and the geometric mean

$$\pi^G(\boldsymbol{\theta}) = \prod\nolimits_{i=1}^{m} \pi_{\theta_i}(\boldsymbol{\theta})^{1/m}.$$

The arithmetic average might seem most natural, since it can be viewed as arising from the hierarchical model in which each π_{θ_i} is assigned probability $1/m$ of being correct. When dealing with improper priors, however, such weights are quite arbitrary because the improper priors can be multiplied by arbitrary constants.

Geometric averaging arises naturally in the definition of reference priors (see *e.g.*, (9.5)), and it can also be seen to result in the optimal prior if one is trying to choose a single prior to minimize the average of the Kullback-Liebler divergences of the prior from the π_{θ_i}'s. And geometric averaging is unaffected by normalizing constants, so the use of geometric averaging with improper priors causes no problem.

Example 12.2 *Bivariate Normal Distribution (continued).* In Example 7.5, the average priors that arise from the two right-Haar reference priors,
$\pi_1^r(\mu_1,\mu_2,\sigma_1,\sigma_2,\rho) = \sigma_1^{-2}(1-\rho^2)^{-1}$ and $\pi_2^r(\mu_1,\mu_2,\sigma_1,\sigma_2,\rho) = \sigma_2^{-2}(1-\rho^2)^{-1}$,
are

$$\pi^A(\mu_1,\mu_2,\sigma_1,\sigma_2,\rho) = \frac{1}{2}\left(\frac{1}{\sigma_1^2(1-\rho^2)} + \frac{1}{\sigma_2^2(1-\rho^2)}\right),$$
$$\pi^G(\mu_1,\mu_2,\sigma_1,\sigma_2,\rho) = \frac{1}{\sigma_1\sigma_2(1-\rho^2)}.$$

Berger & Sun (2008) show that π^A is a substantially worse objective prior than either right-Haar prior alone, while π^G was the recommended overall prior (12.1).

One problem with the averaging approach is that the reference prior for θ_i can depend on the ordering of the other parameters, so there is not just one reference prior corresponding to θ_i.

Example 12.3 *Multinomial example (continued).* The reference prior derived when the parameter of interest is θ_i actually depends on the ordering chosen (in the reference prior derivation) for all the parameters (*e.g.*, $\{\theta_i,\theta_1,\theta_2,\ldots,\theta_{i-1},\theta_{i+1},\ldots,\theta_m\}$); there are thus $(m-1)!$ different reference priors for each parameter of interest and averaging over all these priors can be prohibitive computationally.

There has not been sufficient study of prior averaging (of improper priors) to draw firm conclusions concerning the approach. That said, current evidence clearly supports use of geometric averaging, as opposed to arithmetic averaging.

12.4 Reference Distance Approach

12.4.1 Introduction

We consider a family of candidate prior distributions, $\{\pi(\boldsymbol{\theta}\,|\,\boldsymbol{a}), \boldsymbol{a}\in\mathcal{A}\}$, and choose, as the overall prior, the distribution from this class which yields marginal posteriors that are closest, in an average sense, to the marginal reference posteriors. Thus we want to find $\boldsymbol{a}$ so that the corresponding marginal posteriors $\{\pi(\theta_i\,|\,\boldsymbol{x},\boldsymbol{a}),\ i=1,\ldots,m\}$ are as close as possible to the reference prior marginal posteriors. $\{\pi_{\theta_i}(\theta_i\,|\,\boldsymbol{x}),\ i=1,\ldots,m\}$, arising from the separate reference priors $\{\pi_{\theta_i}(\boldsymbol{\theta}),\ i=1,\cdots,m\}$.

The average distance measure that we will utilize has three components:

(i) As usual, we will measure the 'distance' between $\pi(\theta_i\,|\,\boldsymbol{x},\boldsymbol{a})$ and $\pi_{\theta_i}(\theta_i\,|\,\boldsymbol{x})$ by the KL divergence (1.5), $\kappa\{\pi(\theta_i\,|\,\boldsymbol{x},\boldsymbol{a})\,|\,\pi_{\theta_i}(\theta_i\,|\,\boldsymbol{x})\}$.

(ii) Suppose the relative importance of the θ_i is given by a set of weights $\{w_1, \ldots, w_m\}$, with $0 < w_i < 1$ and $\sum_i w_i = 1$. A natural default value for these is $w_i = 1/m$.

(iii) The distance measure must be apriori, so we average over the data with respect to the reference prior predictives, for $i = 1, \ldots, m$,

$$p_{\theta_i}(\boldsymbol{x}) = \int_{\boldsymbol{\Theta}} p(\boldsymbol{x} \mid \boldsymbol{\theta})\, \pi_{\theta_i}(\boldsymbol{\theta})\, d\boldsymbol{\theta}\,.$$

Definition 12.1 *The best overall prior,* $\pi^o(\boldsymbol{\theta}) \equiv \pi(\boldsymbol{\theta} \mid \boldsymbol{a}^*)$*, within the family* $\{\pi(\boldsymbol{\theta} \mid \boldsymbol{a}), \boldsymbol{a} \in \mathcal{A}\}$*, is defined by* $\boldsymbol{a}^*$*, which is the value of* $\boldsymbol{a}$ *that minimizes (assuming it exists) the* weighted average expected KL divergence

$$d(\boldsymbol{a}) = \sum_{i=1}^{m} w_i \int_{\mathcal{X}} \kappa\{\pi(\boldsymbol{\theta}_i \mid \boldsymbol{x}, \boldsymbol{a}) \mid \pi_{\theta_i}(\boldsymbol{\theta}_i \mid \boldsymbol{x})\}\, p_{\theta_i}(\boldsymbol{x})\, d\boldsymbol{x}. \tag{12.2}$$

Remark 12.1 An alternative useful expression for $d(\boldsymbol{a})$ is

$$d(\boldsymbol{a}) = \sum_{i=1}^{m} w_i \int_{\boldsymbol{\Theta}} \rho_{\theta_i}(\boldsymbol{a} \mid \boldsymbol{\theta})\, \pi_{\theta_i}(\boldsymbol{\theta})\, d\boldsymbol{\theta}\,,$$

$$\rho_{\theta_i}(\boldsymbol{a} \mid \boldsymbol{\theta}) = \int_{\mathcal{X}} \kappa\{\pi(\boldsymbol{\theta}_i \mid \boldsymbol{x}, \boldsymbol{a}) \mid \pi_{\theta_i}(\boldsymbol{\theta}_i \mid \boldsymbol{x})\}\, p(\boldsymbol{x} \mid \boldsymbol{\theta})\, d\boldsymbol{x}\,. \tag{12.3}$$

12.4.2 Multinomial Example

For the multinomial model with m cells and parameters $\boldsymbol{\theta} = (\theta_1, \ldots, \theta_m)$, with $\sum_{i=1}^m \theta_i = 1$, it is natural to consider the Dirichlet $\{\mathrm{Di}(\boldsymbol{\theta} \mid a, \ldots, a) : a \in (0, \infty)\}$ class of priors for $\boldsymbol{\theta}$, because of its conjugacy and reflecting a desire to treat all the θ_i similarly.

The marginal posterior distribution of θ_i, from the $\mathrm{Di}(\boldsymbol{\theta} \mid a, \ldots, a)$ prior, is $\mathrm{Be}(\theta_i \mid x_i + a, n - x_i + (m-1)a)$. The KL divergence of this marginal posterior from the reference marginal posterior, $\mathrm{Be}(\theta_i \mid x_i + \frac{1}{2}, n - x_i + \frac{1}{2})$, is

$$\kappa_i\{a \mid x_i, m, n\} = \kappa_{\mathrm{Be}}\{x_i + a, n - x_i + (m-1)a \mid x_i + \tfrac{1}{2}, n - x_i + \tfrac{1}{2}\}$$

where, with $\psi(\cdot)$ being the digamma function,

$$\begin{aligned}
\kappa_{\mathrm{Be}}\{\alpha_0, \beta_0 \mid \alpha, \beta\} &= \int_0^1 \mathrm{Be}(\theta_i \mid \alpha, \beta) \log\left[\frac{\mathrm{Be}(\theta_i \mid \alpha, \beta)}{\mathrm{Be}(\theta_i \mid \alpha_0, \beta_0)}\right] d\theta_i \\
&= \log\left[\frac{\Gamma(\alpha+\beta)}{\Gamma(\alpha_0+\beta_0)} \frac{\Gamma(\alpha_0)}{\Gamma(\alpha)} \frac{\Gamma(\beta_0)}{\Gamma(\beta)}\right] + (\alpha - \alpha_0)\psi(\alpha) \\
&\quad + (\beta - \beta_0)\psi(\beta) - ((\alpha+\beta) - (\alpha_0+\beta_0))\psi(\alpha+\beta)\,.
\end{aligned}$$

Note that the divergence, $\kappa_i\{a \mid x_i, m, n\}$, between the two posteriors of θ_i depends on the data only through x_i, and the sampling distribution of x_i is

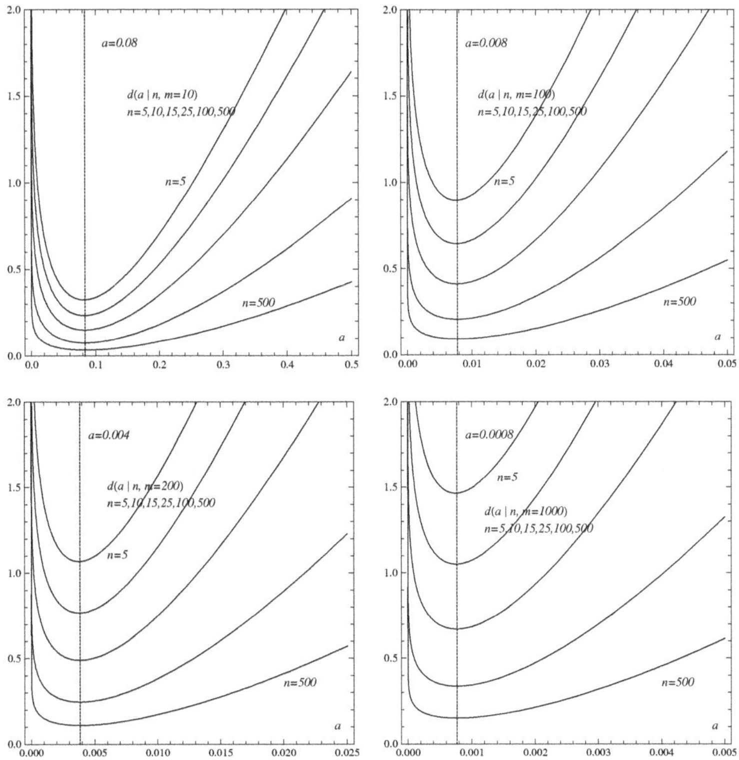

Figure 12.1: Average expected KL divergences, when using a Dirichlet prior with parameter $(a, \ldots, a)$, in a multinomial model with m cells, for sample sizes $n = 5, 10, 25, 100$ and 500. Top panels are for $m = 10$ and $m = 100$; bottom panels for $m = 200$ and $m = 1000$. In all cases, the optimal choice of a is $a^* \approx 0.8/m$.

$\mathrm{Bi}(x_i \mid n, \theta_i)$, which only depends on θ_i. Also, the marginal reference prior for θ_i is $\pi_{\theta_i}(\theta_i) = \mathrm{Be}(\theta_i \mid 1/2, 1/2)$ and, therefore, the corresponding reference prior predictive for x_i is

$$p(x_i \mid n) = \int_0^1 \mathrm{Bi}(x_i \mid n, \theta_i)\, \mathrm{Be}(\theta_i \mid 1/2, 1/2)\, d\theta_i = \frac{1}{\pi} \frac{\Gamma(x_i + \frac{1}{2})\, \Gamma(n - x_i + \frac{1}{2})}{\Gamma(x_i + 1)\, \Gamma(n - x_i + 1)}\,.$$

Hence, using Definition 12.1 with uniform weights, the average expected KL divergence of using the $\mathrm{Di}(\boldsymbol{\theta} \mid a, \ldots, a)$ prior is

$$d(a \mid n, m) = \sum_{x=0}^{n} \kappa\{a \mid x, m, n\}\, p(x \mid n)\,,$$

since the m parameters $(\theta_1, \ldots, \theta_m)$ yield the same average expected divergence.

The function $d(a \mid n, m = 10)$ is graphed in the upper left panel of Figure 12.1 (taken from Berger et al. (2015)) for several values of n. For any n, the function

$d(a \mid n, m)$ has a unique minimum, numerically found to be at $a^* \approx= 0.8/m$. The other three panels also graph $d(a \mid n, m)$, for the same n and $m = 100, 200$, and 1000; in all cases, the minimum is $a^* \approx= 0.8/m$.

It can be concluded that, for all practical purposes when using the reference distance approach, the best global Dirichlet prior, when one is interested in all the parameters of a multinomial model, is that with parameter vector $(1/m, \ldots, 1/m)$ (or $0.8 \times (1/m, \ldots, 1/m)$ to be slightly more precise), yielding approximate marginal reference posteriors for each of the θ_i's

$$\pi^o(\theta_i \mid \boldsymbol{x}) = \pi^o(\theta_i \mid x_i, n, m) = Be(\theta_i \mid x_i{+}1/m, n{-}x_i{+}(m{-}1)/m), \quad i = 1, \ldots, m.$$

The corresponding posterior means are $\mathrm{E}[\theta_i \mid x_i, n] = \hat{\theta}_i = (x_i + 1/m)/(n + 1)$, with posterior variances $\mathrm{Var}[\theta_i \mid x_i, n] = \hat{\theta}_i(1 - \hat{\theta}_i)/(n + 2)$.

Example 12.4 *Multinomial data example (continued).* In Example 4.14, the Jeffreys-rule prior, corresponding to $a = 1/2$, was shown to yield unreasonable posterior means for the θ_i in the situation where $n = 3$, $m = 1000$, $x_{240} = 2$, $x_{876} = 1$, and the other x_i are zero. In particular, the posterior mean for θ_{240} was only 0.005 even though 2/3 of the observations were in that cell; the problem was that the Jeffreys-rule prior effectively adds 1/2 to each cell, so that the cells with $x_i = 0$ overwhelmed the cells with $x_i \neq 0$. Choosing $a = 1/m$ here adds only $1/m = 0.001$ to each cell, so that

$$\mathrm{E}[\theta_i \mid \boldsymbol{x}] = \frac{x_i + 1/m}{\sum_{i=1}^{m}(x_i + 1/m)} = \frac{x_i + 1/m}{n + 1} = \frac{x_i + 0.001}{4}.$$

Thus $\mathrm{E}[\theta_{240} \mid \boldsymbol{x}] \approx 0.5$, $\mathrm{E}[\theta_{876} \mid \boldsymbol{x}] \approx 0.25$, and $\mathrm{E}[\theta_i \mid \boldsymbol{x}] \approx \frac{1}{4000}$ otherwise, all sensible results.

12.5 Hierarchical Reference Approach

12.5.1 Introduction

The hierarchical reference prior approach consists of the following steps.

(i) Choose a class of *proper* priors $\{\pi(\boldsymbol{\theta} \mid \boldsymbol{a}) : \boldsymbol{a} \in \mathcal{A}\}$ reflecting the desired structure of the problem.

(ii) Form the marginal density for the hyperparameter $\boldsymbol{a}$,

$$p(\boldsymbol{x} \mid \boldsymbol{a}) = \int p(\boldsymbol{x} \mid \boldsymbol{\theta})\, \pi(\boldsymbol{\theta} \mid \boldsymbol{a})\, d\boldsymbol{\theta},$$

which is proper because the $\pi(\boldsymbol{\theta} \mid \boldsymbol{a})$ are proper.

(iii) Find the reference prior, $\pi^R(\boldsymbol{a})$, for $\boldsymbol{a}$ in this marginal model.

(iv) Use, as the overall prior for $\boldsymbol{\theta}$,

$$\pi^o(\boldsymbol{\theta}) = \int \pi(\boldsymbol{\theta} \mid \boldsymbol{a})\, \pi^R(\boldsymbol{a})\, d\boldsymbol{a}\,. \tag{12.4}$$

Note that computation with this prior is often easiest if one keeps $\boldsymbol{a}$ as a random variable, rather than integrating it out and working directly with (12.4). This is especially true when $\{\pi(\boldsymbol{\theta} \mid \boldsymbol{a}) : \boldsymbol{a} \in \mathcal{A}\}$ is a conjugate family to the data distribution, so that Gibbs sampling can be employed.

12.5.2 The Multinomial Problem

Derivation of the Hierarchical Prior. The Dirichlet $\{\text{Di}(\boldsymbol{\theta} \mid a, \ldots, a) : a \in (0, \infty)\}$ class of hierarchical priors for $\boldsymbol{\theta}$ is natural here, reflecting a desire to treat all the θ_i similarly. The marginal density is

$$\begin{aligned} p(\boldsymbol{x} \mid a) &= \int \binom{n}{x_1 \ldots x_m} \left(\prod_{i=1}^{m} \theta_i^{x_i} \right) \frac{\Gamma(m\,a)}{\Gamma(a)^m} \prod_{i=1}^{m} \theta_i^{a-1} d\boldsymbol{\theta} \\ &= \binom{n}{x_1 \ldots x_m} \frac{\Gamma(m\,a)}{\Gamma(a)^m} \frac{\prod_{i=1}^{m} \Gamma(x_i + a)}{\Gamma(n + m\,a)}\,. \end{aligned} \tag{12.5}$$

This can be shown to be a regular, one-parameter model, so the reference prior, $\pi^R(a)$, for the hyper parameter a, is the Jeffreys-rule prior for this marginal model. In Berger et al. (2015), this was shown to be given by

$$\pi^R(a) \propto \left[\sum_{j=0}^{n-1} \left(\frac{Q(j \mid a, m, n)}{(a + j)^2} - \frac{m}{(m\,a + j)^2} \right) \right]^{1/2}, \tag{12.6}$$

where $Q(\cdot \mid a, m, n)$, for $j = 0, \ldots, n-1$, is given by

$$Q(j \mid a, m, n) = \sum_{l=j+1}^{n} \binom{n}{l} \frac{\Gamma(l + a)\, \Gamma(n - l + (m-1)a)\, \Gamma(m\,a)}{\Gamma(a)\, \Gamma((m-1)a)\, \Gamma(n + m\,a)}\,.$$

Finally, the overall prior for $\boldsymbol{\theta}$ is

$$\pi^o(\boldsymbol{\theta}) = \int \text{Di}(\boldsymbol{\theta} \mid a)\, \pi^R(a)\, da\,.$$

Propriety of the Reference Prior for a. The reference prior, $\pi^R(a)$, was shown in Berger et al. (2015) to behave as $O(a^{-1/2})$ near $a = 0$ and as $O(a^{-2})$ for large a values. It is thus a proper prior. It is always a surprise when a reference prior turns out to be proper, and this seems to happen when the likelihood does not go to zero at a limit. Indeed, it was shown in Berger et al. (2015) that

$$p(\boldsymbol{x} \mid a) = \begin{cases} O(a^{r_0 - 1}), & \text{as } a \to 0, \\ \binom{n}{\boldsymbol{x}}\, m^{-n}, & \text{as } a \to \infty, \end{cases}$$

where r_0 is the number of nonzero x_i. Thus the likelihood is a nonzero constant at ∞, so that the prior must be proper at infinity for the posterior to exist.

Behavior for Sparse Tables. Recall that m is the number of cells in the table and n is the sample size. Hence, if m/n is relatively large, the table is necessarily a sparse table, *i.e.*, one with mostly zero entries. It was shown in Berger et al. (2015) that, in this case, the reference prior is well approximated by the proper prior

$$\pi^*(a) = \frac{1}{2}\frac{n}{m}\,a^{-1/2}\Big(a+\frac{n}{m}\Big)^{-3/2}. \tag{12.7}$$

Calculus shows that the α-quantile of this distribution is

$$q_\alpha = \frac{n}{m(\alpha^{-2}-1)}\,.$$

For instance, the median of the distribution is $q_{0.5} = n/[3m]$. A 90% prior credible set for a is $\frac{n}{m}(.0025, 9.25)$ and a 50% prior credible set is $\frac{n}{m}(.067, 1.29)$. That a varies here around n/m is interesting; in comparison, the reference distance approach suggested that a should be about $1/m$. Of course, the reference distance approach was not focusing on the sparse table situation.

Computation with the hierarchical reference prior. If a full Bayesian analysis is desired, the obvious MCMC sampler is as follows:

Step 1. Use a Metropolis Hastings move (see Section 1.8.1) to sample from the marginal posterior $\pi^R(a \mid \boldsymbol{x}) \propto \pi^R(a)\,p(\boldsymbol{x} \mid a)$.

Step 2. Given a, sample from the usual beta posteriors $\pi(\theta_i \mid a, \boldsymbol{x})$.

This will be highly efficient if a good proposal distribution for Step 1 can be found. As it is only a one-dimensional distribution, standard techniques should work well.

Even simpler computationally is the use of the approximate reference prior $\pi^*(a \mid m, n)$ in (12.7) because, if there are at least three nonzero x_i, it can be shown that the marginal posterior distribution of a is log-concave. Thus log-concave rejection sampling (Gilks & Wild 1992) can be used to sample from the posterior of a.

Adaptability of the Hierarchical Prior. It is interesting to study the 'large sparse' contingency table scenario, in order to understand how the hierarchical prior relates to various structures in the data, as reflected in the posterior distribution of a. It is easiest to state the key needed result in terms of the transformed variable $v = m\,a$, so let $\pi^R_m(v \mid \boldsymbol{x})$ denote the transformed reference posterior. Then, it is shown in Berger et al. (2015) that, with r_0 being the number of nonzero x_i,

$$\Psi(v) = \lim_{m\to\infty} \pi^R_m(v \mid \boldsymbol{x}) = \frac{\Gamma(v+1)}{\Gamma(v+n)}\, v^{(r_0-\frac{3}{2})} \left[\sum_{i=1}^{n-1} \frac{i}{(v+i)^2}\right]^{1/2}. \tag{12.8}$$

It follows, of course, that a behaves like v/m for large m, where v has the distribution in (12.8). It is very interesting that this 'large m' behavior of the posterior depends on the data only through r_0.

If, in addition, $n \to \infty$ but $n/m \to 0$, it is possible to use (12.8) to determine how the posterior mode of a depends on r_0, in addition to m and n. Indeed Table 12.1, most of which is taken from Berger et al. (2015), shows this, in terms of the growth of r_0 relative to n. Note first that $\hat{a} = 1/m$ (the suggested choice from the reference distance approach) only when r_0 grows like $\log n$, which means that most of the observations are concentrating in a smallish number of cells. At the extreme, where r_0 is a fixed constant C, $\hat{a}$ goes to zero even faster than $1/m$, so the prior will give even less mass than $1/m$ to cells.

At the other extreme, where $r_0 = n$ (i.e., every nonzero x_i is one), $\hat{a}$ is $n^2/[2m]$, much larger than $1/m$. This is particularly interesting because, if one distributes n observations among m cells, they have to go somewhere, so seeing n cells with $x_i = 1$ does not necessarily lead one to think that the corresponding θ_i are particularly large (*e.g.*, observing this is completely compatible with all $\theta_i = 1/m$). The prior is adapting to that automatically, by choosing much larger a, i.e. giving much more mass to the nonzero cells. This adaptivity of hierarchical priors is one of the compelling reasons for their use. For a different hierarchical approach to the problem, see Tuyl (2019).

Table 12.1: The posterior mode of a as $n \to \infty$ and $n/m \to 0$, for various limiting behaviors of the number of non-zero cells r_0 (as a function of n), with c^* being the solution to $c^* \log(1 + \frac{1}{c^*}) = c$.

r_0	n	cn $(0 < c < 1)$	n^b $(0 < b < 1)$	$(\log n)^b$	$\log n$	C
$\hat{a}$	$\frac{n^2}{2m}$	$\frac{c^* n}{m}$	$\frac{n^b}{m(1-b)\log n}$	$\frac{1}{m}(\log n)^{(b-1)}$	$\frac{1}{m}$	$\frac{C}{m \log n}$

12.6 Hierarchical Normal Models

12.6.1 Introduction

Hierarchical modeling and analysis were introduced in Section 1.5. Hierarchical normal models are the basis of much of applied Bayesian statistics, yet there is uncertainty as to which priors to use for hyperparameters (parameters at higher levels of the hierarchical model). The long history of efforts to develop objective hyperpriors, includes Morris (1983), Chaloner (1987), Berger & Strawderman (1996), Daniels & Kass (1999), Everson & Morris (2000), Daniels & Kass (2001), Berger et al. (2005), Gelman (2006), Gustafson et al. (2006), the review article Morris & Lysy (2012), and the recent Berger et al. (2020*b*), on which this section is primarily based.

Formal objective Bayesian approaches, such as the Jeffreys-rule approach or reference prior approach (Fatti (1982), Berger & Bernardo (1992*a*), Ye (1994), Ye (1998), Sun et al. (2001)) are only implementable in relatively simple hierarchical settings, such as the following:

Example 12.5 *Example 1.24 continued.* Recall that the second level model for the normal means μ_i was $\mathrm{N}(\mu_i \mid \xi, \tau^2)$. The formal reference prior approach is that discussed at the beginning of Section 12.5.1. Integrating out the μ_i at the first level model gave the marginal likelihood function for ξ and τ^2 in (1.8) (recalling that σ^2 is here assumed to be known)

$$p(\overline{\boldsymbol{x}} \mid \xi, \tau^2) \propto \frac{1}{(\tau^2 + \sigma^2/n)^{m/2}} \exp\left\{ -\frac{m(\bar{\bar{x}} - \xi)^2 + s^2}{2(\tau^2 + \sigma^2/n)} \right\}. \tag{12.9}$$

The reference prior approach can be applied directly to this model and yields (for either ξ or τ^2 being the parameter of interest) the reference prior in (1.9), namely

$$\pi^R(\xi, \tau^2) = \frac{1}{\tau^2 + \sigma^2/n} \,. \tag{12.10}$$

Applying the formal reference prior approach (or any other formal approach) to complex hierarchical models is simply not feasible, so it has, instead, become common to use adhoc approaches, such as using formal priors from non-hierarchical models in hierarchical settings. The danger of this was illustrated in Example 2.11, however, where using the first level Jeffreys-rule prior at higher levels of the hierarchical model — *e.g.*, $\pi^J(\xi, \tau^2) = 1/\tau^2$ in the above example — results in an improper posterior distribution because of a non-integrable singularity at zero. After this problem became recognized, it became common to use a constant prior for higher level variances or covariances, but the constant prior is much too diffuse, requiring twice as many observations to obtain posterior propriety as is logically needed (see Section 12.6.7).

Berger & Strawderman (1996) and Berger et al. (2005) approached the question of choice of hyperpriors in normal hierarchical models by looking at the frequentist notion of admissibility (see Section 7.6.2 for definition) of resulting estimators. Hyperpriors that are too diffuse result in inadmissible estimators, while hyperpriors that are concentrated enough result in admissible estimators. Hyperpriors that are 'on the boundary of admissibility' thus seem to be sensible choices for objective priors, being as diffuse as possible without resulting in inadmissible procedures, as discussed in Section 7.6.2. In this section and following Berger et al. (2020*b*), we use this approach to recommend hyperpriors for any normal hierarchical model.

12.6.2 The Hierarchical Model Considered

We begin by considering the following two-stage hierarchical model. Suppose that, independently and for $i = 1, \ldots m$,

$$\boldsymbol{x}_i \,|\, \boldsymbol{\theta}_i \sim \mathrm{N}_k(\cdot \,|\, \boldsymbol{\theta}_i, \boldsymbol{\Sigma}_i) \quad \text{and} \quad \boldsymbol{\theta}_i \,|\, \boldsymbol{\beta}, \boldsymbol{V} \sim \mathrm{N}_k(\cdot \,|\, \boldsymbol{Z}_i\boldsymbol{\beta}, \boldsymbol{V})\,, \tag{12.11}$$

where

- the $\boldsymbol{x}_i$ are the $k \times 1$ observation vectors;
- the $\boldsymbol{\Sigma}_i$ are *known* $k \times k$ full rank covariance matrices;
- the $\boldsymbol{\theta}_i$ are $k \times 1$ unknown mean vectors;
- $\boldsymbol{\beta}$ is an $l \times 1$ unknown "hyper-mean" vector;
- the $\boldsymbol{Z}_i$ are $k \times l$ known matrices of covariates of full rank;
- $\boldsymbol{V}$ is an unknown $k \times k$ "hyper-covariance matrix."

The assumption of known $\boldsymbol{\Sigma}_i$ is made to enable the admissibility theory to go through. (Admissibility analysis with unknown first-level covariance matrices is very difficult.) If the $\boldsymbol{\Sigma}_i$ are not known, they either need to be estimated from data and plugged in (resulting in a partly empirical Bayes approach) or incorporated into a larger hierarchical Bayesian analysis using, say, the reference prior for a first level covariance matrix, discussed in Example 7.4.

Example 12.6 (Following the path-breaking Efron & Morris (1975):) Suppose $k = 1$ and, for $i = 1, \ldots, m$, $\hat{b}_i$ is the batting average (in baseball) of player i after one month of the season, the player having batted n_i times. Let b_i be the player's 'true' batting proficiency, as measured (say) by batting average at the end of the season. It is well known that, as long as n_i is not too small, one can make the transformations $x_i = \arcsin(\hat{b}_i)$ and $\theta_i = \arcsin(b_i)$, resulting (approximately) in $x_i \sim \mathrm{N}(x_i \,|\, \theta_i, \sigma_i^2)$ where $\sigma_i^2 = 1/[4n_i]$. The goal is to estimate all the θ_i,

Also available are the (transformed) batting averages, θ_i^o, of the players from the previous year. It is deemed reasonable to regress the current θ_i on the old θ_i^0, resulting in the second stage model $\theta_i \sim \mathrm{N}(\cdot \,|\, \beta_1 + \theta_i^o\beta_2, V)$ so, in the above notation, $\boldsymbol{\beta} = (\beta_1, \beta_2)'$, $\boldsymbol{Z}_i = (1, \theta_i^0)$, and V is real.

Example 12.7 At hospital i, the observations $\boldsymbol{x}_i = (x_{i1}, \ldots, x_{ik})'$ are the sample averages of the costs of k different medical treatments; $\boldsymbol{\theta}_i$ is the corresponding unknown vector of true mean costs of the treatments at the hospital; and $\boldsymbol{\Sigma}_i$ is the covariance matrix associated with these sample averages.

Consider the jth coordinate, θ_{ij}, of each $\boldsymbol{\theta}_i$; this refers to the cost of the jth medical treatment at the ith hospital. It is natural to model this as depending on p hospital characteristics, such as the number of patients receiving the treatment, the average severity of the condition of the patients for the treatments, the average income of the patients, etc. Denoting these characteristics for the jth treatment at the ith hospital as $\boldsymbol{z}_{ij} = (z_{ij1}, \ldots, z_{ijp})$, a reasonable model would be a regression model

$$\theta_{ij} = \boldsymbol{z}_{ij}\boldsymbol{\alpha}_j + \epsilon_{ij}, \quad \text{for } i = 1, \ldots, m; \quad j = 1, \ldots, k\,,$$

where $\boldsymbol{\alpha}_j$ is a $p \times 1$ column vector of weights (specific to each treatment but assumed to be constant across hospitals) determining the effect of hospital characteristics on the cost of treatment j, and ϵ_{ij} is normal error. There would typically be a separate regression of this form for each treatment j, and stacking these regressions vertically leads to equation (12.11), where $l = kp$,

$$\boldsymbol{z}_i = \begin{pmatrix} \boldsymbol{z}_{i1} & \mathbf{0} & \ldots & \mathbf{0} \\ \mathbf{0} & \boldsymbol{z}_{i2} & \ldots & \mathbf{0} \\ \vdots & \vdots & \ddots & \vdots \\ \mathbf{0} & \mathbf{0} & \ldots & \boldsymbol{z}_{ik} \end{pmatrix}, \qquad \boldsymbol{\beta} = \begin{pmatrix} \boldsymbol{\alpha}_1 \\ \boldsymbol{\alpha}_2 \\ \vdots \\ \boldsymbol{\alpha}_k \end{pmatrix}. \tag{12.12}$$

If it were thought that each of the treatment cost means were independent, $\boldsymbol{V}$ could be taken as diagonal, but it is far more likely that there is considerable dependence, so that completely unknown $\boldsymbol{V}$ is reasonable.

In this example, there could be a third level of hierarchial modeling,

$$\boldsymbol{\alpha}_i \sim \mathrm{N}_p(\cdot \mid \boldsymbol{\xi}, \boldsymbol{\Omega})\,. \tag{12.13}$$

Whether or not this is appropriate depends on the precise context, but we will see that the analysis could still be done for this three-level hierarchical model with the methodology proposed herein.

Remark 12.2 Often the data vectors themselves arise from linear models $\boldsymbol{y}_i \sim N_r(\cdot \mid \boldsymbol{B}_i\boldsymbol{\theta}_i, \boldsymbol{\Lambda}_i)$, where the $\boldsymbol{B}_i$ are known design matrices of covariates. In this situation, one can simply transform to the least squares estimates for $\boldsymbol{\theta}_i$,

$$\boldsymbol{x}_i = (\boldsymbol{B}_i'\boldsymbol{\Lambda}_i^{-1}\boldsymbol{B}_i)^{-1}\boldsymbol{B}_i'\boldsymbol{\Lambda}_i^{-1}\boldsymbol{y}_i\,,$$

which will be distributed as in (12.11) with covariance matrix $\boldsymbol{\Sigma}_i = (\boldsymbol{B}_i'\boldsymbol{\Lambda}_i^{-1}\boldsymbol{B}_i)^{-1}$.

12.6.3 The Recommended Prior

As in Berger et al. (2005), we utilize independent priors for the unknown hyperparameters $\boldsymbol{\beta}$ and $\boldsymbol{V}$, i.e., we assume that

$$\pi(\boldsymbol{\beta}, \boldsymbol{V}) = \pi(\boldsymbol{\beta})\pi(\boldsymbol{V})\,. \tag{12.14}$$

The purpose of this section is to justify the following recommendation for choice of $\pi(\boldsymbol{\beta})$ and $\pi(\boldsymbol{V})$:

$$\pi(\boldsymbol{\beta}) \propto \frac{1}{[1+||\boldsymbol{\beta}||^2]^{(l-1)/2}}, \quad \pi(\boldsymbol{V}) = \frac{1}{|\boldsymbol{V}|^{1-1/(2k)} \prod_{i<j}(d_i - d_j)}, \tag{12.15}$$

where $||\boldsymbol{\beta}||^2 = \boldsymbol{\beta}'\boldsymbol{\beta}$ and $d_1 > d_2 \ldots > d_k$ are the ordered eigenvalues of $\boldsymbol{V}$. For $l = 1$, the first prior is $\pi(\beta) = 1$; for $k = 1$, the second is $\pi(V) = 1\sqrt{V}$.

One of the most important properties of this prior is that it can be used at any level of a hierarchical model for the mean parameters and the covariance matrix at that level, as will be discussed in Section 12.6.8. This is not so for other objective priors that have been proposed, such as that in (1.9) which depends on the variance σ^2 at the lower level of the model (and, curiously, on the sample size n at the lower level of the model). The one other prior that can be useable at any level of the hierarchy is the constant prior, but this will be shown, in Section 12.6.7, to be seriously deficient.

12.6.4 Covariate Transformations

For the priors in (12.15) to be reasonable, one must deal with the fact that the covariates (defined by the columns of $\boldsymbol{Z}_i$) often use different units of measurement, units which can be arbitrary, and objective priors should be invariant to the choice of unit of measurement. The easiest way to achieve this is to first rescale the covariates so they are 'unitless' and then proceed with use of the priors in (12.15). Here are two common methods of such re-scaling.

Method 1 - Standard Deviation Rescaling: Define c_j to be the sample standard deviation of all non-zero entries in the jth columns of all the $\boldsymbol{Z}_i$, and rescale to obtain new covariates $\boldsymbol{z}_{ij}^* = \boldsymbol{z}_{ij}/c_j$.

Method 2 - Max Coordinate Rescaling: Define c_j to be the maximum absolute value of all entries in the jth columns of all the $\boldsymbol{Z}_i$, and rescale as above. (This puts everything on the interval from -1 to 1.)

Remark 12.3 If covariate j is rescaled by c_j, for all j, the corresponding hyper-mean is $\boldsymbol{\beta}^* = (c_1\beta_1, \ldots, c_l\beta_l)'$. Then, if the rescaled covariates are used in Bayesian analysis, the analysis will yield the posterior for $\boldsymbol{\beta}^*$, which must then be transformed back to the posterior for $\boldsymbol{\beta}$, if the latter is desired. If only the posterior distributions of the $\boldsymbol{\theta}_i$ are desired, the analysis carried out in the rescaled space is fine.

12.6.5 Computation with the Posterior

Here we consider MCMC sampling from the posterior arising from the hierarchical model and prior. The algorithm given here is from Berger et al. (2020*b*). It has proven to be reasonably efficient for dimensions up to $k = 100$.

Gibbs Sampling Draws of the $\boldsymbol{\theta}_i$. Given $\boldsymbol{\beta}$, $\boldsymbol{V}$, and $\boldsymbol{x}_i$, the conditional posterior distribution of $\boldsymbol{\theta}_i$ is the usual conjugate posterior density

$$\mathrm{N}_k(\cdot \mid (\boldsymbol{\Sigma}_i^{-1} + \boldsymbol{V}^{-1})^{-1}(\boldsymbol{\Sigma}_i^{-1}\boldsymbol{x}_i + \boldsymbol{V}^{-1}\boldsymbol{Z}_i\boldsymbol{\beta})), (\boldsymbol{\Sigma}_i^{-1} + \boldsymbol{V}^{-1})^{-1}),$$

from which Gibbs sampling draws can easily be made.

Gibbs Sampling Draws of $\boldsymbol{\beta}$. To compute with the recommended prior for $\boldsymbol{\beta}$, introduce a latent variable λ, replacing the prior distribution of $\boldsymbol{\beta}$ in (12.15) with the equivalent (easily seen by integrating out over λ)

$$\boldsymbol{\beta} \mid \lambda \sim \mathrm{N}_l(\cdot \mid \boldsymbol{0}, \lambda \boldsymbol{I}_l) \text{ and } \pi(\lambda) \propto \lambda^{-1/2} e^{-1/2\lambda}.$$

- Sample λ from its full conditional, the $\mathrm{Ig}((k-1)/2,\, (1 + ||\boldsymbol{\beta}||^2)/2)$ density. If $l = 1$, this step is not needed as the hyperprior for β is constant.
- Given λ (and $\boldsymbol{V}$ and the $\boldsymbol{\theta}_i$), Gibbs sampling of $\boldsymbol{\beta}$ can be done from its full conditional, which is

$$\mathrm{N}_k\left(\cdot \,\middle|\, \left(\frac{1}{\lambda}\boldsymbol{I} + \sum_{i=1}^m \boldsymbol{Z}_i'\boldsymbol{V}^{-1}\boldsymbol{Z}_i\right)^{-1} \sum_{i=1}^m \boldsymbol{Z}_i'\boldsymbol{V}^{-1}\boldsymbol{\theta}_i,\ \left(\frac{1}{\lambda}\boldsymbol{I} + \sum_{i=1}^m \boldsymbol{Z}_i'\boldsymbol{V}^{-1}\boldsymbol{Z}_i\right)^{-1}\right).$$

  When $l = 1$, set $\lambda = \infty$ above.

Gibbs Sampling Draws of V. Defining $r = \frac{m}{2} + 1 - \frac{1}{2k}$, the conditional posterior density of $\boldsymbol{V}$, for given $(\boldsymbol{\theta}_1, \cdots, \boldsymbol{\theta}_m, \boldsymbol{\beta})$, can be written

$$\pi(\boldsymbol{V} \mid \boldsymbol{\theta}_1, \cdots, \boldsymbol{\theta}_m, \boldsymbol{\beta}) \propto \frac{1}{|\boldsymbol{V}|^r \prod_{i<j}(d_i - d_j)} \mathrm{etr}\left\{ -\frac{1}{2}\boldsymbol{V}^{-1}\boldsymbol{W} \right\}, \tag{12.16}$$

where $\mathrm{etr}(A)$ is short for $\exp(\mathrm{tr}(\boldsymbol{A}))$ for a square matrix $\boldsymbol{A}$, and

$$\boldsymbol{W} = \boldsymbol{W}(\boldsymbol{\theta}, \boldsymbol{\beta}) = \sum_{i=1}^m (\boldsymbol{\theta}_i - \boldsymbol{Z}_i'\boldsymbol{\beta})(\boldsymbol{\theta}_i - \boldsymbol{Z}_i'\boldsymbol{\beta})'. \tag{12.17}$$

Utilize the eigenvalue-vector decomposition $\boldsymbol{V} = \boldsymbol{ODO}'$, where $\boldsymbol{O}$ is orthogonal and $\boldsymbol{D}$ is the diagonal matrix of ordered eigenvalues. For $\boldsymbol{W}$ defined in (12.17), note that $\mathrm{tr}(\boldsymbol{V}^{-1}\boldsymbol{W}) = \mathrm{tr}(\boldsymbol{OD}^{-1}\boldsymbol{O}'\boldsymbol{W}) = \mathrm{tr}(\boldsymbol{D}^{-1}\boldsymbol{O}'\boldsymbol{WO})$, so the conditional density of $\boldsymbol{V}$, given in (12.16), can be transformed to

$$\pi(\boldsymbol{D}, \boldsymbol{O} \mid \boldsymbol{\theta}, \boldsymbol{\beta}) \propto \frac{1}{|\boldsymbol{D}|^r} \mathrm{etr}\left\{ -\frac{1}{2}\boldsymbol{D}^{-1}\boldsymbol{O}'\boldsymbol{WO} \right\} 1_{\{d_1 > \cdots > d_k\}}. \tag{12.18}$$

Gibbs Sampling of $\boldsymbol{D}$. Following Lemma 1 in Berger et al. (2005), i.e.,

$$\int\!\!\int g(\boldsymbol{ODO}') 1_{\{d_1 > \cdots > d_k\}} d\boldsymbol{D} d\boldsymbol{O} = \frac{1}{k!} \int\!\!\int g(\boldsymbol{ODO}') d\boldsymbol{D} d\boldsymbol{O},$$

we can first sample $\boldsymbol{D}$, given $(\boldsymbol{O}, \boldsymbol{W})$, from

$$\pi(\boldsymbol{D} \mid \boldsymbol{O}, \boldsymbol{\theta}, \boldsymbol{\beta}) \propto \frac{1}{\prod_{i=1}^k d_i^r} \mathrm{etr}\Big(-\frac{1}{2}\boldsymbol{D}^{-1}\boldsymbol{O}'\boldsymbol{W}\boldsymbol{O}\Big) = \prod_{i=1}^k \frac{1}{d_i^r} \exp\Big(-\frac{c_i}{d_i}\Big),$$

where c_i is the (i,i) element of $\boldsymbol{O}'\boldsymbol{W}\boldsymbol{O}/2$. Thus we can directly sample d_i independently from $\mathrm{Ig}(r-1, c_i)$ distributions, and then simply rearrange the d_i, so that $d_1 \geq \cdots \geq d_k$.

Gibbs Sampling of $\boldsymbol{O}$. From (12.18), the conditional density of $\boldsymbol{O}$ given $\boldsymbol{D}$ and $\boldsymbol{W}$ is

$$\pi(\boldsymbol{O} \mid \boldsymbol{D}, \boldsymbol{\theta}, \boldsymbol{\beta}) \propto \mathrm{etr}\Big(-\frac{1}{2}\boldsymbol{W}\boldsymbol{O}\boldsymbol{D}^{-1}\boldsymbol{O}'\Big). \tag{12.19}$$

Write $\boldsymbol{W} = \boldsymbol{G}\boldsymbol{U}\boldsymbol{G}'$, where $\boldsymbol{G}$ is the matrix of normalized eigenvectors ($\boldsymbol{G}\boldsymbol{G}' = \boldsymbol{I}_k$), and $\boldsymbol{U} = \mathrm{diag}(u_1, \ldots, u_k)$ is the diagonal matrix of corresponding eigenvalues with $u_1 \geq \cdots \geq u_k$. Denote $\boldsymbol{T} = \boldsymbol{G}'\boldsymbol{O}$. Then the conditional density of $\boldsymbol{T}$ given $\boldsymbol{D}$ and $\boldsymbol{W}$ is

$$\pi(\boldsymbol{T} \mid \boldsymbol{D}, \boldsymbol{\theta}, \boldsymbol{\beta}) \propto \mathrm{etr}\Big(-\frac{1}{2}\boldsymbol{U}\boldsymbol{T}\boldsymbol{D}^{-1}\boldsymbol{T}'\Big). \tag{12.20}$$

Hoff (2009) introduced an MCMC algorithm to sample from the posterior of $\boldsymbol{O}$ by updating two randomly selected columns of $\boldsymbol{O}$. Berger et al. (2020*b*) suggested updating two randomly selected rows of $\boldsymbol{T}$ (essentially equivalent to Hoff's method in full rank cases, but considerably faster in situations of less than full rank). For example, to update the first two rows of $\boldsymbol{T}$, we write $\boldsymbol{T} = \mathrm{diag}(\boldsymbol{\Omega}, \boldsymbol{I}_{k-2})(\boldsymbol{T}_{12}', \boldsymbol{T}_{-12}')'$, where $\boldsymbol{T}_{12}$ is the first 2 rows of $\boldsymbol{T}$, $\boldsymbol{T}_{-12}$ is the remaining $k-2$ rows of $\boldsymbol{T}$, and

$$\boldsymbol{\Omega} = \boldsymbol{D}_\epsilon \boldsymbol{\Omega}_0 = \begin{pmatrix} \epsilon_1 & 0 \\ 0 & \epsilon_2 \end{pmatrix} \begin{pmatrix} \cos\omega & -\sin\omega \\ \sin\omega & \cos\omega \end{pmatrix}.$$

Here $\omega \in (-\frac{\pi}{2}, \frac{\pi}{2}]$ and $\epsilon_i = \pm 1$ for $i = 1, 2$. Write $\boldsymbol{U}_1 = \mathrm{diag}(u_1, u_2)$ and $\boldsymbol{U}_2 = \mathrm{diag}(u_3, \ldots, u_k)$. The conditional posterior density of ω can be shown to be

$$\pi(\omega \mid \boldsymbol{T}_{12}, \boldsymbol{T}_{-12}, \boldsymbol{D}, \boldsymbol{W}) \propto \mathrm{etr}\Big\{-\frac{1}{2}\boldsymbol{U}_1\boldsymbol{\Omega}_0\boldsymbol{T}_{12}\boldsymbol{D}^{-1}\boldsymbol{T}_{12}'\boldsymbol{\Omega}_0'\Big\}. \tag{12.21}$$

Write

$$\boldsymbol{T}_{12}\boldsymbol{D}^{-1}\boldsymbol{T}_{12}' = \begin{pmatrix} \cos\phi & -\sin\phi \\ \sin\phi & \cos\phi \end{pmatrix} \begin{pmatrix} s_1 & 0 \\ 0 & s_2 \end{pmatrix} \begin{pmatrix} \cos\phi & \sin\phi \\ -\sin\phi & \cos\phi \end{pmatrix},$$

where $\phi \in (-\pi/2, \pi/2]$, and $s_1 > s_2$. From (12.21), the conditional posterior of θ is then

$$\pi(\theta \mid \boldsymbol{T}_{-12}, \boldsymbol{D}; \boldsymbol{\theta}, \boldsymbol{\beta}) \propto \exp\{c_0 \cos^2(\omega + \phi)\}, \theta \in (-\pi/2, \pi/2], \tag{12.22}$$

where $c_0 = -\frac{1}{2}(s_1 - s_2)(u_1 - u_2) \leq 0$. Let $\alpha = \cos^2(\omega + \phi)$. Thus the full conditional density of α is

$$\pi(\alpha \mid \boldsymbol{T}_{-12}, \boldsymbol{D}, \boldsymbol{\theta}, \boldsymbol{\beta}) \;\propto\; e^{c_0\alpha}\alpha^{-\frac{1}{2}}(1-\alpha)^{-\frac{1}{2}}, \quad \alpha \in [0,1]. \tag{12.23}$$

Sampling $\alpha \in [0,1]$ can be done by a rejection sampler with a $\mathrm{Be}(\cdot \mid \frac{1}{2}, \frac{1}{2})$ proposal distribution.

12.6.6 Decision-theoretic Justification for the Prior

Suppose we pose the problem as that of estimating $\boldsymbol{\theta}$ by its posterior mean, $\boldsymbol{\delta}^{\pi}(\boldsymbol{x})$, under the quadratic loss

$$L(\boldsymbol{\theta}, \boldsymbol{\delta}^{\pi}) = (\boldsymbol{\theta} - \boldsymbol{\delta}^{\pi})'\boldsymbol{Q}(\boldsymbol{\theta} - \boldsymbol{\delta}^{\pi}), \tag{12.24}$$

where $\boldsymbol{Q}$ is a known positive-definite matrix, with the performance of an estimator being evaluated by the frequentist risk function (integrating over $\boldsymbol{x}$)

$$R(\boldsymbol{\theta}, \boldsymbol{\delta}) = \mathrm{E}_{\boldsymbol{\theta}}[L(\boldsymbol{\theta}, \boldsymbol{\delta}(\boldsymbol{x}))]. \tag{12.25}$$

Theorem 12.1 *For the recommended priors, if $m \geq 2$ and $k \geq 2$, or $m \geq 3$ and $k = 1$, the posterior mean of $\boldsymbol{\theta}$ is admissible and, indeed, is on the 'boundary of admissibility' (see Section 7.6.2 for definitions).*

Proof. For $k \geq 2$ and $m \geq 2$, this was proved in Berger et al. (2020*b*), for the case of no covariates $\boldsymbol{z}_i$ and $l = k$. It was asserted therein that a similar proof would go through in general.

For $k = 1$ and the case of no covariates ($l = 1$), this was proved for $m \geq 3$ in Berger & Strawderman (1996), but the result for $l \geq 2$ was stated therein to be undetermined. We complete the proof for that case here.

On p. 946 of Berger & Strawderman (1996), the last inequality at the bottom of the page sacrificed too much to be able to prove admissibility for our prior, which in the notation of the paper corresponded to $a = 1/2$ and $b = 1/2$. An alternative bound that works for our prior is

$$\begin{aligned} I_4 &\leq C\int_c^{\infty}\int_{c/(1+\sigma_\pi^2)}^{\infty} \frac{\exp\{-C'r^2(1+\sigma_\pi^2)^{-1}[w+(1-w)/(1+v)]\}}{(1+\sigma_\pi^2)^{(p-1)/2}\,(1+v)^{k/2}\,[1+v(1+\sigma_\pi^2)]^{1/2}}\,dv\,d\sigma_\pi^2 \\ &\leq C\int_0^{\infty}\int_0^{\infty} \frac{\exp\{-C'r^2(1+\sigma_\pi^2)^{-1}[w+(1-w)/(1+v)]\}}{(1+\sigma_\pi^2)^{p/2}\,(1+v)^{k/2}\,\sqrt{v}}\,dv\,d\sigma_\pi^2 \,. \end{aligned}$$

Changing variables from σ_π^2 to $y = r^2(1+\sigma_\pi^2)^{-1}[w+(1-w)/(1+v)]$ results in

$$\begin{aligned} I_4 &\leq C\int_0^{\infty}\int_0^{\infty} \frac{\exp\{-C'y\}\,y^{(p-4)/2}}{r^{(p-2)}[w+(1-w)/(1+v)]^{(p-2)/2}(1+v)^{k/2}\sqrt{v}}\,dv\,dy \\ &= \frac{C}{r^{(p-2)}}\int_0^{\infty} \exp\{-C'y\}\,y^{(p-4)/2}\,dy \int_0^{\infty} \frac{(1+v)^{(p-k-2)/2}}{(1+vw)^{(p-2)/2}\sqrt{v}}\,dv \,. \end{aligned}$$

The integral over y is clearly finite if $p \geq 3$ (or $m \geq 3$ in the notation of this section). From Berger & Strawderman (1996), w is a $\mathrm{Be}(\cdot \mid \frac{p-k}{2}, \frac{k}{2})$ random variable, so the expectation of the integral involving v is proportional to

$$\int_0^1 \int_0^\infty \frac{w^{(p-k-2)/2}(1-w)^{(k-2)/2}(1+v)^{(p-k-2)/2}}{(1+vw)^{(p-2)/2}\sqrt{v}} \, dv \, dw \, .$$

Changing variables from v to $z = vw$ results in

$$\int_0^1 \int_0^\infty \frac{(1-w)^{(k-2)/2}(w+z)^{(p-k-2)/2}}{(1+z)^{(p-2)/2}\sqrt{zv}} \, dz \, dw \leq \int_0^1 \int_0^\infty \frac{(1-w)^{(k-2)/2}}{(1+z)^{k/2}\sqrt{zv}} \, dz \, dw \, ,$$

which is clearly finite if $k \geq 2$. (In the notation of this section, that means $l \geq 2$.) Thus we have shown that $I_4 \leq Cr^{-(p-2)}$, which is sufficient to complete the proof of admissibility in Berger & Strawderman (1996). □

12.6.7 Performance of the Recommended Prior

In this section, we compare the performance, from the mean square error (MSE) perspective, of 4 objective hyperpriors created from the product of two objective hyperpriors for $\boldsymbol{\beta}$,

- the constant prior $\pi_1(\boldsymbol{\beta}) = 1$;
- and the recommended prior $\pi_2(\boldsymbol{\beta}) \propto [1 + ||\boldsymbol{\beta}||^2]^{-\frac{(l-1)}{2}}$;

and two objective hyperpriors for $\boldsymbol{V}$,

- the constant prior $\pi_1(\boldsymbol{V}) = 1$;
- and the recommended prior $\pi_2(\boldsymbol{V}) = [|\boldsymbol{V}|^{1-1/(2k)} \prod_{i<j}(d_i - d_j)]^{-1}$.

The recommended prior (12.15) is the prior corresponding to $\pi_2(\boldsymbol{\beta})\pi_2(\boldsymbol{V})$. The commonly used prior today is the constant prior $\pi_1(\boldsymbol{\beta})\pi_1(\boldsymbol{V})$.

Posterior Propriety. The first interesting evaluation of these priors arises from Berger et al. (2005), which discusses the sample sizes needed to have well-defined posterior means; this result is reproduced in Table 12.2.

There are two surprising results from this table. The first is that roughly $m = 2k$ observations are needed for the posterior to be proper when $\pi_1(\boldsymbol{V}) = 1$ is used. For a covariance matrix to be identifiable in classical terms, it is known that roughly $m = k$ observations are needed. That an extra k observations are needed here to make posterior proper is a strong indication that the constant prior is much too diffuse; the first k observations are effectively wasted to correct for the over-diffuse tails of the prior. The second surprise is that only one (vector) observation is needed for the recommended priors to yield a proper posterior; the reason for this is discussed in Berger et al. (2005). Note that, even though the posterior is always proper, it will be extremely vague for most features of $\boldsymbol{V}$ for small m. (There is a suggestion, in Berger et al. (2020*b*), that the posterior for some features of $\boldsymbol{V}$ will be reasonable for small m.)

Table 12.2: Sample sizes, m, needed for the posterior to be proper for the 4 priors, $\pi_i(\boldsymbol{\beta}) \times \pi_j(\boldsymbol{V})$, under consideration. Recall that $\boldsymbol{V}$ is a $k \times k$ matrix.

Prior for $\boldsymbol{\beta}$	Prior for $\boldsymbol{V}$	
	$\pi_1(\boldsymbol{V})$	$\pi_2(\boldsymbol{V})$
$\pi_1(\boldsymbol{\beta})$	$m \geq 2k+2$	$m \geq 3$
$\pi_2(\boldsymbol{\beta})$	$m \geq 2k+1$	$m \geq 1$

Risks of Estimators from the Four Priors. We report (from Berger et al. 2020*b*) the Bayes risks of the posterior means resulting from the 4 priors, in every combination of the following cases:

- $(k_1 = 4, m_1 = 10)$ or $(k_2 = 5, m_2 = 15)$;
- $\boldsymbol{\beta}_1 = \mathbf{1}_k$ or $\boldsymbol{\beta}_2 = 50\mathbf{1}_k$;
- $\boldsymbol{V}_1 = \boldsymbol{I}_k$ or $\boldsymbol{V}_2 = \text{diag}\{8k-7, \ldots, 9, 1\}$.

The choices of $\boldsymbol{\beta}$ and $\boldsymbol{V}$ are to represent 'normal' and 'extreme' hyperparameters, respectively. The moderate choices of k were necessitated because, in the study in Berger et al. (2020*b*), a number of other priors and computational methods were considered that were computationally very inefficient.

Each choice of $(k, m, \boldsymbol{\beta}, \boldsymbol{V})$ specifies a 'true' hierarchical model for the simulation, and the Bayes risks, corresponding to the posterior means for $\boldsymbol{\theta} = (\boldsymbol{\theta}_1', \ldots, \boldsymbol{\theta}_m')'$, given the data $\boldsymbol{x} = (\boldsymbol{x}_1', \ldots, \boldsymbol{x}_m')'$, arising from each of the 4 objective priors, were computed. To simulate these risks, 3000 random $\boldsymbol{\theta}^{(t)} \sim \text{N}_{mk}(\cdot \mid \mathbf{1}_m \otimes \boldsymbol{\beta}, \boldsymbol{I}_m \otimes \boldsymbol{V})$ were generated; from each, $\boldsymbol{x}^{(t)} \sim \text{N}_{mk}(\cdot \mid \boldsymbol{\theta}^{(t)}, \boldsymbol{I}_{mk})$ was drawn. To obtain the posterior means, $\boldsymbol{\delta}^{\pi}(\boldsymbol{x}^{(t)})$, of $\boldsymbol{\theta}^{(t)}$, under the 4 priors, 150,000 MCMC cycles were run after 5,000 burn-in cycles, using the algorithms described above. The Bayes risks were approximated by the average observed mean squared error (MSE)

$$MSE = \frac{1}{3000} \sum_{l=1}^{3000} (\boldsymbol{\theta}^{(t)} - \boldsymbol{\delta}^{\pi}(\boldsymbol{x}^{(t)})'(\boldsymbol{\theta}^{(t)} - \boldsymbol{\delta}^{\pi}(\boldsymbol{x}^{(t)})\,. \tag{12.26}$$

Finally, the *per parameter* MSE, $MSE/(k_i m_i)$, was reported, since there are a total of $k_i m_i$ mean parameters being estimated.

The results are given in Table 12.3. The recommended prior (π_{22}) dominates the others in terms of risk. Using the recommended prior for $\boldsymbol{\beta}$ instead of the constant prior results in only modest gains. However, using the recommended prior for $\boldsymbol{V}$, instead of the constant prior, achieves substantial gains.

Table 12.3: The MSE, in estimating $\boldsymbol{\theta}$, under combinations of $(k_1 = 4, m_1 = 10)$, $(k_2 = 5, m_2 = 15)$; $\boldsymbol{\beta}_1 = \mathbf{1}_k$, $\boldsymbol{\beta}_2 = 50\mathbf{1}_k$; $\boldsymbol{V}_1 = \boldsymbol{I}_k$, $\boldsymbol{V}_2 = diag\{8k-7,\ldots,9,1\}$; using prior π_{ij}, where
For $\boldsymbol{\beta}$: $i = 1$ is the constant prior; $i = 2$ is the recommended prior;
For $\boldsymbol{V}$: $j = 1$ is the constant prior; $j = 2$ is the recommended prior.

	$k_1\beta_1V_1$	$k_1\beta_2V_1$	$k_1\beta_1V_2$	$k_1\beta_2V_2$	$k_2\beta_1V_1$	$k_2\beta_2V_1$	$k_2\beta_1V_2$	$k_2\beta_2V_2$
π_{11}	31.810	33.058	39.329	40.214	53.606	55.379	72.759	73.205
π_{12}	25.082	27.844	36.632	37.344	42.856	45.046	69.321	70.858
π_{21}	30.183	33.048	37.093	38.144	52.227	54.367	68.550	70.204
π_{22}	22.971	24.246	34.199	35.336	42.048	42.195	66.732	68.811

12.6.8 Posterior Propriety and Use for General Hierarchical Models

A primary advantage of the recommended prior, over many other proposed objective priors for hierarchical models, is that it can be used at any level of a hierarchical model. This is clear functionally, as the prior only depends on the parameters at the current level of modeling, and not on lower levels such as the reference prior in (12.9). This is also clear computationally, since the Gibbs sampling steps discussed in Section 12.6.5 will apply to any level of the model. But the properties of the resulting posterior are not clear; in particular it is not obvious that the resulting posterior is proper for any hierarchical model. For a canonical 3-level hierarchical model, posterior propriety, for the recommended priors, was demonstrated in Berger et al. (2020*b*). We suspect that propriety holds in general for normal hierarchical models, but the proof is elusive.

12.7 Discussion and Additional References

The arithmetic averaging approach to obtaining an overall prior seemed highly questionable, while the geometric averaging approach seems to be fine; but there has been no systematic study of this.

The reference distance approach to developing an overall prior is natural, and seems to work well when the reference priors themselves are proper. It also appears to be possible to implement the approach in the case where the reference priors are improper, by operating on suitable large compact sets and showing that the result is not sensitive to the choice of compact set.

The hierarchical reference approach seems excellent (as usual), and can certainly be recommended if one can find a natural hierarchical structure based on a class of proper priors. In particular, the hierarchical reference prior for the contingency table scenario seemed to have extremely interesting adaptations to sparsity.

The suggested priors for *any* normal hierarchical model, while not strictly reference priors, seem to have excellent properties and are computationally feasible.

Chapter 13

Reference Priors with Partial Information

13.1 Introduction

Sometimes one has partial information about $\boldsymbol{\theta}$. For instance, one might know that a normal mean μ is positive. Determining the reference prior, under constraints on the parameter space, is discussed in Section 13.2.

A second type of partial information is knowing some features of the prior distribution, such as some moments or some quantiles. Determining the reference prior, under such constraints, is discussed in Section 13.3.

Sometimes one has partial information about $\boldsymbol{\theta}$ that corresponds to structural knowledge of the prior distribution. For instance, with a $\mathrm{N}(\cdot \mid \mu, \sigma^2)$ model, one might possess strong prior information about μ, but little or no prior information about σ^2. Lavine (1991) considered robust Bayesian inference with specified prior marginals. Sun & Berger (1998) considered the problem of finding reference priors when marginal or conditional prior densities are known. More specifically, suppose $(\boldsymbol{\theta}, \boldsymbol{\lambda})$ is the parameter, and consider the following three situations.

1. Suppose the conditional prior density $\pi^s(\boldsymbol{\lambda} \mid \boldsymbol{\theta})$, for $\boldsymbol{\lambda}$ given $\boldsymbol{\theta}$, is specified and it is desired to complete the prior specification by finding a marginal reference prior $\pi^R(\boldsymbol{\theta})$ for $\boldsymbol{\theta}$. This is discussed in Section 13.4.

2. Suppose the marginal prior density $\pi^s(\boldsymbol{\theta})$, for $\boldsymbol{\theta}$, is specified and it is desired to complete the prior specification by finding a conditional reference prior $\pi^R(\boldsymbol{\lambda} \mid \boldsymbol{\theta})$, for $\boldsymbol{\lambda}$ given $\boldsymbol{\theta}$. This is discussed in Section 13.5.

3. Suppose it is known apriori that $\boldsymbol{\theta}$ and $\boldsymbol{\lambda}$ are independent, and it is desired to find the reference prior subject to this constraint. This is discussed in Section 13.6.

13.2 Constrained Parameter Space

In Section 1.2.8, it was observed how simple it is to deal with a constrained parameter space $\boldsymbol{\Theta}_0$, if one has a prior distribution $\pi(\boldsymbol{\theta})$ on the unconstrained space $\boldsymbol{\Theta}$; one simply restricts $\pi(\boldsymbol{\theta})$ to $\boldsymbol{\Theta}_0$ and renormalizes (if the prior is proper). Thus it would seem that, if one develops a reference prior $\pi^R(\boldsymbol{\theta})$ on $\boldsymbol{\Theta}$, then the reference prior on $\boldsymbol{\Theta}_0$ would simply be the restricted reference prior, *i.e.*, the prior proportional to $\pi^R(\boldsymbol{\theta})1_{\boldsymbol{\Theta}_0}(\boldsymbol{\theta})$. This is established in Proposition 13.1 for the single parameter case. (It also follows from the explicit form for the reference prior given in Proposition 9.7, since the definition applies simultaneously to all compact subsets.)

Proposition 13.1 *Consider the model* $\mathcal{M} \equiv \{p(\boldsymbol{x}\,|\,\theta), \boldsymbol{x} \in \mathcal{X}, \theta \in \Theta \subset \mathbb{R}\}$, *and suppose that* Θ_0 *is an open subset of* Θ. *Under the assumptions of Proposition 9.7, the reference prior which corresponds to the model* $\mathcal{M}_0 \equiv \{p(\boldsymbol{x}\,|\,\theta), \boldsymbol{x} \in \mathcal{X}, \theta \in \Theta_0 \subset \mathbb{R}\}$ *is proportional to the restriction to* Θ_0 *of the reference prior which corresponds to* $\mathcal{M}$.

Proof. Using the notation of Proposition 9.7, the definition of posterior consistency of π^* is that for any $\theta \in \Theta$ and any $\epsilon > 0$,

$$\Pr{}^*(|\tau - \theta| \leq \epsilon\,|\,\boldsymbol{t}_k) \equiv \int_{\{\tau:|\tau-\theta|\leq\epsilon\}} \pi^*(\tau\,|\,\boldsymbol{t}_k)\,d\tau \xrightarrow{P} 1,$$

in probability $p(\boldsymbol{t}_k\,|\,\theta)$ as $k \to \infty$. It is immediate that

$$\Pr{}^*(\Theta_0\,|\,\boldsymbol{t}_k) = \frac{\int_{\Theta_0} p(\boldsymbol{t}_k\,|\,\tau)\pi^*(\tau)\,d\tau}{\int_{\Theta} p(\boldsymbol{t}_k\,|\,\tau)\pi^*(\tau)\,d\tau} \xrightarrow{P} 1$$

with probability $p(\boldsymbol{t}_k\,|\,\theta)$ as $\theta \in \Theta_0$ and $k \to \infty$. Defining f_k^0 as the function f_k but developed on Θ_0, it follows that

$$\frac{f_k^0(\theta)}{f_k(\theta)} = \exp\left\{\int_{\mathcal{X}^k} p(\boldsymbol{t}_k\,|\,\theta)\log[\Pr{}^*(\Theta_0\,|\,\boldsymbol{t}_k)]\right\} \to 1\,,$$

using the monotone convergence theorem (the integrand for each k is negative). The result is immediate. □

In multiparameter models, with, say, $\boldsymbol{\theta} = (\theta_1, \ldots, \theta_m)$ being the parameter ordering, the situation is less clear. When $\boldsymbol{\Theta}_0$ is a rectangle in $\boldsymbol{\Theta}$, it is likely that the usual asymptotic and compact set limiting argument will work, to show that the truncated full space reference prior is actually the reference prior on $\boldsymbol{\Theta}_0$. But, if $\boldsymbol{\Theta}_0$ has boundaries that involve multiple parameters (e.g. $\boldsymbol{\Theta}_0 = \{\boldsymbol{\theta} : \theta_1 < \theta_2 < \ldots < \theta_m\}$), it is possible that the $\boldsymbol{\Theta}_0$ reference prior could be different. As a practical matter, however, it is certainly reasonable to just truncate the full reference prior, in such situations.

13.3 Linear Constraints on the Prior

Linear constraints on the prior $\pi(\boldsymbol{\theta})$ are constraints such as

$$\int_{\boldsymbol{\Theta}} g_k(\boldsymbol{\theta})\pi(\boldsymbol{\theta})d\boldsymbol{\theta} = \mu_k, \quad \text{for } k = 1, \ldots, K, \tag{13.1}$$

where the $g_k(\cdot)$ are specified functions and the μ_k are specified constants. Such constraints typically only make sense for proper priors, so $\pi(\boldsymbol{\theta})$ will be assumed to be proper in this section.

Example 13.1 If θ is a real parameter, choosing $g_k(\theta) = \theta^k$ corresponds to specifying that the kth moment of the prior is μ_k.

Choosing $g_k(\theta) = 1_A(\theta)$ corresponds to specifying that the set $A \subset \Theta$ has prior probability μ_k.

A reference prior, subject to the constraints, is found by restricting the class of distributions accordingly, and then applying the reference prior approach to this restricted class. Proposition 5.26 in Bernardo & Smith (1994) shows that the reference prior, subject to the constraints in (13.1), is given by

$$\pi^{CR}(\boldsymbol{\theta}) = \frac{\exp\{\sum_{k=1}^{K} \lambda_k g_k(\boldsymbol{\theta})\}\pi^R(\boldsymbol{\theta})}{\int \exp\{\sum_{k=1}^{K} \lambda_k g_k(\boldsymbol{\theta})\}\pi^R(\boldsymbol{\theta})d\boldsymbol{\theta}}, \tag{13.2}$$

where $\pi^R(\boldsymbol{\theta})$ is the unconstrained reference prior and the λ_k are constants found by inserting this expression into (13.1) and solving the K equations.

Example 13.2 *Moment Constraints:* If $\Theta = (-\infty, \infty)$, $\pi^R(\theta) = 1$, and $K = 2$ with $g_1(\theta) = \theta$ and $g_2(\theta) = \theta^2$, then (13.2) becomes

$$\pi^{CR}(\theta) = \frac{\exp\{\lambda_1\theta + \lambda_2\theta^2\}}{\int_{-\infty}^{\infty} \exp\{\lambda_1\theta + \lambda_2\theta^2\}d\theta} = \frac{\exp\{\lambda_2(\theta + \frac{\lambda_1}{2\lambda_2})^2\}}{\int_{-\infty}^{\infty} \exp\{\lambda_2(\theta + \frac{\lambda_1}{2\lambda_2})^2\}d\theta},$$

the last equality following since the constants $\exp\{-\lambda_1^2/[4\lambda_2]\}$ cancel in the numerator and denominator. This can be recognized as a normal distribution with mean $\mu_1 = -\lambda_1/[2\lambda_2]$ and variance $\sigma^2 = -1/[2\lambda_2]$. Since $\sigma^2 = (\mu_2 - \mu_1^2)$ (the variance equals the second moment minus the square of the first moment), it follows that $\pi^{CR}(\theta)$ is the $N(\theta \mid \mu_1, (\mu_2 - \mu_1^2))$ distribution. Note that not all values of the μ_k are allowed; here, for instance, μ_2 must be larger than μ_1^2.

Example 13.3 Consider the same situation but with $K = 1$, so that only μ_1 is specified. Then (13.2) becomes $\pi^{CR}(\theta) \propto \exp\{\lambda_1\theta\}$, but this is not a proper distribution on $(-\infty, \infty)$ for any λ_1, so the constrained reference prior does not exist.

The constrained reference prior would exist here if the parameter space were $(0, \infty)$, then being the exponential distribution with mean μ_1. But

$\pi^R(\theta) = 1$ is often not the reference prior when the parameter space is $(0, \infty)$. For instance, if this arose in a scale-parameter problem, the unconstrained reference prior would be $\pi^R(\theta) = 1/\theta$, but (13.2) then becomes

$$\pi^{CR}(\theta) \propto \exp\{\lambda_1 \theta\}/\theta\,, \tag{13.3}$$

which cannot be a proper distribution. If θ were a Poisson mean, on the other hand, the unconstrained reference prior is $\pi^R(\theta) = 1/\sqrt{\theta}$, and the constrained reference prior would be proper; indeed it is the $\mathrm{Ga}(\frac{1}{2}, \frac{1}{2\mu_1})$ distribution.

Interestingly, for a scale-parameter problem and $\pi^R(\theta) = 1/\theta$, (13.2) cannot be a proper prior no matter how many moments are specified, because the prior has a non-integrable singularity at 0. To understand the problem, consider the $K = 1$ case and consider a 'sequence of compact sets' argument, choosing the compact sets to be (ϵ, ∞), with $\epsilon \to 0$. Then (13.3) is proper on each compact set and can be shown to be

$$\pi_\epsilon^{CR}(\theta) \cong \frac{1}{\theta \log \epsilon^{-1}} \exp\left\{-\frac{\theta}{\mu_1 \log \epsilon^{-1}}\right\} 1_{(\epsilon,\infty)}(\theta)\,. \tag{13.4}$$

The problem is that this does not converge to a proper distribution as $\epsilon \to 0$, so that a constrained reference prior cannot be defined. Any of the $\pi_\epsilon^{CR}(\theta)$, for small ϵ, could be called an approximate constrained reference prior, however, and they would typically yield essentially the same answer in posterior analysis.

Example 13.4 *Quantile Constraints:* Suppose $\Theta = (\theta_{min}, \theta_{max})$ is a bounded interval and $g_k(\theta) = 1_{(a_k, a_{k+1})}$, with $\theta_{min} = a_1 < a_2 < \ldots < a_{K+1} = \theta_{max}$. The linear constraints are then

$$\int_{a_k}^{a_{k+1}} \pi(\theta) d\theta = \mu_k\,.$$

Since the intervals (a_k, a_{k+1}) are disjoint, it is easy to see that the constrained reference prior (13.2) can be alternatively expressed as

$$\pi^{CR}(\theta) = \frac{\pi^R(\theta) \sum_{k=1}^K c_k 1_{(a_k, a_{k+1})}(\theta)}{\int_{a_1}^{a_{K+1}} \pi^R(\theta) \sum_{k=1}^K c_k 1_{(a_k, a_{k+1})}(\theta) d\theta}\,,$$

with the constants, c_k, chosen to satisfy the quantile constraints.

If $\pi^R(\theta) = 1$, the constrained reference prior is thus just a step function, with the step heights being $c_k = \mu_k/(a_{k+1} - a_k)$ for $k = 1, \ldots, K$. Note that, if Θ is unbounded, the constrained reference prior will not exist, unless one specifies an infinite number of quantile constraints.

13.4 A Known Conditional Prior

When the conditional density of $\boldsymbol{\lambda}$ given $\boldsymbol{\theta}$, $\pi^s(\boldsymbol{\lambda} \mid \boldsymbol{\theta})$, is known, there are two reasonable options for finding a marginal reference prior, $\pi^R(\boldsymbol{\theta})$.

Option 1. Implementing the base reference prior method. If $\pi^s(\boldsymbol{\lambda} \mid \boldsymbol{\theta})$ is proper, the marginal density of $\boldsymbol{x}$ is

$$p(\boldsymbol{x} \mid \boldsymbol{\theta}) = \int p(\boldsymbol{x} \mid \boldsymbol{\theta}, \boldsymbol{\lambda}) \pi^s(\boldsymbol{\lambda} \mid \boldsymbol{\theta}) d\boldsymbol{\lambda}. \tag{13.5}$$

One can then find the reference prior, $\pi^R(\boldsymbol{\theta})$, for this marginal model and use, as the overall prior,

$$\pi^R(\boldsymbol{\theta}, \boldsymbol{\lambda}) = \pi^s(\boldsymbol{\lambda} \mid \boldsymbol{\theta}) \pi^R(\boldsymbol{\theta}) .$$

Example 13.5 If θ is a real parameter and the marginal density of $\boldsymbol{x}$ is regular, the reference prior for the marginal density will just be the Jeffreys-rule prior, so that

$$\pi^R(\theta, \boldsymbol{\lambda}) = \pi^s(\boldsymbol{\lambda} \mid \theta) \sqrt{I^*(\theta)} ,$$

where $I^*(\theta)$ is the expected Fisher information of θ from the marginal model.

Unfortunately, it can be quite difficult to find the reference prior for the marginal model; even determining the Fisher information can be challenging. Hence, the asymptotic reference prior approach is typically used, instead.

Option 2. Implementing the asymptotic reference prior method. Known, but improper, $\pi^s(\boldsymbol{\lambda} \mid \boldsymbol{\theta})$ primarily arise when one can recognize the actual conditional reference prior, without having to go through the derivation. For instance, one might recognize that, given $\boldsymbol{\theta}$, λ is a scale parameter, and so $\pi^s(\lambda \mid \boldsymbol{\theta}) \propto 1/\lambda$. If it is necessary to compute the asymptotic reference prior, one applies the same process as in Section 10.2.2, except that $\boldsymbol{\theta}$ and $\boldsymbol{\lambda}$ might now be vectors and Step 1 becomes

Step 1. Find the Conditional Reference Prior of $\boldsymbol{\lambda}$ *Given* $\boldsymbol{\theta}$*:*

Case 1. If $\pi^s(\boldsymbol{\lambda} \mid \boldsymbol{\theta})$ is proper, one can proceed directly with this density.

Case 2. If $\pi^s(\boldsymbol{\lambda} \mid \boldsymbol{\theta})$ is improper, constrain $(\boldsymbol{\theta}, \boldsymbol{\lambda})$ to compact subsets $\{\boldsymbol{\Upsilon}_l,\ l = 1, 2, \ldots\}$ of the parameter space $\boldsymbol{\Upsilon}$, such that $\lim_{l \to \infty} \boldsymbol{\Upsilon}_l = \boldsymbol{\Upsilon}$, with associated projections $\{\boldsymbol{\Lambda}_l(\boldsymbol{\theta}), l = 1, 2, \ldots\}$; the constrained conditional reference prior is then

$$\pi_l(\boldsymbol{\lambda} \mid \boldsymbol{\theta}) = \frac{1}{c_l(\boldsymbol{\theta})} \pi^s(\boldsymbol{\lambda} \mid \boldsymbol{\theta}) 1_{\boldsymbol{\Lambda}_l(\boldsymbol{\theta})}(\boldsymbol{\lambda}), \quad c_l(\boldsymbol{\theta}) = \int_{\boldsymbol{\Lambda}_l(\boldsymbol{\theta})} \pi^s(\boldsymbol{\lambda} \mid \boldsymbol{\theta})\, d\boldsymbol{\lambda} . \tag{13.6}$$

One then simply follows Definition 10.1 to define the asymptotic reference prior.

Under aymptotic bi-normality, the reference prior is found, as in Proposition 10.1, replacing θ and λ by $\boldsymbol{\theta}$ and $\boldsymbol{\lambda}$, replacing $V_{11}^{-1/2}$ by $|V_{11}|^{-1/2}$, utilizing $\pi(\boldsymbol{\lambda} \mid \boldsymbol{\theta})$ directly in Case 1, if it is proper, and utilizing (13.6) in Case 2 if it is improper. The following result, from Sun & Berger (1998), gives important special cases; its proof is immediate.

Proposition 13.2 *(a) If $|V_{11}(\boldsymbol{\theta}, \boldsymbol{\lambda})|$ does not depend on $\boldsymbol{\lambda}$, then the marginal and full reference priors are*

$$\pi_1^R(\boldsymbol{\theta}) \propto |V_{11}(\boldsymbol{\theta}, \boldsymbol{\lambda})|^{-1/2}, \quad \pi^R(\boldsymbol{\theta}, \boldsymbol{\lambda}) \propto \pi^s(\boldsymbol{\lambda} \,|\, \boldsymbol{\theta}) |V_{11}(\boldsymbol{\theta}, \boldsymbol{\lambda})|^{-1/2}. \tag{13.7}$$

(b) If $|V_{11}(\boldsymbol{\theta}, \boldsymbol{\lambda})|$ depends only on $\boldsymbol{\lambda}$ and $\pi^s(\boldsymbol{\lambda} \,|\, \boldsymbol{\theta})$ does not depend on $\boldsymbol{\theta}$, then $\pi_1^R(\boldsymbol{\theta}) \propto 1$ and $\pi^R(\boldsymbol{\theta}, \boldsymbol{\lambda}) \propto \pi^s(\boldsymbol{\lambda} \,|\, \boldsymbol{\theta})$.

13.5 A Known Marginal Prior

A known marginal prior density, $\pi^s(\boldsymbol{\theta})$, will virtually always be proper, arising from either a subjective assessment or as the posterior from a previous study. There is no technical advantage obtained from propriety, however, so we will not assume it.

To derive the reference prior, we follow the algorithm given in Section 10.2.1, making changes as needed.

- Step 1 is the same, although note that $\boldsymbol{\lambda}$ and $\boldsymbol{\theta}$ could now be vectors and so $I(\lambda \,|\, \theta)^{1/2}$ would be $|\boldsymbol{I}(\boldsymbol{\lambda} \,|\, \boldsymbol{\theta})|^{1/2}$.

- Steps 2 and 3 are skipped, since their purpose was to find the marginal reference prior for $\boldsymbol{\theta}$, and we instead want to use $\pi^s(\boldsymbol{\theta})$. Interestingly, we are thus able to apply the base reference prior method, and do not have to resort to deriving the asymptotic reference prior.

- Step 4 is essentially the same.

  Case 1. If $c(\boldsymbol{\theta}) < \infty$, the joint reference prior is $\pi(\boldsymbol{\theta}, \boldsymbol{\lambda}) = \pi(\boldsymbol{\lambda} \,|\, \boldsymbol{\theta}) \pi^s(\boldsymbol{\theta})$.
  Case 2. If $c(\boldsymbol{\theta})$ is infinite, the joint reference prior should be computed as

$$\pi(\boldsymbol{\theta}, \boldsymbol{\lambda}) = \lim_{l \to \infty} \frac{\pi_l(\boldsymbol{\lambda} \,|\, \boldsymbol{\theta}) \pi_l^s(\boldsymbol{\theta})}{\pi_l(\boldsymbol{\lambda}_0 \,|\, \boldsymbol{\theta}_0) \pi_l^s(\boldsymbol{\theta}_0)}, \tag{13.8}$$

  where $\pi_l^s(\boldsymbol{\theta})$ is the restriction of $\pi^s(\boldsymbol{\theta})$ to $\boldsymbol{\Upsilon}_l$ and $(\boldsymbol{\theta}_0, \boldsymbol{\lambda}_0)$ is any interior point.

The following theorem, from Sun & Berger (1998), gives a special case under asymptotic normality.

Proposition 13.3 *Assume asymptotic normality of $\boldsymbol{\lambda}$ given $\boldsymbol{\theta}$ and that*

$$|\boldsymbol{I}_{22}(\boldsymbol{\theta}, \boldsymbol{\lambda})| = g_1(\boldsymbol{\theta}) g_2(\boldsymbol{\lambda}), \tag{13.9}$$

for some functions g_1 and g_2. Suppose $g_2(\boldsymbol{\lambda})$ has finite mass or, otherwise, that the compact sets are rectangles in the overall parameter space. Then the reference prior is $\pi^R(\boldsymbol{\theta}, \boldsymbol{\lambda}) \propto [g_2(\boldsymbol{\lambda})]^{\frac{1}{2}} \pi^s(\boldsymbol{\theta})$.

13.6 Knowing Independence of Parameters

Suppose the available information is only that $\boldsymbol{\theta}$ and $\boldsymbol{\lambda}$ are independent, so we restrict consideration to products of priors $\pi_1(\boldsymbol{\theta})\pi_2(\boldsymbol{\lambda})$. If the overall reference prior $\pi^R(\boldsymbol{\theta}, \boldsymbol{\lambda})$ is proper, one can simply find the marginal priors whose product most closely approximates the reference prior.

Proposition 13.4 *If $\pi^R(\boldsymbol{\theta}, \boldsymbol{\lambda})$ is proper, the closest independent prior, in terms of K-L divergence, is $\pi_1^R(\boldsymbol{\theta})\pi_2^R(\boldsymbol{\lambda})$, the product of the marginal densities of $\pi^R(\boldsymbol{\theta}, \boldsymbol{\lambda})$.*

Proof. Simplify notation by writing the independence prior as $f(\boldsymbol{\theta})g(\boldsymbol{\lambda})$. The K-L divergence of this relative to $\pi^R(\boldsymbol{\theta}, \boldsymbol{\lambda})$ is, defining $C = \int \pi^R(\boldsymbol{\theta}, \boldsymbol{\lambda}) \log \pi^R(\boldsymbol{\theta}, \boldsymbol{\lambda})\, d\boldsymbol{\theta} d\boldsymbol{\lambda}$,

$$\kappa\{f(\boldsymbol{\theta})g(\boldsymbol{\lambda}) \mid \pi^R(\boldsymbol{\theta}, \boldsymbol{\lambda})\} = \int \pi^R(\boldsymbol{\theta}, \boldsymbol{\lambda}) \log \left[\frac{\pi^R(\boldsymbol{\theta}, \boldsymbol{\lambda})}{f(\boldsymbol{\theta})g(\boldsymbol{\lambda})}\right] d\boldsymbol{\theta} d\boldsymbol{\lambda}$$

$$= C - \int \pi^R(\boldsymbol{\theta}, \boldsymbol{\lambda})[\log f(\boldsymbol{\theta}) + \log g(\boldsymbol{\lambda})]\, d\boldsymbol{\theta} d\boldsymbol{\lambda}$$

$$= C^* + \int \pi_1^R(\boldsymbol{\theta}) \log \left[\frac{\pi_1^R(\boldsymbol{\theta})}{f(\boldsymbol{\theta})}\right] d\boldsymbol{\theta} + \int \pi_2^R(\boldsymbol{\lambda}) \log \left[\frac{\pi_2^R(\boldsymbol{\lambda})}{g(\boldsymbol{\lambda})}\right] d\boldsymbol{\lambda}, \tag{13.10}$$

where $C^* = C - \int \pi_1^R(\boldsymbol{\theta}) \log \pi_1^R(\boldsymbol{\theta})\, d\boldsymbol{\theta} - \int \pi_2^R(\boldsymbol{\lambda}) \log \pi_2^R(\boldsymbol{\lambda})\, d\boldsymbol{\lambda}$. The integrals in (13.10) are minimized when $f(\boldsymbol{\theta}) = \pi_1^R(\boldsymbol{\theta})$ and $g(\boldsymbol{\lambda}) = \pi_2^R(\boldsymbol{\lambda})$, completing the proof. □

When $\pi^R(\boldsymbol{\theta}, \boldsymbol{\lambda})$ is not proper, there is no obvious way to proceed. In particular, we have not been able to make asymptotic variational arguments work. For the situation of asymptotic bi-normality, Sun & Berger (1998) proposed the following reasonable (but adhoc) iterative scheme, following Proposition 10.1 as closely as possible.

Step 0. Choose any initial nonzero marginal prior density for $\boldsymbol{\lambda}$, say $\pi_2^{(0)}(\boldsymbol{\lambda})$.

Step 1. Define an initial prior density for $\boldsymbol{\theta}$ by

$$\pi_1^{(1)}(\boldsymbol{\theta}) \propto \exp\left\{\int \pi_2^{(0)}(\boldsymbol{\lambda}) \log\left(|V_{11}(\boldsymbol{\theta}, \boldsymbol{\lambda})|^{-1/2}\right) d\boldsymbol{\lambda}\right\}.$$

Step 2. Define an initial prior density for $\boldsymbol{\lambda}$ by

$$\pi_2^{(1)}(\boldsymbol{\lambda}) \propto \exp\left\{\int \pi_1^{(1)}(\boldsymbol{\theta}) \log\left(|V_{22}(\boldsymbol{\theta}, \boldsymbol{\lambda})|^{-1/2}\right) d\boldsymbol{\theta}\right\}.$$

Now iterate, replacing $\pi_2^{(0)}$ in *Step 0* by $\pi_2^{(1)}$ and repeating the steps to obtain $\pi_1^{(2)}$ and $\pi_2^{(2)}$. Continuing in this fashion, two sequences of marginal priors, $\{\pi_1^{(i)}\}_{i\geq 1}$ and $\{\pi_2^{(i)}\}_{i\geq 1}$, are generated. Define the marginal reference priors as the limits $\pi_j^R = \lim_{i\to\infty} \pi_j^{(i)}, j = 1, 2$, assuming the limits exist. In applying the iterative algorithm, it may be necessary to operate on compact sets, and then let the sets grow.

We do not know the extent to which this algorithm converges. We have studied several specific situations, and convergence was achieved quickly. For instance, in the two-parameter Weibull model, the equations iterate to the usual reference prior given in Sun (1997).

For many important situations, it is possible to state the result of the above algorithm. Here are two sufficient conditions under which this can be done, the proofs being immediate. For simplicity we only state the version that does not require utilization of a sequence of compact sets.

Proposition 13.5 *(a) If $|V_{11}(\boldsymbol{\theta},\boldsymbol{\lambda})|$ does not depend on $\boldsymbol{\lambda}$, then the marginal reference priors are*

$$\pi_1^R(\boldsymbol{\theta}) \propto |V_{11}(\boldsymbol{\theta},\boldsymbol{\lambda})|^{-1/2}, \tag{13.11}$$

$$\pi_2^R(\boldsymbol{\lambda}) \propto \exp\Big\{\int \pi_1^R(\boldsymbol{\theta}) \log(|V_{22}(\boldsymbol{\theta},\boldsymbol{\lambda})|^{-1/2}) d\boldsymbol{\theta}\Big\}. \tag{13.12}$$

(b) If $|V_{22}(\boldsymbol{\theta},\boldsymbol{\lambda})|$ does not depend on $\boldsymbol{\theta}$, then the marginal reference priors are

$$\pi_2^R(\boldsymbol{\lambda}) \propto |V_{22}(\boldsymbol{\theta},\boldsymbol{\lambda})|^{-1/2}, \tag{13.13}$$

$$\pi_1^R(\boldsymbol{\theta}) \propto \exp\Big\{\int \pi_2^R(\boldsymbol{\lambda}) \log(|V_{11}(\boldsymbol{\theta},\boldsymbol{\lambda})|^{-1/2}) d\boldsymbol{\lambda}\Big\}. \tag{13.14}$$

The following result gives a condition under which they are the same. The proof is obvious, and is omitted.

Proposition 13.6 *If the Fisher information matrix of $(\boldsymbol{\theta},\boldsymbol{\lambda})$ is of the form*

$$\boldsymbol{I}(\boldsymbol{\theta},\boldsymbol{\lambda}) = diag\{g_1(\boldsymbol{\theta})h_1(\boldsymbol{\lambda}),\ g_2(\boldsymbol{\theta})h_2(\boldsymbol{\lambda})\},$$

then the independent marginal reference priors are

$$\pi_1^R(\boldsymbol{\theta}) \propto [g_1(\boldsymbol{\theta})]^{\frac{1}{2}} \text{ and } \pi_2^R(\boldsymbol{\lambda}) \propto [h_2(\boldsymbol{\lambda})]^{\frac{1}{2}}. \tag{13.15}$$

Under the conditions of the proposition, when either $\boldsymbol{\theta}$ or $\boldsymbol{\lambda}$ is the parameter of interest, the unconstrained reference prior has the same form: $\pi^R(\boldsymbol{\theta},\boldsymbol{\lambda}) \propto [g_1(\boldsymbol{\theta})h_2(\boldsymbol{\lambda})]^{\frac{1}{2}}$ (*cf.* Datta & Ghosh (1995*e*)). Therefore, the assumption of independence did not affect the results.

Example 13.6 *Bivariate binomial distribution.* Crowder & Sweeting (1989) consider the bivariate binomial distribution, whose probability density is given by

$$f(r,s\,|\,p,q) = \binom{m}{r} p^R (1-p)^{m-r} \binom{r}{s} q^s (1-q)^{r-s},$$

where $0 < p, q < 1$, and s and r are nonnegative integers satisfying $0 \le s \le r \le m$. The Fisher information matrix for (p, q) is given by

$$\boldsymbol{I}(p, q) = m \ \mathrm{diag}[\{p(1-p)\}^{-1},\ p\{q(1-q)\}^{-1}].$$

Clearly, the Jeffreys-rule prior is proportional to $\{(1-p)q(1-q)\}^{-\frac{1}{2}}$. Based on the assumptions that p and q are independent, that $\theta = pq$ and $\phi = p(1-q)(1-pq)^{-1}$ are independent, and some invariance considerations, Crowder & Sweeting (1989) derived the noninformative prior, $\pi^{CS}(p, q) \propto \{p(1-p)q(1-q)\}^{-1}$. Polson & Wasserman (1990) derived, as the reference prior when either p or q is the parameter of interest, $\pi^{R}(p, q) \propto \{p(1-p)q(1-q)\}^{-\frac{1}{2}}$. From Proposition 13.6, this is also the reference prior under the assumption of independence of p and q.

13.7 Discussion and Additional References

13.7.1 Maximum Entropy Priors

There has been considerable effort in defining objective priors, subject to linear constraints, by maximizing the entropy of the prior distribution subject to the constraints; *cf.* Jaynes (1968), Jaynes (1982) and Jaynes (2003). Entropy is a measure of the uncertainty in a random variable. A distribution with maximum entropy would thus appear to be a reasonable candidate for an objective prior.

For a discrete parameter space $\Theta = \{\theta_1, \ldots, \theta_m\}$ (*i.e.*, θ can assume only m possible values), the entropy of a density $\pi(\theta)$ is measured by

$$\mathcal{E}(\pi) = -\sum_i \pi(\theta_i) \log \pi(\theta_i)\,.$$

This can be shown to be maximized by the uniform density on Θ, i.e.

$$-\sum_{i=1}^{m} \pi(\theta_i) \log \pi(\theta_i) \le -\sum_{i=1}^{m} \frac{1}{m} \log \frac{1}{m} = \log m\,,$$

so that the uniform density on Θ is the maximum entropy distribution, and a very reasonable objective prior, as we have seen.

For a continuous parameter $\boldsymbol{\theta} \in \boldsymbol{\Theta}$, the entropy of $\pi(\boldsymbol{\theta})$ is defined as

$$\mathcal{E}(\pi) = -\int_{\boldsymbol{\Theta}} \pi(\boldsymbol{\theta}) \log\left(\frac{\pi(\boldsymbol{\theta})}{\pi_0(\boldsymbol{\theta})}\right) d\boldsymbol{\theta},$$

where π_0 is a base or reference density. This immediately indicates one of the problems with the maximum entropy approach to objective Bayesian analysis: one needs to start with π_0, and there is no guidance in the approach as to how this should be chosen. (Jaynes (1982) and Jaynes (2003) would typically utilize invariance arguments to choose π_0, if invariance is present.)

A second problem with the maximum entropy approach is that the maximum entropy distribution typically does not exist, unless one places constraints on the distribution, such as moment constraints. Thus maximum entropy is primarily useful when one already has a base objective prior $\pi_0(\boldsymbol{\theta})$ (say, the reference prior) along with some constraints on the prior, and wants to be as objective as possible relative to this information. The result of the maximum entropy analysis is typically the constrained reference prior discussed in Section 13.3.

For further information on maximum entropy and its use in Bayesian statistics see Akaike (1977), Csiszar (1985), Csiszár (1991), Erickson & Smith (1988), Skilling (1989), Fougere (1990), Clarke & Barron (1994), Grünwald & Dawid (2004), and references therein.

Chapter 14

Models with Special Structures

Objective priors have been developed for many specific classes of models, too many to present in this book. In this chapter we illustrate such developments by considering four such classes of models: a special two parameter exponential family, models with Fisher information matrices that have certain patterns, star-shaped graphical models, and sequential models.

14.1 A Two Parameter Exponential Family

Bar-Lev & Reiser (1982) considered the following density function of the generic two-parameter exponential family:

$$f(x \mid \theta_1, \theta_2) = a(x)\exp\{\theta_1 U_1(x) - \theta_1 G_2'(\theta_2) U_2(x) - \psi(\theta_1, \theta_2)\}, \qquad (14.1)$$

where $\theta_1 < 0$, $\theta_2 = \mathrm{E}[U_2(X) \mid \theta_1, \theta_2]$, $G_i(\cdot), (i = 1, 2)$ are infinitely differentiable functions satisfying $G_i'' > 0$, and

$$\psi(\theta_1, \theta_2) = -\theta_1[\theta_2 G_2'(\theta_2) - G_2(\theta_2)] + G_1(\theta_1).$$

This class of distributions includes many popular distributions, such as the normal, inverse Gaussian, gamma, and inverse gamma.

Table 14.1, reproduced from Sun (1994), indicates how each distribution arises. The Fisher information matrix of (θ_1, θ_2), per observation, is

$$\mathbf{I}(\theta_1, \theta_2) = \begin{pmatrix} G_1''(\theta_1) & 0 \\ 0 & -\theta_1 G_2''(\theta_2) \end{pmatrix}.$$

Table 14.1: Special cases of Bar-Lev & Reiser (1982) two parameter exponential family, where $h(\theta_1) = -\theta_1 + \theta_1 \log(-\theta_1) + \log(\Gamma(-\theta_1))$.

	$U_1(x)$	$U_2(x)$	θ_1	θ_2	$G_1(\theta_1)$	$G_2(\theta_2)$
$N(\mu, \sigma^2)$	x^2	x	$\frac{-1}{2\sigma^2}$	μ	$-\frac{1}{2}\log(-2\theta_1)$	θ_2^2
Inverse Gaussian	$\frac{1}{x}$	x	$-\frac{\alpha}{2}$	$\sqrt{\frac{\alpha}{\mu}}$	$-\frac{1}{2}\log(-2\theta_1)$	$\frac{1}{\theta_2}$
Gamma	$-\log x$	x	$-\alpha$	μ	$h(\theta_1)$	$-\log \theta_2$
Inverse Gamma	$\log x$	$\frac{1}{x}$	$-\alpha$	μ	$h(\theta_1)$	$-\log \theta_2$

Thus the Jeffreys-rule prior for (θ_1, θ_2) is

$$\pi^J(\theta_1, \theta_2) \propto \sqrt{|\theta_1|}\sqrt{G_1''(\theta_1)G_2''(\theta_2)}.$$

When θ_1 or θ_2 is the parameter of interest and the other is the nuisance parameter, it can be shown (Sun & Ye (1996)) that the one-at-a-time reference priors are the same and are given by

$$\pi^R(\theta_1, \theta_2) \propto \sqrt{G_1''(\theta_1)G_2''(\theta_2)}. \tag{14.2}$$

The first order matching priors for θ_1 and θ_2 can be shown to be, respectively,

$$\pi_1^M(\theta_1, \theta_2) \propto \sqrt{G_1''(\theta_1)f_2(\theta_2)}, \quad \pi_2^M(\theta_1, \theta_2) \propto f_1(\theta_1)\sqrt{G_2''(\theta_2)}, \tag{14.3}$$

where f_1 and f_2 are arbitrary positive functions. Since (14.2) satisfies (14.3), the one at-a-time reference prior is a first order matching prior for both θ_1 and θ_2.

14.2 Fisher Information of Certain Patterns

If the multivariate posterior distribution is asymptotically normal, the asymptotic reference prior only depends on the Fisher information matrix. The derivation of such reference priors can be substantially simplified under some specific assumptions on the structure of the information matrix, as first shown in Berger & Bernardo (1992*a*) and Datta & Ghosh (1995*d*).

Suppose that the p-dimensional unknown parameter $\boldsymbol{\theta}$ can be decomposed into m sub-vectors, $\boldsymbol{\theta} = (\boldsymbol{\theta}_1, \ldots, \boldsymbol{\theta}_m)$, where $\boldsymbol{\theta}_i = (\theta_{i1}, \cdots, \theta_{ip_i}) \subset \boldsymbol{\Theta}_i \subset \mathbb{R}^{p_i}$, for $i = 1, \cdots, m$, and $\boldsymbol{\Theta} = \boldsymbol{\Theta}_1 \times \cdots \times \boldsymbol{\Theta}_m$, of dimension $p_1 + \cdots + p_m = p$. Typically, either all the $p_i = 1$, in which case one is in the situation of the one-at-a-time reference prior discussed in Section 10.3, or one is in the two group case, discussed in Section 10.3.5, where $\boldsymbol{\theta}_1 = \theta_1$ is the real parameter of interest and $\boldsymbol{\theta}_2$ is the vector of all remaining nuisance parameters. The theorems discussed herein were developed for the general case, in which one iterated the reference prior algorithm over all the $\boldsymbol{\theta}_i$, and so will be presented in that generality.

The first theorem is from Liseo et al. (2009), which generalized a theorem from Datta & Ghosh (1995*d*).

Theorem 14.1 *Suppose there exists a nested sequence* $\{\boldsymbol{\Theta}_i^l\}$ *of compact subsets of* $\boldsymbol{\Theta}_i$, *not depending on* $\boldsymbol{\theta}_i$ *such that* $\bigcup_{l=1}^{\infty} \boldsymbol{\Theta}_i^l = \boldsymbol{\Theta}_i$, $i = 1, \ldots, m$. *Suppose the following two conditions hold:*

Condition A1. *The Fisher information matrix,* $\boldsymbol{I}(\boldsymbol{\theta})$, *can be decomposed as* $\boldsymbol{I}(\boldsymbol{\theta}) = \boldsymbol{G}(\boldsymbol{\theta})\boldsymbol{I}^*\boldsymbol{G}(\boldsymbol{\theta})$, *where* $\boldsymbol{I}^*$ *is a positive definite matrix that does not depend on* $\boldsymbol{\theta}$, *and* $\boldsymbol{G}(\boldsymbol{\theta}) = Diag\{\boldsymbol{G}_1(\boldsymbol{\theta}), \cdots, \boldsymbol{G}_m(\boldsymbol{\theta})\}$, *where* $\boldsymbol{G}_i(\boldsymbol{\theta})$ *is a* $p_i \times p_i$ *positive definite matrix,* $i = 1, \cdots, m$.

Condition A2. $|\boldsymbol{G}_i(\boldsymbol{\theta})| = g_{1i}(\boldsymbol{\theta}_i) g_{2i}(\boldsymbol{\theta}_i^c)$, *where* $\boldsymbol{\theta}_i^c = (\boldsymbol{\theta}_1, \cdots, \boldsymbol{\theta}_{i-1}, \boldsymbol{\theta}_{i+1}, \cdots, \boldsymbol{\theta}_m)$. *Then the* m*-group reference prior for* $\boldsymbol{\theta}$, *for the ordering* $\{\boldsymbol{\theta}_1, \ldots, \boldsymbol{\theta}_m\}$, *is*

$$\pi^R(\boldsymbol{\theta}) \propto \prod_{i=1}^{m} g_{1i}(\boldsymbol{\theta}_i). \tag{14.4}$$

Example 14.1 *Two independent Weibull populations with a common shape parameter.* For $i = 1, 2$, suppose x_i are independent draws from the Weibull(η_i, β) density $p(x \mid \eta_i, \beta) = \beta x^{\beta-1} \eta_i^{-\beta} \exp\{-(x/\eta_i)^{\beta}\}$. It follows from Sun et al. (1998) that the Fisher information matrix for (η_1, η_2, β) is

$$\boldsymbol{I}(\eta_1, \eta_2, \beta) = \begin{pmatrix} \frac{\beta^2}{\eta_1^2} & 0 & -\frac{\gamma}{\eta_1} \\ 0 & \frac{\beta^2}{\eta_2^2} & -\frac{\gamma}{\eta_2} \\ -\frac{\gamma}{\eta_1} & -\frac{\gamma}{\eta_2} & \frac{2(\gamma^2+\gamma^*)}{\beta^2} \end{pmatrix},$$

where $\gamma = 1 + \int_0^{\infty} (\log z) e^{-z} dz$, ($1 - \gamma$ is Euler's constant) and $\gamma^* = 1 + \int_0^{\infty} (\log z)^2 e^{-z} dz - \{\int_0^{\infty} (\log z) e^{-z} dz\}^2$. $\boldsymbol{I}(\eta_1, \eta_2, \beta)$ can be re-written as

$$\boldsymbol{I}(\eta_1, \eta_2, \beta) = \boldsymbol{G}(\eta_1, \eta_2, \beta)\, \boldsymbol{I}^*\, \boldsymbol{G}(\eta_1, \eta_2, \beta),$$

where

$$\boldsymbol{G}(\eta_1, \eta_2, \beta) = \mathrm{Diag}\Big(\frac{\beta}{\eta_1}, \frac{\beta}{\eta_2}, \frac{1}{\beta}\Big), \quad \boldsymbol{I}^* = \begin{pmatrix} 1 & 0 & -\gamma \\ 0 & 1 & -\gamma \\ -\gamma & -\gamma & 2(\gamma^2 + \gamma^*) \end{pmatrix}.$$

Applying Theorem 14.1, it follows that for, every ordering of (η_1, η_2, β), the reference prior is given by

$$\pi^R(\eta_1, \eta_2, \beta) \propto g_{11}(\eta_1) g_{12}(\eta_2) g_{13}(\beta) \propto \frac{1}{\eta_1 \eta_2 \beta}.$$

Liseo et al. (2009) developed a second result depending on a different structure for the information matrix.

Theorem 14.2 *Suppose there exists a nested sequence $\{\boldsymbol{\Theta}_i^l\}$ of compact subsets of $\boldsymbol{\Theta}_i$ not depending on $\boldsymbol{\theta}$ such that $\bigcup_{l=1}^{\infty} \boldsymbol{\Theta}_i^l = \boldsymbol{\Theta}_i$, $i = 1, \ldots, m$. Suppose the following two conditions hold:*

Condition B1. *The Fisher information matrix, $\boldsymbol{I}(\boldsymbol{\theta})$, can be decomposed as $\boldsymbol{I}(\boldsymbol{\theta}) = \boldsymbol{G}(\boldsymbol{\theta}_{i^*}^c)\boldsymbol{I}^*(\boldsymbol{\theta}_{i^*})\boldsymbol{G}(\boldsymbol{\theta}_{i^*}^c)$, with $\boldsymbol{I}^*(\boldsymbol{\theta}_{i^*})$ positive definite and depending only on $\boldsymbol{\theta}_{i^*}$ (one of the $\boldsymbol{\theta}_i$) and $\boldsymbol{G}(\boldsymbol{\theta}_{i^*}^c) = Diag\Big(\boldsymbol{G}_1(\boldsymbol{\theta}_{i^*}^c), \cdots, \boldsymbol{G}_m(\boldsymbol{\theta}_{i^*}^c)\Big)$, where the $\boldsymbol{G}_i(\boldsymbol{\theta}_{i^*}^c)$ are $p_i \times p_i$ matrices depending only on $\boldsymbol{\theta}_{i^*}^c$.*

Condition B2. *For $i \neq i^*$, $|\boldsymbol{G}_i(\boldsymbol{\theta}_{i^*}^c)| = g_{1i}(\boldsymbol{\theta}_i)g_{2i}(\boldsymbol{\theta}_{(i,i^*)}^C)$, where $\boldsymbol{\theta}_{(i,i^*)}^C$ is the vector $\boldsymbol{\theta}$, with $\boldsymbol{\theta}_i$ and $\boldsymbol{\theta}_{i^*}$ omitted.*

Then the m-group reference prior for $\boldsymbol{\theta}$, for the ordering $\{\boldsymbol{\theta}_1, \boldsymbol{\theta}_2, \cdots, \boldsymbol{\theta}_m\}$, is

$$\pi^R(\boldsymbol{\theta}) \propto \frac{|\boldsymbol{I}_{i^*}^*(\boldsymbol{\theta}_{i^*})|^{\frac{1}{2}}}{|\boldsymbol{I}_{i^*+1}^*(\boldsymbol{\theta}_{i^*})|^{\frac{1}{2}}} \prod_{i \neq i^*}^{m} g_{1i}(\boldsymbol{\theta}_i), \tag{14.5}$$

where $\boldsymbol{I}_j^*(\boldsymbol{\theta}_{i^*})$ is the lower-right $(p_j + \cdots + p_m) \times (p_j + \cdots + p_m)$ submatrix of $\boldsymbol{I}^*(\boldsymbol{\theta}_{i^*})$, $j = 1, \ldots, m$, and, if $i^* = m$, $|\boldsymbol{I}_{m+1}^*(\boldsymbol{\theta}_{i^*})|$ is set equal to 1.

Example 14.2 *Fieller's problem.* Let x and y be two independent normal variables with unknown means α and β, respectively, and unknown *common* variance σ^2. The parameter of interest is $\theta = \alpha/\beta$, the ratio of the means. With the parametrization $(\theta, \beta, \sigma^2)$, the Fisher information matrix is (see, for instance, Bernardo (1977*a*))

$$\boldsymbol{I}(\theta, \beta, \sigma^2) = \frac{1}{\sigma^4} \begin{pmatrix} \beta^2 & \theta\beta & 0 \\ \theta\beta & 1+\theta^2 & 0 \\ 0 & 0 & 1 \end{pmatrix},$$

which can be re-expressed as

$$\boldsymbol{I}(\theta, \beta, \sigma^2) = \boldsymbol{G}(\beta, \sigma^2)\boldsymbol{I}^*(\theta)\boldsymbol{G}(\beta, \sigma^2),$$

where $\boldsymbol{G}(\beta, \sigma^2) = (\sigma^2)^{-1}\text{Diag}(\beta, 1, 1)$ and

$$\boldsymbol{I}^*(\theta) = \begin{pmatrix} 1 & \theta & 0 \\ \theta & 1+\theta^2 & 0 \\ 0 & 0 & 1 \end{pmatrix}.$$

From Theorem 14.2, the reference prior for $(\theta, \beta, \sigma^2)$, in the ordering $\{\theta, \beta, \sigma^2\}$ or $\{\theta, \sigma^2, \beta\}$, is

$$\pi^R(\theta, \beta, \sigma^2) \propto \frac{|I_1^*(\theta)|^{\frac{1}{2}}}{|I_2^*(\theta)|^{\frac{1}{2}}} g_{12}(\beta)g_{13}(\sigma^2) \propto \frac{1}{(1+\theta^2)^{1/2}\sigma^2}.$$

14.3 Star-Shaped Models

14.3.1 Definition of Star-Shaped Models

There is a large literature on estimating the covariance matrix and precision matrix of a multivariate normal distribution when the matrices are of full rank;

see Berger et al. (2020*a*) for references. Often, however, only matrices with special structure are considered. One important case, to be considered here, is where variables can be conditionally independent, given other variables.

For the multivariate normal distribution, conditional independence relationships correspond to the presence of zeros in the entries of the precision matrix (the inverse of the covariance matrix). See Dempster (1972), Whittaker (1990), and Lauritzen (1996) for general discussion. Here we consider one important example, namely *star-shaped models*; these arise as follows.

Assume that $\boldsymbol{x} \sim N_p(0, \boldsymbol{\Sigma})$. The vector $\boldsymbol{x}$ is partitioned into k groups, that is, $\boldsymbol{x} = (\boldsymbol{x}_1', \boldsymbol{x}_2', \cdots, \boldsymbol{x}_k')'$, where $\boldsymbol{x}_i$ is p_i-dimensional and $\sum_{i=1}^k p_i = p$. Assume that, given $\boldsymbol{x}_1$, the subvectors $\boldsymbol{x}_2, \cdots, \boldsymbol{x}_k$ are mutually conditionally independent. From Whittaker (1990) and Lauritzen (1996), the precision matrix $\boldsymbol{\Omega} = \boldsymbol{\Sigma}^{-1}$ then has the following special structure:

$$\boldsymbol{\Omega} = \begin{pmatrix} \boldsymbol{\Omega}_{11} & \boldsymbol{\Omega}_{12} & \boldsymbol{\Omega}_{13} & \cdots & \boldsymbol{\Omega}_{1k} \\ \boldsymbol{\Omega}_{21} & \boldsymbol{\Omega}_{22} & \mathbf{0} & \cdots & \mathbf{0} \\ \boldsymbol{\Omega}_{31} & \mathbf{0} & \boldsymbol{\Omega}_{33} & \cdots & \mathbf{0} \\ \vdots & \vdots & \vdots & \ddots & \vdots \\ \boldsymbol{\Omega}_{k1} & \mathbf{0} & \mathbf{0} & \cdots & \boldsymbol{\Omega}_{kk} \end{pmatrix}. \tag{14.6}$$

Here are some examples.

Example 14.3 Let x_1 be the prime interest rate in the United States, which is a national variable, and $x_2, \cdots, x_k$ be the median house prices in $k-1$ cities in different states, which are local variables. It is often assumed that $x_2, \cdots, x_k$ are conditionally independent, given x_1, and are normally distributed. Then the precision matrix of $(x_1, \cdots, x_k)$ would have the structure (14.6), with the $\boldsymbol{\Omega}_{ij}$ now being real variables.

Example 14.4 This conditional independence assumption is very common in graphical models. The case $k = 3$ was considered in detail in Whittaker (1990), wherein it was called the 'butterfly model.' For $k \geq 3$, a normal model with precision matrix as in (14.6) is called a *star-shaped* model because the graphical shape of the relationship among the variables when $k \geq 4$ looks like a star. The cases $k = 3, 4, 7$ are illustrated in Figure 14.1.

The star-shaped model is a special case of lattice conditional independence models, introduced by Andersson & Perlman (1993). These have been widely used in graph theory and social network models.

14.3.2 Objective Priors for Star-Shaped Models

Invariance Priors. As discussed in Section 6.4.4, invariance priors for matrices are typically found by using the Cholesky decomposition, here $\boldsymbol{\Omega} = \boldsymbol{\Psi}'\boldsymbol{\Psi}$, where $\boldsymbol{\Psi}$ is a $p \times p$ lower-triangular matrix with positive diagonal entries, and then finding the invariance priors for $\boldsymbol{\Psi}$. (This, of course, induces a prior on $\boldsymbol{\Omega}$.)

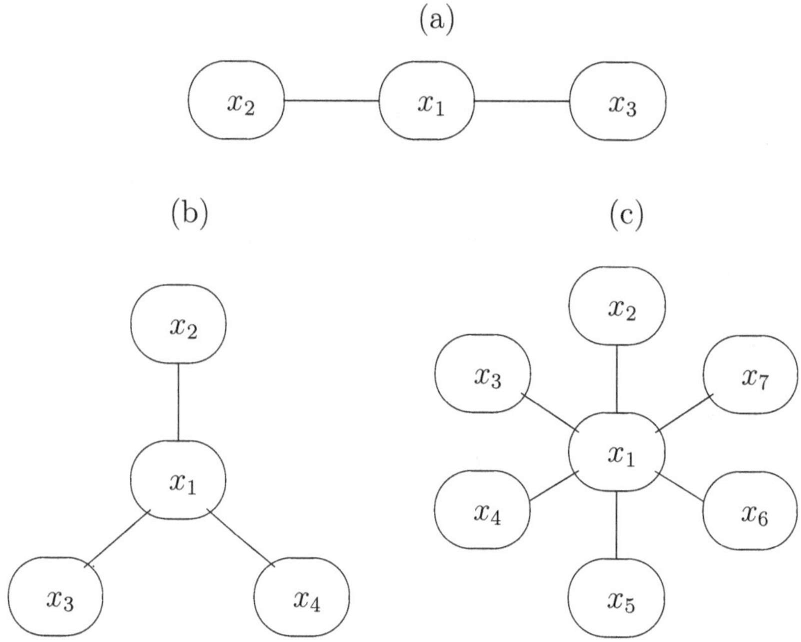

Figure 14.1: Graphs corresponding to the conditional independence assumption in (14.6): (a) $k = 3$; (b) $k = 4$, and (c) $k = 7$.

For the structure of $\mathbf{\Omega}$ given by (14.6), one can show that $\mathbf{\Psi}$ has the following block structure:

$$\mathbf{\Psi} = \begin{pmatrix} \mathbf{\Psi}_{11} & \mathbf{0} & \mathbf{0} & \cdots & \mathbf{0} \\ \mathbf{\Psi}_{21} & \mathbf{\Psi}_{22} & \mathbf{0} & \cdots & \mathbf{0} \\ \mathbf{\Psi}_{31} & \mathbf{0} & \mathbf{\Psi}_{33} & \cdots & \mathbf{0} \\ \vdots & \vdots & \vdots & \ddots & \vdots \\ \mathbf{\Psi}_{k1} & \mathbf{0} & \mathbf{0} & \cdots & \mathbf{\Psi}_{kk} \end{pmatrix}, \tag{14.7}$$

where each $\mathbf{\Psi}_{ii}$ is a $p_i \times p_i$ lower-triangular matrix with positive diagonal elements and there are no restrictions on the $\mathbf{\Psi}_{ij}$ $(i > j)$. We apply the invariance theory from Chapter 6 to find the right-Haar and left-Haar priors for $\mathbf{\Psi}$.

Proposition 14.1 *The right-Haar and left-Haar priors for* $\mathbf{\Psi}$ *are, respectively,*

$$\pi^r(\mathbf{\Psi}) = \left[\prod_{j=1}^{p_1} \psi_{1jj}^{j} \cdot \prod_{i=2}^{k}\prod_{j=1}^{p_i} \psi_{ijj}^{p_1+j}\right]^{-1}, \tag{14.8}$$

$$\pi^l(\mathbf{\Psi}) = \left[\prod_{j=1}^{p_1} \psi_{1jj}^{p-j+1} \cdot \prod_{i=2}^{k}\prod_{j=1}^{p_i} \psi_{ijj}^{p_i-j+1}\right]^{-1}, \tag{14.9}$$

where, for $i = 1, 2, \cdots, k$,

$$\boldsymbol{\Psi}_{ii} = \begin{pmatrix} \psi_{i11} & 0 & \cdots & 0 \\ \psi_{i21} & \psi_{i22} & \cdots & 0 \\ \vdots & \vdots & \ddots & \vdots \\ \psi_{ip_i1} & \psi_{ip_i1} & \cdots & \psi_{ip_ip_i} \end{pmatrix}.$$

Proof. The general proof is given in Sun & Sun (2005). Here we only present the proof for the case where $p = 3$ and the $p_i = 1$, since this shows the key ideas. Thus assume that $\boldsymbol{x} = (x_1, x_2, x_3) \sim N_3(\boldsymbol{0}, \boldsymbol{\Omega}^{-1})$, where the precision matrix $\boldsymbol{\Omega}$ is of the form

$$\boldsymbol{\Omega} = \boldsymbol{\Psi}'\boldsymbol{\Psi} = \begin{pmatrix} \omega_{11} & \omega_{21} & \omega_{31} \\ \omega_{21} & \omega_{22} & 0 \\ \omega_{31} & 0 & \omega_{33} \end{pmatrix}.$$

It is easy to see that $Cov(x_2, x_3 \mid x_1) = \omega_{23} = 0$. So, given x_1, x_2 and x_3 are independent and we have the 'butterfly model.' The parameter space for

$$\boldsymbol{\Psi} = \begin{pmatrix} \psi_{11} & 0 & 0 \\ \psi_{21} & \psi_{22} & 0 \\ \psi_{31} & 0 & \psi_{33} \end{pmatrix}$$

is the following subset of the 3×3 lower triangular matrices:

$$\boldsymbol{\Theta} = \{\boldsymbol{C} = (c_{ij})_{3\times 3}, c_{32} = 0, c_{ij} = 0, i < j, c_{ii} > 0\}.$$

For any $\boldsymbol{C} \in \boldsymbol{\Theta}$, consider a linear transformation $g_{\boldsymbol{C}}(\boldsymbol{x}) = \boldsymbol{C}\boldsymbol{x} \sim N_3(\boldsymbol{0}, \boldsymbol{C}\boldsymbol{\Omega}^{-1}\boldsymbol{C}')$. It is easy to show that such transformations form a group, with the unit element $e = g_{\boldsymbol{I}_3}$ (with $\boldsymbol{C} = \boldsymbol{I}_3$) and $g_{\boldsymbol{C}}^{-1} = g_{\boldsymbol{C}^{-1}}$. To determine the group action on $\boldsymbol{\Psi}$, note that the precision matrix corresponding to $g_{\boldsymbol{C}}(\boldsymbol{x})$ is

$$(\boldsymbol{C}\boldsymbol{\Omega}^{-1}\boldsymbol{C}')^{-1} = (\boldsymbol{\Psi}\boldsymbol{C}^{-1})'(\boldsymbol{\Psi}\boldsymbol{C}^{-1}).$$

Thus $\bar{g}_{\boldsymbol{C}}(\boldsymbol{\Psi}) = \boldsymbol{\Psi}\boldsymbol{C}^{-1}$ and, for $\boldsymbol{C}^*, \boldsymbol{C} \in \boldsymbol{\Theta}$,

$$\bar{g}_{\boldsymbol{C}^*} \circ \bar{g}_{\boldsymbol{C}}(\boldsymbol{\Psi}) = \boldsymbol{\Psi}(\boldsymbol{C}^*\boldsymbol{C})^{-1}, \quad \boldsymbol{\Psi} \in \boldsymbol{\Theta},$$

with

$$\begin{aligned} \boldsymbol{C}^*\boldsymbol{C} &= \begin{pmatrix} c_{11}^* & 0 & 0 \\ c_{21}^* & c_{22}^* & 0 \\ c_{31}^* & 0 & c_{33}^* \end{pmatrix} \begin{pmatrix} c_{11} & 0 & 0 \\ c_{21} & c_{22} & 0 \\ c_{31} & 0 & c_{33} \end{pmatrix} \\ &= \begin{pmatrix} c_{11}^*c_{11} & 0 & 0 \\ c_{21}^*c_{11} + c_{22}^*c_{21} & c_{22}^*c_{22} & 0 \\ c_{31}^*c_{11} + c_{33}^*c_{31} & 0 & c_{33}^*c_{33} \end{pmatrix}. \end{aligned}$$

As in Section 6.4.2, the right-Haar measure for this group action is given by

$$\begin{aligned} J_{\bar{g}}^r(\bar{g}^*) &= \frac{\partial(\boldsymbol{C}^*\boldsymbol{C})}{\partial(c_{11}^*, c_{21}^*, c_{22}^*, c_{31}^*, c_{33}^*)} \\ &= \begin{vmatrix} c_{11} & 0 & 0 & 0 & 0 \\ 0 & c_{11} & c_{21} & 0 & 0 \\ 0 & 0 & c_{22} & 0 & 0 \\ 0 & 0 & 0 & c_{11} & c_{31} \\ 0 & 0 & 0 & 0 & c_{33} \end{vmatrix} = c_{11}^3 c_{22} c_{33}. \end{aligned}$$

Similarly,

$$\boldsymbol{C}\boldsymbol{C}^* = \begin{pmatrix} c_{11}c_{11}^* & 0 & 0 \\ c_{21}c_{11}^* + c_{22}c_{21}^* & c_{22}c_{22}^* & 0 \\ c_{31}c_{11}^* + c_{33}c_{31}^* & 0 & c_{33}c_{33}^* \end{pmatrix},$$

and the left-Haar measure is given by

$$\begin{aligned} J_{\bar{g}}^l(\bar{g}^*) &= \frac{\partial(\boldsymbol{C}\boldsymbol{C}^*)}{\partial(c_{11}^*, c_{21}^*, c_{22}^*, c_{31}^*, c_{33}^*)} \\ &= \begin{vmatrix} c_{11} & 0 & 0 & 0 & 0 \\ c_{21} & c_{22} & 0 & 0 & 0 \\ 0 & 0 & c_{22} & 0 & 0 \\ c_{31} & 0 & 0 & c_{33} & 0 \\ 0 & 0 & 0 & 0 & c_{33} \end{vmatrix} = c_{11}c_{22}^2c_{33}^2. \end{aligned}$$

We now have to transform these Haar measures on the group to priors on the parameter space, as in Section 6.4.2. Define $\mathcal{I} = \{\bar{g}_C(\boldsymbol{I}_3) = \boldsymbol{C}^{-1}, \boldsymbol{C} \in \boldsymbol{\Theta}\}$ and define $\boldsymbol{C}^{-1} = \boldsymbol{\Psi}$, or

$$\boldsymbol{C} = \boldsymbol{\Psi}^{-1} = \begin{pmatrix} \frac{1}{\psi_{11}} & 0 & 0 \\ \frac{-\psi_{21}}{\psi_{11}\psi_{22}} & \frac{1}{\psi_{22}} & 0 \\ \frac{-\psi_{31}}{\psi_{11}\psi_{33}} & 0 & \frac{1}{\psi_{33}} \end{pmatrix}.$$

The Jacobian of $\boldsymbol{C} = \boldsymbol{\Psi}^{-1}$, $J(\boldsymbol{\Psi}) = \frac{\partial(c_{11}, c_{21}, c_{22}, c_{31}, c_{33})}{\partial(\psi_{11}, \psi_{21}, \psi_{22}, \psi_{31}, \psi_{33})}$, is

$$J(\boldsymbol{\Psi}) = \begin{vmatrix} -\frac{1}{\psi_{11}^2} & 0 & 0 & 0 & 0 \\ \frac{\psi_{21}}{\psi_{11}^2\psi_{22}} & \frac{-1}{\psi_{11}\psi_{22}} & \frac{\psi_{21}}{\psi_{11}\psi_{22}^2} & 0 & 0 \\ 0 & 0 & -\frac{1}{\psi_{22}^2} & 0 & 0 \\ \frac{\psi_{31}}{\psi_{11}^2\psi_{33}} & 0 & 0 & \frac{-1}{\psi_{11}\psi_{33}} & \frac{\psi_{31}}{\psi_{11}\psi_{33}^2} \\ 0 & 0 & 0 & 0 & -\frac{1}{\psi_{33}^2} \end{vmatrix} = \frac{1}{\psi_{11}^4\psi_{22}^3\psi_{33}^3}.$$

Following (6.4), the right and left Haar priors for $\boldsymbol{\Psi}$ are, respectively,

$$\begin{aligned} \pi^r(\boldsymbol{\Psi}) &= \frac{J(\boldsymbol{\Psi})}{J_{\bar{g}}^r(\bar{g}^*)} = \frac{\psi_{11}^3\psi_{22}\psi_{33}}{\psi_{11}^4\psi_{22}^3\psi_{33}^3} = \frac{1}{\psi_{11}\psi_{22}^2\psi_{33}^2}, \\ \pi^l(\boldsymbol{\Psi}) &= \frac{J(\boldsymbol{\Psi})}{J_{\bar{g}}^l(\bar{g}^*)} = \frac{\psi_{11}\psi_{22}^2\psi_{33}^2}{\psi_{11}^4\psi_{22}^3\psi_{33}^3} = \frac{1}{\psi_{11}^3\psi_{22}\psi_{33}}. \end{aligned} \tag{14.10}$$

□

The Jeffreys-Rule and Reference Priors for Star-Shaped Models.

Proposition 14.2 *Consider the star-shaped model satisfying (14.6).*
(a) *The Jeffreys-rule prior, $\pi^J(\boldsymbol{\Psi})$, for $\boldsymbol{\Psi}$ is the same as the left-Haar prior, given by (14.9).*

(b) *Suppose that each* $\Psi_{ij} = \psi_{ij}$ *is real, with* ψ_{11} *being the parameter of interest, and that we reparameterize all the off-diagonal nuisance parameters in* $\boldsymbol{\Psi}$ *as* $\beta_{j1} = \psi_{j1}/\psi_{jj}$, $j = 2, \ldots, k$. *Then the one-at-a-time reference prior, corresponding to the parameter ordering* $(\psi_{11}, \beta_{21}, \psi_{22}, \ldots, \beta_{k1}, \psi_{kk})$ *(but with the prior transformed back to* $\boldsymbol{\Psi} = (\psi_{11}, \psi_{21}, \psi_{22}, \ldots, \psi_{k1}, \psi_{kk})$*), is given by* $\pi^R(\boldsymbol{\Psi}) \propto \psi_{11}^{-1} \prod_{j=2}^{k} \psi_{jj}^{-2}$, *which is the same as the right-Haar prior in (14.8).*

Proof. Again, we only present the proof for the case where $p = 3$ and the $p_i = 1$ (the 'butterfly model') and use the notation from the proof of Proposition 14.1.

To find the Jeffreys-rule prior and reference priors for $\boldsymbol{\Psi}$, we need the Fisher information matrix for $\boldsymbol{\Psi}$. Noting that

$$f(\boldsymbol{x} \mid \boldsymbol{\Psi}) = \frac{|\boldsymbol{\Psi}|}{(2\pi)^{3/2}} \exp\left\{ -\frac{1}{2}\boldsymbol{x}'\boldsymbol{\Psi}'\boldsymbol{\Psi}\boldsymbol{x} \right\},$$

the Fisher information matrix for $\boldsymbol{\psi} = (\psi_{11}, \psi_{21}, \psi_{22}, \psi_{31}, \psi_{33})$ can be computed to be

$$\boldsymbol{I}(\boldsymbol{\psi}) = \mathrm{Diag}\left(\frac{2}{\psi_{11}^2}, \begin{pmatrix} \frac{1}{\psi_{11}^2} & \frac{-\psi_{21}}{\psi_{11}^2\psi_{22}} \\ \frac{-\psi_{21}}{\psi_{11}^2\psi_{22}} & \frac{\psi_{21}^2}{\psi_{11}^2\psi_{22}^2} + \frac{2}{\psi_{22}^2} \end{pmatrix}, \begin{pmatrix} \frac{1}{\psi_{11}^2} & \frac{-\psi_{31}}{\psi_{11}^2\psi_{33}} \\ \frac{-\psi_{31}}{\psi_{11}^2\psi_{33}} & \frac{\psi_{31}^2}{\psi_{11}^2\psi_{33}^2} + \frac{2}{\psi_{33}^2} \end{pmatrix} \right).$$

The determinant of $\boldsymbol{I}(\boldsymbol{\psi})$ is $|\boldsymbol{I}(\boldsymbol{\psi})| = 6/[\psi_{11}^6\psi_{22}^2\psi_{33}^2]$, so the Jeffreys prior is

$$\pi^J(\boldsymbol{\psi}) \propto \frac{1}{\psi_{11}^3\psi_{22}\psi_{33}}.$$

A one-at-a-time reference prior for ψ_{11} is most easily found by reparameterizing two of the nuisance parameters to be $\beta_{21} = \psi_{21}/\psi_{22}$ and $\beta_{31} = \psi_{31}/\psi_{33}$. The Fisher information matrix for $(\psi_{11}, \beta_{21}, \psi_{22}, \beta_{31}, \psi_{33})$ is

$$\boldsymbol{I} = diag\left(\frac{2}{\psi_{11}^2}, \frac{\psi_{22}^2}{\psi_{11}^2}, \frac{2}{\psi_{22}^2}, \frac{\psi_{33}^2}{\psi_{11}^2}, \frac{2}{\psi_{33}^2} \right).$$

It follows from Theorem 14.1, that the one-at-a-time reference prior for $(\psi_{11}, \beta_{21}, \psi_{22}, \beta_{31}, \psi_{33})$, with any ordering, is

$$\pi^R(\psi_{11}, \beta_{21}, \psi_{22}, \beta_{31}, \psi_{33}) = \frac{1}{\psi_{11}\psi_{22}\psi_{33}}.$$

Note that, for $j = 2, 3$, the Jacobians are

$$\frac{\partial(\psi_{j1}, \psi_{jj})}{\partial(\beta_{j1}, \psi_{jj})} = \begin{vmatrix} \psi_{jj} & \beta_{j1} \\ 0 & 1 \end{vmatrix} = \psi_{jj}\,,$$

so transforming back to $\boldsymbol{\Psi}$, yields the result. Note that this prior agrees with the right-Haar prior for $\boldsymbol{\Psi}$ in (14.10). □

14.3.3 Posterior Distributions and Estimates

Let $\boldsymbol{y}_1, \boldsymbol{y}_2, \cdots, \boldsymbol{y}_n$ be a random sample from $N_p(0, \boldsymbol{\Omega}^{-1})$, where $\boldsymbol{\Omega}$ satisfies (14.6). (We use $\boldsymbol{y}_i$ for the observations, instead of $\boldsymbol{x}_i$, since the $\boldsymbol{x}_i$ were earlier used to denote subvectors of a single observation to describe the conditional independencies.) Let $\boldsymbol{S} = \sum_{i=1}^n \boldsymbol{y}_i\boldsymbol{y}_i'$ and assume that $n > p+1$. Then $\boldsymbol{S}$ is a sufficient statistic for $\boldsymbol{\Omega}$ and has a Wishart distribution with parameters n and $\boldsymbol{\Omega}^{-1}$. Write $\boldsymbol{S} = (\boldsymbol{S}_{ij})$, where $\boldsymbol{S}_{ij}$ is the $p_i \times p_j$ sub-matrix, and define $\boldsymbol{S}_{ii\cdot 1} = \boldsymbol{S}_{ii} - \boldsymbol{S}_{i1}\boldsymbol{S}_{11}^{-1}\boldsymbol{S}_{i1}$, $i = 2, \cdots, k$.

Consider the class of priors for $\boldsymbol{\Psi}$:

$$\pi(\boldsymbol{\Psi}) \propto \prod_{i=1}^{k}\prod_{j=1}^{p_i} \psi_{ijj}^{\alpha_{ij}} \exp(-\beta_{ij}\psi_{ijj}^2), \tag{14.11}$$

where $\beta_{ij} \geq 0, j = 1, \cdots, p_i, i = 1, \cdots, k$. This class includes the right-Haar prior $\pi^r(\boldsymbol{\Psi})$, the left-Haar prior $\pi^l(\boldsymbol{\Psi})$ (also the Jeffreys-rule prior $\pi^J(\boldsymbol{\Psi})$), and the reference prior $\pi^R(\boldsymbol{\Psi})$. The following theorem is straightforward to establish.

Theorem 14.3 *For the star-shaped model, the posterior $\pi(\boldsymbol{\Psi} \mid \boldsymbol{S})$, under the prior $\pi(\boldsymbol{\Psi})$ in (14.11), has the following properties:*

(a) *$\pi(\boldsymbol{\Psi} \mid \boldsymbol{S})$ is proper iff $n + \alpha_{ij} + 1 > 0$, $j = 1, \cdots, p_i, i = 1, \cdots, k$.*

(b) *$\boldsymbol{\Psi}_{11}, (\boldsymbol{\Psi}_{21}, \boldsymbol{\Psi}_{22}), \cdots, (\boldsymbol{\Psi}_{k1}, \boldsymbol{\Psi}_{kk})$ are mutually independent;*

(c) *For $i = 2, \cdots, k$, the conditional distribution of $\boldsymbol{\Psi}_{i1}$ given $\boldsymbol{\Psi}_{ii}$ is $N_{p_i+p_1}(-\boldsymbol{\Psi}_{ii}\boldsymbol{S}_{i1}\boldsymbol{S}_{11}^{-1}, \boldsymbol{I}_{p_i} \otimes \boldsymbol{S}_{11}^{-1})$;*

(d) *For $i = 1, \cdots, k$, the marginal posteriors of $\boldsymbol{\Psi}_{ii}$ are independent, with densities*

$$\pi(\boldsymbol{\Psi}_{ii} \mid \boldsymbol{S}) \propto \exp\{-\tfrac{1}{2}tr(\boldsymbol{\Psi}_{ii}\boldsymbol{S}_{ii\cdot 1}\boldsymbol{\Psi}_{ii}')\}\textstyle\prod_{j=1}^{p_i} \psi_{ijj}^{n+\alpha_{ij}} \exp(-\beta_{ij}\psi_{ijj}^2).$$

Rather than using the posterior mean as an estimate of the precision matrix $\boldsymbol{\Omega}$, it is common to use the Bayesian estimate under the entropy (or Stein) loss

$$L(\hat{\boldsymbol{\Omega}}, \boldsymbol{\Omega}) = tr(\hat{\boldsymbol{\Omega}}\boldsymbol{\Omega}^{-1}) - \log|\hat{\boldsymbol{\Omega}}\boldsymbol{\Omega}^{-1}| - p. \tag{14.12}$$

The following theorem, whose proof can be found in Sun & Sun (2005), gives the Bayes estimate for this loss under the right-Haar prior (also the reference prior for suitable parameter orderings). It was also shown in Sun & Sun (2005) that this estimator is markedly superior to other estimators, such as the maximum likelihood estimate, the unbiased estimate, and the Bayes estimate arising from the left-Haar (also Jeffreys-rule) prior.

Theorem 14.4 *Under the entropy loss, the Bayes estimate of the precision matrix $\boldsymbol{\Omega}$, under the posterior arising from the right-Haar prior $\pi^r(\boldsymbol{\Psi})$, is*

$$\hat{\boldsymbol{\Omega}} = \boldsymbol{R}'\boldsymbol{B}^{-1}\boldsymbol{R},$$

where

$$
\boldsymbol{R} = \begin{pmatrix}
\boldsymbol{T}_{11}^{-1} & \boldsymbol{0} & \boldsymbol{0} & \cdots & \boldsymbol{0} \\
-\boldsymbol{T}_{22}^{-1}\boldsymbol{S}_{21}\boldsymbol{S}_{11}^{-1} & \boldsymbol{T}_{22}^{-1} & \boldsymbol{0} & \cdots & \boldsymbol{0} \\
-\boldsymbol{T}_{33}^{-1}\boldsymbol{S}_{31}\boldsymbol{S}_{11}^{-1} & \boldsymbol{0} & \boldsymbol{T}_{33}^{-1} & \cdots & \boldsymbol{0} \\
\vdots & \vdots & \vdots & \ddots & \vdots \\
-\boldsymbol{T}_{kk}^{-1}\boldsymbol{S}_{k1}\boldsymbol{S}_{11}^{-1} & \boldsymbol{0} & \boldsymbol{0} & \cdots & \boldsymbol{T}_{kk}^{-1}
\end{pmatrix},
$$

with $\boldsymbol{T}_{ii}$ *being the Cholesky decomposition of* $\boldsymbol{S}_{ii\cdot 1}$ (i.e., $\boldsymbol{S}_{ii\cdot 1} = \boldsymbol{T}_{ii}\boldsymbol{T}_{ii}'$), $\boldsymbol{B} = Diag(\boldsymbol{B}_1, \boldsymbol{B}_2, \cdots, \boldsymbol{B}_k)$ *and* $\boldsymbol{B}_i = Diag(b_{i1}, \cdots, b_{ip_i})$ *with*

$$
b_{ij} = \begin{cases} \frac{n-1}{(n-j-1)(n-j)}, & \textit{if } i = 1,\ j = 1, \cdots, p_1, \\ \frac{n-1}{(n-p_1-j-1)(n-p_1-j)}, & \textit{if } i = 2, \cdots, k,\ j = 1, \cdots, p_i. \end{cases}
$$

14.4 Reference Priors for Sequential Experiments

Statistical experiments are often conducted sequentially, with data arriving over a period of time. See, for example, Ghosh et al. (1997). Bartholomew (1965) and Geisser (1979) introduced the notion that objective priors for a sequential experiment should depend on the expected stopping time. Ye (1993) derived the reference prior for sequential experiments when the expected stopping time depends only on the parameter of interest. Sun & Berger (2008) generalized Ye's result in various directions, and provided some computational tools for use with priors that depend on expected stopping times.

14.4.1 Objective Priors with a Given Stopping Rule

Let $x_1, x_2, \ldots,$ be an iid sequence of random variables with density $p(x \mid \boldsymbol{\theta})$, and assume that the data arrive sequentially (i.e. x_1 is first observed, then x_2, etc.); here $\boldsymbol{\theta}$ is an m-vector of unknown parameters. Let N denote a *proper stopping time* for the sequential experiment, namely a series of functions

$$\{N_1(x_1), N_2(x_1, x_2), \ldots N_j(x_1, x_2, \ldots, x_j), \ldots\}$$

where $N_j(x_1, x_2, \ldots, x_j)$ gives the conditional probability (usually zero or one) of stopping the experiment, based on the data obtained to that point and given that the experiment did not previously stop; the word 'proper' indicates that, with probability one, sampling must stop eventually. For a formal definition, see Govindarajulu (1981), which also proves the following well-known lemma.

Lemma 14.1 *Let* $\boldsymbol{I}(\boldsymbol{\theta})$ *be the Fisher information matrix based on* x_1 *in a fixed sample side experiment. If the stopping time,* N*, has finite expectation* $E_{\boldsymbol{\theta}}[N]$*, the Fisher information matrix corresponding to the data* $(x_1, \ldots, x_N)$ *is*

$$\boldsymbol{I}^*(\boldsymbol{\theta}) = E_{\boldsymbol{\theta}}[N]\, \boldsymbol{I}(\boldsymbol{\theta}). \tag{14.13}$$

It follows from the above lemma that the Jeffreys-rule prior for $\boldsymbol{\theta}$ is

$$\pi^J(\boldsymbol{\theta}) = |\boldsymbol{I}^*(\boldsymbol{\theta})|^{1/2} = [\mathrm{E}_{\boldsymbol{\theta}}[N]]^{m/2}|\boldsymbol{I}(\boldsymbol{\theta})|^{1/2}. \tag{14.14}$$

Example 14.5 Let $x_1, x_2, \ldots$ be a sequence of Bernoulli random variables with unknown success probability θ. For a fixed positive integer r, define a stopping time, N_r, as follows:

$$N_r = \inf\{n \geq 1 : x_1 + \cdots + x_n = r\}.$$

The probability density of N_r can be shown to be

$$p(N_r = k) = \binom{k-1}{r-1}\theta^r(1-\theta)^{k-r}, \quad \text{for } k = r, r+1, \cdots,$$

which is recognizable as the density of the negative binomial distribution $\mathrm{Nb}(r, \theta)$. Thus we know that $\mathrm{E}_\theta[N_r] = r/\theta$. We have seen that the Fisher information for a Bernoulli random variable is $I(\theta) = 1/[\theta(1-\theta)]$, so (14.14) yields that the Jeffreys rule prior for the negative binomial distribution is $\pi^J(\theta) \propto (r/\theta)^{1/2}I(\theta)^{1/2} \propto 1/(\theta\sqrt{1-\theta})$. This, of course, also follows from a direct computation with the negative binomial distribution.

Here is an example of a continuous stopping time.

Example 14.6 Let $\{Z(t) : t > 0\}$ be a Brownian motion with constant drift θ and variance 1 per unit time, so $Z(t) \sim N(\theta t, t)$. Let $-\infty < a < 0 < b < \infty$, and let N_{ab} denote the random stopping time

$$N_{ab} = \inf\{t > 0 : Z(t) \leq a \text{ or } Z(t) \geq b\}.$$

It can be shown (see *e.g.*, Hall (1992)) that

$$\mathrm{E}_\theta[N_{ab}] = \begin{cases} \dfrac{1}{\theta}\left\{b - (b-a)\dfrac{e^{2b\theta}-1}{e^{2(b-a)\theta}-1}\right\}, & \text{if } \theta \neq 0, \\ -ab, & \text{if } \theta = 0. \end{cases}$$

The Jeffreys-rule prior when stopping at a fixed time is constant (Polson & Roberts (1993), Sivaganesan & Lingham (2002)), from which it follows that the Jeffreys-rule or reference prior for this situation is $\pi(\theta) = \sqrt{\mathrm{E}_\theta[N_{ab}]}$.

This is of additional interest because of the result from Brown (1988), that the commonly used estimate $Z(N_{ab})/N_{ab})$ of θ, which is the posterior mean under a constant prior, is inadmissible under estimation with squared error loss. Brown (1988) suggested that prior distributions behaving like $|\theta|^{-1}$ as $|\theta| \to \infty$ were optimal. The Jeffreys-rule and reference prior above can be shown to behave like $|\theta|^{-1/2}$ as $|\theta| \to \infty$, which is only part way to the optimal prior. But admissibility is dependent on the loss function used and it is likely that the Jeffreys-rule/reference prior would be admissible for an appropriate weighted squared-error loss.

14.4.2 Reference Priors under Asymptotic Normality

As in Section 14.2, suppose that the p-dimensional unknown parameter $\boldsymbol{\theta}$ can be decomposed into m sub-vectors, $\boldsymbol{\theta} = (\boldsymbol{\theta}_1, \ldots, \boldsymbol{\theta}_m)$, where $\boldsymbol{\theta}_i = (\theta_{i1}, \cdots, \theta_{ip_i})$. A sequential version of Theorem 14.1 is as follows. This is a slight generalization of a theorem in Sun & Berger (2008).

Theorem 14.5 *Consider the situation of Theorem 14.1, where the fixed sample size Fisher information matrix satisfies Conditions A1 and A2. Suppose, instead, that the data is collected sequentially, with a stopping time that satisfies*

$$E_{\boldsymbol{\theta}}[N] = \prod_{i=1}^{m} h_i(\boldsymbol{\theta}_i). \tag{14.15}$$

Then the reference prior for $\boldsymbol{\theta}$, in the sequential experiment, is

$$\pi^{R*}(\boldsymbol{\theta}) \propto \prod_{i=1}^{m} [h_i(\boldsymbol{\theta}_i)]^{p_i/2} \pi^R(\boldsymbol{\theta}), \tag{14.16}$$

where $\pi^R(\boldsymbol{\theta})$ is the fixed sample size reference prior given in (14.4).

Proof. Using the notation from Theorem 14.1, it is clear that the Fisher information matrix for the sequential experiment can be written as $\boldsymbol{G}^*(\boldsymbol{\theta})\boldsymbol{I}^*\boldsymbol{G}^*(\boldsymbol{\theta})$, where $\boldsymbol{G}^*(\boldsymbol{\theta}) = \sqrt{\mathrm{E}_{\boldsymbol{\theta}}[N]}\,\boldsymbol{G}(\boldsymbol{\theta})$. Under (14.15), it is clear that the assumptions of Theorem 14.1 are satisfied for this new $\boldsymbol{G}^*(\boldsymbol{\theta})$, and computation yields (14.16) as the reference prior. □

Corollary 14.1 *Suppose that the conditions of Theorem 14.5 hold and $\pi^R(\boldsymbol{\theta})$ is as in (14.4).*

(a) If $p_1 = \cdots = p_m = 1$, the resulting one-at-a-time reference prior for $\boldsymbol{\theta}$, in the sequential experiment, is

$$\pi^{R*}(\boldsymbol{\theta}) \propto \sqrt{E_{\boldsymbol{\theta}}[N]}\pi^R(\boldsymbol{\theta}).$$

(b) If $p_1 = \cdots = p_m = 2$, the resulting one-at-a-time reference prior for $\boldsymbol{\theta}$ in, the sequential experiment, is

$$\pi^{R*}(\boldsymbol{\theta}) \propto E_{\boldsymbol{\theta}}[N]\pi^R(\boldsymbol{\theta}).$$

Example 14.7 Let $x_1, x_2, \cdots$ be a sequence of random variables from (14.1). Bose & Boukai (1993) considered inference about $\theta_2 = \mathrm{E}[U_2(x_1)]$, in sequential experimentation, with the following stopping time:

$$N_a = \inf\Big\{n \geq m_0 : y_n < nG_1'\Big(-\frac{a^2}{n^2}\Big)\Big\}, \quad a \geq 0, \tag{14.17}$$

where $y_n = n^{-1}\sum_{i=1}^n U_1(x_i) - G_2\{n^{-1}\sum_{i=1}^n U_2(x_i)\}$ and $m_0 \geq 2$ is an initial sample size. Theorem 2 of Bose & Boukai (1993) shows that

$$\lim_{a\to\infty} \frac{N_a}{a} = \frac{1}{\sqrt{|\theta_1|}} \quad \text{and} \quad \lim_{a\to\infty} \mathrm{E}_{\boldsymbol{\theta}}\Big(\frac{N_a}{a}\Big) = \frac{1}{\sqrt{|\theta_1|}} \quad a.s.. \tag{14.18}$$

Bar-Lev & Reiser (1982) showed that the distribution of y_n does not depend on θ_2. Thus condition (14.15) is approximately satisfied, when either θ_1 or θ_2 is the parameter of interest and a is large. It follows from Corollary 14.1, (14.1), and (14.2), that, for large a, the Jeffreys-rule prior and one-at-a-time reference prior are, approximately,

$$\pi^{J*}(\theta_1, \theta_2) \propto \sqrt{G_1''(\theta_1)G_2''(\theta_2)}, \quad \pi^{R*}(\theta_1, \theta_2) \propto |\theta_1|^{-\frac{1}{4}}\pi^{J*}(\theta_1, \theta_2). \tag{14.19}$$

As a specific example, if the data arise from the $\mathrm{N}(x \mid \mu, \sigma^2)$ distribution, Table 14.1 gives $\theta_1 = -1/2\sigma^2, \theta_2 = \mu$, $G_1'(\theta_1) = -1/2\theta_1$, and $y_n = \sum_{i=1}^{n}(x_i - \overline{x}_n)^2$. The stopping rule (14.17) becomes

$$N_a = \inf\Big\{n \geq m_0 : \frac{1}{n}\sum_{i=1}^{n}(x_i - \overline{x}_n)^2 < \frac{n^2}{2a^2}\Big\}.$$

So the priors in (14.1), (14.2), and (14.19) are, respectively,

$$\pi^J(\mu, \sigma^2) \propto \frac{1}{(\sigma^2)^{3/2}}, \ \pi^R(\mu, \sigma^2) \propto \frac{1}{\sigma^2}, \ \pi^{J*}(\mu, \sigma^2) \propto \frac{1}{\sigma^2}, \ \pi^{R*}(\mu, \sigma^2) \propto \frac{1}{(\sigma^2)^{3/4}}\,.$$

14.4.3 Computation

If $\mathrm{E}_{\boldsymbol{\theta}}[N]$ is available in closed form, as in the previous examples in this chapter, computation with any of the sequential objective priors can be done using the MCMC techniques discussed in Section 1.8. Hence we now consider the case where $\mathrm{E}_{\boldsymbol{\theta}}[N]$ can only computed numerically, while the fixed sample size posterior is available in closed form.

Brute Force Computation. The Jeffreys-rule and reference priors that have been discussed for a sequential experiment are of the form $h(\mathrm{E}_{\boldsymbol{\theta}}[N])\pi^F(\boldsymbol{\theta})$, where h is some positive function and π^F is the corresponding prior for the fixed sample size experiment. The posterior density of $\boldsymbol{\theta}$ is then

$$\pi(\boldsymbol{\theta} \mid \boldsymbol{x}_N) \propto h(\mathrm{E}_{\boldsymbol{\theta}}[N])\pi^F(\boldsymbol{\theta}) \prod_{i=1}^{N} p(x_i \mid \boldsymbol{\theta}), \tag{14.20}$$

where $\boldsymbol{x}_N = (x_1, \cdots, x_N)$ is the observed data.

A brute force method for simulating from this posterior, when $\pi^F(\boldsymbol{\theta})$ and $\prod_{i=1}^{N} p(x_i \mid \boldsymbol{\theta})$ are available in closed form, is the following Metropolis algorithm:

Step 1. Sample a proposed $\boldsymbol{\theta}^*$ from the fixed sample size posterior density of $\boldsymbol{\theta}$, which is proportional to $\pi^F(\boldsymbol{\theta})\prod_{i=1}^{N} p(x_i \mid \boldsymbol{\theta})$. (This may also require MCMC methodology; if so, iterate this MCMC enough times to achieve near independence of successive proposals, as the steps of this MCMC are typically much cheaper than *Step 2* below.)

Step 2. Numerically estimate $\mathrm{E}_{\boldsymbol{\theta}^*}[N]$. For instance, one could sample N from its distribution, given $\boldsymbol{\theta}^*$, by repeatedly simulating the sequential experiment for the given $\boldsymbol{\theta}^*$, observing the N that results from each simulation, and averaging these N to obtain the estimate $\widehat{\mathrm{E}_{\boldsymbol{\theta}^*}[N]}$.

Step 3. Perform a Metropolis step by sampling $u \sim \mathrm{Un}(u \mid 0,1)$ and, with $\boldsymbol{\theta}$ denoting the previous sampled value of the parameter, accepting the new proposed $\boldsymbol{\theta}^*$ if

$$u \le \min\left\{1, \frac{h(\widehat{\mathrm{E}_{\boldsymbol{\theta}^*}[N]})}{h(\widehat{\mathrm{E}_{\boldsymbol{\theta}}[N]})}\right\}, \tag{14.21}$$

and setting $\boldsymbol{\theta}^*$ equal to the previous $\boldsymbol{\theta}$ otherwise.

The Two-Dimensional Case. If using the Jeffreys-rule prior in a two-dimensional problem or the reference prior in the situation of Corollary 14.1 (b), the posterior distribution is

$$\pi(\boldsymbol{\theta} \mid \boldsymbol{x}_N) \propto \mathrm{E}_{\boldsymbol{\theta}}[N]\pi^F(\boldsymbol{\theta}) \prod_{i=1}^{N} p(x_i \mid \boldsymbol{\theta}) = \sum_{\tilde{N}=1}^{\infty} \tilde{N} p(\tilde{N} \mid \boldsymbol{\theta}) \pi^F(\boldsymbol{\theta}) \prod_{i=1}^{N} p(x_i \mid \boldsymbol{\theta}) \tag{14.22}$$

where the integrand of the expectation has been changed to $\tilde{N}$ to avoid confusion, and $p(\tilde{N} \mid \boldsymbol{\theta})$ is the same density, given $\boldsymbol{\theta}$, as that of N.

This allows a wonderful simplification in the computation, because the summation sign in (14.22) can be dropped, with $\tilde{N}$ being treated as a latent variable and MCMC used to generate all unknowns from the extended posterior

$$\pi(\tilde{N}, \boldsymbol{\theta} \mid \boldsymbol{x}_N) \propto \tilde{N}\, p(\tilde{N} \mid \boldsymbol{\theta})\, \pi^F(\boldsymbol{\theta}) \prod_{i=1}^{N} p(x_i \mid \boldsymbol{\theta})\,. \tag{14.23}$$

Here is a Metropolis algorithm for sampling from (14.23).

Step 1. Sample a proposed $\boldsymbol{\theta}^*$, from the fixed sample size posterior density of $\boldsymbol{\theta}$, which is proportional to $\pi^F(\boldsymbol{\theta}) \prod_{i=1}^{N} p(x_i \mid \boldsymbol{\theta})$ (using MCMC, if necessary).

Step 2. Sample a proposed $\tilde{N}^*$ from $p(\tilde{N} \mid \boldsymbol{\theta}^*)$. This can always be done by simulating the sequential experiment once, given $\boldsymbol{\theta}^*$, *i.e.*, generating data until the stopping rule says stop, with $\tilde{N}^*$ being the resulting sample size.

Step 3. Perform a Metropolis step: sample $u \sim \mathrm{Un}(u \mid 0,1)$ and, letting $(\tilde{N}, \boldsymbol{\theta})$ denote the previous values of the parameters, accept $(\tilde{N}^*, \boldsymbol{\theta}^*)$ if

$$u \le \min\left\{1, \frac{\tilde{N}^*}{\tilde{N}}\right\}, \tag{14.24}$$

and set $(\tilde{N}^*, \boldsymbol{\theta}^*)$ equal to the previous $(\tilde{N}, \boldsymbol{\theta})$ otherwise. (Clearly, if $\tilde{N}^* \ge \tilde{N}$, one would accept the candidate.)

This is much more efficient than the brute force algorithm because one only simulates a single draw of $\tilde{N}^*$ in *Step 2*, whereas a huge number of draws would typically be needed for the brute force algorithm to estimate $\widehat{\mathrm{E}_{\boldsymbol{\theta}^*}[N]}$.

14.4.4 Modified Reference Priors

The recommended prior is the one-at-a-time reference prior, given in Corollary 14.1 (a), resulting in the posterior distribution

$$\pi^{R*}(\boldsymbol{\theta} \mid \boldsymbol{x}_N) \propto \sqrt{\mathrm{E}_{\boldsymbol{\theta}}[N]}\, \pi^R(\boldsymbol{\theta}) \prod_{i=1}^{N} p(x_i \mid \boldsymbol{\theta})\,. \tag{14.25}$$

While the latent variable approach is not directly available for sampling from this distribution, it is often the case that

$$\sqrt{\mathrm{E}_{\boldsymbol{\theta}}[N]} \approx \mathrm{E}_{\boldsymbol{\theta}}[\sqrt{N}]\,, \tag{14.26}$$

leading to the *modified reference prior*

$$\pi^{MR}(\boldsymbol{\theta} \mid \boldsymbol{x}_N) \propto \mathrm{E}_{\boldsymbol{\theta}}[\sqrt{N}]\, \pi^R(\boldsymbol{\theta}) \prod_{i=1}^{N} p(x_i \mid \boldsymbol{\theta})\,. \tag{14.27}$$

The latent variable approach can be directly applied to the modified reference prior; simply replace $\tilde{N}/\tilde{N}^*$ in the Metropolis step (14.24) with $\sqrt{\tilde{N}/\tilde{N}^*}$.

Example 14.8 (Example 14.7, continued). For the stopping time N_a defined in (14.17), it follows from (14.18) that $\lim_{a\to\infty} \mathrm{E}_{\boldsymbol{\theta}}[\sqrt{N_a/a}] = 1/|\theta_1|^{1/4}$ $= \lim_{a\to\infty} \sqrt{\mathrm{E}_{\boldsymbol{\theta}}[N_a/a]}$, so that (14.26) holds for large a.

Example 14.9 (Example 14.5, continued). Recall that N_r has the $\mathrm{Nb}(r,\theta)$ distribution and that $\mathrm{E}_\theta[N_r] = r/\theta$, so that $\mathrm{E}_\theta[N_r/r] = \sqrt{\theta}$. It can be shown that $\lim_{r\to\infty} \mathrm{E}_\theta[\sqrt{N_r/r}\,] = \sqrt{\theta}$, so that (14.26) holds for large r.

To study the difference between $\mathrm{E}_\theta[\sqrt{N_r/r}]$ and $\sqrt{\mathrm{E}_\theta[N_r/r]}$ for moderate r, the two functions of θ are plotted in Figure 14.2, for $r = 1$ and $r = 10$. For $r = 10$, the functions are essentially indistinguishable; even for the minimal $r = 1$ they are close.

It is also of interest to compare the posterior distributions of θ here. In Figure 14.3, the posterior densities are plotted for three priors, the non-sequential Jeffreys prior $\pi^J(\theta) \propto [\theta(1-\theta)]^{-1/2}$, the sequential reference prior $\pi^{R*}(\theta) \propto 1/(\theta\sqrt{1-\theta})$ and the modified reference prior $\pi^{MR}(\theta) \propto \mathrm{E}_\theta[\sqrt{N_r}][\theta(1-\theta)]^{-1/2}$. Even for the very small $r = 2$, the posterior densities under the reference and modified reference priors are quite close, yet differ substantially from the posterior under π^J. For the moderate $r = 10$, the reference and modified reference posteriors are almost identical.

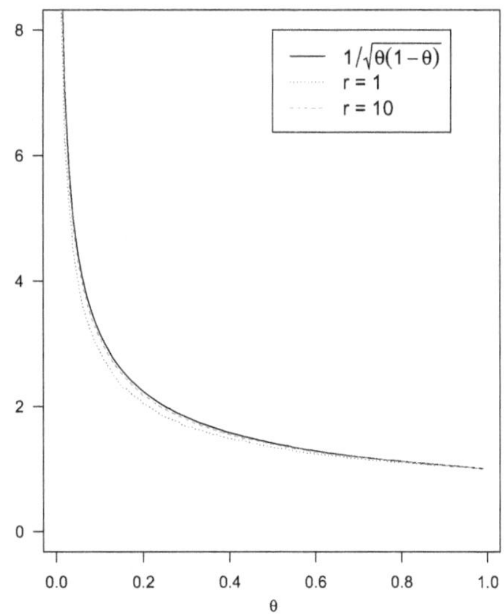

Figure 14.2: Negative binomial example: comparison of $\sqrt{\mathrm{E}_\theta[N_r/r]} = 1/\sqrt{\theta(1-\theta)}$ and $\mathrm{E}_\theta[\sqrt{N_r/r}]$ for $r = 1$ and $r = 10$.

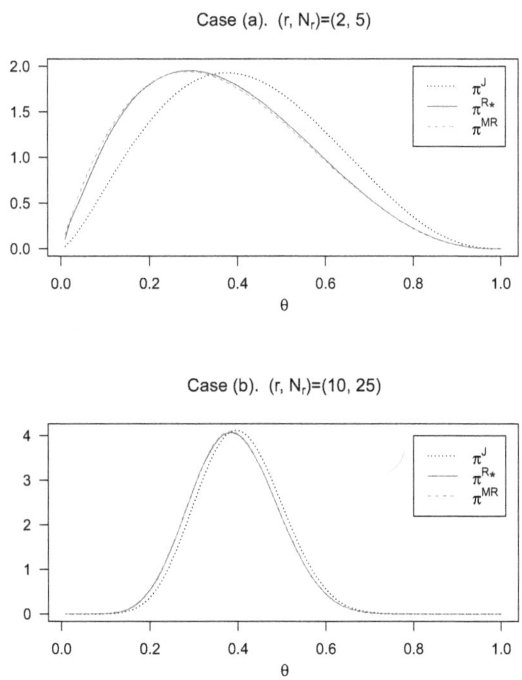

Figure 14.3: Posterior densities of θ based on the priors π^J, π^{R*}, and π^{MR}, when (a) $(r, N_r) = (2, 5)$; (b) $(r, N_r) = (10, 25)$.

Table 14.2: Coverage probability of one-sided 5% and 95% (in parentheses) Bayesian credible intervals, under the three priors π^J, π^{R*}, and π^{MR}.

r	θ	π^J	π^{R*}	π^{MR}
2	0.1	.1142(.9738)	.0516(.9511)	.0487(.9509)
2	0.5	.0002(.9652)	.0010(.9381)	.0008(.9455)
2	0.9	.0001(.9724)	.0003(.9700)	.0000(.9729)
8	0.1	.0751(.9642)	.0474(.9498)	.0465(.9534)
8	0.5	.0552(.9688)	.0522(.9536)	.0568(.9517)
8	0.9	.0000(.9307)	.0001(.9310)	.0002(.9339)
30	0.1	.0617(.9571)	.0508(.9497)	.0516(.9523)
30	0.5	.0556(.9594)	.0512(.9495)	.0525(.9503)
30	0.9	.0426(.9369)	.0438(.9410)	.0442(.9368)

As a final indication of the similarity of the reference and modified reference priors in this example, and to see the value of using the sequential reference priors, we compare the frequentist coverage probabilities that result from their use in obtaining confidence intervals for θ. Table 14.2 considers the frequentist coverage of one-sided 5% and 95% Bayesian credible intervals, based on π^{MR}, π^{R*}, and π^J, for various combination of r and θ. The fixed sample size Jeffreys prior performs worse then the other two, indicating the value of using the sequential versions, while the reference prior and the modified reference prior are essentially equally good.

14.5 Discussion and Additional References

Reference priors have been developed for discrete graphical models by Consonni & Leucari (2006). Consonni & Massam (2012) showed that reparameterization could be important and interesting in multinomial decomposable graphical models. Related work can be found for discrete hierarchical log-linear models by Massam et al. (2009). Peterson et al. (2009) proposed Bayesian inference of multiple Gaussian graphical models.

Chapter 15

A Catalog of Objective Priors

Creation of a catalog of objective priors was initiated in Yang & Berger (1997). Thought was given to including the catalog in this book, but the catalog should be a living repository, able to be augmented and updated as new objective priors are developed. The current version of the catalog is available at the website for this book, `www.worldscientific.com/worldscibooks/10.1142/13640#t=suppl`.

The catalog is organized around statistical models, with the models being listed in alphabetical order. Each model entry is kept as self contained as possible. Listed for each are (i) the model density; (ii) various objective priors that have been considered for the model; and (iii) some of the resulting posteriors and their properties, including computational implementations. The latter information is often limited to the recommended prior (usually the reference prior).

Appendix A

Common Distributions

This appendix consists of a set of tables which record the notation, parameter range, variable range, definition, and first two moments of the probability distributions (discrete and continuous, univariate and multivariate) used in this book.

A.1 Univariate Discrete Distributions

$\text{Br}(x \mid \theta)$	*Bernoulli*
$0 < \theta < 1$	$x = 0, 1$
$p(x) = \theta^x (1-\theta)^{1-x}$	
$\text{E}[x] = \theta$	$\text{Var}[x] = \theta(1-\theta)$

$\text{Bi}(x \mid \theta, n)$	*Binomial*
$0 < \theta < 1,\ n = 1, 2, \dots$	$x = 0, 1, \dots, n$
$p(x) = \binom{n}{x} \theta^x (1-\theta)^{n-x}$	
$\text{E}[x] = n\theta$	$\text{Var}[x] = n\theta(1-\theta)$

$\text{BiBe}(x \mid \alpha, \beta, n)$	*Binomial-Beta*
$\alpha > 0,\ \beta > 0,\ n = 1, 2, \dots$	$x = 0, 1, \dots, n$
$p(x) = c \binom{n}{x} \Gamma(\alpha + x)\Gamma(\beta + n - x)$	$c = \dfrac{\Gamma(\alpha+\beta)}{\Gamma(\alpha)\Gamma(\beta)\Gamma(\alpha+\beta+n)}$
$\text{E}[x] = n\frac{\alpha}{\alpha+\beta}$	$\text{Var}[x] = \dfrac{n\alpha\beta}{(\alpha+\beta)^2} \dfrac{(\alpha+\beta+n)}{(\alpha+\beta+1)}$

$\text{Hy}(x \mid N, M, n)$	*Hypergeometric*
$N = 1, 2, \dots$	$x = a, a+1, \dots, b$
$M = 1, 2, \dots$	$a = \max(0, n - M)$
$n = 1, \dots, N + M$	$b = \min(n, N)$
$p(x) = c \binom{N}{x} \binom{M}{n-x}$	$c = \binom{N+M}{n}^{-1}$
$E[x] = n\dfrac{N}{N+M}$	$V[x] = \dfrac{nNM}{(N+M)^2} \dfrac{N+M-n}{N+M-1}$

$\text{Nb}(x \mid \theta, r)$	*Negative-Binomial*
$0 < \theta < 1,\ r = 1, 2, \dots$	$x = 0, 1, 2, \dots$
$p(x) = c \binom{r+x-1}{r-1} (1-\theta)^x$	$c = \theta^r$
$\text{E}[x] = \dfrac{r(1-\theta)}{\theta}$	$\text{Var}[x] = \dfrac{r(1-\theta)}{\theta^2}$

$\text{NeBiBe}(x \mid \alpha, \beta, r)$	*Negative-Binomial-Beta*
$\alpha > 0,\ \beta > 0,\ r = 1, 2 \dots$	$x = 0, 1, 2, \dots$
$p(x) = c \binom{r+x-1}{r-1} \dfrac{\Gamma(\beta+x)}{\Gamma(\alpha+\beta+r+x)}$	$c = \dfrac{\Gamma(\alpha+\beta)\Gamma(\alpha+r)}{\Gamma(\alpha)\Gamma(\beta)}$
$\text{E}[x] = \dfrac{r\beta}{\alpha-1}$	$\text{Var}[x] = \dfrac{r\beta}{\alpha-1} \left[\dfrac{\alpha+\beta+r-1}{(\alpha-2)} + \dfrac{r\beta}{(\alpha-1)(\alpha-2)} \right]$

Univariate Discrete Distributions (continued)

$\mathrm{Pn}(x \mid \lambda)$	*Poisson*
$\lambda > 0$	$x = 0, 1, 2, \ldots$
$p(x) = c\dfrac{\lambda^x}{x!}$	$c = e^{-\lambda}$
$\mathrm{E}[x] = \lambda$	$\mathrm{Var}[x] = \lambda$

A.2 Univariate Continuous Distributions

$\mathrm{Pn}(x \mid \alpha, \beta, n)$	*Poisson-Gamma*
$\alpha > 0,\ \beta > 0,\ \gamma > 0$	$x = 0, 1, 2, \ldots$
$p(x) = c\dfrac{\Gamma(\alpha + x)}{x!}\dfrac{\gamma^x}{(\beta + \gamma)^{\alpha + x}}$	$c = \dfrac{\beta^\alpha}{\Gamma(\alpha)}$
$\mathrm{E}[x] = \gamma\dfrac{\alpha}{\beta}$	$\mathrm{Var}[x] = \dfrac{\gamma\alpha}{\beta}\left[1 + \dfrac{\gamma}{\beta}\right]$

$\mathrm{Be}(x \mid \alpha, \beta)$	*Beta*
$\alpha > 0,\ \beta > 0$	$0 < x < 1$
$p(x) = c\, x^{\alpha - 1}(1 - x)^{\beta - 1}$	$c = \dfrac{\Gamma(\alpha + \beta)}{\Gamma(\alpha)\Gamma(\beta)}$
$\mathrm{E}[x] = \dfrac{\alpha}{\alpha + \beta}$	$\mathrm{Var}[x] = \dfrac{\alpha\beta}{(\alpha + \beta)^2(\alpha + \beta + 1)}$

$\mathrm{Un}(x \mid a, b)$	*Uniform*
$b > a$	$a < x < b$
$p(x) = c$	$c = (b - a)^{-1}$
$\mathrm{E}[x] = \frac{1}{2}(a + b)$	$\mathrm{Var}[x] = \frac{1}{12}(b - a)^2$

$\mathrm{Ga}(x \mid \alpha, \beta)$	*Gamma*
$\alpha > 0,\ \beta > 0$	$x > 0$
$p(x) = c\, x^{\alpha - 1}e^{-\beta x}$	$c = \dfrac{\beta^\alpha}{\Gamma(\alpha)}$
$\mathrm{E}[x] = \alpha\beta^{-1}$	$\mathrm{Var}[x] = \alpha\beta^{-2}$

$\mathrm{Ex}\ (x \mid \theta)$	*Exponential*
$\theta > 0$	$x > 0$
$p(x) = c\, e^{-\theta x}$	$c = \theta$
$\mathrm{E}[x] = 1/\theta$	$\mathrm{Var}[x] = 1/\theta^2$

Univariate Continuous Distributions (continued)

$\mathrm{Gg}(x \mid \alpha, \beta, n)$	*Gamma-Gamma*
$\alpha > 0,\ \beta > 0,\ n > 0$	$x > 0$
$p(x) = c\,\dfrac{x^{n-1}}{(\beta + x)^{\alpha+n}}$	$c = \dfrac{\beta^\alpha}{\Gamma(\alpha)}\ \dfrac{\Gamma(\alpha+n)}{\Gamma(n)}$
$\mathrm{E}[x] = n\dfrac{\beta}{\alpha-1}$	$\mathrm{Var}[x] = \dfrac{\beta^2(n^2 + n(\alpha-1))}{(\alpha-1)^2(\alpha-2)}$
$\chi^2(x \mid \nu) = \chi^2_\nu$	*Chi-squared*
$\nu > 0$	$x > 0$
$p(x) = c\,x^{(\nu/2)-1}e^{-x/2}$	$c = \dfrac{(1/2)^{\nu/2}}{\Gamma(\nu/2)}$
$\mathrm{E}[x] = \nu$	$\mathrm{Var}[x] = 2\nu$
$\chi^2(x \mid \nu, \lambda)$	*Non-central Chi-squared*
$\nu > 0,\ \lambda > 0$	$x > 0$
$p(x) = \sum_{i=0}^{\infty} \mathrm{Pn}\ (i \mid \frac{\lambda}{2})\chi^2(x \mid \nu + 2i)$	
$\mathrm{E}[x] = \nu + \lambda$	$\mathrm{Var}[x] = 2(\nu + 2\lambda)$
$\mathrm{Ig}(x \mid \alpha, \beta)$	*Inverse Gamma*
$\alpha > 0,\ \beta > 0$	$x > 0$
$p(x) = c\,x^{-(\alpha+1)}e^{-\beta/x}$	$c = \dfrac{\beta^\alpha}{\Gamma(\alpha)}$
$\mathrm{E[x]}=\dfrac{\beta}{\alpha-1}$	$\mathrm{Var[x]}=\dfrac{\beta^2}{(\alpha-1)^2(\alpha-2)}$
$\chi^{-1}(x \mid \nu)$	*Inverse Chi-squared*
$\nu > 0$	$x > 0$
$p(x) = c\ x^{-(\nu/2+1)}e^{-1/2x^2}$	$c = \dfrac{(1/2)^{\nu/2}}{\Gamma(\nu/2)}$
$\mathrm{E}[x] = \dfrac{1}{\nu-2}$	$Var[x] = \dfrac{2}{(\nu-2)^2(\nu-4)}$
$\mathrm{Ga}^{-1/2}(x \mid \alpha, \beta)$	*Square-root Inverse Gamma*
$\alpha > 0,\ \beta > 0$	$x > 0$
$p(x) = c\ x^{-(2\alpha+1)}e^{-\beta/x^2}$	$c = \dfrac{2\beta^\alpha}{\Gamma(\alpha)}$
$\mathrm{E}[x] = \dfrac{\sqrt{\beta}\,\Gamma(\alpha-1/2)}{\Gamma(\alpha)}$	$\mathrm{Var}[x] = \dfrac{\beta}{\alpha-1} - \mathrm{E}[x]^2$
$\mathrm{Pa}(x \mid \alpha, \beta)$	*Pareto*
$\alpha > 0,\ \beta > 0$	$\beta \le x < +\infty$
$p(x) = c\,x^{-(\alpha+1)}$	$c = \alpha\beta^\alpha$
$\mathrm{E}[x] = \beta\alpha(\alpha-1)^{-1}$	$\mathrm{Var}[x] = \beta^2\alpha(\alpha-1)^{-2}(\alpha-2)^{-1}$

Univariate Continuous Distributions (continued)

$\mathrm{Ip}(x \mid \alpha, \beta)$	*Inverse Pareto*
$\alpha > 0, \quad \beta > 0$	$0 < x < \beta^{-1}$
$p(x) = c\, x^{\alpha-1}$	$c = \alpha\beta^{\alpha}$
$\mathrm{E}[x] = \beta^{-1}\alpha(\alpha+1)^{-1}$	$\mathrm{Var}[x] = \beta^{-2}\alpha(\alpha+1)^{-2}(\alpha+2)^{-1}$
$\mathrm{N}(x \mid \mu, \sigma^2)$	*Normal*
$-\infty < \mu < +\infty,\ \sigma > 0$	$-\infty < x < +\infty$
$p(x) = c\ \exp\left\{-\frac{1}{2\sigma^2}(x-\mu)^2\right\}$	$c = (2\pi\sigma^2)^{-1/2}$
$\mathrm{E}[x] = \mu$	$\mathrm{Var}[x] = \sigma^2$
$\mathrm{St}(x \mid \mu, \lambda, \alpha)$ *Student t*	
$-\infty < \mu < +\infty,\ \lambda > 0,\ \alpha > 0$	$-\infty < x < +\infty$
$p(x) = c\left[1 + \frac{1}{\alpha}\left(\frac{x-\mu}{\sigma}\right)^2\right]^{-(\alpha+1)/2}$	$c = \frac{\Gamma\left(\frac{1}{2}(\alpha+1)\right)}{\Gamma(\frac{1}{2}\alpha)}\left(\frac{1}{\alpha\pi\sigma^2}\right)^{1/2}$
$\mathrm{E}[x] = \mu$	$\mathrm{Var}[x] = \frac{\alpha\sigma^2}{\alpha-2}$
$\mathrm{Lo}(x \mid \alpha, \beta)$	*Logistic*
$-\infty < \alpha < +\infty,\ \beta > 0$	$-\infty < x < +\infty$
$p(x) = \frac{\exp\{-(x-\alpha)/\beta\}}{\beta\,[1+\exp\{-(x-\alpha)/\beta\}]^2}$	
$\mathrm{E}[x] = \alpha$	$Var[x] = \beta^2\pi^2/3$

A.3 Multivariate Discrete Distributions

$\mathrm{Mu}_k(\boldsymbol{x} \mid \boldsymbol{\theta}, n)$	*Multinomial*
$\boldsymbol{\theta} = (\theta_1, \ldots, \theta_k)$	$\boldsymbol{x} = (x_1, \ldots, x_k)$
$0 < \theta_i < 1, \quad \sum_{\ell=1}^{k} \theta_\ell \leq 1$	$x_i = 0, 1, 2, \ldots, n, \ \sum_{\ell=1}^{k} x_\ell \leq n, \quad n = 1, 2, \ldots$
$p(\boldsymbol{x}) = \dfrac{n!}{\prod_{\ell=1}^{k+1} x_\ell!} \prod_{\ell=1}^{k+1} \theta^{x_\ell},$	$\theta_{k+1} = 1 - \sum_{\ell=1}^{k} \theta_\ell, \quad x_{k+1} = n - \sum_{\ell=1}^{k} x_\ell$
$\mathrm{E}[x_i] = n\theta_i$	$\mathrm{Var}[x_i] = n\theta_i(1 - \theta_i), \ \ C[x_i, x_j] = -n\theta_i\theta_j$

$\mathrm{Md}_k(\boldsymbol{x} \mid \boldsymbol{\theta}, n)$	*Multinomial-Dirichlet*
$\boldsymbol{\alpha} = (\alpha_1, \ldots, \alpha_{k+1})$	$\boldsymbol{x} = (x_1, \ldots, x_k)$
$\alpha_i > 0$	$x_i = 0, 1, \ldots, n, \ \sum_{\ell=1}^{n} x_l \leq n, \quad n = 1, 2, \ldots$
$p(\boldsymbol{x}) = c \prod_{\ell=1}^{k+1} \dfrac{\alpha_\ell^{[x_\ell]}}{x_\ell!}$	$c = \dfrac{n!}{\left(\sum_{\ell=1}^{k+1} \alpha_\ell\right)^{[n]}}$
$\alpha^{[s]} = \prod_{\ell=1}^{s} (\alpha + \ell - 1)$	$x_{k+1} = n - \sum_{\ell=1}^{k} x_\ell$
$\mathrm{E}[x_i] = np_i$	$\mathrm{Var}[x_i] = \dfrac{n + \sum_{\ell=1}^{k+1} \alpha_\ell}{1 + \sum_{\ell=1}^{k+1} \alpha_\ell} \, np_i(1 - p_i)$
$p_i = \dfrac{\alpha_i}{\sum_{\ell=1}^{k+1} \alpha_\ell}$	$C[x_i, x_j] = -\dfrac{n + \sum_{\ell=1}^{k+1} \alpha_\ell}{1 + \sum_{\ell=1}^{k+1} \alpha_\ell} \, np_ip_j$

A.4 Multivariate Continuous Distributions

$\mathrm{Di}_k(\boldsymbol{x} \mid \boldsymbol{\alpha})$	*Dirichlet*
$\boldsymbol{\alpha} = (\alpha_1, \ldots, \alpha_{k+1})$	$\boldsymbol{x} = (x_1, \ldots, x_k)$
$\alpha_i > 0$	$0 < x_i < 1, \quad \sum_{\ell=1}^{k} x_\ell \leq 1$
$p(\boldsymbol{x}) = c\left(1 - \sum_{\ell=1}^{k} x_\ell\right)^{\alpha_{k+1}-1} \prod_{\ell=1}^{k} x_\ell^{\alpha_\ell - 1}$	$c = \dfrac{\Gamma(\sum_{\ell=1}^{k+1} \alpha_\ell)}{\prod_{\ell=1}^{k+1} \Gamma(\alpha_\ell)}$
$\mathrm{E}[x_i] = \dfrac{\alpha_i}{\sum_{\ell=1}^{k+1} \alpha_\ell}$	$\mathrm{Var}[x_i] = \dfrac{\mathrm{E}[x_i](1 - \mathrm{E}[x_i])}{1 + \sum_{\ell=1}^{k+1} \alpha_\ell}$
	$C[x_i, x_j] = \dfrac{-\mathrm{E}[x_i]\mathrm{E}[x_j]}{1 + \sum_{\ell=1}^{k+1} \alpha_\ell}$

$\mathrm{Ng}(x, y \mid \mu, \lambda, \alpha, \beta)$	*Normal-Gamma*
$\mu \in \mathbb{R},\ \lambda > 0,\ \alpha > 0,\ \beta > 0,$	$x \in \mathbb{R},\ y > 0$
$p(x, y) = N(x \mid \mu, \lambda y)\ \mathrm{Ga}(y \mid \alpha, \beta)$	
$\mathrm{E}[x] = \mu$	$\mathrm{Var}[x] = \beta\lambda^{-1}(\alpha - 1)^{-1}$
$\mathrm{E}[y] = \alpha\beta^{-1}$	$\mathrm{Var}[y] = \alpha\beta^{-2}$
$p(x) = \mathrm{St}\ (x \mid \mu, \alpha\beta^{-1}\lambda, 2\alpha)$	

$\mathrm{N}_k(\boldsymbol{x} \mid \boldsymbol{\mu}, \boldsymbol{\Sigma})$	*Multivariate Normal*
$\boldsymbol{\mu} = (\mu_1, \ldots, \mu_k) \in \mathbb{R}^k$	$\boldsymbol{x} = (x_1, \ldots, x_k) \in \mathbb{R}^k$
$\boldsymbol{\Sigma}:$ *symmetric positive-definite*	
$p(\boldsymbol{x}) = c\ \exp\left\{-\frac{1}{2}(\boldsymbol{x} - \boldsymbol{\mu})\boldsymbol{\Sigma}^{-1}(\boldsymbol{x} - \boldsymbol{\mu})'\right\}$	$c = (2\pi)^{-k/2}\lVert\boldsymbol{\Sigma}\rVert^{-1/2}$
$\mathrm{E}[\boldsymbol{x}] = \boldsymbol{\mu}$	$\mathrm{Var}[\boldsymbol{x}] = \boldsymbol{\Sigma}$

$\mathrm{Pa}_2(x, y \mid \alpha, \beta_0, \beta_1)$	*Bilateral Pareto*
$(\beta_0, \beta_1) \in \mathbb{R}^2,\ \beta_0 < \beta_1,\ \alpha > 0$	$(x, y) \in \mathbb{R}^2,\ x < \beta_0,\ y > \beta_1$
$p(x, y) = c\ (y - x)^{-(\alpha+2)}$	$c = \alpha(\alpha + 1)(\beta_1 - \beta_0)^\alpha$
$\mathrm{E}[x] = \dfrac{\alpha\beta_0 - \beta_1}{\alpha - 1}\ \mathrm{E}[y] = \dfrac{\alpha\beta_1 - \beta_0}{\alpha - 1}$	$\mathrm{Var}[x] = Var[y] = \dfrac{\alpha(\beta_1 - \beta_0)^2}{(\alpha - 1)^2(\alpha - 2)}$

$\mathrm{Ng}_k(\boldsymbol{x}, \boldsymbol{y} \mid \boldsymbol{\mu}, \boldsymbol{\Lambda}, \alpha, \beta)$	*Multivariate Normal-Gamma*
$-\infty < \mu_i < +\infty,\ \alpha > 0,\ \beta > 0$	$\boldsymbol{x} = (x_1, \ldots, x_k)$
$\boldsymbol{\Lambda}:$ *symmetric positive-definite*	$-\infty < x_i < \infty, \quad y > 0$
$p(\boldsymbol{x}, y) = N_k(\boldsymbol{x} \mid \boldsymbol{\mu}, \boldsymbol{\Lambda} y)\ \ \mathrm{Ga}(y \mid \alpha, \beta)$	$\mathrm{Var}[\boldsymbol{x}] = (\alpha - 1)^{-1}\beta\boldsymbol{\Lambda}^{-1}$
$\mathrm{E}[\boldsymbol{x}, y] = (\boldsymbol{\mu},\ \alpha\beta^{-1})$	$\mathrm{Var}[y] = \alpha\beta^{-2}$
$p(\boldsymbol{x}) = \mathrm{St}_k(\boldsymbol{x} \mid \boldsymbol{\mu}, \boldsymbol{\Lambda}\alpha\beta^{-1}, 2\alpha)$	$p(y) = \mathrm{Ga}(y \mid \alpha, \beta)$

Multivariate Continuous Distributions (continued)

$\mathrm{Nw}_k(\boldsymbol{x}, \boldsymbol{Y} \mid \boldsymbol{\mu}, \lambda, \alpha, \boldsymbol{\Omega})$	*Multivariate Normal-Wishart*
$-\infty < \mu_i < +\infty,\ \lambda > 0$	$\boldsymbol{x} = (x_1, \ldots, x_k)$
$2\alpha > k - 1$	$-\infty < x_i < +\infty$
$\boldsymbol{\Omega}$: *symmetric non-singular*	$\boldsymbol{Y}$: *symmetric positive-definite*
$p(\boldsymbol{x}, \boldsymbol{Y}) = \mathrm{N}_k(\boldsymbol{x} \mid \boldsymbol{\mu}, \lambda\boldsymbol{Y})\ \ \mathrm{Wi}_k(\boldsymbol{Y} \mid \alpha, \boldsymbol{\Omega})$	
$\mathrm{E}[\boldsymbol{x}, \boldsymbol{Y}] = \{\boldsymbol{\mu}, \alpha\boldsymbol{\Omega}^{-1}\}$	$\mathrm{Var}[\boldsymbol{x}] = (\alpha - 1)^{-1}\boldsymbol{\Omega}\lambda^{-1}$
$p(\boldsymbol{x}) = \mathrm{St}_k(\boldsymbol{x} \mid \boldsymbol{\mu}, \lambda\alpha\boldsymbol{\Omega}^{-1}, 2\alpha)$	$p(\boldsymbol{Y}) = \mathrm{Wi}_k(\boldsymbol{Y} \mid \alpha, \boldsymbol{\Omega})$
$\mathrm{St}_k(\boldsymbol{x} \mid \boldsymbol{\mu}, \boldsymbol{\Sigma}, \alpha)$	*Multivariate Student*
$\boldsymbol{\mu} = (\mu_1, \cdots, \mu_k),\ \mu_i \in \mathbb{R},\ \ \alpha > 0$	$\boldsymbol{x} = (x_1, \ldots, x_k)$
$\boldsymbol{\Lambda}$: *symmetric positive-definite*	$-\infty < x_i < +\infty$
$p(\mathrm{x}) = c\Big[1 + \dfrac{1}{\alpha}(\boldsymbol{x} - \boldsymbol{\mu})^t\boldsymbol{\Sigma}^{-1}(\boldsymbol{x} - \boldsymbol{\mu})\Big]^{-(\alpha+k)/2}$	$c = \dfrac{\Gamma\left(\frac{1}{2}(\alpha + k)\right)}{\Gamma(\frac{1}{2}\alpha)(\alpha\pi)^{k/2}}\lvert\boldsymbol{\Sigma}\rvert^{-1/2}$
$\mathrm{E}[\boldsymbol{x}] = \boldsymbol{\mu}$	$\mathrm{Var}[\boldsymbol{x}] = \frac{\alpha}{\alpha-2}\boldsymbol{\Sigma}$
$\mathrm{Wi}_k(\boldsymbol{X} \mid \boldsymbol{\Sigma}, \alpha)$	*Wishart*
$\boldsymbol{\Sigma}$: *symmetric non-singular*	$\boldsymbol{X}$: *symmetric positive-definite*
$\alpha \geq k$	
$p(\boldsymbol{X}) = c\,\lvert\boldsymbol{X}\rvert^{(\alpha-k-1)/2}\exp\{-\frac{1}{2}tr(\boldsymbol{\Sigma}^{-1}\boldsymbol{X})\}$	$c = \dfrac{2^{-k\alpha/2}\pi^{-k(k-1)/4}}{\lvert\boldsymbol{\Sigma}\rvert^{\alpha/2}\prod_{i=1}^{k}\Gamma(\frac{1}{2}(\alpha + 1 - i))}$
$\mathrm{E}[\boldsymbol{X}] = \alpha\boldsymbol{\Sigma}$	$\mathrm{E}[\boldsymbol{X}^{-1}] = \frac{1}{\alpha-k-1}\boldsymbol{\Sigma}^{-1}$
$\mathrm{IWi}_k(\boldsymbol{X} \mid \boldsymbol{\Omega}, \alpha)$	*Inverted Wishart*
$\boldsymbol{\Omega}$: *symmetric non-singular*	$\boldsymbol{X}$: *symmetric positive-definite*
$\alpha \geq k$	
$p(\boldsymbol{X}) = c\,\lvert\boldsymbol{X}\rvert^{(\alpha+k+1)/2}\exp\{-\frac{1}{2}tr(\boldsymbol{\Omega}\boldsymbol{X}^{-1})\}$	$c = \dfrac{2^{-k\alpha/2}\pi^{-k(k-1)/4}\lvert\boldsymbol{\Omega}\rvert^{\alpha/2}}{\prod_{i=1}^{k}\Gamma(\frac{1}{2}(\alpha + 1 - i))}$
$\mathrm{E}[\boldsymbol{X}] = \frac{1}{\alpha-k-1}\boldsymbol{\Omega}$	$\mathrm{E}[\boldsymbol{X}^{-1}] = \alpha\boldsymbol{\Omega}^{-1}$

List special functions used in the book.

Bibliography

Abdullahi, S., Pradhan, B. & Jebur, M. N. (2015), 'Gis-based sustainable city compactness assessment using integration of MCDM, Bayes theorem and RADAR technology', *Geocarto International* **30**, 365–387.

Abe, S. (2014), 'Conditional maximum-entropy method for selecting prior distributions in Bayesian statistics', *Europhysics Letters* **108**, 40008.

Abe, S. (2018), Bayesian approach to extreme-value statistics based on conditional maximum-entropy method, *in* 'Journal of Physics: Conference Series', Vol. 1113, p. 012001.

Abramowitz, M. & Stegun, I. A. (1964), *Handbook of Mathematical Functions*, Mineola, NY: Dover.

Abul-Magd, A., Harney, H. L., Simbel, M. & Weidenmüller, H. A. (2004), 'Statistics of 2+ levels in even–even nuclei', *Physics Letters B* **579**, 278–284.

Abul-Magd, A., Harney, H. L., Simbel, M. & Weidenmüller, H. A. (2006), 'Statistical analysis of composite spectra', *Annals of Physics* **321**, 560–580.

Achcar, J. & Bolfarine, H. (1986), 'The log-linear model with a generalized gamma distribution for the error: a Bayesian approach', *Statistics & Probability Letters* **4**, 325–332.

Achcar, J., Moala, F. A. & Boleta, J. (2015), 'Generalized exponential distribution: A Bayesian approach using mcmc methods', *International Journal of Industrial Engineering Computations* **6**, 1–14.

Agliari, A. & Parisetti, C. C. (1988), 'A g-reference informative prior: A note on Zellner's g-prior', *The Statistician* **37**, 271–275.

Agresti, A. & Min, Y. (2005), 'Frequentist performance of Bayesian confidence intervals for comparing proportions in 2×2 contingency tables', *Biometrics* **61**, 515–523.

Aguilar, O. & West, M. (2000), 'Bayesian dynamic factor models and portfolio allocation', *J. Business Econ. Statist.* **18**, 338–357.

Aitchison, J. (1964), 'Bayesian tolerance regions', *J. Roy. Statist. Soc. B* **26**, 161–175.

Aitkin, M. (1991), 'Posterior Bayes factors', *J. Roy. Statist. Soc. B* **53**, 111–128 (with discussion).

Aitkin, M. (2018), 'Likelihood, credible and confidence intervals for parameters in complex models', *Metron* **76**, 305–320.

Akaike, H. (1977), On entropy maximization principle, *in* '*Applications of Statistics* (P. R. Krishnaiah, ed)', Amsterdam: North-Holland, pp. 27–41.

Akaike, H. (1980*a*), 'Ignorance prior distribution of a hyperparameter and Stein's estimator', *Ann. Inst. Statist. Math.* **32**, 171–178.

Akaike, H. (1980*b*), 'The interpretation of improper prior distributions as limits of data dependent proper prior distributions', *J. Roy. Statist. Soc. B* **42**, 46–52.

Akaike, H. (1983), 'On minimum information prior distributions', *Ann. Inst. Statist. Math.* **35**, 139–149.

Albers, C. J., Kiers, H. A. L., van Ravenzwaaij, D. & Savalei, V. (2018), 'Credible confidence: A pragmatic view on the frequentist vs Bayesian debate', *Collabra: Psychology* **4**, 108–119.

Albert, J. H. & Chib, S. (1993), 'Bayesian analysis of binary and polychotomous response data', *J. Amer. Statist. Assoc.* **88**, 669–679.

Albert, J. H. & Pepple, P. A. (1989), 'A Bayesian approach to some overdispersion models', *Canad. J. Statist.* **17**, 333–344.

Aldrich, J. (2000), 'Fisher's "inverse probability"of 1930', *International Statistical Review* **68**, 155–172.

Allais, M. (1953), 'Le comportement de l'homme rational devant le risque: Critique des postulats et axiomes de l'école américaine', *Econometrica* pp. 503–546.

Allais, M. (1990), Allais paradox, *in* 'Utility and Probability', Berlin: Springer, pp. 3–9.

Alston, C. L., Mengersen, K. & Pettitt, A. N. (2012), *Case Studies in Bayesian Statistical Modelling and Analysis*, Chichester: Wiley.

Amarante, M. (2009), 'Foundations of neo-Bayesian statistics', *J. Economic Theory* **144**, 2146–2173.

Andersson, S. A. & Perlman, M. D. (1993), 'Lattice models for conditional independence in a multivariate normal distribution', *The Annals of Statistics* **21**, 1318–1358.

Andrews, R. W., Berger, J. O. & Smith, M. H. (1993), 'Bayesian estimation of manufacturing effects in a fuel economy model', *J. Applied Econometrics* **8**, 5–18.

Armero, C. (1985), Bayesian analysis of M/M/1/∞/FIFO queues, *in* '*Bayesian Statistics 2* (J. M. Bernardo, M. H. DeGroot, D. V. Lindley and A. F. M. Smith, eds). Amsterdam: North-Holland', pp. 613–618.

Armero, C. & Bayarri, M. J. (1997), 'A Bayesian analysis of a queuing system with unlimited service', *J. Statist. Planning and Inference* **58**, 241–261.

Asakawa, M., Nakahara, Y. & Hatsuda, T. (2001), 'Maximum entropy analysis of the spectral functions in lattice qcd', *Progress in Particle and Nuclear Physics* **46**, 459–508.

Aslan, M. (2002), 'Reference prior for the parameters of the Rao-Kupper model', *Proc. Pakistan Acad. Sci.* **39**, 215–224.

Atay-Kayis, A. & Massam, H. (2005), 'The marginal likelihood for decomposable and non-decomposable graphical Gaussian models', *Biometrika* **92**, 317–335.

Azriel, D. (2019), 'The conditionality principle in high-dimensional regression', *Biometrika* **106**, 702–707.

Balamurugan, S. A. A. & Rajaram, R. (2009), 'Effective and efficient feature selection for large-scale data using Bayes' theorem', *Internat J. Automation Computing* **6**, 62–71.

Bar-Lev, S. K. & Reiser, B. (1982), 'An exponential subfamily which admits UMPU test based on a single test statistic', *Ann. Statist.* **10**, 979–989.

Barger, K. & Bunge, J. (2008), 'Bayesian estimation of the number of species using noninformative priors', *Biometrical J.* **50**, 1064–1076.

Barger, K. & Bunge, J. (2010), 'Objective Bayesian estimation for the number of species', *Bayesian Analysis* **5**, 765–785.

Barnard, G. A. (1951), 'The theory of information', *J. Roy. Statist. Soc. B* **13**, 46–64.

Barnard, G. A. (1980), Pivotal inference and the Bayesian controversy, *in* '*Bayesian Statistics* (J. M. Bernardo, M. H. DeGroot, D. V. Lindley and A. F. M. Smith, eds). Valencia: University Press', pp. 295–318 (with discussion).

Barndorff-Nielsen, O. E. (1986), 'Inference on full or partial parameters based on the standardized signed log likelihood ratio', *Biometrika* **73**, 307–322.

Barnes III, T. G., Jefferys, W. H., Berger, J. O., Mueller, P. J., Orr, K. & Rodriguez, R. (2003), 'A Bayesian analysis of the Cepheid distance scale', *Astrophysical J.* **592**, 539.

Barron, A. R. (1999), Information-theoretic characterization of Bayes performance and the choice of priors in parametric and nonparametric problems, *in* '*Bayesian Statistics 6* (J. M. Bernardo, J. O. Berger, A. P. Dawid and A. F. M. Smith, eds). Oxford: University Press', pp. 27–52 (with discussion).

Barron, A. R., Rissanen, J. & Yu, B. (1998), 'The minimum description length principle in coding and modelling', *IEEE Trans. Information Theory* **44**, 2743–2760.

Bartholomew, D. J. (1965), 'A comparison of some Bayesian and frequentist inferences', *Biometrika* **52**, 19–35.

Basu, D. (1975), 'Statistical information and likelihood', *Sankhyā A* **37**, 1–71.

Basu, D. (1992), Learning statistics from counter examples: Ancillary statistics, *in* '*Bayesian Analysis in Statistics and Econometrics: Essays in Honor of Arnold Zellner* (D. A. Berry, K. M. Chaloner and J. K. Geweke, eds). Hoboken, NJ: Wiley', pp. 217–223.

Basu, S. & Ebrahimi, N. (2001), 'Bayesian capture-recapture methods for error detection and estimation of population size: heterogeneity and dependence', *Biometrika* **2001**, 269–279.

Bayarri, M. J. (1981), 'Inferencia Bayesiana sobre el coeficiente de correlación de una población normal bivariante', *Trab. Estadist.* **32**, 18–31.

Bayarri, M. J. (1985), 'Bayesian inference on the parameters of a beta distribution', *Statistics and Decisions* **2**, 17–22.

Bayarri, M. J. & Berger, J. O. (2004), 'The interplay of Bayesian and frequentist analysis', *Statistical Science* **19**, 58–80.

Bayes, T. (1763), 'An essay towards solving a problem in the doctrine of chances', *Phil. Trans. Roy. Soc. London* **53** and **54**, 370–416 and 370–416. Reprinted in *Biometrika*, **45** (1958), 293–315, with a biographical note by G. A. Barnard. Reproduced in Press (1989), 185–217.

Belzile, E. & Angers, J.-F. (1995), 'Choice of noninformative priors for the variance components of an unbalanced one-way random model', *Comm. Statist. Theory and Methods* **24**, 1325–1341.

Benford, F. (1938), 'The law of anomalous numbers', *Proc. Amer. Philos. Soc.* **78**, 551–572.

Berger, J. O. (1984), The robust Bayesian viewpoint, *in* '*Studies in Bayesian Econometrics*, (A. Zellner and J. B. Kadane, eds)', Amsterdam: North-Holland, pp. 63–144 (with discussion).

Berger, J. O. (1985), *Statistical Decision Theory and Bayesian Analysis*, Second edn, New York: Springer.

Berger, J. O. (1994), 'An overview of robust Bayesian analysis)', *Test* **3**, 5–124 (with discussion).

Berger, J. O. (2014), Conditioning is the issue, *in* X. Lin, C. Genest, D. L. Banks, G. Molenberghs, D. W. Scott & W. J.-Li., eds, 'Past, Present and Future of Statistical Science', London: Chapman and Hall, pp. 253–266.

Berger, J. O. & Bernardo, J. M. (1989), 'Estimating a product of means: Bayesian analysis with reference priors', *J. Amer. Statist. Assoc.* **84**, 200–207.

Berger, J. O. & Bernardo, J. M. (1992*a*), On the development of reference priors, *in* '*Bayesian Statistics 4* (J. M. Bernardo, J. O. Berger, A. P. Dawid and A. F. M. Smith, eds). Oxford: University Press', pp. 35–60 (with discussion).

Berger, J. O. & Bernardo, J. M. (1992*b*), 'Ordered group reference priors, with applications to multinomial problems', *Biometrika* **79**, 25–37.

Berger, J. O. & Bernardo, J. M. (1992*c*), Reference priors in a variance components problem, *in* '*Bayesian Analysis in Statistics and Econometrics* (P. K. Goel and N. S. Iyengar, eds). Berlin: Springer', pp. 323–340.

Berger, J. O., Bernardo, J. M. & Mendoza, M. (1989), On priors that maximize expected information, *in* '*Recent Developments in Statistics and their Applications* (J. P. Klein and J. C. Lee, eds)', Seoul: Freedom Academy, pp. 1–20.

Berger, J. O., Bernardo, J. M. & Sun, D. (2009*a*), 'The formal definition of reference priors', *Ann. Statist.* **37**, 905–938.

Berger, J. O., Bernardo, J. M. & Sun, D. (2009*b*), 'Natural induction: an objective Bayesian approach', *Rev. R. Acad. Cienc. Exactas Fis. Nat. A* **103**, 125–135.

Berger, J. O., Bernardo, J. M. & Sun, D. (2012), 'Objective priors for discrete parameter spaces', *J. Amer. Statist. Assoc.* **107**, 636–648.

Berger, J. O., Bernardo, J. M. & Sun, D. (2015), 'Overall objective priors', *Bayesian Analysis* **10**, 189–221 (with discussion).

Berger, J. O. & Berry, D. A. (1988), 'Statistical analysis and the illusion of objectivity', *American Scientist* **76**, 159–165.

Berger, J. O., Betro, B., Moreno, E., Pericchi, L. R., Ruggeri, F. & Salinetti, G., eds (1996), *Bayesian Robustness*, Hayward, CA: IMS.

Berger, J. O. & Chen, M.-H. (1993), 'Predicting retirement patterns: Prediction for a multinomial distribution with constrained parameter space', *The Statistician* **42**, 427–443.

Berger, J. O., De Oliveira, V. & Sansó, B. (2001), 'Objective Bayesian analysis of spatially correlated data', *J. Amer. Statist. Assoc.* **96**(456), 1361–1374.

Berger, J. O., Liseo, B. & Wolpert, R. L. (1999), 'Integrated likelihood methods for eliminating nuisance parameters', *Statistical Sci.* **14**, 1–28 (with discussion).

Berger, J. O., Philippe, A. & Robert, C. P. (1998), 'Estimation of quadratic functions: Noninformative priors for non-centrality parameters', *Statistica Sinica* **8**, 359–376.

Berger, J. O. & Strawderman, W. E. (1996), 'Choice of hierarchical priors: Admissibility in estimation of normal means', *Ann. Statist.* **24**, 931–951.

Berger, J. O., Strawderman, W. E. & Tang, D. (2005), 'Posterior propriety and admissibility of hyperpriors in normal hierarchical models', *Ann. Statist.* **33**, 606–646.

Berger, J. O. & Sun, D. (1993), 'Bayesian analysis for the poly-Weibull distribution', *J. Amer. Statist. Assoc.* **88**, 1412–1418.

Berger, J. O. & Sun, D. (2006), 'Objective priors for a bivariate normal model with multivariate generalizations', *Duke University ISDS Tech. Rep. 2007* **6**.

Berger, J. O. & Sun, D. (2008), 'Objective priors for the bivariate normal model', *Ann. Statist.* **36**, 963–982.

Berger, J. O., Sun, D. & Song, C. (2020*a*), 'Bayesian analysis of covariance matrix of multivariate normal distribution with a new class of priors', *Ann. Statist.* **48**, 2381–2403.

Berger, J. O., Sun, D. & Song, C. (2020*b*), 'An objective prior for hyperparameters in normal hierarchical models', *J. Multivariate Analysis* **178**, 104606.

Berger, J. O. & Wolpert, R. L. (1988), *The Likelihood Principle*, 2nd edn, Hayward, CA: IMS.

Berger, J. O. & Yang, R. (1994), 'Noninformative priors and Bayesian testing for the AR(1) model', *Econometric Theory* **10**, 461–482.

Berk, R. H. (1966), 'Limiting behavior of posterior distributions when the model is incorrect', *Ann. Math. Statist.* **37**, 51–58. Corr: Vol **37**, 745.

Berk, R. H. (1970), 'Consistency a posteriori', *Ann. Math. Statist.* **41**, 894–906.

Bernard, J.-M. (1996), 'Bayesian interpretation of frequentist procedures for a bernoulli process', *Amer. Statist.* **50**, 7–13.

Bernardo, J. M. (1977*a*), Inferences about the ratio of normal means: A Bayesian approach to the Fieller-Creasy problem, *in* '*Recent Developments in Statistics.* (J. R. Barra *et al.* eds). Amsterdam: North-Holland', pp. 345–350.

Bernardo, J. M. (1977*b*), 'Inferencia bayesiana sobre el coeficiente de variación: una solución a la paradoja de marginalización', *Trab. Estadist.* **28**, 23–80.

Bernardo, J. M. (1978), 'Unacceptable implications of the left Haar measure in a standard normal theory inference problem', *Trab. Estadist.* **29**, 3–9.

Bernardo, J. M. (1979*a*), 'Expected information as expected utility', *Ann. Statist.* **7**, 686–690.

Bernardo, J. M. (1979*b*), 'Reference posterior distributions for Bayesian inference', *J. Roy. Statist. Soc. B* **41**, 113–147 (with discussion). Reprinted in *Bayesian Inference* **1** (G. C. Tiao and N. G. Polson, eds). Oxford: Edward Elgar, 1995, 229–263.

Bernardo, J. M. (1981), 'Reference decisions', *Symposia Mathematica* **25**, 85–94.

Bernardo, J. M. (1982), 'Contraste de modelos probabilísticos desde una perspectiva Bayesiana', *Trab. Estadist.* **33**, 16–30.

Bernardo, J. M. (1984), 'Monitoring the 1982 Spanish socialist victory: A Bayesian analysis', *J. Amer. Statist. Assoc.* **79**, 510–515.

Bernardo, J. M. (1985*a*), 'Análisis Bayesiano de los contrastes de hipótesis paramétricos', *Trab. Estadist.* **36**, 45–54.

Bernardo, J. M. (1985*b*), 'On a famous problem of induction', *Trab. Estadist.* **36**, 24–30.

Bernardo, J. M. (1994), Optimizing prediction with hierarchical models: Bayesian clustering, *in* '*Aspects of Uncertainty: a Tribute to D. V. Lindley* (P. R. Freeman and A. F. M. Smith, eds). Chichester: Wiley', pp. 67–76.

Bernardo, J. M. (1997), 'Non-informative priors do not exist', *J. Statist. Planning and Inference* **65**, 159–189 (with discussion).

Bernardo, J. M. (1999*a*), 'Model-free objective Bayesian prediction', *Rev. Acad. Ciencias Madrid* **93**, 295–302.

Bernardo, J. M. (1999*b*), Nested hypothesis testing: The Bayesian reference criterion, *in 'Bayesian Statistics 6* (J. M. Bernardo, J. O. Berger, A. P. Dawid and A. F. M. Smith, eds). Oxford: University Press', pp. 101–130 (with discussion).

Bernardo, J. M. (2005*a*), 'Intrinsic credible regions: An objective Bayesian approach to interval estimation', *Test* **14**, 317–384.

Bernardo, J. M. (2005*b*), Reference analysis, *in 'Handbook of Statistics* **25** (D. K. Dey and C. R. Rao, eds). Amsterdam: Elsevier', pp. 17–90.

Bernardo, J. M. (2006), Intrinsic point estimation of the normal variance, *in* S. K. Upadhyay, U. Singh & D. K. Dey, eds, *'Bayesian Statistics and its Applications'*, New Delhi: Anamaya, pp. 110–121.

Bernardo, J. M. & Juárez, M. (2003), Intrinsic estimation, *in 'Bayesian Statistics 7* (J. M. Bernardo, M. J. Bayarri, J. O. Berger, A. P. Dawid, D. Heckerman, A. F. M. Smith and M. West, eds). Oxford: University Press', pp. 465–476.

Bernardo, J. M. & Ramón, J. M. (1998), 'An introduction to Bayesian reference analysis: Inference on the ratio of multinomial parameters', *The Statistician* **47**, 101–135.

Bernardo, J. M. & Smith, A. F. M. (1994), *Bayesian Theory*, Chichester: Wiley.

Bhapkar, V. P. (1989), 'Conditioning on ancillary statistics and loss of information in the presence of nuisance parameters', *J. Statist. Planning and Inference* pp. 139–160.

Bickel, P. J. & Ghosh, J. K. (1990), 'A decomposition for the likelihood ratio statistic and the Bartlett correction – A Bayesian argument', *Ann. Statist.* **18**, 1070–1090.

Bickel, P. J., Götze, F. & Van Zwet, W. R. (1985), A simple analysis of third order efficiency of estimates, *in 'Proc. Berkeley Conf. in Honor of Jerzy Neyman and Jack Kiefer*, (L. Le Cam and R. A. Olshen, eds). Pacific Drove, CA: Wadsworth', Vol. 2, pp. 749–768.

Birnbaum, A. (1961), 'A unified theory of estimation, I', *Ann. Math. Statist.* **32**, 112–135.

Birnbaum, A. (1962), 'On the foundations of statistical inference', *J. Amer. Statist. Assoc.* **57**, 269–326 (with discussion).

Birnbaum, A. (1972), 'More on concepts of statistical evidence', *J. Amer. Statist. Assoc.* **67**, 858–861.

Bjornstad, J. (1996), 'On the generalization of the likelihood function and the likelihood principle', *J. Amer. Statist. Assoc.* **91**, 791–806.

Bolstad, W. M. (2010), *Understanding Computational Bayesian Statistics*, Hoboken, NJ: Wiley.

Bondar, J. V. & Milnes, P. (1981), 'Amenability: A survey for statistical applications of Hunt-Stein and related conditions on groups', *Zeits. Wahrscheinlichkeitstheorie verwandte Gebiete* **57**, 103–128.

Boole, G. (1854), *The Laws of Thought*, London: Walton and Maberly. Reprinted in 1957, Mineola, NY: Dover.

Borowska, A., Giurghita, D. & Husmeier, D. (2021), 'Gaussian process enhanced semi-automatic approximate Bayesian computation: parameter inference in a stochastic differential equation system for chemotaxis', *Journal of Computational Physics* **429**, 109999.

Bose, A. & Boukai, B. (1993), 'Sequential estimation results for a two-parameter exponential family of distribution', *Ann. Statist.* **21**, 484–502.

Box, G. E. P. (1980), 'Sampling and Bayes inference in scientific modelling and robustness', *J. Roy. Statist. Soc. A* **143**, 383–430 (with discussion).

Box, G. E. P. & Cox, D. R. (1964), 'An analysis of transformations', *J. Roy. Statist. Soc. B* **26**, 211–252 (with discussion).

Box, G. E. P., Leonard, T. H. & Wu, C.-F., eds (1983), *Scientific Inference, Data Analysis, and Robustness*, Cambridge, MA: Academic Press.

Box, G. E. P. & Tiao, G. C. (1965), 'Multiparameter problems from a Bayesian point of view', *Ann. Math. Statist.* **36**, 1468–1482.

Box, G. E. P. & Tiao, G. C. (1968), 'A Bayesian approach to some outlier problems', *Biometrika* **55**, 119–129.

Box, G. E. P. & Tiao, G. C. (1973), *Bayesian Inference in Statistical Analysis*, Reading, MA: Addison-Wesley.

Branco, M. D., Genton, M. G. & Liseo, B. (2013), 'Objective Bayesian analysis of skew-t distributions', *Scandinavian Journal of Statistics* **40**, 63–85.

Bridgman, P. W. (1927), *The Logic of Modern Physics*, London: Macmillan.

Brillinger, D. R. (1962), 'Examples bearing on the definition of fiducial probability with a bibliography', *Ann. Math. Statistics* **33**, 1349–1355.

Brillinger, D. R. (1963), 'Necessary and sufficient conditions for a statistical problem to be invariant under a lie group', *The Annals of Mathematical Statistics* pp. 492–500.

Broad, C. D. (1918), 'On the relation between induction and probability', *Mind* **27**, 389–404.

Broeckx, F., Goovaerts, M. & Van den Broeck, J. (1974), 'On the prior density functions proposed by Jeffreys and Haldane, in a Bayesian framework', *Belgian J. Oper. Res., Stat., and Computer Science* **14**, 19–33.

Broeckx, F., Goovaerts, M. & Van den Broeck, J. (1977), 'A Bayesian approach to some specific business problems', *Belgian J. Oper. Res., Stat., and Computer Science* **17**, 1–14.

Broemeling, L. D. (1985), *Bayesian Analysis of Linear Models*, New York: Marcel Dekker.

Brown, L. D. (1978), 'A contribution to Kiefer's theory of conditional confidence procedures', *Ann. Statist.* **6**, 59–71.

Brown, L. D. (1986), *Foundations of Exponential Families*, Hayward, CA: IMS.

Brown, L. D. (1988), The differential inequality of a statistical estimation problem, *in* '*Statistical Decision Theory and Related Topics IV* **1** (S. S. Gupta and J. O. Berger, eds). New York: Springer', pp. 299–324.

Brown, L. D., Cai, T. T. & DasGupta, A. (2001*a*), 'Confidence intervals for a binomial proportion and asymptotic expansions', *Ann. Statist.* **30**, 160–201.

Brown, L. D., Cai, T. T. & DasGupta, A. (2001*b*), 'Interval estimation of a binomial proportion', *Statistical Sci.* **16**, 101–133 (with discussion).

Bucephala, J. P. & Gatsonis, C. A. (1988), 'Bayesian inference for ratios of coefficients in a linear model', *Biometrics* **44**, 87–101.

Buehler, R. J. (1976), 'Coherent preferences', *Ann. Statist.* **4**, 1051–1064.

Bunouf, P. & Lecoutre, B. (2006), 'Bayesian priors in sequential binomial design', *Comptes Rendus Mathematique* **343**, 339–344.

Burch, B. D. & Harris, I. R. (1999), 'Bayesian estimators of the intraclass correlation coefficient in the one-way random effects model', *Comm. Statist. Theory and Methods* **28**, 1247–1272.

Butler, R. W. (1986), 'Predictive likelihood inference with applications', *J. Roy. Statist. Soc. B* **48**, 1–38.

Byers, S. N. & Roberts, C. A. (2003), 'Bayes' theorem in paleopathological diagnosis', *Amer. J. Physical Anthropology* **121**, 1–9.

Cabras, S. (2013), Default priors based on pseudo-likelihoods for the poisson-gpd model, *in* N. Torelli, F. Pesarin & A. Bar-Hen, eds, 'Advances in Theoretical and Applied Statistics', Berlin: Springer, pp. 3–12.

Cabras, S., Castellanos, M. E. & Gamerman, D. (2011), 'A default Bayesian approach for regression on extremes', *Statistical Modelling* **11**, 557–580.

Carnap, R. (1962), *Logical Foundations of Probability*, 2nd edn, Chicago, IL: University of Chicago Press.

Carroll, J. & Lombard, F. (1985), 'A note on n estimators for the binomial distribution', *J. Amer. Statist. Assoc.* **80**, 423–426.

Casadei, D. (2012), 'Reference analysis of the signal+ background model in counting experiments', *Journal of Instrumentation* **7**(01), P01012.

Casella, G., Robert, C. P. & Wells, M. T. (2004), 'Mixture models, latent variables and partitioned importance sampling', *Statistical Methodology* **1**, 1–18.

Casella, G., Robert, C. P., Wells, M. T. et al. (2004), Generalized accept-reject sampling schemes, *in* '*A Festschrift for Herman Rubin* (H. Rubin and A. DasGupta, eds)', Hayward, CA: IMS, pp. 342–347.

Catling, D. C., Krissansen-Totton, J., Kiang, N. Y., Crisp, D., Robinson, T. D., DasSarma, S., Rushby, A. J., Del Genio, A., Bains, W. & Domagal-Goldman, S. (2018), 'Exoplanet biosignatures: a framework for their assessment', *Astrobiology* **18**, 709–738.

Chaloner, K. (1987), 'A Bayesian approach to the estimation of variance components for the unbalanced one-way random model', *Technometrics* **29**, 323–337.

Chaloner, K. (1991), 'Bayesian residual analysis in the presence of censoring', *Biometrika* **78**, 637–644.

Chang, L.-Y., Toghiani, S., Ling, A., Hay, E. H., Aggrey, S. E. & Rekaya, R. (2017), 'Analysis of multiple binary responses using a threshold model', *Journal of Agricultural, Biological and Environmental Statistics* **22**, 640–651.

Chang, T. & Eaves, D. (1990), 'Reference priors for the orbit in a group model', *Ann. Statist.* **18**, 1595–1614.

Chang, T. & Villegas, C. (1986), 'On a theorem of Stein relating Bayesian and classical inferences in group models', *Canad. J. Statist.* **14**, 289–296.

Chao, J. C. & Phillips, P. C. B. (1998), 'Posterior distributions in limited information analysis of the simultaneous equations model using the Jeffreys prior', *Journal of Econometrics* **87**, 49–86.

Chao, J. C. & Phillips, P. C. B. (2002), 'Jeffreys prior analysis of the simultaneous equations model in the case with $n + 1$ endogenous variables', *J. Econometrics* **111**, 251–283.

Chen, B. & Varshney, P. K. (2002), 'A Bayesian sampling approach to decision fusion using hierarchical models', *IEEE Transactions on signal processing* **50**, 1809–1818.

Chen, C.-F. (1985), 'On asymptotic normality of limiting density functions with Bayesian implications', *J. Roy. Statist. Soc. B* **97**, 540–546.

Chernoff, H. (1956), 'Large-sample theory: Parametric case', *Ann. Math. Statist.* **27**, 1–22.

Cho, J. S. & Baek, S. U. (2002), 'Development of matching priors for the ratio of the failure rate in the Burr model', *Far East J. Theor. Statist.* **8**, 79–87.

Chung, Y. & Dey, D. K. (1998), 'Bayesian approach to estimation in intraclass correlation using reference priors', *Comm. Statist. Theory and Methods* **27**, 2241–2255.

Chung, Y. & Dey, D. K. (2002), 'Model determination for the variance component model using reference priors', *J. Statist. Planning and Inference* **100**, 49–65.

Cifarelli, D. & Regazzini, E. (1987), Priors for exponential families which maximize the association between past and future observations, *in* '*Probability and Bayesian Statistics* (R. Viertl, ed). London: Plenum', pp. 83–95.

Clarke, B. (1996), 'Implications of reference priors for prior information and for sample size', *J. Amer. Statist. Assoc.* **91**, 173–184.

Clarke, B. & Barron, A. R. (1990), 'Information-theoretic asymptotics of Bayes methods', *IEEE Trans. Information Theory* **36**, 453–471.

Clarke, B. & Barron, A. R. (1994), 'Jeffreys' prior is asymptotically least favorable under entropy risk', *J. Statist. Planning and Inference* **41**, 37–60.

Clarke, B. & Sun, D. (1997), 'Reference priors under the chi-squared distance', *Sankhyā A* **59**, 215–231.

Clarke, B. & Wasserman, L. (1993), 'Noninformative priors and nuisance parameters', *J. Amer. Statist. Assoc.* **88**, 1427–1432.

Coelho-Barros, E. A., Mazucheli, J., Achcar, J., Barco, K. V. P. & Tovar Cuevas, J. R. (2018), 'The inverse power Lindley distribution in the presence of left-censored data', *Journal of Applied Statistics* **45**, 2081–2094.

Consonni, G., Fouskakis, D., Liseo, B. & Ntzoufras, I. (2018), 'Prior Distributions for Objective Bayesian Analysis', *Bayesian Analysis* **13**(2), 627–679.
URL: *https://doi.org/10.1214/18-BA1103*

Consonni, G. & Leucari, V. (2006), 'Reference priors for discrete graphical models', *Biometrika* **93**, 23–40.

Consonni, G. & Massam, H. (2012), 'Parametrizations and reference priors for multinomial decomposable graphical models', *Journal of Multivariate Analysis* **105**(1), 380–396.

Consonni, G. & Veronese, P. (1989*a*), 'A note on coherent invariant distributions as non-informative priors for exponential and location-scale families', *Comm. Statist. Theory and Methods* **18**, 2883–2907.

Consonni, G. & Veronese, P. (1989*b*), 'Some remarks on the use of improper priors for the analysis of exponential regression models', *Biometrika* **76**, 101–106.

Consonni, G. & Veronese, P. (1992), Bayes factors for linear models and improper priors, *in* '*Bayesian Statistics 4* (J. M. Bernardo, J. O. Berger, A. P. Dawid and A. F. M. Smith, eds). Oxford: University Press', pp. 587–594.

Consonni, G. & Veronese, P. (1993), 'Unbiased Bayes estimates and improper priors', *Ann. Inst. Statist. Math.1* **45**, 303–315.

Consonni, G. & Veronese, P. (2003), 'Enriched conjugate and reference priors for the Wishart family on symmetric cones', *Ann. Statist.* **31**, 1491–1516.

Consonni, G., Veronese, P. & Gutiérrez-Peña, E. (2004), 'Reference priors for exponential families with simple quadratic variance function', *J. Multivariate Analysis* **88**, 335–364.

Contreras-Cristán, A., Gutiérrez-Peña, E. & O'Reilly, F. (2003), 'Inference using latent variables for mixtures of distributions for censored data with partial identification', *Communications in Statistics-Theory and Methods* **32**, 749–774.

Cook, J. & Lewandowsky, S. (2016), 'Rational irrationality: Modeling climate change belief polarization using Bayesian networks', *Topics Cognitive Sci.* **8**, 160–179.

Cornfield, J. (1969), 'The Bayesian outlook and its application', *Biometrics* **25**, 617–657 (with discussion).

Cox, D. R. (1958), 'Some problems connected with statistical inference', *Ann. Math. Statist.* **29**, 357–372.

Cox, D. R. (1988), 'Some aspects of conditional and asymptotic inference: A review', *Sankhyā A* **50**, 314–337.

Cox, D. R. (2006), Frequentist and Bayesian statistics: A critique, *in* 'Statistical Problems in Particle Physics, Astrophysics and Cosmology', World Scientific, pp. 3–6.

Cox, D. R. & Hinkley, D. V. (1974), *Theoretical Statistics*, London: Chapman and Hall.

Cox, D. R. & Reid, N. (1987), 'Parameter orthogonality and approximate conditional inference', *J. Roy. Statist. Soc. B* **49**, 1–39 (with discussion).

Cox, R. T. (1946), 'Probability, frequency and reasonable expectation', *American J. Physics* **14**, 1–13.

Cox, R. T. (1961), *The Algebra of Probable Inference*, Baltimore: John Hopkins University Press.

Crowder, M. (1992), 'Bayesian priors based on a parameter transformation using the distribution function', *Ann. Inst. Statist. Math.* **44**, 405–416.

Crowder, M. & Sweeting, T. J. (1989), 'Bayesian inference for a bivariate binomial distribution', *Biometrika* **76**, 599–603.

Csiszar, I. (1985), An extended maximum entropy principle and a Bayesian justification, *in* '*Bayesian Statistics 2* (J. M. Bernardo, M. H. DeGroot, D. V. Lindley and A. F. M. Smith, eds). Amsterdam: North-Holland', pp. 83–98 (with discussion).

Csiszár, I. (1991), 'Why least squared and maximum entropy? An axiomatic approach to inference for linear inverse problems', *Ann. Statist.* **19**, 2032–2066.

Cui, Y. & Hannig, J. (2019), 'Nonparametric generalized fiducial inference for survival functions under censoring', *Biometrika* **106**, 501–518.

Cui, Y. & Hannig, J. (2022), 'Demystifying inferential models: A fiducial perspective'. arXiv:2205.05612[stat.ME]

Dale, A. I. (2012), *A History of Inverse Probability: From Thomas Bayes to Karl Pearson*, Berlin: Springer.

Daniels, M. J. (1999), 'A prior for the variance in hierarchical models', *Canad. J. Statist.* **27**, 567–578.

Daniels, M. J. (2005), 'A class of shrinkage priors for the dependence structure in longitudinal data', *J. Statist. Planning and Inference* **127**, 119–130.

Daniels, M. J. & Kass, R. E. (1999), 'Nonconjugate Bayesian estimation of covariance matrices and its use in hierarchical models', *J. Amer. Statist. Assoc.* **94**, 1254–1263.

Daniels, M. J. & Kass, R. E. (2001), 'Shrinkage estimators for covariance matrices', *Biometrics* **57**, 1173–1184.

Daniels, M. J. & Pourahmadi, M. (2002), 'Bayesian analysis of covariance matrices and dynamic models for longitudinal data', *Biometrika* **89**, 553–566.

DasGupta, A. (1985), 'Bayes minimax estimation in multiparameter families when the parameter space is restricted to a bounded convex set', *Sankhyā A* **47**, 326–332.

DasGupta, A. (1991), 'Diameter and volume minimizing confidence sets in Bayes and classical problems', *Ann. Statist.* **19**, 1225–1243.

DasGupta, A. & Rubin, H. (2005), 'Estimation of binomial parameters when both n, p are unknown', *J. Statist. Planning and Inference* **130**, 391–404.

Datta, G. S. (1996), 'On priors providing frequentist validity of Bayesian inference for multiple parametric functions', *Biometrika* **83**, 287–298.

Datta, G. S. & Ghosh, J. K. (1995*a*), 'Noninformative priors for maximal invariant parameter in group models', *Test* **4**, 95–114.

Datta, G. S. & Ghosh, J. K. (1995*b*), 'On priors providing frequentist validity for Bayesian inference', *Biometrika* **82**, 37–45.

Datta, G. S. & Ghosh, J. K. (1995*c*), 'On priors providing frequentist validity for Bayesian inference', *Biometrika* **82**, 37–45.

Datta, G. S. & Ghosh, M. (1995*d*), 'Hierarchical Bayes estimators of the error variance in one-way ANOVA models', *J. Statist. Planning and Inference* **45**, 399–411.

Datta, G. S. & Ghosh, M. (1995*e*), 'Some remarks on noninformative priors', *J. Amer. Statist. Assoc.* **90**, 1357–1363.

Datta, G. S. & Ghosh, M. (1996), 'On the invariance of noninformative priors', *Ann. Statist.* **24**, 141–159.

Datta, G. S., Ghosh, M. & Kim, Y.-H. (2002), 'Probability matching priors for one-way unbalanced random effect models', *Statistics and Risk Modeling* **20**, 29–52.

Datta, G. S., Ghosh, M. & Mukerjee, R. (2000), 'Some new results on probability matching priors', *Calcutta Statist. Assoc. Bull.* **50**, 179–192. Corr: Vol **51**, 125.

Datta, G. S. & Mukerjee, R. (2004), *Probability Matching Priors: Higher Order Asymptotics*, New York: Springer.

Datta, G. S., Mukerjee, R., Ghosh, M. & Sweeting, T. J. (2000), 'Bayesian prediction with approximate frequentist validity', *Ann. Statist.* **28**, 1414–1426.

Datta, G. S., Mukerjee, R., Ghosh, M., Sweeting, T. J. et al. (2000), 'Bayesian prediction with approximate frequentist validity', *Ann. Statist.* **28**, 1414–1426.

Datta, G. S. & Smith, D. D. (2003), 'On propriety of posterior distributions of variance components in small area estimation', *J. Statist. Planning and Inference* **112**, 175–183.

Datta, G. S. & Sweeting, T. J. (2005), 'Probability matching priors', *Handbook of statistics* **25**, 91–114.

David, H. A. & Edwards, A. W. F. (2001), *Annotated Readings in the History of Statistics*, New York: Springer.

Dawid, A. P. (1977), 'Invariant distributions and analysis of variance models', *Biometrika* **64**, 291–298.

Dawid, A. P. (1980), A Bayesian look at nuisance parameters, *in* '*Bayesian Statistics* (J. M. Bernardo, M. H. DeGroot, D. V. Lindley and A. F. M. Smith, eds). Valencia: University Press', pp. 167–203 (with discussion).

Dawid, A. P. (1983*a*), Invariant prior distributions, *in* '*Encyclopedia of Statistical Sciences* **1**. (S. Kotz, C. B. Read, N. Balakrishnan, B. Vidakovic, and N. L. Johnson, eds). Hoboken, NJ: Wiley', pp. 228–236.

Dawid, A. P. (1983*b*), Statistical inference: I, *in* '*Encyclopedia of Statistical Sciences* **1**. (S. Kotz, C. B. Read, N. Balakrishnan, B. Vidakovic, and N. L. Johnson, eds). Hoboken, NJ: Wiley', pp. 89–105.

Dawid, A. P. (1984), 'Present position and potential developments: Some personal views. Statistical theory – The prequential approach', *J. Roy. Statist. Soc. A* **147**, 278–292 (with discussion).

Dawid, A. P. (1991), 'Fisherian inference in likelihood and prequential frames of reference', *J. Roy. Statist. Soc. B* **53**, 79–109 (with discussion).

Dawid, A. P. (1992), Prequential analysis, stochastic complexity and Bayesian inference, *in* '*Bayesian Statistics 4* (J. M. Bernardo, J. O. Berger, A. P. Dawid and A. F. M. Smith, eds). Oxford: University Press', pp. 109–125 (with discussion).

Dawid, A. P. & Stone, M. (1972), 'Expectation consistency of inverse probability distributions', *Biometrika* **59**, 486–489.

Dawid, A. P. & Stone, M. (1973), 'Expectation consistency and generalized Bayes inference', *Ann. Statist.* **1**, 478–485.

Dawid, A. P. & Stone, M. (1982), 'The functional-model basis of fiducial inference', *Ann. Statist.* **10**, 1054–1067.

Dawid, A. P., Stone, M. & Zidek, J. V. (1973), 'Marginalization paradoxes in Bayesian and structural inference', *J. Roy. Statist. Soc. B* **35**, 189–233 (with discussion).

de Cristofaro, R. (2002), 'The inductive reasoning in statistical inference', *Communications in Statistics-Theory and Methods* **31**, 1079–1089.

de Cristofaro, R. (2008), 'A new formulation of the principle of indifference', *Synthese* **163**, 329–339.

de Finetti, B. (1937), 'La prévision: ses lois logiques, ses sources subjectives', *Ann. Inst. Henri Poincaré* **7**, 1–68. Reprinted in *Studies in Subjective Probability* (H. E. Kyburg and H. E. Smokler, eds), Chichester: Wiley, 1980, 93–158.

de Finetti, B. (1970), *Teoria delle Probabilità* **1**, Turin: Einaudi. English translation as *Theory of Probability* **1**, Chichester: Wiley, 1974.

de la Horra, J. (1987), 'Generalized estimators: A Bayesian decision-theoretic view', *Statistics and Decisions* **5**, 347–352.

de la Horra, J. (1988), Parametric estimation with L^1 distance, *in* '*Bayesian Statistics 3* (J. M. Bernardo, M. H. DeGroot, D. V. Lindley and A. F. M. Smith, eds). Oxford: University Press', pp. 579–583.

de Melo, A. C. & Sanchez, A. J. (2008), 'Software maintenance project delays prediction using Bayesian networks', *Expert Systems with Applications* **34**, 908–919.

de Morgan, A. (1838), *An Essay on Probabilities*, London: Longman.

De Oliveira, V. (2012), 'Bayesian analysis of conditional autoregressive models', *Annals of the Institute of Statistical Mathematics* **64**, 107–133.

De Oliveira, V. & Song, J. J. (2008), 'Bayesian analysis of simultaneous autoregressive models', *Sankhyā B* pp. 323–350.

De Santis, F., Mortera, J. & Nardi, A. (2001), 'Jeffreys priors for survival models with censored data', *Journal of Statistical Planning and Inference* **99**, 193–209.

de Waal, D. J. & Groenewald, P. C. N. (1989), 'On measuring the amount of information from the data in a Bayesian analysis', *South African Statist. J.* **23**, 23–62.

DeGroot, M. H. (1970), *Optimal Statistical Decisions*, New York: McGraw-Hill.

Delampady, M., DasGupta, A., Casella, G., Rubin, H. & Strawderman, W. E. (2001), 'A new approach to default priors and robust Bayesian methodology', *Canad. J. Statist.* **29**, 437–450.

Demortier, L., Jain, S. & Prosper, H. B. (2010), 'Reference priors for high energy physics', *Physical Review D* **82**(3), 034002.

Dempster, A. M. (1972), 'Covariance selection', *Biometrics* **28**, 157–175.

Dempster, A. P. (1968), 'A generalization of Bayesian inference', *J. Roy. Statist. Soc. B* **30**, 205–247 (with discussion).

DeRobertis, L. & Hartigan, J. A. (1981), 'Bayesian inference using intervals of measures', *Ann. Statist.* **9**, 235–244.

Devroye, L. (1984), 'A simple algorithm for generating random variates with a log-concave density', *Computing* **33**, 247–257.

Dey, D. K. & Peng, F. (1993), 'On the choice of prior for the Bayes estimation in accelerated life testing', *J. Statist. Computation and Simulation* **48**, 207–217.

Dey, D. K. & Peng, F. (1995), 'Elimination of nuisance parameters using information measure', *Parisankhyan Samikkha* **2**, 9–29.

Diaconis, P. & Freedman, D. (1986*a*), 'On inconsistent Bayes estimates of location', *Ann. Statist.* **14**, 68–87.

Diaconis, P. & Freedman, D. (1986*b*), 'On the consistency of Bayes estimates', *Ann. Statist.* **14**, 1–67 (with discussion).

Diaconis, P. & Ylvisaker, D. (1979), 'Conjugate priors for exponential families', *Ann. Statist.* **7**, 269–281.

Díaz, R. F. (2018), Modelling light and velocity curves of exoplanet hosts, *in* T. L. Campante, N. C. Santos & M. J. Monteiro, eds, 'Asteroseismology and Exoplanets: Listening to the Stars and Searching for New Worlds', Berlin: Springer, pp. 199–224.

Diciccio, T. J., Kuffner, T. A. & Young, G. A. (2012), 'Objective Bayes, conditional inference and the signed root likelihood ratio statistic', *Biometrika* **99**, 675–686.

Dickey, J. M. (1976), 'Approximate posterior distribution', *J. Amer. Statist. Assoc.* **71**, 680–689.

Dittrich, D., Leenders, R. T. A. J. & Mulder, J. (2017), 'Bayesian estimation of the network autocorrelation model', *Social Networks* **48**, 213–236.

du Plessis, J. L., Van der Merwe, A. J. & Groenewald, P. C. N. (1995), 'Reference priors for the multivariate calibration problem', *South African Statist. J.* **29**, 155–168.

Durbin, J. (1970), 'On Birnbaum's theorem on the relation between sufficiency, conditionality and likelihood', *J. Amer. Statist. Assoc.* **65**, 395–403 (with discussion).

Eaton, M. L. (1982), A method for evaluating improper prior distributions, *in* '*Statistical Decision Theory and Related Topics III* **1** (S. S. Gupta and J. O. Berger, eds). Cambridge, MA: Academic Press', pp. 329–352.

Eaton, M. L. (1989), *Group Invariance Applications in Statistics*, Hayward, CA: IMS.

Eaton, M. L. & Freedman, D. A. (2004), 'Dutch book against some 'objective' priors', *Bernoulli* **10**, 861–872.

Eaton, M. L. & Sudderth, W. D. (2002), 'Group invariant inference and right Haar measure', *J. Statist. Planning and Inference* **103**, 87–99.

Eaton, M. L. & Sudderth, W. D. (2004), 'Properties of right Haar predictive inference', *Sankhyā A* **66**, 487–512.

Eaves, D. & Chang, T. (1992), 'Priors for ordered conditional variance and vector partial correlation', *J. Multivariate Analysis* **41**, 43–55.

Eaves, D. M. (1983*a*), 'Minimally informative prior analysis of a non-linear model', *The Statistician* **32**, 117–117.

Eaves, D. M. (1983*b*), 'On Bayesian nonlinear regression with an enzyme example', *Biometrika* **70**, 373–379.

Edgeworth, F. Y. (1883), 'The method of least squares', *London, Edinburgh and Dublin Phil. Mag. and J. of Science* **16**, 360–375.

Edwards, A. W. F. (1992), *Likelihood*, 2nd edn, Baltimore: J. Hopkins Univ. Press.

Edwards, A. W. F. (1997), 'What did Fisher mean by "inverse probability" in 1912–1922?', *Statistical Science* **12**, 177–184.

Efron, B. (1985), 'Bootstrap confidence intervals for a class of parametric problems', *Biometrica* **76**, 45–58.

Efron, B. (1986), 'Why isn't everyone a Bayesian?', *Amer. Statist.* **40**, 1–11 (with discussion). Reprinted in *The Science of Bradley Efron.* (C. N. Morris and R. Tibshirani, eds) Ch. 13. New York: Springer, 2008.

Efron, B. (1993), 'Bayes and likelihood calculations from confidence intervals', *Biometrika* **80**, 3–26.

Efron, B. (2013), 'Bayes theorem in the 21st century', *Science* **340**, 1177–1178.

Efron, B. (2015), 'Frequentist accuracy of Bayesian estimates', *J. Roy. Statist. Soc. B* **77**, 617.

Efron, B. & Morris, C. N. (1972), 'Empirical Bayes on vector observations: An extension of Stein's method', *Biometrika* **59**, 335–347.

Efron, B. & Morris, C. N. (1975), 'Data analysis using Stein's estimator and its generalizations', *J. Amer. Statist. Assoc.* **70**, 311–319.

Elhor, A. & Pensky, M. (2000), 'Bayesian estimation of location of lightning events', *Sankhyā B* **62**, 202–206.

Ellsberg, D. (1961), 'Risk, ambiguity, and the Savage axioms', *Quarterly J. Economics* pp. 643–669.Reply in vol 77, 336–342.

Eno, D. R. & Ye, K. (2000), 'Bayesian reference prior analysis for polynomial calibration models', *Test* **9**, 191–202.

Eno, D. R. & Ye, K. (2001), 'Probability matching priors for an extended statistical calibration problem', *Canad. J. Statist.* **29**, 19–35.

Eric, Z. & Peter, C. B. P. (1994), 'A Bayesian analysis of trend determination in economic time series', *Econometric Reviews* **13**, 291–336.

Erickson, G. J. & Smith, C. R., eds (1988), *Maximum-Entropy and Bayesian Methods in Science and Engineering*, Dordrecht: Kluwer.

Evans, I. G. & M., N. A. (1980), 'Bayesian prediction for two parameter Weibull lifetime models', *Comm. Statist. Theory and Methods* **9**, 649–658.

Everson, P. J. & Morris, C. N. (2000), 'Inference for multivariate normal hierarchical models', *J. Roy. Statist. Soc. B* **62**, 399–412.

Fan, T.-H. (2001), 'Noninformative Bayesian estimation for the optimum in a single factor quadratic response model', *Test* **10**, 225–240.

Fan, T.-H. & Berger, J. O. (2000), 'Robust Bayesian displays for standard inferences concerning a normal mean', *J. Comp. Statist. and Data Analysis* **33**, 381–399.

Farrell, R. H. (1964), 'Estimators of a location parameter in the absolutely continuous case', *Ann. Math. Statist.* **35**, 949–998.

Fatti, L. P. (1982), 'Predictive discrimination under the random effect model', *South African Statist. J.* **16**, 55–77.

Feibleman, J. (1945), 'Pragmatism and inverse probability', *Philosophy and Phenomenological Research* **5**, 309–319.

Fenton, N., Neil, M. & Berger, D. (2016), 'Bayes and the law', *Ann. Rev. Statist. Application* **3**, 51–77.

Ferguson, T. S. & Phadia, E. G. (1979), 'Bayesian nonparametric estimation based on censored data', *Ann. Statist.* **7**, 163–186.

Fernández, C., Osiewalski, J. & Steel, M. (1997), 'On the use of panel data in stochastic frontier models with improper priors', *J. Econometrics* **79**, 169–193.

Fernández, C. & Steel, M. (1998), 'Reference priors for the non-normal two-sample problems', *Test* **7**, 179–205.

Fernández, C. & Steel, M. (1998b), 'On Bayesian modelling of fat tails and skewness', *J. Amer. Statist. Assoc.* **93**, 359–371.

Fernández, C. & Steel, M. (1999a), On the dangers of modelling through continuous distributions: A Bayesian perspective, *in* '*Bayesian Statistics 6* (J. M. Bernardo, J. O. Berger, A. P. Dawid and A. F. M. Smith, eds). Oxford: University Press', pp. 213–238 (with discussion).

Fernández, C. & Steel, M. (1999b), 'Reference priors for the general location-scale model', *Statistics and Probab. Letters* **43**, 377–384.

Feroze, N. & Aslam, M. (2017), 'On selection of a suitable prior for the posterior analysis of censored mixture of topp leone distribution', *Communications in Statistics-Simulation and Computation* **46**, 5184–5211.

Ferrándiz, J. R. (1982), 'Una solución bayesiana a la paradoja de Stein', *Trab. Estadist.* **33**, 31–46.

Ferrándiz, J. R. (1985), Bayesian inference on Mahalanobis distance: An alternative approach to Bayesian model testing, *in* '*Bayesian Statistics 2* (J. M. Bernardo, M. H. DeGroot, D. V. Lindley and A. F. M. Smith, eds). Amsterdam: North-Holland', pp. 645–654.

Ferreira, M. A. R. & Salazar, E. (2014), 'Bayesian reference analysis for exponential power regression models', *Journal of Statistical Distributions and Applications* **1**, 1–20.

Fienberg, S. E. (1997), 'Introduction to RA Fisher on inverse probability and likelihood', *Statistical Science* **12**, 161–161.

Fienberg, S. E. (2006), 'When did Bayesian inference become Bayesian?', *Bayesian Analysis* **1**, 1–40.

Fienberg, S. E. & Holland, P. W. (1972), 'On the choice of flattening constants for estimating multinomial probabilities', *Journal of Multivariate Analysis* **2**(1), 127–134.

Fienberg, S. E. & Holland, P. W. (1973), 'Simultaneous estimation of multinomial cell probabilities', *Journal of the American Statistical Association* **68**(343), 683–691.

Fishburn, P. C. (1981), 'Subjective expected utility: A review of normative theories', *Theory and Decision* **13**, 139–199.

Fishburn, P. C. (1982), *The Foundations of Expected Utility*, Dordrecht: Reidel.

Fishburn, P. C. (1986), 'The axioms of subjective probability', *Statistical Sci.* **1**, 335–358 (with discussion).

Fisher, R. A. (1930), 'Inverse probability', *Proc. Cambridge Philos. Soc.* **26**, 528–535.

Fisher, R. A. (1932), 'Inverse probability and the use of likelihood', *Math. Proc. Cambridge Phil. Soc.* **28**, 257–261.

Fisher, R. A. (1933), 'The concepts of inverse probability and fiducial probability referring to unknown parameters', *Proc. Royal Soc. of London. Series A* **139**, 343–348.

Fisher, R. A. (1935), 'The fiducial argument in statistical inference', *Ann. Eugenics* **6**, 391–398.

Fisher, R. A. (1939), 'A note on fiducial inference', *Ann. Math. Statist.* **10**, 383–388.

Fisher, R. A. (1973), *Statistical Methods and Scientific Inference* (3rd edition), Edinburgh: Oliver and Boyd. Reprinted within *Statistical Methods, Experimental Design and Scientific Inference* (J. H. Bennett, ed). Oxford: University Press, 1990.

Florens, J.-P. (1978), 'Measures à priori et invariance dans une expèrience bayèsienne', *Pub. Inst. Statist. Univ. Paris* **23**, 29–55.

Florens, J.-P. (1982), 'Expèriences bayèsiennes invariantes', *Ann. Inst. H. Poincaré* **18**, 305–317.

Fonseca, T. C. O., Ferreira, M. A. R. & Migon, H. S. (2008), 'Objective Bayesian analysis for the student-t regression model', *Biometrika* **95**, 325–333.

Fougere, P. F., ed. (1990), *Maximum Entropy and Bayesian Methods*, Dordrecht: Kluwer.

Fraser, D. A. S. (1961), 'On fiducial inference', *Ann. Math. Statist.* **32**, 661–676.

Fraser, D. A. S. (1963), 'On the sufficiency and likelihood principles', *J. Amer. Statist. Assoc.* **58**, 641–647.

Fraser, D. A. S. (1968), *The Structure of Inference*, Hoboken, NJ: Wiley.

Fraser, D. A. S. (1972), 'Bayes, likelihood, or structural', *Ann. Math. Statist.* **43**, 777–790.

Fraser, D. A. S. (1979), *Inference and Linear Models*, New York: McGraw-Hill.

Fraser, D. A. S. (2004), Structural inference, *in* '*Encyclopedia of Statistical Sciences* **13**. (S. Kotz, C. B. Read, N. Balakrishnan, B. Vidakovic, and N. L. Johnson, eds). Hoboken, NJ: Wiley'.

Fraser, D. A. S. (2011), 'Is Bayes posterior just quick and dirty confidence?', *Stat. Sci.* **26**, 299–316.

Fraser, D. A. S., McDunnough, P. & Taback, N. (1997), Improper priors, posterior asymptotic normality, and conditional inference, *in* '*Advances in the Theory and Practice of Statistics: A Volume in Honor of Samuel Kotz* (N. L. Johnson and N. Balakrishnan, eds)', Hoboken, NJ: Wiley, pp. 563–569.

Fraser, D. A. S., Monette, G. & Ng, K.-W. (1985), Marginalization, likelihood and structured models, *in* '*Multivariate Analysis* **6** (D. K. Dey and C. R. Rao, eds). Amsterdam: North-Holland', pp. 209–217.

Fraser, D. A. S. & Reid, N. (1989), 'Adjustments to profile likelihood', *Biometrika* **76**, 477–488.

Fraser, D. A. S. & Reid, N. (1996), Bayes posteriors for scalar interest parameters, *in* '*Bayesian Statistics 5* (J. M. Bernardo, J. O. Berger, A. P. Dawid and A. F. M. Smith, eds). Oxford: University Press', pp. 581–585.

Fraser, D. A. S. & Reid, N. (2002), 'Strong matching of frequentist and Bayesian parametric inference', *J. Statist. Planning and Inference* **103**, 263–285.

Fraser, D. A. S., Reid, N., Marras, E. & Yi, G. Y. (2010), 'Default priors for Bayesian and frequentist inference', *J. Roy. Statist. Soc. B* **72**, 631–654.

Fraser, D. A. S., Reid, N., Wong, A. & Yi, G. Y. (2003), Direct Bayes for interest parameters, *in* '*Bayesian Statistics 7* (J. M. Bernardo, M. J. Bayarri, J. O. Berger, A. P. Dawid, D. Heckerman, A. F. M. Smith and M. West, eds). Oxford: University Press', pp. 529–534.

Fraser, D. A. S., Reid, N. & Wu, J. (1999), 'A simple general formula for tail probabilities for frequentist and Bayesian inference', *Biometrika* **86**, 249–264.

Fraser, D. A. S. & Yi, G. Y. (2002), 'Location reparametrization and default priors for statistical analysis', *J. Iranian Statist. Soc.* **1**, 55–78.

Freedman, D. A. & Purves, R. A. (1969), 'Bayes' method for bookies', *Ann. Math. Statist.* **40**, 1177–1186.

Freedman, D. & Diaconis, P. (1983), 'On inconsistent Bayes estimates in the discrete case', *Ann. Statist.* **11**, 1109–1118.

French, S. (1982), 'On the axiomatisation of subjective probabilities', *Theory and Decision* **14**, 19–33.

Fu, J. C. & Kass, R. E. (1988), 'The exponential rates of convergence of posterior distributions', *Ann. Inst. Statist. Math.* **40**, 683–691.

Fuglstad, G.-A., Simpson, D., Lindgren, F. & Rue, H. (2019), 'Constructing priors that penalize the complexity of gaussian random fields', *J. Amer. Statist. Assoc.* **114**, 445–452.

Gamerman, D. & Migon, H. S. (1993), 'Dynamic hierarchical models', *J. Roy. Statist. Soc. B* **55**, 628–642.

Gart, J. J. & Nam, J. (1988), 'Approximate interval estimation of the ratio of binomial parameters: A review and corrections for skewness', *Biometrics* **44**, 323–338.

Garvan, C. W. & Ghosh, M. (1997), 'Noninformative priors for dispersion models', *Biometrika* **84**, 976–982.

Garvan, C. W. & Ghosh, M. (1999), 'On the property of posteriors for dispersion models', *J. Statist. Planning and Inference* **78**, 229–241.

Geisser, S. (1965), 'Bayesian estimation in multivariate analysis', *Ann. Math. Statist.* **36**, 150–159.

Geisser, S. (1974), 'A predictive approach to the random effect model', *Biometrika* **61**, 101–107.

Geisser, S. (1979), 'Comments on "Reference posterior distributions for Bayesian inference" by J. M. Bernardo', *J. Roy. Statist. Soc. B* **41**, 136–137.

Geisser, S. (1984), 'On prior distributions for binary trials', *Amer. Statistician* **38**, 244–247.

Geisser, S. (1993), *Predictive Inference. An introduction*, London: Chapman and Hall.

Geisser, S. & Cornfield, J. (1963), 'Posterior distributions for multivariate normal parameters', *J. Roy. Statist. Soc. B* **25**, 368–376.

Gelfand, A. E. & Smith, A. F. M. (1990), 'Sampling-based approaches to calculating marginal densities', *J. Amer. Statist. Assoc.* **85**, 398–409.

Gelman, A. (2006), 'Prior distributions for variance parameters in hierarchical models (comment on article by browne and Draper)', *Bayesian Analysis* **1**, 515–534.

Gelman, A., Carlin, J. B., Stern, H. S., Dunson, D. B., Vehtari, A. & Rubin, D. B. (2013), *Bayesian data analysis*, CRC press.

Geman, S. & Geman, D. (1984), 'Stochastic relaxation, Gibbs distributions and the Bayesian restoration of images', *IEEE Trans. Patt. Anal. Mach. Intelligence* **6**, 721–740.

George, E. I. & McCulloch, R. E. (1993), 'On obtaining invariant prior distributions', *J. Statist. Planning and Inference* **37**, 169–179.

Ghosal, S. (1997), 'Reference priors in multiparameter nonregular cases', *Test* **6**, 159–186.

Ghosal, S. (1999), 'Probability matching priors for non-regular cases', *Biometrika* **86**, 956–964.

Ghosal, S., Ghosh, J. K. & Ramamoorthi, R. V. (1997), Non-informative priors via sieves and packing numbers, *in 'Advances in Decision Theory and Applications* (S. Panchpakesan and N. Balakrisnan, eds). Boston: Birkhauser', pp. 119–132.

Ghosal, S. & Samanta, T. (1997), 'Expansion of Bayes risk for entropy loss and reference prior in nonregular cases', *Statistics and Decisions* **15**, 129–140.

Ghosal, S. & Van der Vaart, A. (2017), *Fundamentals of nonparametric Bayesian inference*, Cambridge University Press, Cambridge, UK.

Ghosh, J. K. (1994), *Higher Order Asymptotics*, Hayward, CA: IMS.

Ghosh, J. K., Ghosal, S. & Samanta, T. (1994), Stability and convergence of the posterior in non-regular problems, *in 'Statistical Decision Theory and Related Topics V* (S. S. Gupta and J. O. Berger, eds). New York: Springer, Berlin: Springer, pp. 183–199.

Ghosh, J. K. & Mukerjee, R. (1991), 'Characterization of priors under which Bayesian and frequentist Bartlett corrections are equivalent in the multiparameter case', *J. Multivariate Analysis* **38**, 385–393.

Ghosh, J. K. & Mukerjee, R. (1992*a*), 'Bayesian and frequentist Bartlett corrections for likelihood ratio and conditional likelihood ratio tests', *J. Roy. Statist. Soc. B* **54**, 867–875.

Ghosh, J. K. & Mukerjee, R. (1992*b*), Non-informative priors, *in 'Bayesian Statistics 4* (J. M. Bernardo, J. O. Berger, A. P. Dawid and A. F. M. Smith, eds). Oxford: University Press', pp. 195–210 (with discussion).

Ghosh, J. K. & Mukerjee, R. (1993*a*), 'Frequentist validity of highest posterior density regions in the multiparameter case', *Amer. Statistician* **45**, 293–302; Corr: 602.

Ghosh, J. K. & Mukerjee, R. (1993*b*), 'On priors that match posterior and frequentist distribution functions', *Canad. J. Statist.* **21**, 89–96.

Ghosh, J. K. & Mukerjee, R. (1995*a*), 'Frequentist validity of highest posterior density regions in the presence of nuisance parameters', *Statistics and Decisions* **13**, 131–139.

Ghosh, J. K. & Mukerjee, R. (1995*b*), 'On perturbed ellipsoidal and highest posterior density regions with approximate frequentist validity', *J. Roy. Statist. Soc. B* **57**, 761–769.

Ghosh, J. K. & Mukerjee, R. (1998), Recent developments on probability matching priors, *in* S. E. Ahmed, M. Ashanullah & B. K. Sinha, eds, *'Applied Statistical Science III'*, Hauppauge, NY: Nova Science, pp. 227–252.

Ghosh, J. K., Sinha, B. K. & Joshi, S. M. (1982), Expansions for posterior probability and integrated Bayes risk, *in 'Statistical Decision Theory and Related Topics III* **1** (S. S. Gupta and J. O. Berger, eds). Cambridge, MA: Academic Press', pp. 403–456.

Ghosh, M. (1991), Hierarchical and empirical Bayes sequential estimation, *in* B. K. Ghosh & P. K. Sen, eds, *'Handbook of Sequential Analysis'*, New York: Marcel Dekker, pp. 441–458.

Ghosh, M. (1992*a*), 'Constrained Bayes estimation with applications', *J. Amer. Statist. Assoc.* **87**, 533–540.

Ghosh, M. (1992*b*), Hierarchical and empirical Bayes multivariate estimation, *in 'Current Issues in Statistical Inference: Essays in Honor of D. Basu.* (M. Ghosh and P. K. Pathak eds). Hayward, CA: IMS', pp. 151–177.

Ghosh, M., Carlin, B. P. & Srivastava, M. S. (1995), 'Probability matching priors for linear calibration', *Test* **4**, 333–357.

Ghosh, M., Chen, M.-H., Ghosh, A. & Agresti, A. (2000), 'Noninformative priors for one-parameter item response models', *J. Statist. Planning and Inference* **88**, 99–115.

Ghosh, M. & Heo, J. (2003*a*), 'Default Bayesian priors for regression models with first-order autoregressive residuals', *J. Time Series Analysis* **24**, 269–282.

Ghosh, M. & Heo, J. (2003*b*), 'Noninformative priors, credible sets and Bayesian hypothesis testing for the intraclass model', *J. Statist. Planning and Inference* **112**, 133–146.

Ghosh, M. & Kim, D. H. (2001), 'The Behrens-Fisher problem revisited: a Bayes-frequentist synthesis', *Canad. J. Statist.* **29**, 5–17.

Ghosh, M., Rousseau, J. & Kim, D. H. (2001), 'Noninformative priors for the bivariate Fieller-Creasy problem', *Statistics and Decisions* **19**, 277–288.

Ghosh, M., Sen, P. K. & Mukhopadhyay, N. (1997), *Sequential Estimation*, Hoboken, NJ: Wiley.

Ghosh, M. & Yang, M.-C. (1996), 'Noninformative priors for the two sample normal problem', *Test* **5**, 145–157.

Ghosh, M., Yin, M. & Kim, Y.-H. (2003), 'Objective Bayesian inference for ratios of regression coefficients in linear models', *Statistica Sinica* **13**, 409–422.

Gibbs, J. W. (1876), *The Scientific Papers of J. W. Gibbs, 1876–1878*, Reprinted in 1961, Mineola, NY: Dover.

Gilks, W. R. & Wild, P. (1992), 'Adaptive rejection sampling for Gibbs sampling', *Applied Statistics* **41**, 337–348.

Giocoli, N. (2013), 'From Wald to Savage: homo economicus becomes a Bayesian statistician', *J. History of Behavioural Sci.* pp. 63–95.

Girón, F. J., Martınez, M. & Moreno, E. (2003), 'Bayesian analysis of matched pairs', *Journal of statistical planning and inference* **113**, 49–66.

Gleser, L. J. & Hwang, J. T. (1987), 'The nonexistence of $100(1-\alpha)$ percent confidence sets of finite expected diameter in errors-in-variables and related models', *Ann. Statist.* **15**, 1351–1362.

Godambe, V. P. (1980), 'On sufficiency and ancillarity in the presence of a nuisance parameter', *Biometrika* **67**, 155–162.

Godambe, V. P. (1984), 'On ancillarity and Fisher information in the presence of a nuisance parameter', *Biometrika* pp. 626–629.

Gokhale, D. V. & Press, S. J. (1982), 'Assessment of a prior distribution for the correlation coefficient in a bivariate normal distribution', *J. Roy. Statist. Soc. A* **145**, 237–249.

Good, I. J. (1952), 'Rational decisions', *J. Roy. Statist. Soc. B* **14**, 107–114.

Good, I. J. (1960), 'Weight of evidence, corroboration, explanatory power, information, and the utility of experiments', *J. Roy. Statist. Soc. B* **22**, 319–331. Corr: **30** (1968) 203.

Good, I. J. (1963), 'Maximum entropy for hypothesis formulation, especially for multidimensional contingency tables', *Ann. Math. Statist.* **34**, 911–934.

Good, I. J. (1966), 'A derivation of the probabilistic explication of information', *J. Roy. Statist. Soc. B* **28**, 578–581.

Good, I. J. (1969), What is the use of a distribution?, *in* '*Multivariate Analysis* **2** (D. K. Dey and C. R. Rao, eds). Amsterdam: North-Holland', pp. 183–203.

Good, I. J. (1980), Some history of the hierarchical Bayesian methodology, *in* '*Bayesian Statistics* (J. M. Bernardo, M. H. DeGroot, D. V. Lindley and A. F. M. Smith, eds). Valencia: University Press', pp. 489–519 (with discussion).

Good, I. J. (1981), 'Flexible priors for estimating a normal distribution', *J. Statist. Computation and Simulation* **13**, 151–153.

Good, I. J. (1982), Degrees of belief, *in* '*Encyclopedia of Statistical Sciences* **1**. (S. Kotz, C. B. Read, N. Balakrishnan, B. Vidakovic, and N. L. Johnson, eds). Hoboken, NJ: Wiley', pp. 287–293.

Good, I. J. (1983), *Good Thinking: The Foundations of Probability and its Applications*, Minneapolis: Univ. of Minnesota Press.

Good, I. J. (1986), 'Invariant hyperpriors', *J. Statist. Computation and Simulation* **24**, 323–324.

Good, I. J. (1992), 'The Bayes/non-Bayes compromise: A brief review', *J. Amer. Statist. Assoc.* **87**, 597–606.

Good, I. J. & Crook, J. F. (1974), 'The Bayes/non-Bayes compromise and the multinomial distribution', *J. Amer. Statist. Assoc.* **69**, 711–720.

Gordon, R. D. (1940), 'Inverse probability and modern statisticians', *Philosophy of Science* **7**, 389–399.

Gosset, W. S. (1908), 'The probable error of a mean', *Biometrika* **6**, 1–25.

Goudie, I. B. J. & Goldie, C. M. (1981), 'Initial size estimation for the pure death process', *Biometrika* **68**, 543–550.

Goudie, I. B. J. & Goldie, C. M. (1994), 'On the erratic behavior of estimators of n in the binomial (n, p) distribution', *J. Amer. Statist. Assoc.* **89**, 344–352.

Govindarajulu, Z. (1981), *The Sequential Statistical Analysis of Hypothesis Testing, Point and Interval Estimation, and Decision Theory*, Columbus, OH: American Science Press.

Goyal, P. (2005), Prior probabilities: An information-theoretic approach, *in* 'AIP Conference Proceedings', Vol. 803, American Institute of Physics, pp. 366–373.

Grientschnig, D. (2015), 'Reference priors for poisson processes involving nuisance parameters', *Communications in Statistics-Theory and Methods* **44**(23), 4893–4911.

Grientschnig, D. & Lira, I. (2011), 'Reassessment of a calibration model by Bayesian reference analysis', *Metrologia* **48**(1), L7.

Grünwald, P. D. & Dawid, A. P. (2004), 'Game theory, maximum entropy, minimum discrepancy and robust Bayesian decision theory', *Ann. Statist.* **32**, 1367–1433.

Gu, C. (1992), 'Penalized likelihood regression: A Bayesian analysis', *Statistica Sinica* **2**, 255–264.

Gubler, P. & Oka, M. (2010), 'A Bayesian approach to qcd sum rules', *Progress of Theoretical Physics* **124**, 995–1018.

Gustafson, P., Hossain, S. & MacNab, Y. C. (2006), 'Conservative prior distributions for variance parameters in hierarchical models', *Canad. J. Statist.* **34**, 377–390.

Gutiérrez-Peña, E. (1992), Expected logarithmic divergence for exponential families, *in* '*Bayesian Statistics 4* (J. M. Bernardo, J. O. Berger, A. P. Dawid and A. F. M. Smith, eds). Oxford: University Press', pp. 669–674.

Gutiérrez-Peña, E. & Rueda, R. (2003), 'Reference priors for exponential families', *J. Statist. Planning and Inference* **110**, 35–54.

Gutiérrez-Redomero, E., Alonso, C., Romero, E. & Galera, V. (2008), 'Variability of fingerprint ridge density in a sample of spanish caucasians and its application to sex determination', *Forensic Sci. Internat.* **180**, 17–22.

Hacking, I. (1965), *Logic of Statistical Inference*, Cambridge, UK: Cambridge University Press.

Hadjicostas, P. (1998), 'Improper and proper posteriors with improper priors in a hierarchical model with a beta-binomial likelihood', *Comm. Statist. Theory and Methods* **27**, 1905–1914.

Hadjicostas, P. & Berry, S. M. (1999), 'Improper and proper posteriors with improper priors in a Poisson-Gamma hierarchical model', *Test* **8**, 147–166.

Haff, L. R. (1980), 'Empirical Bayes estimation of the multivariate normal covariance matrix', *The Annals of Statistics* **8**, 586–597.

Hajec, A. (2019), Interpretations of probability, *in* E. N. Zalta, ed., 'Stanford Encyclopedia of Philosophy', Stanford University, Stanford, CA.

Hald, A. (1998), *A History of Mathematical Statistics from 1750 to 1930*, Hoboken, NJ: Wiley.

Hald, A. (1999), 'On the history of maximum likelihood in relation to inverse probability and least squares', *Statistical Science* **14**, 214–222.

Haldane, J. B. S. (1932), 'A note on inverse probability', *Math. Proc. Cambridge Phil. Soc.* **28**, 55–61.

Haldane, J. B. S. (1948), 'The precision of observed values of small frequencies', *Biometrika* **35**, 297–303.

Hall, W. J. (1992), *A Course in Sequential Analysis*, Rochester, NY: University of Rochester.

Halmos, P. R. (1950), *Measure Theory*, New York: Van Nostrand.

Hamelryck, T. (2009), 'Probabilistic models and machine learning in structural bioinformatics', *Statistical methods in medical research* **18**, 505–526.

Hampel, F. (2006), The proper fiducial argument, *in* '*General Theory of Information Transfer and Combinatorics.* (R. Ahlswede *et al.* eds)', New York: Springer, pp. 512–526.

Hand, D. J. & Yu, K. (2001), 'Idiot's Bayes – not so stupid after all?', *Internat. Statist. Rev.* **69**, 385–398.

Hannig, J. (2009), 'On generalized fiducial inference', *Statistica Sinica* **19**, 491–544.

Harney, H. L. (2003), Beyond form invariance: The geometric prior, *in* 'Bayesian Inference', Berlin: Springer, pp. 71–80.

Hartigan, J. A. (1964), 'Invariant prior distributions', *Ann. Math. Statist.* **35**, 836–845.

Hartigan, J. A. (1965), 'The asymptotically unbiased prior distribution', *Ann. Math. Statist.* **36**, 1137–1152.

Hartigan, J. A. (1966), 'Note on the confidence-prior of Welch and Peers', *J. Roy. Statist. Soc. B* **28**, 55–56.

Hartigan, J. A. (1967), 'The likelihood and invariance principles', *J. Roy. Statist. Soc. B* **29**, 533–539.

Hartigan, J. A. (1971), Similarity and probability, *in* '*Foundations of Statistical Inference* (V. P. Godambe and D. A. Sprott, eds). Toronto: Holt, Rinehart and Winston', pp. 305–313 (with discussion).

Hartigan, J. A. (1975), 'Necessary and sufficient conditions for asymptotic joint normality of a statistic and its subsample values', *Ann. Statist.* **3**, 573–580.

Hartigan, J. A. (1983), *Bayes Theory*, New York: Springer.

Hartigan, J. A. (1996), 'Locally uniform prior distributions', *Ann. Statist.* **24**, 160–173.

Hartigan, J. A. (1998), 'The maximum likelihood prior', *Ann. Statist.* **26**, 2083–2103.

Hartigan, J. A. (2004), 'Uniform priors on convex sets improve risk', *Statistics and Probab. Letters* **67**, 285–288.

Harvey, J. & Van der Merwe, A. J. (2012), 'Bayesian confidence intervals for means and variances of lognormal and bivariate lognormal distributions', *Journal of Statistical Planning and Inference* **142**, 1294–1309.

Harvey, J. & Van der Merwe, A. J. (2018), 'A Bayesian approach to inference on the variance of lognormal data', *South African Statistical Journal* **52**, 29–55.

Hastings, W. K. (1970), 'Monte Carlo sampling methods using Markov chains and their applications', *Biometrika* **57**, 97–109.

Hattori, K., Valluri, M., Loebman, S. R. & Bell, E. F. (2017), 'Reliability of the measured velocity anisotropy of the milky way stellar halo', *The Astrophysical Journal* **841**, 91.

Haussler, D. & Opper, M. (1997), 'Mutual information, metric entropy, and cumulative entropy risk', *Ann. Statist.* **25**, 2451–2492.

He, C. Z. (2003), 'Bayesian modelling of age-specific survival in bird nesting studies under irregular visits', *Biometrics* **59**, 962–973.

He, C. Z. (2011), 'Bayesian analysis for the multinomial data under noninformative priors', *Tech. Rep.*, University of Missouri .

Heath, D. C. & Sudderth, W. D. (1972), 'On a theorem of de finetti, oddsmaking, and game theory', *Ann. Math. Statist.* pp. 2072–2077.

Heath, D. C. & Sudderth, W. D. (1976), 'De Finetti's theorem on exchangeable variables', *Amer. Statistician* **30**, 188–189.

Heath, D. C. & Sudderth, W. D. (1978), 'On finitely additive priors, coherence, and extended admissibility', *Ann. Statist.* **6**, 333–345.

Heath, D. C. & Sudderth, W. D. (1989), 'Coherent inference from improper priors and from finitely additive priors', *Ann. Statist.* **17**, 907–919.

Hee, S., Vázquez, J. A., Handley, W. J., Hobson, M. P. & Lasenby, A. N. (2017), 'Constraining the dark energy equation of state using Bayes theorem and the Kullback–Leibler divergence', *Monthly Notices Roy. Astronom. Soci.* **466**, 369–377.

Helland, I. S. (1995), 'Simple counterexamples against the conditionality principle', *The American Statistician* **49**, 351–356.

Heyde, C. C. & Johnstone, I. M. (1979), 'On asymptotic posterior normality for stochastic processes', *J. Roy. Statist. Soc. B* **41**, 184–189.

Hill, B. M. (1965), 'Inference about variance components in the one-way model', *J. Amer. Statist. Assoc.* **60**, 806–825.

Hill, B. M. (1987), 'The validity of the likelihood principle', *Amer. Statistician* **41**, 95–100.

Hills, S. E. (1987), 'Reference priors and identifiability problems in non-linear models', *The Statistician* **36**, 235–240.

Hills, S. E. & Smith, A. F. M. (1992), Parameterization issues in Bayesian inference, *in* '*Bayesian Statistics 4* (J. M. Bernardo, J. O. Berger, A. P. Dawid and A. F. M. Smith, eds). Oxford: University Press', pp. 227–246 (with discussion).

Hinkley, D. (1979), 'Predictive likelihood', *Ann. Statist.* **7**, 718–728. Corr: **8**, 694.

Hipp, C. (1974), 'Sufficient statistics and exponential families', *Ann. Statist.* **2**, 1283–1292.

Hjort, N. L., Holmes, C., Müller, P. & Walker, S. G. (2010), *Bayesian Nonparametrics*, Cambridge: University Press.

Ho, K.-W. (2012), 'The use of Jeffreys priors for the student-t distribution', *Journal of Statistical Computation and Simulation* **82**, 1015–1021.

Hoadley, B. (1970), 'A Bayesian look at inverse linear regression', *J. Amer. Statist. Assoc.* **65**, 356–369.

Hoff, P. D. (2009), 'A hierarchical eigenmodel for pooled covariance estimation', *J. Roy. Statist. Soc. B* **71**, 971–992.

Hooper, M. (2013), 'Richard Price, Bayes' theorem, and God', *Significance* **10**, 36–39.

Hora, R. B. & Buehler, R. J. (1966), 'Fiducial theory and invariant estimation', *Ann. Math. Statist.* **37**, 643–656.

Hörmann, W., Leydold, J. & Derflinger, G. (2004), *Automatic Nonuniform Random Variate Generation*, New York: Springer. Ch. 15.

Howlader, H. A. & Weiss, G. (1988), 'Bayesian reliability estimation of a two parameter Cauchy distribution', *Biometrical J.* **30**, 329–337.

Howson, C. (1997), 'A logic of induction', *Philosophy of Science* **64**, 268–290.

Hu, G.-Q., Liu, Y., Wei, X.-C. & Chen, T.-S. (2011), 'Joint pp and ps avo inversion based on Bayes theorem', *Appl. Geophysics* **8**, 293–302.

Hunt, G. & Stein, C. (1946), 'Most stringent tests of statistical hypotheses', *Tech. Rep.*, Stanford University.

Ibragimov, I. A. & Khasminskii, R. Z. (1973), On the information in a sample about a parameter, *in* '*Proc. 2nd Internat. Symp. Information Theory.* (B. N. Petrov and F. Csaki, eds). Budapest: Akademiai Kiadó', pp. 295–309.

Ibrahim, J. G. (1997), 'On properties of predictive priors in linear models', *Amer. Statistician* **51**, 333–337.

Ibrahim, J. G. & Chen, M.-H. (1998), 'Prior distributions and Bayesian computation for proportional hazards models', *Sankhyā B* **60**, 48–64.

Ibrahim, J. G. & Laud, P. W. (1991), 'On Bayesian analysis of generalized linear models using Jeffreys's prior', *J. Amer. Statist. Assoc.* **86**, 981–986.

Inaba, T. (1984), 'Non-informative prior distribution in a simultaneous equations model', *J. Japan Statist. Soc.* **14**, 93–103.

Irony, T. Z. (1992), 'Bayesian estimation for discrete distributions', *J. Appl. Statist.* **19**, 533–549.

Isserlis, L. (1936), 'Inverse probability', *Journal of the Royal Statistical Society* **99**, 130–137.

Jaynes, E. T. (1958), *Probability Theory in Science and Engineering*, Dallas, TX: Mobil Oil Co.

Jaynes, E. T. (1968), 'Prior probabilities', *IEEE Trans. Systems Science Cybernetics* **4**, 227–291.

Jaynes, E. T. (1976), Confidence intervals vs. Bayesian intervals, *in 'Foundations of Probability Theory, Statistical inference, and Statistical Theories of Science* (I. J. Good, W. L. Harper, and C. A. Hooker, eds). Dordrecht: Reidel', pp. 175–257 (with discussion).

Jaynes, E. T. (1982), 'On the rationale of maximum-entropy methods', *Proc. of the IEEE* **70**, 939–952.

Jaynes, E. T. (1985), Highly informative priors, *in 'Bayesian Statistics 2* (J. M. Bernardo, M. H. DeGroot, D. V. Lindley and A. F. M. Smith, eds). Amsterdam: North-Holland', pp. 329–359.

Jaynes, E. T. (2003), *Probability Theory: The Logic of Science*, Cambridge: University Press.

Jeffrey, R. (1983), *The Logic of Decision*, 2nd edn, Chicago, IL: University of Chicago Press.

Jeffreys, H. (1938), 'Maximum likelihood, inverse probability and the method of moments', *Annals of Eugenics* **8**, 146–151.

Jeffreys, H. (1946), 'An invariant form for the prior probability in estimation problems', *Proc. Roy. Soc. A* **186**, 453–461.

Jeffreys, H. (1955), 'The present position in probability theory', *Brit. J. Philos. Sci.* **5**, 275–289.

Jeffreys, H. (1961), *Theory of Probability*, 3rd edn, Oxford: University Press.

Johnson, R. A. (1970), 'Asymptotic expansions associated with posterior distributions', *Ann. Math. Statist.* **41**, 851–864.

Jones, G. T. & Yarn, D. H. (2003), 'Evaluation dispute resolution under uncertainty: An empirical look at Bayes' theorem and the expected value of perfect information', *J. Disp. Resol.* p. 427.

Joshi, S. & Shah, M. (1991), 'Estimating the mean of an inverse Gaussian distribution with known coefficient of variation', *Comm. Statist. Theory and Methods* **20**, 2907–2912.

Joshi, V. M. (1983), Likelihood principle, *in 'Encyclopedia of Statistical Sciences* **1**. (S. Kotz, C. B. Read, N. Balakrishnan, B. Vidakovic, and N. L. Johnson, eds). Hoboken, NJ: Wiley', pp. 644–647.

Jowett, C. (2001), 'Lies, damned lies, and DNA statistics: DNA match testing, Bayes' theorem, and the criminal courts', *Medicine, Science and the Law* **41**, 194–205.

Joyce, J. (2003), Bayes theorem, *in* E. N. Zalta, ed., 'Stanford Encyclopedia of Philosophy', Stanford University, Stanford, CA.

Juárez, M. A. (2005), 'Objective Bayes estimation and hypothesis testing'. *Tech. Rep.*, Univ. Warwick.

Jung, B.-C., Biscontin, G. & Gardoni, P. (2009), 'Bayesian updating of a unified soil compression model', *Georisk* **3**, 87–96.

Justice, J. H., ed. (1986), *Maximum Entropy and Bayesian Methods in Applied Statistics*, Cambridge: University Press.

Kadane, J. B. & O'Hagan, A. (1995), 'Using finitely additive probability: Uniform distributions on the natural numbers', *J. Amer. Statist. Assoc.* **90**, 626–631.

Kadane, J. B., Schervish, M. J. & Seidenfeld, T. (1986), Statistical implications of finitely additive probability, *in* '*Bayesian Inference and Decision Techniques: Essays in Honor of Bruno de Finetti* (P. K. Goel and A. Zellner, eds). Amsterdam: North-Holland', pp. 59–76.

Kadane, J. B., Schervish, M. J. & Seidenfeld, T., eds (1999), *Rethinking the Foundations of Statistics*, Cambridge: University Press.

Kahn, W. D. (1987), 'A cautionary note for Bayesian estimation of the binomial parameter *n*', *Amer. Statistician* **41**, 38–39.

Kahneman, D., Slovic, S. P., Slovic, P. & Tversky, A., eds (1982), *Judgment under Uncertainty: Heuristics and Biases*, Cambridge: University Press.

Kalbfleisch, J. D. (1975), 'Sufficiency and conditionality', *Biometrika* **62**, 251–259.

Kalbfleisch, J. D. & Sprott, D. A. (1973), 'Marginal and conditional likelihoods', *Sankhyā A* **35**, 311–328.

Kaplan, E. H. & Barnett, A. (2003), 'A new approach to estimating the probability of winning the presidency', *Operations Research* **51**, 32–40.

Kaplan, M. (1998), *Decision Theory as Philosophy*, Cambridge, UK: Cambridge University Press.

Kappenman, R. F., Geisser, S. & Antle, C. E. (1970), 'Bayesian and fiducial solutions for the Fieller-Creasy problem', *Sankhyā B* **32**, 331–340.

Kardaun, O., Salomé, D., Schaafsma, W., Steerneman, A., Willems, J. & Cox, D. R. (2003), 'Reflections on fourteen cryptic issues concerning the nature of statistical inference', *International Statistical Review* pp. 277–303.

Kashyap, R. L. (1971), 'Prior probability and uncertainty', *IEEE Trans. Information Theory* **4**, 641–650.

Kass, R. E. (1989), 'The geometry of asymptotic inference', *Statistical Sci.* **4**, 188–234 (with discussion).

Kass, R. E. (1990), 'Data-translated likelihood and Jeffreys rules', *Biometrika* **77**, 107–114.

Kass, R. E. & Slate, E. H. (1992), Reparameterization and diagnostics of posterior non-normality, *in* '*Bayesian Statistics 4* (J. M. Bernardo, J. O. Berger, A. P. Dawid and A. F. M. Smith, eds). Oxford: University Press', pp. 289–305 (with discussion).

Kass, R. E., Tierney, L. & Kadane, J. B. (1990), The validity of posterior expansions based on Laplace's method, *in* '*Bayesian and Likelihood Methods in Statistics and Econometrics: Essays in Honor of George A. Barnard* (S. Geisser, J. S. Hodges, S. J. Press and A. Zellner, eds). Amsterdam: North-Holland', pp. 473–488.

Kass, R. E., Tierney, L. & Kadane, J. B. (1991), 'Laplace's method in Bayesian analysis', *Contemporary Mathematics* **115**, 89–99.

Kass, R. E. & Wasserman, L. (1996), 'The selection of prior distributions by formal rules', *J. Amer. Statist. Assoc.* **91**, 1343–1370. Corr: **93**, 412.

Kazianka, H. (2012), 'Objective Bayesian analysis for the normal compositional model', *Computational Statistics & Data Analysis* **56**, 1528–1544.

Kazianka, H. (2013), 'Objective Bayesian analysis of geometrically anisotropic spatial data', *Journal of Agricultural, Biological, and Environmental Statistics* **18**, 514–537.

Kazianka, H. & Pilz, J. (2012), 'Objective Bayesian analysis of spatial data with uncertain nugget and range parameters', *Canadian Journal of Statistics* **40**, 304–327.

Keefe, M. J., Ferreira, M. A. R. & Franck, C. T. (2019), 'Objective Bayesian analysis for gaussian hierarchical models with intrinsic conditional autoregressive priors', *Bayesian Analysis* **14**, 181–209.

Kelly, N. E., Javed, A., Shimoda, Y., Zastepa, A., Watson, S., Mugalingam, S. & Arhonditsis, G. B. (2019), 'A Bayesian risk assessment framework for microcystin violations of drinking water and recreational standards in the bay of quinte, lake ontario, canada', *Water research* **162**, 288–301.

Kershaw, J., Ardekani, B. A. & Kanno, I. (1999), 'Application of Bayesian inference to fmri data analysis', *IEEE transactions on Medical imaging* **18**, 1138–1153.

Kershaw, J., Kashikura, K., Zhang, X., Abe, S. & Kanno, I. (2001), 'Bayesian technique for investigating linearity in event-related bold fmri', *Magnetic Resonance in Medicine* **45**, 1081–1094.

Keynes, J. M. (1929), *A Treatise on Probability*, 2nd edn, London: Macmillan. Reprinted in 1962. New York: Harper and Row.

Kiefer, J. (1957), 'Invariance, minimax sequential estimation, and continuous time processes', *Ann. Math. Statist.* **28**, 573–601.

Kiefer, J. (1966), Multivariate optimality results, *in* '*Multivariate Analysis* (P. R. Krishnaiah, ed). Cambridge, MA: Academic Press', pp. 255–274.

Kiefer, J. (1975), Conditional confidence approach in multi decision problems, *in* '*Multivariate Analysis* **4** (D. K. Dey and C. R. Rao, eds). Amsterdam: North-Holland', pp. 143–158.

Kiefer, J. (1976), 'Admissibility of conditional confidence procedures', *Ann. Statist.* **4**, 836–865.

Kiefer, J. (1977), 'Conditional confidence statements and confidence estimators', *J. Amer. Statist. Assoc.* **72**, 789–827 (with discussion).

Kim, B. H., Chang, I. H. & Kang, C. K. (1961), 'Bayesian estimation for the reliability in Weibull stress-strength systems using noninformative priors', *Far East J. Theor. Statist.* **5**, 299–315.

Kim, D. H., Kang, S. G. & Cho, J. S. (2000), 'Noninformative priors for stress-strength system in the Burr-Type X model', *J. Korean Statist. Soc.* **29**, 17–28.

Kim, D. H., Kang, S. G. & Lee, W. D. (2001), 'Noninformative priors for intraclass correlation coefficient in familial data', *Far East J. Theor. Statist.* **5**, 51–65.

Kim, D.-H., Kang, S.-G. & Lee, W.-D. (2002), 'Noninformative priors for the power law process', *Journal of the Korean Statistical Society* **31**, 17–31.

Kim, S. W. & Ibrahim, J. G. (2000), 'On Bayesian inference for proportional hazards models using noninformative priors', *Lifetime Data Analysis* **6**, 331–341.

Kinderman, A. J. & Monahan, J. F. (1977), 'Computer generation of random variables using the ratio of uniform deviates', *ACM Trans. Math. Software* **3**, 257–260.

Kolmogórov, A. N. (1956), *Grundbegriffe der Wahrscheinlichkeitsrechnung*, 2nd edn, Berlin: Springer. English translation as *Foundations of the Theory of Probability*, Mineola, NY: Dover, 2018.

Koopman, B. O. (1940), 'The axioms and algebra of intuitive probability', *Ann. Mathematics* pp. 269–292.

Kopp, B. (2007), The p300 component of the event-related brain potential and Bayes' theorem, *in* M.-K. Sun, ed., 'Cognitive Sciences at the Leading Edge', New York, NY: Nova Science, pp. 113–125.

Kramer, M. & Starr, N. (1990), 'Optimal stopping in a size dependent search', *Sequential Anal.* **9**, 59–80.

Krichevskiy, R. E. (1998), 'Laplace's law of succession and universal encoding', *IEEE Transactions on Information Theory* **44**(1), 296–303.

Krishnamoorthy, K. & Gupta, A. K. (1989), 'Improved minimax estimation of a normal precision matrix', *The Canadian Journal of Statistics* **17**, 91–102.

Kuboki, H. (1998), 'Reference priors for prediction', *J. Statist. Planning and Inference* **69**, 295–317.

Kullback, S. (1968), *Information Theory and Statistics,* 2nd edition, Mineola, NY: Dover. Reprinted in 1978, Gloucester, MA: Peter Smith.

Kullback, S. & Leibler, R. A. (1951), 'On information and sufficiency', *Ann. Math. Statist.* **22**, 79–86.

Lafferty, J. & Wasserman, L. (2001), Iterative Markov chain Monte Carlo computation of reference priors and minimax risk, *in* J. S. Breese & D. Koller, eds, '*Proc. 17th Conf. Uncertainty in Artificial Intelligence*', San Francisco, CA: Morgan Kaufman, pp. 293–300.

Lane, D. A. & Sudderth, W. D. (1984), 'Coherent predictive inference', *Sankhyā A* **46**, 166–185.

Laplace, P. S. (1774), Mémoire sur la probabilité des causes par les événements, *in* '*Oeuvres Complètes* **8**', Paris: Gauthier-Villars, pp. 27–68.

Laplace, P. S. (1812), *Théorie Analytique des Probabilités*, Paris: Courcier. Reprinted in *Oeuvres Complètes de Laplace* **7**, 1878–1912. Paris: Gauthier-Villars.

Laplace, P. S. (1825), *Essai Philosophique sur les Probabilitiés*, Paris: Courcier (5th ed). English translation in 1952 as *Philosophical Essay on Probabilities.* New York: Dover.

Lauretto, M., Pereira, C. A. B., Stern, J. M. & Zacks, S. (2003), 'Comparing parameters of two bivariate normal distributions using the invariant FBST, Full Bayesian Significance Test', *Brazilian J. Probab. Statist.* **17**, 147–168.

Lauritzen, S. L. (1996), *Graphical models*, Oxford University Press.

Lavine, M. (1991), 'Sensitivity in Bayesian statistics: The prior and the likelihood', *J. Amer. Statist. Assoc.* **86**, 396–399.

Lavine, M. (2020), 'Assessing procedures vs. assessing evidence', *Notices Amer. Math. Soc.* **67**(7).

Lawless, J. F. (1982), *Statistical Models and Methods for Lifetime Data*, Hoboken, NJ: Wiley.

Lawley, D. (1956), 'A general method for approximating to the distribution of the likelihood ratio criteria', *Biometrika* **43**, 295–303.

Le Cam, L. M. (1986), *Asymptotic Methods in Statistical Decision Theory*, New York: Springer.

Lecoutre, B., Derzko, G. & ElQasyr, K. (2010), 'Frequentist performance of Bayesian inference with response-adaptive designs', *Statistics in Medicine* **29**, 3219–3231.

Lee, G. (1998), 'Development of matching priors for $P(X < Y)$ in exponential distributions', *J. Korean Statist. Soc.* **27**, 421–433.

Lee, H. K. H. (2003), 'A noninformative prior for neural networks', *Machine Learning* **50**, 197–212.

Lee, H. K. H. (2004), 'Priors for neural networks', *Classification, Clustering, and Data Mining Applications* pp. 141–150.

Lee, H. K. H. (2007), 'Default priors for neural network classification', *Journal of classification* **24**, 53–70.

Lee, J. & Berger, J. O. (2001), 'Semiparametric Bayesian analysis of selection models', *J. Amer. Statist. Assoc.* **96**, 1397–1409.

Lee, J. C. & Hwang, R. C. (2000), 'On estimation and prediction for temporally correlated longitudinal data', *J. Statist. Planning and Inference* **87**, 87–104.

Lee, J. J. & Shin, W. S. (1990), 'Prior distributions using the entropy principles', *Korean J. Appl. Statist.* **3**, 91–104.

Leisen, F., Rossini, L. & Villa, C. (2018), 'Objective Bayesian analysis of the yule–simon distribution with applications', *Computational Statistics* **33**, 99–126.

Leisen, F., Villa, C. & Walker, S. G. (2020), 'On a class of objective priors from scoring rules (with discussion)', *Bayesian Analysis* **15**, 1345–1423.

Leonard, T. H. (1977), 'A Bayesian approach to some multinomial estimation and pretesting problems', *J. Amer. Statist. Assoc.* **72**, 869–874.

Leonard, T. H. & Hsu, J. S. J. (1992), 'Bayesian inference for a covariance matrix', *Ann. Statist.* **20**, 1669–1696.

Leonard, T. H. & Hsu, J. S. J. (1994), The Bayesian analysis of categorical data: A selective review, *in* '*Aspects of Uncertainty: a Tribute to D. V. Lindley* (P. R. Freeman and A. F. M. Smith, eds). Chichester: Wiley', pp. 283–310.

Leucari, V. & Consonni, G. (2003), Compatible priors for causal Bayesian networks, *in* '*Bayesian Statistics 7* (J. M. Bernardo, M. J. Bayarri, J. O. Berger, A. P. Dawid, D. Heckerman, A. F. M. Smith and M. West, eds). Oxford: University Press', pp. 597–606.

Levine, R. & Tribus, M., eds (1978), *The Maximum Entropy Formalism*, Cambridge, MA: The MIT Press.

Ley, E. & Steel, M. F. J. (2009), 'On the effect of prior assumptions in Bayesian model averaging with applications to growth regression', *J. Applied Econometrics* **24**, 651–674.

Leydold, J. (2003), 'Short universal generators via generalized ratio-of-uniforms method', *Mathematics of Computation* **72**, 1453–1471.

Li, S. & Wang, M. (2017), 'Bayesian estimation of the generalized lognormal distribution using objective priors', *Journal of Statistical Computation and Simulation* **87**, 1323–1341.

Liang, K.-Y. (1984), 'The asymptotic efficiency of conditional likelihood methods', *Biometrika* **71**, 305–313.

Liang, Y. & Sun, D. (2012), 'Objective priors for generative star-shape models', *Statistics & Probability Letters* **82**, 991–997.

Lindley, D. V. (1956), 'On a measure of information provided by an experiment', *Ann. Math. Statist.* **27**, 986–1005.

Lindley, D. V. (1957), 'A statistical paradox', *Biometrika* **44**, 187–192.

Lindley, D. V. (1958), 'Fiducial distribution and Bayes' theorem', *J. Roy. Statist. Soc. B* **20**, 102–107.

Lindley, D. V. (1961), The use of prior probability distributions in statistical inference and decisions, *in 'Proc. Fourth Berkeley Symp.'* (J. Neyman and E. L. Scott, eds). Berkeley: Univ. California Press, pp. 453–468.

Lindley, D. V. (1964), 'The Bayesian analysis of contingency tables', *Ann. Math. Statist.* **35**, 1622–1643.

Lindley, D. V. (1965), *Introduction to Probability and Statistics from a Bayesian Viewpoint*, Cambridge: University Press.

Lindley, D. V. (1982*a*), Coherence, *in 'Encyclopedia of Statistical Sciences* **1**. (S. Kotz, C. B. Read, N. Balakrishnan, B. Vidakovic, and N. L. Johnson, eds). Hoboken, NJ: Wiley', pp. 29–31.

Lindley, D. V. (1982*b*), 'Scoring rules and the inevitability of probability', *Internat. Statist. Rev.* **50**, 1–26 (with discussion).

Lindley, D. V. (1988), Statistical inference concerning Hardy-Weinberg equilibrium, *in 'Bayesian Statistics 3* (J. M. Bernardo, M. H. DeGroot, D. V. Lindley and A. F. M. Smith, eds). Oxford: University Press', pp. 307–326 (with discussion).

Lindley, D. V. & Smith, A. F. M. (1972), 'Bayes estimates for the linear model', *J. Roy. Statist. Soc. B* **34**, 1–41 (with discussion).

Lindqvist, B. H. & Taraldsen, G. (2018), 'On the proper treatment of improper distributions', *J. Statist. Planning and Inference* **195**, 93–104.

Lindsay, B. G. & Roeder, K. (1987), 'A unified treatment of integer parameter models', *J. Amer. Statist. Assoc.* **82**, 758–764.

Lira, I. & Grientschnig, D. (2011), 'Assignment of a non-informative prior when using a calibration function', *Measurement Science and Technology* **23**, 015001.

Liseo, B. (1993), 'Elimination of nuisance parameters with reference priors', *Biometrika* **80**, 295–304.

Liseo, B. (2003), 'Bayesian and conditional frequentist analyses of the Fieller's problem. A critical review', *Metron* **61**, 133–152.

Liseo, B. (2005), The elimination of nuisance parameters, *in 'Handbook of Statistics* **25** (D. K. Dey and C. R. Rao, eds). Amsterdam: Elsevier', pp. 193–219.

Liseo, B. & Loperfido, N. (2006), 'A note on reference priors for the scalar skew-normal distribution', *J. Statist. Planning and Inference* **136**, 373–389.

Liseo, B. & Macaro, C. (2013), 'Objective priors for causal AR (p) with partial autocorrelations', *Journal of Statistical Computation and Simulation* **83**(9), 1613–1628.

Liseo, B., Solari, F. & Sun, D. (2009), 'Noninformative priors under special patterns of the Fisher information matrix', *Sankhyā A* **71**, 284–297.

Liu, J. S. (2001), *Monte Carlo Strategies in Scientific Computing*, New York: Springer.

Liu, L.-M. (1981), 'Estimation of random coefficient regression models', *Journal of Statistical Computation and Simulation* **13**, 27–39.

Liu, L.-M. & Hanssens, D. M. (1981), 'A Bayesian approach to time-varying cross-sectional regression models', *Journal of Econometrics* **15**, 341–356.

Long, A. J., Raveri, M., Hu, W. & Dodelson, S. (2018), 'Neutrino mass priors for cosmology from random matrices', *Physical Review D* **97**, 043510.

Luce, R. D. (2012), *Individual Choice Behavior: A Theoretical Analysis*, Mineola, NY: Dover.

Lüroth, J. (1876), 'Vergelichung von zwei Werthen des wahrscheinlichen Fehlers', *Astronomische Nachrichten* **87**, 209.

Ly, A., Marsman, M. & Wagenmakers, E.-J. (2018), 'Analytic posteriors for pearson's correlation coefficient', *Statistica Neerlandica* **72**(1), 4–13.

Machina, M. J. (1992), Choice under uncertainty: Problems solved and unsolved, *in* G. Dionne & S. E. Harrington, eds, 'Foundations of Insurance Economics', Berlin: Springer, pp. 49–82.

Manderson, A. A., Murray, K. & Turlach, B. A. (2020), Flexible regression modelling under shape constraints, *in* Y. Fan, D. Nott, M. Smith & J.-L. Dortet-Bernadet, eds, 'Flexible Bayesian Regression Modelling', Elsevier, pp. 251–279.

Maneerat, P., Niwitpong, S.-A. & Niwitpong, S. (2020), 'Bayesian confidence intervals for the difference between variances of delta-lognormal distributions', *Biometrical Journal* **62**, 1769–1790.

Martin, R. & Liu, C. (2015), *Inferential Models: Reasoning with Uncertainty*, CRC Press.

Massam, H., Liu, J. S. & Dobra, A. (2009), 'A conjugate prior for discrete hierarchical log-linear models', *The Annals of Statistics* **37**, 3431–3467.

Massam, H. & Neher, E. (1998), 'Estimation and testing for lattice conditional independence models on Euclidean Jordan algebras', *Ann. Statist.* **26**, 1051–1082.

McCullagh, P. & Han, H. (2011), 'On Bayes theorem for improper mixtures', *Ann. Statist.* pp. 2007–2020.

McCulloch, R. E., Polson, N. G. & Rossi, P. E. (2000), 'A Bayesian analysis of the multinomial probit model with fully identified parameters', *J. Econometrics* **99**, 173–193.

McGrayne, S. B. (2011), *The Theory That Would Not Die*, New Haven, CT: Yale University Press.

Meeden, G. (1990), 'Admissible contour credible sets', *Statistics and Decisions* **8**, 1–10.

Meeden, G. & Vardeman, S. (1991), 'A noninformative Bayesian approach to interval estimation in finite population sampling', *J. Amer. Statist. Assoc.* **86**, 972–980.

Mendoza, M. (1990), 'A Bayesian analysis of the slope ratio bioassay', *Biometrika* **46**, 1059–1069.

Mendoza, M. (1994), 'Asymptotic normality under transformations. A result with Bayesian applications', *Test* **3**, 173–180.

Mendoza, M. & Gutiérrez-Peña, E. (1999), 'Bayesian inference for the ratio of the means of two normal populations with unequal variances', *Biometrical J.* **41**, 133–147.

Mendoza, M. & Gutiérrez-Peña, E. (2000), 'Bayesian conjugate analysis of the galton-watson process', *Test* **9**, 149–171.

Meng, X.-L. & Zaslavsky, A. M. (2002), 'Single observation unbiased priors', *Ann. Statist.* **30**, 1345–1375.

Metropolis, N., Rosenbluth, A. W., Rosenbluth, M. N., Teller, A. H. & Teller, E. (1953), 'Equation of state calculations by fast computing machines', *J. Chemical Physics* **21**, 1087–1092.

Michaeli, R. & Simon, L. (2008), 'An illustration of Bayes' theorem and its use as a decision-making aid for competitive intelligence and marketing analysts', *European J. Marketing* **42**, 804–813.

Millar, R. B. (2002), 'Reference priors for Bayesian fisheries models', *Canad. J. Fish. Aquat. Sci.* **59**, 1492–1502.

Miller, R. B. (1980), 'Bayesian analysis of the two-parameter Gamma distribution', *Technometrics* **22**, 65–69.

Milne, P. (1983), 'A note on scale invariance', *British J./ Philos./ Science* **34**, 49–55.

Moala, F. A. & Dey, S. (2018), 'Objective and subjective prior distributions for the gompertz distribution', *Anais da Academia Brasileira de Ciências* **90**, 2643–2661.

Moala, F. A. & Garcia, L. M. (2013), 'A Bayesian analysis for the parameters of the exponential-logarithmic distribution', *Quality Engineering* **25**, 282–291.

Moala, F. A., Rodrigues, J. & Tomazella, V. L. D. (2009), 'A note on the prior distributions of weibull parameters for the reliability function', *Communications in StatisticsTheory and Methods* **38**(7), 1041–1054.

Mohsin, M., Kazianka, H., Pilz, J. & Gebhardt, A. (2014), 'A new bivariate exponential distribution for modeling moderately negative dependence', *Statistical Methods & Applications* **23**, 123–148.

Molitor, J. & Sun, D. (2002), 'Bayesian analysis under ordered functions of parameters', *Environ. Ecological Statist.* **9**, 179–193.

Moreno, E. & Girón, F. J. (1998), 'Estimating with incomplete count data: A Bayesian approach', *J. Statist. Planning and Inference* **66**, 147–159.

Morris, C. N. (1983), 'Parametric empirical Bayes inference: theory and applications', *J. Amer. Statist. Assoc.* **78**, 47–55.

Morris, C. N. (1988), Approximating posterior distributions and posterior moments, *in* '*Bayesian Statistics 3* (J. M. Bernardo, M. H. DeGroot, D. V. Lindley and A. F. M. Smith, eds). Oxford: University Press', pp. 327–344.

Morris, C. N. & Lysy, M. (2012), 'Shrinkage estimation in multilevel normal models', *Statistical Sci.* **27**, 115–134.

Mortera, J. (1986), 'Bayesian forecasting', *Metron* **44**, 277–296.

Mossman, D. (2000), 'The meaning of malingering data: Further applications of Bayes' theorem', *Behavioral Sci. and the Law* **18**(6), 761–779.

Mossman, D. & Berger, J. O. (2001), 'Intervals for post-test probabilities: A comparison of 5 methods', *Medical Decision Making* **21**, 498.

Mukerjee, R. & Chen, Z. (2003), 'On expected lengths of predictive intervals', *Scandinavian J. Statist.* **30**, 757–766.

Mukerjee, R. & Dey, D. K. (1993), 'Frequentist validity of posterior quantiles in the presence of a nuisance parameter: Higher order asymptotics', *Biometrika* **80**, 499–505.

Mukerjee, R. & Ghosh, M. (1997), 'Second-order probability matching priors', *Biometrika* **84**, 970–975.

Mukerjee, R. & Reid, N. (1999), 'On a property of probability matching priors: Matching the alternative coverage probabilities', *Biometrika* **86**, 333–340.

Mukerjee, R. & Reid, N. (2001), 'Second-order probability matching priors for a parametric function with application to Bayesian tolerance limits', *Biometrika* **88**, 587–592.

Mukhopadhyay, S. & Ghosh, M. (1995), 'On the uniform approximation of Laplace's prior by t-priors in location problems', *J. Multivariate Analysis* **54**, 284–294.

Muré, J. (2019), 'Optimal compromise between incompatible conditional probability distributions, with application to objective Bayesian kriging', *ESAIM: Probability and Statistics* **23**, 271–309.

Nadarajah, S., Bityukov, S. & Krasnikov, N. (2015), 'Confidence distributions: A review', *Statistical Methodology* **22**, 23–46.

Natarajan, R. (2001), 'On the propriety of a modified Jeffreys's prior for variance components in binary random effects models', *Statistics & Probability letters* **51**, 409–414.

Natarajan, R. & Kass, R. E. (2000), 'Reference Bayesian methods for generalized linear mixed models', *J. Amer. Statist. Assoc.* **95**, 227–237.

Neal, R. M. (2003), 'Slice sampling', *Ann. Statist.* **31**, 705–741.

Neto, E. C. & Branco, M. E. (2007), 'Reference analysis for the p-dimensional calibration problem', *Brazilian J. Prob. Statist.* **21**, 121–139.

Neyman, J. (1934), 'On the two different aspects of representative method: The method of stratified sampling and the method of purpose selection', *J. Roy. Statist. Soc. A* **97**, 558–625.

Neyman, J. (1937), 'Outline of a theory of statistical estimation based on the classical theory of probability', *Philos. Trans. Roy. Soc. A.* **237**, 333–380.

Neyman, J. & Scott, E. L. (1948), 'Consistent estimates based on partially consistent observations', *Econometrica* **16**, 1–32.

Ni, S. & Sun, D. (2003), 'Noninformative priors and frequentist risks of Bayesian estimators of vector-autoregressive models', *J. Econometrics* **115**, 159–197.

Nicolaou, A. (1993), 'Bayesian intervals with good frequentist behaviour in the presence of nuisance parameters', *J. Roy. Statist. Soc. B* **55**, 377–390.

Nogarotto, D. C., Azevedo, C. L. N. & Bazán, J. L. (2020), 'Bayesian modeling and prior sensitivity analysis for zero–one augmented beta regression models with an application to psychometric data', *Brazilian Journal of Probability and Statistics* **34**, 304–322.

Nosek, D. & Nosková, J. (2016), 'On Bayesian analysis of on–off measurements', *Nuclear Instruments and Methods in Physics Research A* **820**, 23–33.

Novick, M. R. (1969), 'Multiparameter Bayesian indifference procedure', *J. Roy. Statist. Soc. B* **31**, 29–64 (with discussion).

Novick, M. R. & Hall, W. J. (1965), 'A Bayesian indifference procedure', *J. Amer. Statist. Assoc.* **60**, 1104–1117.

Oaksford, M. & Chater, N. (2007), *Bayesian Rationality: The Probabilistic Approach to Human Reasoning*, Oxford: University Press.

Ogata, Y. (1990), 'A Monte Carlo method for an objective Bayesian procedure', *Ann. Inst. Statist. Math.1* **42**, 403–433.

Oh, M.-S. & Kim, Y. (2000), 'Jeffreys noninformative prior in Bayesian conjoint analysis', *J. Korean Statist. Soc.* **29**, 137–153.

O'Hagan, A. (1994), *Kendall's Advanced Theory of Statistics. Volume 2B: Bayesian Inference*, London: Edward Arnold.

Ortega, F. J. & Basulto, J. (2016), 'A generalization of Jeffreys' rule for non regular models', *Communications in Statistics-Theory and Methods* **45**, 4433–4444.

Pal, N. (1993), 'Estimating the normal dispersion matrix and the precision matrix from a decision-theoretic point of view: A review', *Statistical Papers [Formerly: Statistischen Hefte]* **34**, 1–26.

Palmer, J. L. & Pettit, L. I. (1996), 'Risks of using improper priors with Gibbs sampling and autocorrelated errors', *J. Comp. Graphical Statist.* **5**, 245–249.

Paulo, R. (2005), 'Default priors for Gaussian processes', *The Annals of Statistics* **33**, 556–582.

Pawlak, Z. (2003), Decision algorithms, Bayes' theorem and flow graphs, *in* L. Rutkowski, ed., 'Neural Networks and Soft Computing', Berlin: Springer, pp. 18–24.

Pearl, J. (1988), *Probabilistic Reasoning in Intelligent Systems: Networks of Plausible Inference*, San Mateo, CA: Morgan Kaufmann.

Pearson, E. S. (1925), 'Bayes' theorem, examined in the light of experimental sampling', *Biometrika* **17**, 388–442.

Pearson, E. S., ed. (1978), *The History of Statistics in the 17th and 18th Centuries. Lectures by Karl Pearson, 1921-1933*, London: Macmillan.

Peers, H. W. (1965), 'On confidence points and Bayesian probability points in the case of several parameters', *J. Roy. Statist. Soc. B* **27**, 9–16.

Peers, H. W. (1968), 'Confidence properties of Bayesian interval estimates', *J. Roy. Statist. Soc. B* **30**, 535–544.

Pericchi, L. R. (1981), 'A Bayesian approach to transformations to normality', *Biometrika* **68**, 35–43.

Pericchi, L. R. & Nazaret, W. A. (1988), On being imprecise at the higher levels of a hierarchical linear model, *in* '*Bayesian Statistics 3* (J. M. Bernardo, M. H. DeGroot, D. V. Lindley and A. F. M. Smith, eds). Oxford: University Press', pp. 361–375 (with discussion).

Pericchi, L. R. & Walley, P. (1991), 'Robust Bayesian credible intervals and prior ignorance', *Internat. Statist. Rev.* **59**, 1–23.

Perks, W. (1947), 'Some observations on inverse probability, including a new indifference rule', *J. Inst. Actuaries* **73**, 285–334 (with discussion).

Pete Blair, J. & Rossmo, D. K. (2010), 'Evidence in context: Bayes' theorem and investigations', *Police Quarterly* **13**, 123–135.

Peterson, C., Stingo, F. C. & Vannucci, M. (2009), 'Bayesian inference of multiple gaussian graphical models', *The Annals of Statistics* **110**, 159–170.

Philippe, A. & Robert, C. P. (1998), 'A note on the confidence properties of reference priors for the calibration model', *Test* **7**, 147–160.

Phillips, P. C. B. (1991), 'To criticize the critics: An objective Bayesian analysis of stochastic trends', *J. Applied Econometrics* **6**, 333–473 (with discussion).

Piccinato, L. (1973), 'Un metodo per determinare distribuzioni iniziali relativamente non-informative', *Metron* **31**, 124–156.

Piccinato, L. (1978), Predictive distributions and non-informative priors, *in* '*Trans. 7th. Prague Conf. Information Theory* (M. Uldrich, ed). Prague: Czech. Acad. Sciences', pp. 399–407.

Pilz, J., Kazianka, H. & Spöck, G. (2012), 'Some advances in Bayesian spatial prediction and sampling design', *Spatial Statistics* **1**, 65–81.

Poirier, D. J. (1994), 'Jeffreys prior for logit models', *J. Econometrics* **63**, 327–339.

Pole, A. & West, M. (1989), 'Reference analysis of the dynamic linear model', *J. Time Series Analysis* **10**, 131–147.

Pole, A., West, M. & Harrison, J. (1994), *Applied Bayesian Forecasting and Time Series Analysis*, London: Chapman and Hall.

Polson, N. G. (1991), 'A representation of the posterior mean for a location model', *Biometrika* **78**, 426–430.

Polson, N. G. & Roberts, G. O. (1993), 'A utility based approach to information for stochastic differential equations', *Stochastic Proc. Appl.* **48**, 341–356.

Polson, N. G. & Wasserman, L. (1990), 'Prior distributions for the bivariate binomial', *Biometrika* **77**, 901–904.

Popper, K. R. (1957), 'Probability magic or knowledge out of ignorance', *Dialectica* **11**, 354–374.

Pratt, J. W. (1961), 'Length of confidence intervals', *J. Amer. Statist. Assoc.* **56**, 549–567.

Pratt, J. W. (1965), 'Bayesian interpretation of standard inference statements', *J. Roy. Statist. Soc. B* **27**, 169–203 (with discussion).

Press, S. J. (1982), *Applied Multivariate Analysis: Using Bayesian and Frequentist Methods of Inference* (2nd edition), Mineola, NY: Dover.

Press, S. J. (1989), *Bayesian Statistics: Principles, Models, and Applications*, Hoboken, NJ: Wiley.

Press, S. J. & Zellner, A. (1978), 'Posterior distribution for the multiple correlation coefficient with fixed regressors', *J. Econometrics* **8**, 307–321.

Pretorius, A. L. & Van der Merwe, A. J. (2002), 'Reference and probability-matching priors in Bayesian analysis of mixed linear models', *J. Anim. Breed. Genet.* **119**, 311–324.

Price, R. M. & Bonett, D. G. (2000), 'Estimating the ratio of two Poisson rates', *Comput. Statist. Data Anal.* **34**, 345–356.

Pulcini, A. & Liseo, B. (2015), Approximate Bayesian quantile regression for panel data, *in* A. M. Paganoni & P. Secchi, eds, 'Advances in Complex Data Modeling and Computational Methods in Statistics', Berlin: Springer, pp. 173–189.

Rabena, M. T. (1998), 'Deriving reference decisions', *Test* **7**, 161–177.

Raftery, A. E. (1988*a*), 'Analysis of a simple debugging model', *Appl. Statist.* **37**, 12–22.

Raftery, A. E. (1988*b*), 'Inference for the binomial n parameter: A hierarchical Bayes approach', *Biometrika* **75**, 223–228.

Raftery, A. E. & Akman, V. E. (1986), 'Bayesian analysis of a Poisson process with a change-point', *Biometrika* **73**, 85–89.

Raftery, A. E. & Schweder, T. (1993), 'Inference about the ratio of two parameters, with application to whale censusing', *Amer. Statistician* **47**, 259–264.

Rainey, C. (2016), 'Dealing with separation in logistic regression models', *Political Analysis* **24**, 339–355.

Ramos, P. L., Dey, D. K., Louzada, F. & Ramos, E. (2021), 'On posterior properties of the two parameter gamma family of distributions', *Annals of the Brazilian Academy of Sciences* **93**, 4.

Ramos, P. L., Louzada, F. & Ramos, E. (2015), 'Posterior properties of the nakagami-m distribution using non-informative priors and applications in reliability', *IEEE Trans Reliability* **14**, 1–13.

Ramos, P. L., Louzada, F. & Ramos, E. (2017), 'Posterior properties of the nakagami-m distribution using noninformative priors and applications in reliability', *IEEE Transactions on Reliability* **67**, 105–117.

Ramsey, F. P. (1931), Truth and probability, *in* R. B. Braithwaite, ed., '*The Foundations of Mathematics and Other Logical Essays*', London: Kegan Paul, pp. 156–198. Reprinted in 1980 in *Studies in Subjective Probability* (H. E. Kyburg and F. E. Smokler, eds) New York: Dover, 61–92.

Rao, C. R. & Mukerjee, R. (1995), 'On posterior credible sets based on the score statistic', *Statistica Sinica* **5**, 781–791.

Rao, K. A. & D'Cunha, J. G. (2016), 'Bayesian inference for median of the lognormal distribution', *Journal of Modern Applied Statistical Methods* **15**, 32.

Reichenbach, H. (1949), *The Theory of Probability*, Berkeley, CA: University of California Press.

Reid, D. H. (1995*a*), 'Operational time and a fundamental problem of insurance in a data-rich environment', *Appl. Stochastic Models and Data Analysis* **11**, 257–269.

Reid, N. (1995*b*), 'The roles of conditioning in inference', *Statistical Science* **10**, 138–157.

Reid, N. (1996), Likelihood and Bayesian approximation methods, *in* '*Bayesian Statistics 5* (J. M. Bernardo, J. O. Berger, A. P. Dawid and A. F. M. Smith, eds). Oxford: University Press', pp. 351–368 (with discussion).

Ren, C. (2013), 'Objective Bayesian analysis of SAR models', *Spatial Statistics* **4**, 68–78.

Ren, C., Ke, W.-M. & Lu, H.-T. (2009), 'Objective Bayesian analysis for spatio-temporal models', *International Journal of Intelligent Technologies and Applied Statistics* **2**, 121–134.

Ren, C. & Sun, D. (2013), 'Objective Bayesian analysis for CAR models', *Annals of the Institute of Statistical Mathematics* **65**, 457–472.

Ren, C. & Sun, D. (2014), 'Objective Bayesian analysis for autoregressive models with nugget effects', *Journal of Multivariate Analysis* **124**, 260–280.

Ren, C., Sun, D. & He, C. Z. (2012), 'Objective Bayesian analysis for a spatial model with nugget effects', *Journal of Statistical Planning and Inference* **142**, 1933–1946.

Ren, C., Sun, D. & Sahu, S. K. (2013), 'Objective Bayesian analysis of spatial models with separable correlation functions', *Canadian Journal of Statistics* **41**, 488–507.

Rényi, A. (1955), 'On a new axiomatic theory of probability', *Acta Math. Acad. ScientHungaricae* **11**, 285–385.

Rényi, A. (1961), On measures of entropy and information, *in* '*Proc. Fourth Berkeley Symp.* **1** (J. Neyman and E. L. Scott, eds). Berkeley: Univ. California Press', pp. 547–561.

Ríos, D. & Ruggeri, F., eds (2012), *Robust Bayesian Analysis*, Berlin: Springer.

Rissanen, J. (1983), 'A universal prior for integers and estimation by minimum description length', *Ann. Statist.* **11**, 416–431.

Rissanen, J. (1986), 'Stochastic complexity and modelling', *Ann. Statist.* **14**, 1080–1100.

Rissanen, J. (1987), 'Stochastic complexity', *J. Roy. Statist. Soc. B* **49**, 223–262 (with discussion).

Rissanen, J. (1988), Minimum description length principle, *in* '*Encyclopedia of Statistical Sciences* **5**. (S. Kotz, C. B. Read, N. Balakrishnan, B. Vidakovic, and N. L. Johnson, eds). Hoboken, NJ: Wiley', pp. 523–527.

Rissanen, J. (1989), *Stochastic Complexity in Statistical Enquiry*, Singapore: World Scientific Pub.

Robert, C. P. (1993), 'A note on Jeffreys-Lindley paradox', *Statistica Sinica* **3**, 601–608.

Robert, C. P. (1994), *The Bayesian Choice: A Decision-Theoretic Motivation*, Berlin: Springer.

Robert, C. P. (1996), 'Intrinsic loss functions', *Theory and Decision* **40**, 192–214.

Robert, C. P. & Casella, G. (2004), *Monte Carlo Statistical Methods*, 2nd edn, Berlin: Springer.

Robert, C. P., Hwang, J. T. G. & Strawderman, W. E. (1993), 'Is Pitman closeness a reasonable criterion?', *J. Amer. Statist. Assoc.* **88**, 57–76 (with discussion).

Roberts, G. O. & Rosenthal, J. S. (2001), 'Infinite hierarchies and prior distributions', *Bernoulli* **7**, 453–471.

Robins, J. & Wasserman, L. (2000), 'Conditioning, likelihood, and coherence: A review of some foundational concepts', *J. Amer. Statist. Assoc.* **95**, 1340–1346.

Rodriguez, C. C. (1988), Understanding ignorance, *in* E. G. J. & S. C. R., eds, 'Maximum-Entropy and Bayesian Methods in Science and Engineering', Berlin: Springer, pp. 189–204.

Rodríguez, C. C. (1991), From Euclid to entropy, *in* '*Maximum Entropy and Bayesian Methods* (W. T. Grandy and L. H. Schick eds)', Dordrecht: Kluwer, pp. 343–348.

Rosa, G. J. M. & Gianola, D. (2001*a*), 'Inferences about the coefficient of correlation in the standard bivariate normal distribution', *South African Statist. J.* **35**, 69–93.

Rosa, G. J. M. & Gianola, D. (2001*b*), 'Inferences about the coefficient of correlation in the standard bivariate normal distribution: theory and methods', *South African Statistical Journal* **35**, 69–93.

Rosenkrantz, R. D., ed. (1989), *E. T. Jaynes: Papers on Probability, Statistics and Statistical Physics*, 2nd edn, Dordrecht: Kluwer.

Rouder, J. N. & Morey, R. D. (2019), 'Teaching Bayes' theorem: Strength of evidence as predictive accuracy', *Amer. Statistician* **73**, 186–190.

Rousseau, J. (2000), 'Coverage properties of one-sided intervals in the discrete case and application to matching priors', *Ann. Inst. Statist. Math.1* **52**, 28–42.

Rousseau, J. (2016), 'On the frequentist properties of Bayesian nonparametric methods', *Annual Review of Statistics and its Application* **3**, 211–231.

Roverato, A. & Consonni, G. (2004), 'Compatible prior distributions for DAG models', *J. Roy. Statist. Soc. B* **66**, 47–61.

Rubin, D. B. (1984), 'Bayesianly justifiable and relevant frequency calculations for the applied statistician', *Ann. Statist.* **12**, 1151–1172.

Rubio, F. J. & Liseo, B. (2014), 'On the independence Jeffreys prior for skew-symmetric models', *Statistics & Probability Letters* **85**, 91–97.

Rubio, F. J. & Steel, M. F. J. (2014), 'Inference in two-piece location-scale models with Jeffreys priors', *Bayesian Analysis* **9**, 1–22.

Ruli, E., Sartori, N. & Ventura, L. (2016), 'Improved Laplace approximation for marginal likelihoods', *Electronic Journal of Statistics* **10**, 3986–4009.

Russell, B. (1948), *Human Knowledge: Its Scope and Limits*, Reprinted in 2009, Milton Park, UK: Routledge.

Salazar, E., Ferreira, M. A. R. & Migon, H. S. (2012), 'Objective Bayesian analysis for exponential power regression models', *Sankhya B* **74**, 107–125.

Salo, B., Sirén, J., Corander, J., Zappalà, A., Bosco, D., Mokros, A. & Santtila, P. (2013), 'Using Bayes' theorem in behavioural crime linking of serial homicide', *Legal and criminological psychology* **18**, 356–370.

Samaniego, F. J. (2010), *A Comparison of the Bayesian and Frequentist Approaches to Estimation*, Springer.

Sansó, B. & Pericchi, L. R. (1992), 'Near ignorance classes of log-concave priors for the location model', *Test* **1**, 39–46.

Sansó, B. & Pericchi, L. R. (1994), 'On near ignorance classes', *Rev. Brasilera Prob. Estatist.* **8**, 119–126.

Sareen, S. (2003), 'Reference Bayesian inference in nonregular models', *J. Econometrics* **113**, 265–288.

Savage, I. R. (1980), On not being rational, *in* '*Bayesian Statistics* (J. M. Bernardo, M. H. DeGroot, D. V. Lindley and A. F. M. Smith, eds). Valencia: University Press', pp. 321–328 and 339–346 (with discussion).

Savage, L. J. (1954), *The Foundations of Statistics*, Hoboken, NJ: Wiley. 2nd edition in 1972, New York: Dover.

Savage, L. J. (1961), The foundations of statistics reconsidered, *in* '*Proc. First Berkeley Symp.* (J. Neyman ed). Berkeley: Univ. California Press, pp. 575–586.

Savage, L. J. (1970*a*), 'Comments on a weakened principle of conditionality', *Journal of the American Statistical Association* **65**, 399–401.

Savage, L. J. (1970*b*), 'Reading suggestions for the foundations of statistics', *Amer. Statistician* **24**, 23–27.

Schervish, M. J. (1995), *Theory of Statistics*, New York: Springer.

Schweder, T. & Hjort, N. L. (2002), 'Confidence and likelihood', *Scandinavian J. Statist.* **29**, 309–332.

Schweder, T. & Hjort, N. L. (2003), Frequentist analogues of priors and posteriors, *in* B. Stigum, ed., '*Econometrics and the Philosophy of Economics*', Princeton, NJ: University Press, pp. 285–387.

Schweder, T. & Hjort, N. L. (2016), *Confidence, Likelihood, Probability*, Cambridge: University Press.

Scott, J. G. & Berger, J. O. (2010), 'Bayes and empirical-Bayes multiplicity adjustment in the variable-selection problem', *Ann. Statist.* **38**, 2587–2619.

Scricciolo, C. (1999), 'Probability matching priors: A review', *J. Italian Statist. Soc.* **8**, 83–100.

Sendra, M. (1982), 'Distribucion final de referencia para el problema de Fieller-Creasy', *Trab. Estadist.* **33**, 55–72.

Severini, T. A., Mukerjee, R. & Ghosh, M. (2002), 'On an exact probability matching property of right-invariant priors', *Biometrika* **89**, 953–957.

Shafer, G. (1981), 'Constructive probability', *Synthese* **48**, 1–60.

Shafer, G. (1982*a*), 'Belief functions and parametric models', *J. Roy. Statist. Soc. B* **44**, 322–352 (with discussion).

Shafer, G. (1982*b*), 'Lindleys paradox', *J. Amer. Statist. Assoc.* **77**, 325–351 (with discussion).

Shannon, C. E. (1948), 'A mathematical theory of communication', *Bell System Tech. J.* **27**, 379–423 and 623–656. Reprinted in *The Mathematical Theory of Communication* (Shannon, C. E. and Weaver, W., eds, 1964), 29–125; Urbana, IL: Univ. Illinois Press.

Sharma, S. (2017), 'Markov chain monte carlo methods for Bayesian data analysis in astronomy', *Annual Review of Astronomy and Astrophysics* **55**, 213–259.

Shemyakin, A. (2014), 'Hellinger distance and non-informative priors', *Bayesian Analysis* **9**, 923–938.

Shemyakin, A. et al. (2012), 'A new approach to construction of objective priors: Hellinger information', *Applied Econometrics* **28**, 124–137.

Shieh, G. & Lee, J. C. (2002), 'Bayesian prediction analysis for growth curve model using noninformative priors', *Ann. Inst. Statist. Math.1* **54**, 324–337.

Siegmund, D. (1988), *Sequential Analysis: Tests and Confidence Intervals*, New York: Springer.

Silliman, N. P. (1997), 'Hierarchical selection models with applications in meta-analysis', *J. Amer. Statist. Assoc.* **92**, 926–936.

Singh, K. & Xie, M. (2012), Cd-posterior - combining prior and data through confidence distributions, *in* D. Fourdrinier, E. Marchand & A. L. Rukhin, eds, '*Contemporary Developments in Bayesian Analysis and Statistical Decision Theory: A Festschrift for William E. Strawderman*', Hayward, CA: IMS, pp. 200–214.

Singh, K., Xie, M. & Strawderman, W. E. (2005), 'Combining information from independent sources through confidence distributions', *Ann. Statist.* **33**, 159–183.

Singh, K., Xie, M. & Strawderman, W. E. (2007), 'Confidence distribution (cd)-distribution estimator of a parameter', *IMS Lecture Notes Series* **54**, 132–150.

Singpurwalla, N. D. & Wilson, P. (2004), 'When can finite testing ensure infinite trustworthiness?', *J. Iranian Statist. Soc.* **3**, 1–37 (with discussion).

Sinha, B. K. & Ghosh, M. (1987), 'Inadmissibility of the best equivariant estimators of the variance-covariance matrix, the precision matrix, and the generalized variance under entropy loss', *Statistics & Decisions* **5**, 201–227.

Siu, N. O. & Kelly, D. L. (1998), 'Bayesian parameter estimation in probabilistic risk assessment', *Reliability Engineering & System Safety* **62**, 89–116.

Sivaganesan, S. & Lingham, R. T. (2002), 'Bayes factors for model selection with diffusion processes under improper priors', *Ann. Inst. Statist. Math.* **54**, 500–516.

Skilling, J., ed. (1989), *Maximum Entropy and Bayesian Methods*, Dordrecht: Kluwer.

Slater, P. B. (1995), 'Quantum coin-tossing in a Bayesian Jeffreys framework', *Physics Letters A* **206**, 66–72.

Slater, P. B. (1996), 'Bayesian inference for complex and quaternionic two-level quantum systems', *Physica A: Statistical Mechanics and its Applications* **223**, 167–174.

Slater, P. B. (2000), 'High-temperature expansions of Bures and Fisher information priors', *Physical Review E* **61**, 6087.

Slovic, P. & Tversky, A. (1974), 'Who accepts Savage's axioms?', *Behavioural Sci.* pp. 368–378.

Smith, A. F. M. (1973*a*), 'Bayes estimates in one-way and two-way models', *Biometrika* **60**, 319–329.

Smith, A. F. M. (1973*b*), 'A general Bayesian linear model', *J. Roy. Statist. Soc. B* **35**, 67–75.

Smith, A. F. M. (1991), 'Bayesian computational methods', *Phil. Trans. Roy. Soc. London A* **337**, 369–386.

Smith, A. F. M. & Gelfand, A. E. (1992), 'Bayesian statistics without tears: A sampling-resampling perspective', *Amer. Statistician* **46**, 84–88. Corr: **47**, 58.

Smith, A. F. M. & Roberts, G. O. (1993), 'Bayesian computation via the Gibbs sampler and related Markov chain Monte Carlo methods', *J. Roy. Statist. Soc. B* **55**, 3–102 (with discussion).

Smith, R. L. (1999), Bayesian and frequentist approaches to parametric predictive inference, *in* '*Bayesian Statistics 6* (J. M. Bernardo, J. O. Berger, A. P. Dawid and A. F. M. Smith, eds). Oxford: University Press', pp. 589–612 (with discussion).

Smith, R. L. & Naylor, J. C. (1987), 'A comparison of maximum likelihood and Bayesian estimators for the three-parameter Weibull distribution', *Appl. Statist.* **36**, 358–369.

Solari, F., Liseo, B. & Sun, D. (2008), 'Some remarks on Bayesian inference for one-way anova models', *Annals of the Institute of Statistical Mathematics* **60**, 483–498.

Song, C., Sun, D., Fan, K. & Mu, R. (2020), 'Posterior propriety of an objective prior in a 4-level normal hierarchical model', *Mathematical Problems in Engineering* **2020**.

Sono, S. (1983), 'On a noninformative prior distribution for Bayesian inference of multinomial distribution's parameters', *Ann. Inst. Statist. Math.* **35**, 167–174.

Soyer, R. (2009), 'Bayesian networks and probabilistic inference in forensic science', *Technometrics* **51**, 99.

Spall, J. C. & Hill, S. D. (1990), 'Least-informative Bayesian prior distributions for finite samples based on information theory', *IEEE Trans. Automatic Control* **35**, 580–583.

Spiegelhalter, D. J. (1985), Exact Bayesian inference on the parameters of a Cauchy distribution with vague prior information, *in* '*Bayesian Statistics 2* (J. M. Bernardo, M. H. DeGroot, D. V. Lindley and A. F. M. Smith, eds). Amsterdam: North-Holland', pp. 743–749.

Stäel von Holstein, C. A. S., ed. (1974), *The Psychological Concept of Subjective Probability: A Measurement-Theoretic View*, Berlin: Springer.

Starr, N. (1974), 'Optimal and adaptive stopping based on capture times', *J. Appl. Probability* **11**, 294–301.

Steel, M. F. J. (1991), 'A Bayesian analysis of simultaneous equation models by combining recursive analytical and numerical approaches', *Journal of Econometrics* **48**, 83–117.

Stein, C. (1956), Inadmissibility of the usual estimator for the mean of a multivariate normal distribution, *in* '*Proc. First Berkeley Symp.* (J. Neyman ed). Berkeley: Univ. California Press, pp. 197–206.

Stein, C. (1959), 'An example of wide discrepancy between fiducial and confidence intervals', *Ann. Math. Statist.* **30**, 877–880.

Stein, C. (1965), Approximation of improper prior measures by prior probability measures, *in* J. Neyman & L. M. Le Cam, eds, '*Bernoulli 1713, Bayes 1763, Laplace 1813*', New York: Springer, pp. 217–240.

Stein, C. (1985), On the coverage probability of confidence sets based on a prior distribution, *in* R. Zielisnski, ed., '*Sequential Methods in Statistics*', Varsaw: Polish Scientific Publications, pp. 485–514.

Stewart, W. E. (1987), 'Multiresponse parameter estimation with a new and noninformative prior', *Biometrika* **74**, 557–562.

Stigler, S. M. (1982), 'Thomas Bayes's Bayesian inference', *J. Roy. Statist. Soc. A* **145**, 250–258.

Stigler, S. M. (1986*a*), 'Laplace's 1774 memoir on inverse probability', *Statistical Science* pp. 359–363.

Stigler, S. M. (1986*b*), *The History of Statistics: The Measurement of Uncertainty Before 1900*, Cambridge, MA: Harvard University Press.

Stone, M. (1959), 'Application of a measure of information to the design and comparison of experiments', *Ann. Math. Statist.* **30**, 55–70.

Stone, M. (1963), 'The posterior t distribution', *Ann. Math. Statist.* **34**, 568–573.

Stone, M. (1965), 'Right Haar measure for convergence in probability to quasi posterior distributions', *Ann. Math. Statist.* **36**, 440–453.

Stone, M. (1969), 'The role of experimental randomization in Bayesian statistics: Finite sampling and two Bayesians', *Biometrika* **56**, 681–683.

Stone, M. (1970), 'Necessary and sufficient condition for convergence in probability to invariant posterior distributions', *Ann. Math. Statist.* **41**, 1349–1353.

Stone, M. (1974), 'Cross-validatory choice and assessment of statistical predictions', *J. Roy. Statist. Soc. B* **36**, 111–147 (with discussion). Corr: Vol **38** (1976), 102.

Stone, M. (1976), 'Strong inconsistency from uniform priors', *J. Amer. Statist. Assoc.* **71**, 114–125 (with discussion).

Stone, M. & Dawid, A. P. (1972), 'Un-Bayesian implications of improper Bayes inference in routine statistical problems', *Biometrika* **59**, 369–375.

Stone, M. & Springer, B. G. F. (1965), 'A paradox involving quasi prior distributions', *Biometrika* **52**, 623–627.

Sudderth, W. D. (1980), 'Finitely additive priors, coherence and the marginalization paradox', *J. Roy. Statist. Soc. B* **42**, 339–341.

Sugiura, N. & Ishibayashi, H. (1997), 'Reference prior Bayes estimator for bivariate normal covariance matrix with risk comparison', *Comm. Statist. Theory and Methods* **26**, 2203–2221.

Sun, D. (1994), 'Integrable expansions for posterior distributions for a two-parameter exponential family', *Ann. Statist.* **22**, 1808–1830.

Sun, D. (1997), 'A note on noninformative priors for Weibull distributions', *J. Statist. Planning and Inference* **61**, 319–338.

Sun, D. & Berger, J. O. (1994), 'Bayesian sequential reliability for Weibull and related distributions', *Ann. Inst. Statist. Math.* **46**, 221–249.

Sun, D. & Berger, J. O. (1998), 'Reference priors with partial information', *Biometrika* **85**, 55–71.

Sun, D. & Berger, J. O. (2007*a*), 'Objective Bayesian analysis for the multivariate normal model', *Bayesian Statistics* **8**, 525–562.

Sun, D. & Berger, J. O. (2007*b*), Objective Bayesian analysis for the multivariate normal model, *in* '*Bayesian Statistics 8* (J. M. Bernardo, M. J. Bayarri, J. O. Berger, A. P. Dawid, D. Heckerman, A. F. M. Smith and M. West, eds). Oxford: University Press', Vol. 8, pp. 525–562, (with discussion).

Sun, D. & Berger, J. O. (2008), Objective priors under the sequential experimentation, *in* '*Pushing the Limits of Contemporary Statistics: Contributions in Honor of Jayanta K. Ghosh* (B. Clarke and S. Ghosal, eds', Hayward, CA: IMS, pp. 19–32.

Sun, D., Ghosh, M. & Basu, A. P. (1998), 'Bayesian analysis for a stress-strength system under noninformative priors', *Canad. J. Statist.* **26**, 323–332.

Sun, D. & Sun, X. (2005), 'Estimation of the multivariate normal precision and covariance matrices in a star-shape model', *Annals of the Institute of Statistical Mathematics* **57**, 455–494.

Sun, D., Tsutakawa, R. K. & He, C. Z. (2001), 'Propriety of posteriors with improper priors in hierarchical linear mixed models', *Statistica Sinica* **11**, 77–95.

Sun, D. & Ye, K. (1995), 'Reference prior Bayesian analysis for normal mean products', *J. Amer. Statist. Assoc.* **90**, 589–597.

Sun, D. & Ye, K. (1996), 'Frequentist validity of posterior quantiles for a two-parameter exponential family', *Biometrika* **83**, 55–65.

Sun, D. & Ye, K. (1999), 'Reference prior Bayesian analysis for normal mean products', *Canad. J. Statist.* **27**, 97–103.

Suppes, P. (1973), New foundations of objective probability: Axioms for propensities, *in* P. Suppes, L. Henkin, A. Joja & G. C. Moisil, eds, 'Logic, Methodology and Philosophy of Science IV', Amsterdam: North-Holland, pp. 515–529.

Sweeting, T. J. (1984), 'On the choice of prior distribution for the Box-Cox transformed linear model', *Biometrika* **71**, 127–134.

Sweeting, T. J. (1985), Consistent prior distributions for transformed models, *in* '*Bayesian Statistics 2* (J. M. Bernardo, M. H. DeGroot, D. V. Lindley and A. F. M. Smith, eds). Amsterdam: North-Holland', pp. 755–762.

Sweeting, T. J. (1992), On asymptotic posterior normality in the multiparameter case, *in* '*Bayesian Statistics 4* (J. M. Bernardo, J. O. Berger, A. P. Dawid and A. F. M. Smith, eds). Oxford: University Press', pp. 825–835.

Sweeting, T. J. (1995*a*), 'A Bayesian approach to approximate conditional inference', *Biometrika* **82**, 25–36.

Sweeting, T. J. (1995*b*), Consistent prior distributions for transformed models, *in* '*Bayesian Statistics 2* (J. M. Bernardo, M. H. DeGroot, D. V. Lindley and A. F. M. Smith, eds). Amsterdam: North-Holland', pp. 755–762.

Sweeting, T. J. (1995*c*), 'A framework for Bayesian and likelihood approximations in statistics', *Biometrika* **82**, 1–24.

Sweeting, T. J. (1996), Approximate Bayesian computation based on signed roots of log-density ratios, *in* '*Bayesian Statistics 5* (J. M. Bernardo, J. O. Berger, A. P. Dawid and A. F. M. Smith, eds). Oxford: University Press', pp. 427–444 (with discussion).

Sweeting, T. J. (2001), 'Coverage probability bias, objective Bayes and the likelihood principle', *Biometrika* **88**, 657–675.

Szabó, B., Van der Vaart, A., van Zanten, J. H. et al. (2015), 'Frequentist coverage of adaptive nonparametric Bayesian credible sets', *Ann. Statist.* **43**, 1391–1428.

Szökefalvi-Nagy, B. (1985), 'Alfred Haar (1885–1933)', *Results in Mathematics* **8**, 194–196.

Talbott, W. J. (2001), Bayesian epistemology, *in* E. N. Zalta, ed., 'Stanford Encyclopedia of Philosophy', Stanford University, Stanford, CA.

Taraldsen, G. & Lindqvist, B. H. (2010), 'Improper priors are not improper', *Amer. Statistician* **64**, 154–158.

Tardella, L. (2002), 'A new Bayesian method for nonparametric capture-recapture models in presence of heterogeneity', *Biometrika* **89**, 807–817.

Tiao, G. C. & Box, G. E. P. (1968), 'Bayesian estimation of means for the random effect model', *J. Amer. Statist. Assoc.* **63**, 174–181.

Tiao, G. C. & Box, G. E. P. (1975), Some comments on Bayes estimators, *in* '*Studies in Bayesian Econometrics and Statistics: in Honor of Leonard J. Savage* (S. E. Fienberg and A. Zellner, eds). Amsterdam: North-Holland', pp. 619–626.

Tibshirani, R. (1989), 'Noninformative priors for one parameter of many', *Biometrika* **76**, 604–608.

Tiwari, R. C., Chib, S. & Jammalamadaka, S. R. (1989), 'Bayes estimation of the multiple correlation coefficient', *Comm. Statist. Theory and Methods* **18**, 1401–1413.

Torgersen, E. N. (1981), 'Measures of information based on comparison with total information and with total ignorance', *Ann. Statist.* **9**, 638–657.

Tran, T. D., Lesaffre, E., Verbeke, G. & Duyck, J. (2021), 'Latent ornstein-uhlenbeck models for Bayesian analysis of multivariate longitudinal categorical responses', *Biometrics* **77**, 689–701.

Tsutakawa, R. K. (1992), 'Prior distribution for item response curves', *British J. Math. Statist. Philosophy* **45**, 51–74.

Tuyl, F. (2019), 'A method to handle zero counts in the multinomial model', *The American Statistician* **73**(2), 151–158.

Tuyl, F., Gerlach, R. & Mengersen, K. (2008), 'A comparison of Bayes–Laplace, Jeffreys, and other priors: the case of zero events', *The American Statistician* **62**, 40–44.

Tuyl, F., Gerlach, R. & Mengersen, K. (2009), 'Posterior predictive arguments in favor of the Bayes-Laplace prior as the consensus prior for binomial and multinomial parameters', *Bayesian Analysis* **4**, 151–158.

Vallejos, C. A. & Steel, M. F. J. (2015), 'Objective Bayesian survival analysis using shape mixtures of log-normal distributions', *J. Amer. Statist. Assoc.* **110**, 697–710.

Van den Boogaart, K. G., Egozcue, J. J. & Pawlowsky-Glahn, V. (2014), 'Bayes hilbert spaces', *Australian and New Zealand J. Statist.* **56**, 171–194.

van der Linde, A. (2000), 'Reference priors for shrinkage and smoothing parameters', *J. Statist. Planning and Inference* **90**, 245–274.

Van der Merwe, A. J. & Chikobvu, D. (2004), 'Bayesian analysis of the process capability index c_{pk}', *South African Statist. J.* **38**, 97–117.

van Zwet, E. (2019), 'A default prior for regression coefficients', *Statistical Methods in Medical Research* **28**, 3799–3807.

Vaurio, J. K. (1992), 'Objective prior distributions and Bayesian updating', *Reliability Eng. System Safety* **35**, 55–59.

Venn, J. (1866), *The Logic of Chance*, Reprinted in 2019. London: Books on Demand.

Veronese, P. & Melilli, E. (2015), 'Fiducial and confidence distributions for real exponential families', *Scandinavian Journal of Statistics* **42**, 471–484.

Veronese, P. & Melilli, E. (2018), 'Fiducial, confidence and objective Bayesian posterior distributions for a multidimensional parameter', *Journal of Statistical Planning and Inference* **195**, 153–173.

Villa, C. & Rubio, F. J. (2018), 'Objective priors for the number of degrees of freedom of a multivariate t distribution and the t-copula', *Computational Statistics & Data Analysis* **124**, 197–219.

Villa, C. & Walker, S. G. (2014), 'Objective prior for the number of degrees of freedom of at distribution', *Bayesian Analysis* **9**, 197–220.

Villegas, C. (1969), 'On the a priori distribution of the covariance matrix', *Ann. Math. Statist.* **40**, 1098–1099.

Villegas, C. (1971), On Haar priors, *in* '*Foundations of Statistical Inference* (V. P. Godambe and D. A. Sprott, eds). Toronto: Holt, Rinehart and Winston', pp. 409–414 (with discussion).

Villegas, C. (1977*a*), 'Inner statistical inference', *J. Amer. Statist. Assoc.* **72**, 453–458.

Villegas, C. (1977*b*), 'On the representation of ignorance', *J. Amer. Statist. Assoc.* **72**, 651–654.

Villegas, C. (1981), 'Inner statistical inference II', *Ann. Statist.* **9**, 768–776.

Villegas, C. (1990), 'Bayesian inference in models with Euclidean structures', *J. Amer. Statist. Assoc.* **85**, 1159–1164.

Wagenmakers, E.-J., Lee, M., Lodewyckx, T. & Iverson, G. J. (2008), Bayesian versus frequentist inference, *in* 'Bayesian Evaluation of Informative Hypotheses', Berlin: Springer, pp. 181–207.

Wahba, G. (1978), 'Improper priors, spline smoothing and the problem of guarding against model errors in regression', *J. Roy. Statist. Soc. B* **40**, 364–372.

Wakefield, J. (2013), *Bayesian and Frequentist Regression Methods*, Springer.

Wakefield, J. C., Gelfand, A. E. & Smith, A. F. M. (1991), 'Efficient generation of random variates via the ratio-of-uniforms method', *Statistics and Computing* **1**, 129–133.

Walker, S. G. & Gutiérrez-Peña, E. (1999), Robustifying Bayesian procedures, *in* '*Bayesian Statistics 6* (J. M. Bernardo, J. O. Berger, A. P. Dawid and A. F. M. Smith, eds). Oxford: University Press', pp. 685–710, (with discussion).

Walker, S. G. & Muliere, P. (1999), 'A characterization of a neutral to the right prior via an extension of Johnson's sufficientness postulate', *Ann. Statist.* **27**, 589–599.

Walley, P. (1991), *Statistical Reasoning with Imprecise Probabilities*, London: Chapman and Hall.

Wallstrom, T. C. (2007), The marginalization paradox and the formal Bayes law, *in* 'AIP Conference Proceedings **954**', American Institute of Physics, pp. 93–100.

Wang, G. & Sivaganesan, S. (2013), 'Objective priors for parameters in a normal linear regression with measurement error', *Communications in Statistics-Theory and Methods* **42**(15), 2694–2713.

Wang, M. & Yang, M. (2016), 'Posterior property of student-t linear regression model using objective priors', *Statistics & Probability Letters* **113**, 23–29.

Wang, X., Berger, J. O. & Burdick, D. S. (2013), 'Bayesian analysis of dynamic item response models in educational testing', *The Annals of Applied Statistics* **7**, 126–153.

Wang, Y., Huang, K. & Cao, Z. (2014), 'Bayesian identification of soil strata in london clay', *Géotechnique* **64**, 239–246.

Wasserman, L. (1989), 'A robust Bayesian interpretation of likelihood regions', *Ann. Statist.* **17**, 1387–1393.

Wasserman, L. (1990*a*), 'Belief functions and statistical inference', *Canad. J. Statist.* **18**, 183–196.

Wasserman, L. (1990*b*), 'Prior envelopes based on belief functions', *Ann. Statist.* **18**, 454–464.

Wasserman, L. (1992), 'Invariance properties of density ratio priors', *Ann. Statist.* **20**, 2177–2182.

Wasserman, L. (1996), 'The conflict between improper priors and robustness', *J. Statist. Planning and Inference* **52**, 1–15.

Wasserman, L. (2000), 'Asymptotic inference for mixture models using data-dependent priors', *J. Roy. Statist. Soc. B* **62**, 159–180.

Wasserman, L. & Clarke, B. (1995), 'Information tradeoff', *Test* **4**, 19–38.

Welch, B. L. (1965), 'On comparisons between confidence point procedures in the case of a single parameter', *J. Roy. Statist. Soc. B* **27**, 1–8.

Welch, B. L. & Peers, H. W. (1963), 'On formulae for confidence points based on intervals of weighted likelihoods', *J. Roy. Statist. Soc. B* **25**, 318–329.

Wertheimer, A. (1939), 'A note on confidence intervals and inverse probability', *Ann. Math. Statist.* **10**, 74–76.

West, M. & Harrison, J. (1989), *Bayesian Forecasting and Dynamic Models*, New York: Springer.

Whittaker, J. (1990), *Graphical models in applied multivariate statistics*, John Wiley & Sons.

Wijsman, R. A. (1990), *Invariant Measures on Groups and their Uses in Statistics*, Hayward, CA: IMS.

Wollert, R. (2006), 'Low base rates limit expert certainty when current actuarials are used to identify sexually violent predators: An application of Bayes's theorem', *Psychology, Public Policy, and Law* **12**, 56–85.

Wong, F., Carter, C. & Kohn, R. (2002), Efficient estimation of covariance selection modles, *in* 'Bayesian Statistics 7. Proceedings of the Seventh Valencia International Meeting', pp. 182–217.

Woodroofe, M. (1978), 'Large deviations of the likelihood ratio statistic with applications to sequential testing', *Ann. Statist.* **6**, 72–84.

Woodroofe, M. (1982), *Nonlinear Renewal Theory in Sequential Analysis*, Philadelphia, PA: SIAM.

Wrinch, D. & Jeffreys, H. (1919), 'On some aspects of the theory of probability', *Phil. Mag. Ser. 6* **38**, 715–731.

Wrinch, D. & Jeffreys, H. (1921), 'On certain fundamental principles of scientific inquiry', *Phil. Mag. Ser. 6* **42**, 369–390.

Xie, M.-G. & Singh, K. (2013), 'Confidence distribution, the frequentist distribution estimator of a parameter: a review', *Internat. Statist. Rev.* **81**, 3–39.

Xu, G., Ding, Y., Zhao, J., Hu, L. & Fu, X. (2013), 'A novel artificial bee colony approach of live virtual machine migration policy using Bayes theorem', *Scient. World J.* **2013**, 1–13.

Yang, R. (1995), 'Invariance of the reference prior under reparametrization', *Test* **4**, 83–94.

Yang, R. & Berger, J. O. (1994), 'Estimation of a covariance matrix using the reference prior', *Ann. Statist.* **22**, 1195–1211.

Yang, R. & Berger, J. O. (1997), *A Catalogue of Noninformative Priors*, Durham, NC: Duke University, ISDS 97-42.

Yang, R. & Chen, M.-H. (1995), 'Bayesian analysis for random coefficient regression models using noninformative priors', *J. Multivariate Analysis* **55**, 283–311.

Yanuar, F., Yozza, H. & Rescha, R. V. (2019), 'Comparison of two priors in Bayesian estimation for parameter of weibull distribution', *Science and Technology Indonesia* **4**, 82–87.

Ye, K. (1993), 'Reference priors when the stopping rule depends on the parameter of interest', *J. Amer. Statist. Assoc.* **88**, 360–363.

Ye, K. (1994), 'Bayesian reference prior analysis on the ratio of variances for the balanced one-way random effect model', *J. Statist. Planning and Inference* **41**, 267–280.

Ye, K. (1995), 'Selection of the reference priors for a balanced random effects model', *Test* **4**, 1–17.

Ye, K. (1998), 'Estimating a ratio of the variances in a balanced one-way random effects model', *Statistics and Decisions* **16**, 163–180.

Ye, K. & Berger, J. O. (1991), 'Noninformative priors for inferences in exponential regression models', *Biometrika* **78**, 645–656.

Yosboonruang, N. & Niwitpong, S.-A. (2021), 'Confidence intervals for the coefficient of quartile variation of a zero-inflated lognormal distribution', *Emerging Science Journal* **5**, 457–470.

Yu, B. (2010), 'A Bayesian MCMC approach to survival analysis with doubly-censored data', *Comput. Statist. Data Anal.* **54**, 1921–1929.

Yuan, A. & Clarke, B. (1999), 'A minimally informative likelihood for decision analysis: Illustration and robustness', *Canad. J. Statist.* **27**, 649–665.

Yuan, A. et al. (2009), 'Bayesian frequentist hybrid inference', *Ann. Statist.* **37**(5), 2458–2501.

Zabell, S. (1989), 'R. A. Fisher on the history of inverse probability', *Statistical Science* pp. 247–256.

Zabell, S. (1992), 'R. A. Fisher and fiducial argument', *Statistical Sci.* **7**, 369–387.

Zaffalon, M. & Miranda, E. (2013), 'Probability and time', *Artificial Intelligence* **198**, 1–51.

Zellner, A. (1971), *An Introduction to Bayesian Inference in Econometrics*, Hoboken, NJ: Wiley. Reprinted in 1987, Melbourne, FL: Krieger.

Zellner, A. (1977), Maximal data information prior distributions, *in* '*New Developments in the Applications of Bayesian Methods* (A. Ayka M and C. Brumat, eds). Amsterdam: North-Holland', pp. 211–232.

Zellner, A. (1983), 'Applications of Bayesian analysis in econometrics', *The Statistician* **32**, 23–34.

Zellner, A. (1985), 'Bayesian econometrics', *Econometrica* **53**, 253–269.

Zellner, A. (1986), 'Bayesian estimation and prediction using asymmetric loss functions', *J. Amer. Statist. Assoc.* **81**, 446–451.

Zellner, A. (1991), Bayesian methods and entropy in economics and econometrics, *in* '*Maximum Entropy and Bayesian Methods* (W. T. Grandy and L. H. Schick eds)', Dordrecht: Kluwer, pp. 17–31.

Zhang, W., Chaloner, K., Cowles, M. K., Zhang, Y. & Stapleton, J. T. (2008), 'A Bayesian analysis of doubly censored data using a hierarchical Cox model', *Statistics in Medicine* **27**, 529–542.

Zhang, Z. (1994), *Discrete Noninformative Priors*, PhD Thesis, University of Yale, USA.

Zhou, R., Sivaganesan, S. & Longla, M. (2014), 'An objective Bayesian estimation of parameters in a log-binomial model', *Journal of Statistical Planning and Inference* **146**, 113–121.

Zhu, M. & Lu, A. Y. (2004), 'The counter-intuitive non-informative prior for the bernoulli family', *Journal of Statistics Education* **12**.

Zidek, J. V. (1969), 'A representation of Bayes invariant procedures in terms of Haar measure', *Ann. Inst. Statist. Math.* **21**, 291–308.

Author Index

Index

www.ingramcontent.com/pod-product-compliance
Lightning Source LLC
LaVergne TN
LVHW060023210926
848481LV00013B/3773

* 9 7 8 9 8 1 9 8 3 3 1 8 4 *